AF356151

# Applications of Finite Element Modeling for Mechanical and Mechatronic Systems

*Editors*
Marek Krawczuk
Department of Mechatronics and High Voltage
Engineering, Faculty of Electrical and Control
Engineering, Gdańsk University of Technology
Poland

Magdalena Palacz
Department of Production Engineering,
Faculty of Organisation and Management,
Silesian University of Technology
Poland

*Editorial Office*
MDPI
St. Alban-Anlage 66
4052 Basel, Switzerland

This is a reprint of articles from the Special Issue published online in the open access journal *Journal Not Specified* (ISSN ) (available at: https://www.mdpi.com/journal/applsci/special_issues/FEM).

For citation purposes, cite each article independently as indicated on the article page online and as indicated below:

LastName, A.A.; LastName, B.B.; LastName, C.C. Article Title. *Journal Name* **Year**, *Volume Number*, Page Range.

**ISBN 978-3-0365-1555-7 (Hbk)**
**ISBN 978-3-0365-1556-4 (PDF)**

# Contents

# About the Editors

**Marek Krawczuk** graduated in 1987 from the Gdańsk University of Technology. He obtained his doctorate in 1991 at the Institute of Fluid-Flow Machinery of the Polish Academy of Sciences. In 1995, he defended his habilitation thesis, and, in 2003, he was nominated as Professor from the President of Poland. The main areas of his scientific interest are numerical methods applied to the analysis of statics and dynamics of structures with failures. For many years, he has also been dealing with the issue of using vibration signals to detect structural damage. He also deals with issues in the areas of smart materials and biomedical engineering. He is the author of over 100 papers from the Web of Sciences database, which have been cited over 3000 times. His Hirsh index is currently 31.

**Magdalena Palacz** Ph.D., D.Sc.Eng., Assoc. Prof. at Silesian University of Technology, Faculty of Organization and Management, lecturer and researcher. She specializes in effective numerical tools for modeling dynamic parameters of structural elements for their use in technical diagnostics as well as experimental verification. Her interests cover the mechanical properties of materials with controllable physical parameters in terms of their practical application. She has been involved in the realization of various national and international scientific projects, which have benefitted from publications such as 31 articles in indexed international journals, 20 international conference papers, 2 monographs (in Polish). Her current Hirsh index is 14.

 *applied sciences*

*Editorial*

# Special Issue "Applications of Finite Element Modeling for Mechanical and Mechatronic Systems"

**Marek Krawczuk** [1,†] and **Magdalena Palacz** [2,*,†]

1. Department of Mechatronics and High Voltage Engineering, Faculty of Electrical and Control Engineering, Gdańsk University of Technology, Narutowicza 11/12, 80-233 Gdańsk, Poland; marek.krawczuk@pg.edu.pl
2. Department of Production Engineering, Faculty of Organisation and Management, Silesian University of Technology, Roosevelta 26, 41-806 Zabrze, Poland
* Correspondence: magdalena.palacz@polsl.pl; Tel.: +48-32-237-7393
† M. Krawczuk and M. Palacz contributed equally to this work.

**Abstract:** Modern engineering practice requires advanced numerical modeling because, among other things, it reduces the costs associated with prototyping or predicting the occurrence of potentially dangerous situations during operation in certain defined conditions. Different methods have so far been used to implement the real structure into the numerical version. The most popular have been variations of the finite element method (FEM). The aim of this Special Issue has been to familiarize the reader with the latest applications of the FEM for the modeling and analysis of diverse mechanical problems. Authors are encouraged to provide a concise description of the specific application or a potential application of the Special Issue.

**Keywords:** finite element method; numerical modeling; mechanical parameters; damage detection

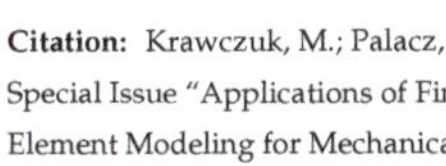

**Citation:** Krawczuk, M.; Palacz, M. Special Issue "Applications of Finite Element Modeling for Mechanical and Mechatronic Systems". *Appl. Sci.* **2021**, *11*, 5170. https://doi.org/10.3390/app11115170

Received: 28 May 2021
Accepted: 31 May 2021
Published: 2 June 2021

**Publisher's Note:** MDPI stays neutral with regard to jurisdictional claims in published maps and institutional affiliations.



Modern mechatronic systems first appeared about 50 years ago. A mechatronic system is currently defined as a structure implementing tasks from the area of four subsystems related to sensors (collect data about the operation of the system), control (a system processing data collected from sensors and regulating the operation of the executive system, whose task is to control the device), executive system (regulating the operation of the actuators) and a linking system (transmitting data between the above-mentioned subsystems). The design and state analysis of such complex systems is undoubtedly a demanding challenge for engineers and designers, as it requires the use of expertise from various engineering areas.

The main problem that engineers encounter is the use of well-known analytical mathematical methods, which are inadequate for more complex systems and can lead to erroneous results. For this reason, modern engineering in principle already depends on numerical modelling. This makes it possible not only to reduce the cost of prototyping designed mechatronic systems, but also to predict potentially dangerous situations resulting from operation. Among the various methods of numerical modelling which can be used in modern engineering, the most common is the finite element method due to its universal nature and the relatively simple way of preparing the model. Despite the fact that the method appeared commercially in the 1950s, its popularity is still increasing and the mathematical core is constantly being improved. It is used in the modelling of almost all areas of science and technology. In this book it was possible to gather examples of the applications of the finite element method in many areas of technology, carried out by various research centres. One general statement may be derived here—FEM is a viable tool.

In the group of papers devoted to the analysis of complex mechatronic systems, it is necessary to notice studies from the area of automotive and aerospace industry, in particular [1–6]. Among papers devoted especially to the analysis of complex mechatronic systems the following ones can be listed. The first listed [1] presents a methodology for conducting a study of the behaviour of a disabled driver during a crash by the use of FEM. Interesting conclusions have been derived, the reader is encouraged to follow them. In the [2] a direction for crash model development in Multi Body Systems programs to consider a varied 3D body space zones stiffness related to the structure of the car body and the internal car elements instead of modeling the car body as a solid with an average stiffness has been indicated. Such an approach would provide an alternative improvement to Finite Element Method (FEM) conventional modeling. Following [3] presented a methodology of optimising the weight of a bell crank, sourced from a Louis Christen Road Racing F1 Sidecar by developing a 3D model in ABAQUS (one of leading worldwide known commercial FEM based software). The model has been verified through the experimental measurements of dynamic characteristics. After that the model was used for topology optimisation procedure and then converted back to a 3D model and then fabricated to produce a physical prototype for design verification and validation by means of FE analysis and laboratory experiments and then compared with the original part. The proposed procedure is applicable and effective in topology optimization to obtain a lightweight (approximately 3% weight saving) and structurally strong design. The authors of [4] have used FEM for the structural analysis of the front single-sided swing-arm of a new three-wheel electric motorcycle, recently designed to meet the challenges of the vehicle electrification era. A dedicated Computer Aided Engineering (CAE) software has been used for the structural evaluation of different swing-arm designs, through a series of finite element analysis simulations. A topology optimization procedure has been also implemented to assist the redesign effort and reduce the weight of the final design. Simulation results in the worst-case loading conditions, have indicated that the proposed structure is effective and promising for actual prototyping. A direct comparison of results for the initial and final swing-arm design revealed that a 23.2% weight reduction has been achieved. In this article [5] the authors developed a model of all-steel radial tire that has been expanded to include the non-linear stress–strain relationship for textile cord and its thermal shrinkage. Variable cord density and cord angle in the cord-rubber bias tire composite are the major challenges in pneumatic tire modeling. The available FEM code with implemented user subroutines in MSC, have allowed the description of the tire specific properties. The distinguishing feature of the developed model from other ones is the exact determination of the cord angles in a vulcanized tire and the possibility of simulation with the tire mounting on the rim and with cord thermal shrinkage taken into account. The model may be an effective tool in bias tire design. Finally in a paper from aerospace industrial application the paper [6] shows an application of FEM for the analysis of thin-layer composites used for designed load-bearing structures. Due to proposed software enabling quick evaluation of such structure a facilitation of the initial concept has been possible. The proposed procedure used FEM for verification and improvement of the composite component.

Further examples of the application of FEM for the analysis of complex mechatronic issues present in modern production engineering industry is the paper [7] describing the compensation of strain deviation in the machine direction of a web in the roll-to-roll process for optimal mass-production of flexible devices at low cost. According to the results of a comparative experiment conducted to confirm the correcting performance of the optimized roller, the strain deviation in the machine direction decreased by approximately 48% with the proposed roller compared to that of the conventional roller. An industrial solution of engineering problem has been performed in [8]. The aim of the study presented here has been to predict properly the influence of high peening coverage on the Almen intensity and residual comprehensive stress. Therefore a quantitative description of the peening coverage has been developed. Based on the quantitative description, the finite element simulation and Almen test have been carried out. The simulation results of arc height and saturation curve have agreed well with that of the Almen test, by which the effectiveness of the quantitative description and FE simulation have been proved. The further study indicated that in shot peening processes, the excessive peening coverage does not improved Almen intensity and residual compressive stress. Following, an important aspect of industrial safety analysis that is of key importance to the mining industry an interesting FEM application for mechatronic system has been presented by [9] where a thorough analysis has been performed in order to determine how the tensile strength of roof rocks influences the extent of the zone with a particularly high risk of spontaneous coal combustion (endogenous fires) in caving goaves of the long-walls ventilated with the Y-type system. To achieve this goal, model-based tests have been conducted for a region of the long-wall mined with caving and ventilated with the Y-type system. The results obtained indicate that the type of rocks forming the caving affects its permeability and the extent of the risk zone for spontaneous coal combustion. The effectiveness of these measures significantly may improve the safety of mining exploitation.

The complex nature of the analysis of issues related to the safety of machinery and equipment, and in particular the detection and observation of fatigue damage, is the subject of the following articles [10–12]. In this study, [10], the technique of a digital holographic microscope and a digital height correlation method has been applied in combination with finite element analysis using a 2D and 3D model simulating the turbine blades to ensure their safety against damage. Analysis performed have clarified that the change in the surface properties under a small load varied according to the presence or absence of a crack, and elucidated the strain distribution that caused the difference in the change. The difference in the change in the top surface height distribution of the materials with and without a crack was directly proportional to the crack length. The authors here [11] have analysed the effects of impact loading acting on composite panels made out of epoxy resin as a matrix with carbon fabrics reinforced with aramid. The numerical calculations have been performed by the use of FEM and each reinforcement layer has been modeled as an independent part. The performed numerical and microscopy tests allowed to determine some destruction mechanisms of the panels depending on the geometry of the striker. Interesting conclusions have been derived regarding the striker shape and final delamination area properties. Finally in this study [12], a finite element, fully three-dimensional solid modeling method has been used to study the mechanical response of a steel-cored aluminum strand with a mid-phase jumper under a bending deformation. The analysis showed that the swing of the mid-phase jumper in the east–west direction caused a greater bending moment at the lower area of the mid-phase jumper, which led to the stress concentration appearing near the outlet of the tension clamp. This explained why the actual mid-phase jumper breakage occurred at the outlet of the tension clamp. Although this modeling method has been applied to the stress and deformation analysis of a mid-phase jumper in this study, it can be used to study the bending deformation of rope structures with a complex geometry and the main bending deformation. The finite element modeling method of a mid-phase jumper presented in this paper can be implemented in any general finite element software.

This book includes also papers from which it is clear that increasing interest in the scientific community is being shown for problems whose solution requires cooperation of many scientific disciplines [13,14]. Furthermore, so the problem of active control in certain structures has been analysed in [13]. Due to the fact that modern technology allows to design dedicated structures of specific features the authors have performed numerical research on a beam element built of periodically arranged elementary cells, with active piezoelectric elements. The control of parameters of this structure enables one for active damping of vibrations in a specific band in the beam spectrum. For this analysis, the authors propose numerical models based on the finite element method (FEM) and the spectral finite element methods defined in the frequency domain (FDSFEM) and the time domain (TDSFEM). An interesting example of another multidisciplinary application of FEM has been presented in [14], where an application of FEM analysis for printed circuit analysis, especially to determine the stress by changing the misalignment to below 30% of the primary channel width. The motivation for the research is based on commonly known fact, that printed circuit heat exchangers (PCHEs), which are used for thermal heat storage and power generation, are often subject to severe pressure and temperature differences between primary and secondary channels, which causes mechanical integrity problems. After the analysis performed it has been concluded that for this particular application the mechanical integrity of the PCHE with low-pressure molten salt or liquid metal and a high-pressure steam channel is acceptable in terms of utilization factor.

Mechatronic systems are not only advanced industrial systems, but also less complex mechanical systems requiring the determination of important parameters resulting from specific operating conditions. In this group may be placed papers [15–17]. In the study [15] the authors applied nonuniform-rational-basis-spline (NURBS) curves for the design of torsion springs, analyzed the displacements of these springs using finite-element analysis, and verified the design of these springs through experimentation. A method was proposed for deriving the coordinates of a control point for shifted elements by applying the inverse method on the basis of data derived through finite-element analysis. In addition, the relationship between the movement of the control point and stiffness matrix was identified and formulated by varying the torsion-spring parameters. In this paper, [16], according to the needs of large-diameter core drilling, a core barrel joint has been designed with an outer diameter of diameter 135 mm and a trapezoidal thread profile. Subsequently, a three-dimensional simulation model of the joint has been established. The influence of the external load, connection state and thread structure on the stress distribution in the joint has been analyzed through simulations, from which the optimal thread structure was determined. An example of FEM implementation for the analysis the designed metrology system may be found here [17]. In this paper, simulations have been performed to compare the distribution of temperatures developed in the measurement system for thermal conductivity of solid materials, which operates under a condition of permanent heat flow. The radial heat leaks, which affect the measurement parameters for an aluminium bar has also been analysed. The authors used copper bars as reference material. On that bases the authors implemented a thermal conductivity measurement system for solid materials limited in its operating intervals to measurements of maximum 300 °C for solid conductive materials.

In spite of the fact that FEM is very popular and most of the commercially available software allows obtaining reliable results there are occasional situations where the quality of obtained results is affected by potential numerical errors, therefore it is especially appreciated that this book contains works that address this issue [18,19]. In the first of them [18] may be found the algorithm of transition piezoelectric elements for adaptive analysis of electro-mechanical systems with analysis of numerical effectiveness of the models and their approximations in the contexts of: ability to remove high stress gradients between the basic and transition models, and convergence of the numerical solutions for the model problems of piezoelectrics with and without the proposed transition elements. In the second [19]—proposed detection algorithms assigned for the hpq-adaptive finite element analysis of the solid mechanics problems affected by the locking phenomena. The algorithms have been combined with the M- and hpq-adaptive finite element method, where M is the element model, h denotes the element size parameter, and p and q stand for the longitudinal and transverse approximation orders within an element.

In conclusion, the editors would like to thank all the authors whose careful work and dedication has resulted in so many interesting articles. The editors also hope that after reading this book, the reader will find answers to the questions in the area of FEM applications to the analysis of mechatronic systems.

**Author Contributions:** Conceptualization, M.P. and M.K.; validation, M.P. and M.K.; formal analysis, M.P. and M.K.; writing—original draft preparation, M.P.; writing—review and editing, M.P. and M.K. All authors have read and agreed to the published version of the manuscript.

**Funding:** This research received no external funding.

**Institutional Review Board Statement:** Not applicable.

**Conflicts of Interest:** The authors declare no conflict of interest.

## References


1. Sybilski, K.; Małachowski, J. Impact of Disabled Driver's Mass Center Location on Biomechanical Parameters during Crash. *Appl. Sci.* **2021**, *11*, 1427. [CrossRef]
2. Aleksandrowicz, P. Modeling Head-On Collisions: The Problem of Identifying Collision Parameters. *Appl. Sci.* **2020**, *10*, 6212. [CrossRef]
3. Pang, T.Y.; Fard, M. Reverse Engineering and Topology Optimization for Weight-Reduction of a Bell-Crank. *Appl. Sci.* **2020**, *10*, 8568. [CrossRef]
4. Spanoudakis, P.; Christenas, E.; Tsourveloudis, N.C. Design and Structural Analysis of a Front Single-Sided Swingarm for an Electric Three-Wheel Motorcycle. *Appl. Sci.* **2020**, *10*, 6063. [CrossRef]
5. Pelc, J. Bias Truck Tire Deformation Analysis with Finite Element Modeling. *Appl. Sci.* **2020**, *10*, 4326. [CrossRef]
6. Skarka, W.; Jałowiecki, A. Automation of a Thin-Layer Load-Bearing Structure Design on the Example of High Altitude Long Endurance Unmanned Aerial Vehicle (HALE UAV). *Appl. Sci.* **2021**, *11*, 2645. [CrossRef]
7. Kang, Y.; Jeon, Y.; Ji, H.; Kwon, S.; Kim, G.E.; Lee, M.G. Optimizing Roller Design to Improve Web Strain Uniformity in Roll-to-Roll Process. *Appl. Sci.* **2020**, *10*, 7564. [CrossRef]
8. Yang, Z.; Lee, Y.; He, S.; Jia, W.; Zhao, J. Analysis of the Influence of High Peening Coverage on Almen Intensity and Residual Compressive Stress. *Appl. Sci.* **2020**, *10*, 105. [CrossRef]
9. Tutak, M.; Brodny, J. The Impact of the Strength of Roof Rocks on the Extent of the Zone with a High Risk of Spontaneous Coal Combustion for Fully Powered Longwalls Ventilated with the Y-Type System—A Case Study. *Appl. Sci.* **2019**, *9*, 5315. [CrossRef]
10. Sakamoto, J.; Tada, N.; Uemori, T.; Kuniyasu, H. Finite Element Study of the Effect of Internal Cracks on Surface Profile Change due to Low Loading of Turbine Blade. *Appl. Sci.* **2020**, *10*, 4883. [CrossRef]
11. Sławski, S.; Szymiczek, M.; Kaczmarczyk, J.; Domin, J.; Duda, S. Experimental and Numerical Investigation of Striker Shape Influence on the Destruction Image in Multilayered Composite after Low Velocity Impact. *Appl. Sci.* **2020**, *10*, 288. [CrossRef]
12. Ma, P.; Li, Y.; Han, J.; He, C.; Xiao, W. Finite Element Modeling and Stress Analysis of a Six-Splitting Mid-Phase Jumper. *Appl. Sci.* **2020**, *10*, 644. [CrossRef]
13. Waszkowiak, W.; Krawczuk, M.; Palacz, M. Finite Element Approaches to Model Electromechanical, Periodic Beams. *Appl. Sci.* **2020**, *10*, 1992. [CrossRef]
14. Simanjuntak, A.P.; Lee, J.Y. Mechanical Integrity Analysis of a Printed Circuit Heat Exchanger with Channel Misalignment. *Appl. Sci.* **2020**, *10*, 2169. [CrossRef]
15. Kim, Y.S.; Song, Y.J.; Jeon, E.S. Numerical and Experimental Analysis of Torsion Springs Using NURBS Curves. *Appl. Sci.* **2020**, *10*, 2629. [CrossRef]
16. Wang, Y.; Qian, C.; Kong, L.; Zhou, Q.; Gong, J. Design Optimization for the Thin-Walled Joint Thread of a Coring Tool Used for Deep Boreholes. *Appl. Sci.* **2020**, *10*, 2669. [CrossRef]
17. Gonzalez Duran, J.E.E.; González-Rodríguez, O.J.; Zamora-Antuñano, M.A.; Rodríguez-Reséndiz, J.; Méndez-Lozano, N.; Gómez Meléndez, D.J.; García García, R. Finite Element Method and Cut Bar Method-Based Comparison Under 150°, 175° and 310 °C for an Aluminium Bar. *Appl. Sci.* **2020**, *10*, 296. [CrossRef]
18. Zboiński, G.; Zielińska, M. 3D-Based Transition hpq/hp-Adaptive Finite Elements for Analysis of Piezoelectrics. *Appl. Sci.* **2021**, *11*, 4062. [CrossRef]
19. Miazio, Ł.; Zboiński, G. A Posteriori Detection of Numerical Locking in hpq-Adaptive Finite Element Analysis. *Appl. Sci.* **2020**, *10*, 8247. [CrossRef]

*applied sciences*

MDPI

*Article*

# Impact of Disabled Driver's Mass Center Location on Biomechanical Parameters during Crash

**Kamil Sybilski * and Jerzy Małachowski**

Faculty of Mechanical Engineering, Military University of Technology, Gen. Sylwestra Kaliskiego Street 2, 00-908 Warsaw, Poland; jerzy.malachowski@wat.edu.pl
* Correspondence: kamil.sybilski@wat.edu.pl; Tel.: +48-261-83-96-83

**Abstract:** Adapting a car for a disable person involves adding additional equipment to compensate for the driver's disability. During this process, the change in the driver's position and kinematics and their impact on safety levels during crash is not considered. There is also a lack of studies in the literature on this problem. This paper describes a methodology for conducting a study of the behavior of a disabled driver during a crash using the finite element method, based on an explicit time integration method. A validated car model and a commercial dummy model were used. The results show that the use of a handle on the steering wheel and a hand control unit causes dangerous lateral displacements relative to the seat. Amputation of the left leg or right arm causes significant shoulder rotations, amputation of the left leg causes increased thoracic loads. Amputation or additional equipment have no significant impact on head injuries.

**Keywords:** finite element method; numerical simulation; biomechanics; head injury; safety; injury criteria; disability; driver

**Citation:** Sybilski, K.; Małachowski, J. Impact of Disabled Driver's Mass Center Location on Biomechanical Parameters during Crash. *Appl. Sci.* **2021**, *11*, 1427. https://doi.org/10.3390/app11041427

Received: 31 December 2020
Accepted: 26 January 2021
Published: 4 February 2021

**Publisher's Note:** MDPI stays neutral with regard to jurisdictional claims in published maps and institutional affiliations.



## 1. Introduction

Over a billion people live with some form of disability, which represents 15% of the world's population [1]. Between 110–190 million adults have very significant difficulties in functioning. Rate of disability is increasing. According to [2], approximately 2% of road accidents in Spain result in moderate, serious or total disability. The authors of [2] point out that the acquisition of a disability is associated with a reduced ability to work, greater functional dependence, greater need for assistance, and the need for family support.

Technological and medical developments make it possible to extend and improve quality of life. A great deal of attention has been paid in recent years to activating older people and people with disabilities (DP—disabled persons). Researchers from all over the world carry out research related to different aspects of DPs' lives, looking, for example, at how they spend their leisure time [3], how well the infrastructure fits their needs [4–6], or their preferences when making choices [7]. These people in many cases possess valuable skills. Understanding the specific requirements of this social group allows the development of technical solutions that remove barriers that prevent them from active functioning, socially and professionally. This will enable the public to benefit from these skills.

Among the many factors influencing the professional and social activation of people with disabilities, aspects related to the mobility of DPs and the adaptation of existing systems of transportation to the needs of older and disabled people are often mentioned [5,8]. While in large cities, DPs can count on public transport adapted to their needs, in smaller towns and villages DPs are practically dependent on having individual means of transport or on third parties to provide them. One possible solution of this problem is the individual adaptation of a car to the needs of the particular person with disabilities [9]. However, it should be considered to ensure that any structural changes made do not result in reduced road safety, other people, and disabled driver.

Among many causes of road accidents, the human factor is indicated as one of the main ones [10]. It is difficult to eliminate all the imperfections and limitations of the physical driver, but thanks to technological developments, the driver has more and more systems to support his actions. Among these are constantly developed active safety systems. Their use, combined with appropriate training of drivers to operate them, can significantly reduce the number of road accidents [11,12].

New solutions for use in the automotive industry must be tested in accordance with product standards. The most extensive testing applies to new vehicles, as each must pass very stringent tests before it can be put into road traffic. Of course, cars on the market have varying levels of safety (depending, among other things, on the number of active driver assistance systems), but any new car that does not meet the minimum requirements cannot be put on the road. Another point worth noting is that new cars are designed with the average consumer in mind, and tests are carried out for the chosen body configuration and weight of the occupants. Therefore, modification of the vehicle by changing the steering equipment or adjusting the car for a person with a physique significantly different from the one assumed during design, requires additional tests [13].

There are now very many methods in use for analysing dangerous situations that may occur on the road. Experimental research is undoubtedly the most important of these. Their disadvantage is the very high unit cost of each test and restrictions on carrying out certain measurements. Therefore, a very popular method of verifying the operation of technical objects is the numerical analysis [14–18]. Testing in virtual space allows for evaluating the structure in a short period of time in order to check the compliance with many standards [19,20], and for predicting the structure's behaviour in different load scenarios. The lack of the need to physically build new prototypes and prepare experimental research also allows for significant financial savings. An additional advantage of computer simulations is the possibility to record more data than in the case of experimental studies [21].

Many different numerical tools are currently available for analyzing dangerous traffic situations [22]. One of the fastest are calculations using analytical formulae that take into account, among other things, the velocity, weight and stiffness of vehicles [23–25]. They allow many scenarios to be analyzed in a very short time and, when supplemented with a reliable vehicle database, make it possible to assess a real accident. An additional advantage of analytical methods is the possibility of transferring the loads acting on the driver to 3D models and further local analysis of his behavior.

Another fast and accurate method based on Reduced Order Dynamic Model [26], in which discretization of vehicle's perimeter takes place only in a 2D environment. It reduces the number of equation and thus reduce time calculation.

Analytical and 2D models do not allow an accurate analysis of driver behavior. Therefore, if the aim of the study is to determine e.g., injuries to the driver, methods based on multibody analyzes [19] or FEM are used [27]. The big disadvantage of these methods is the long calculation time. To reduce it, it is possible to use different strategies. One of them is an approach in which a global collision is analyzed using analytical or 2D models and the results are transferred to 3D models and further analysis of only a selected area. A second solution is to use multi-stage analyses, in which selected aspects of a hazardous road event are investigated independently.

## 2. Materials and Methods

### 2.1. Numerical Research Strategy

The main objective of this article is to examine the impact of the change in the position of the driver's mass, caused by various types of disabilities or the use of specialist equipment, on the driver's biomechanical parameters. Additional objectives are to show how to model selected elements of car safety systems and to draw attention to possible changes in the driver's biomechanics related to the adaptation of the car to the needs of the disabled.

This article is a continuation and development of models described in the work [28]. The paper uses a three-stage scheme of numerical tests (Figure 1), in which the first stage is to carry out an analysis of the frontal impact of the validated full car model [28]. Based on these analyses, the change of velocity of the vehicle interior, which is used in the third stage, is determined. The second stage involves subsidence a dummy on the seat while resting its limbs on the floor, the steering wheel or a special handle mounted on the steering wheel and the manual gas and brake control unit. The subsidence is a very important process. It involves the dummy falling under the gravity load on the interior elements of the vehicle. This deforms the structure of e.g., the seat and causes forces between them, which are transferred to the third stage. The magnitude of these forces directly affects the magnitude of the friction forces and thus the behaviour of the dummy during the impact.

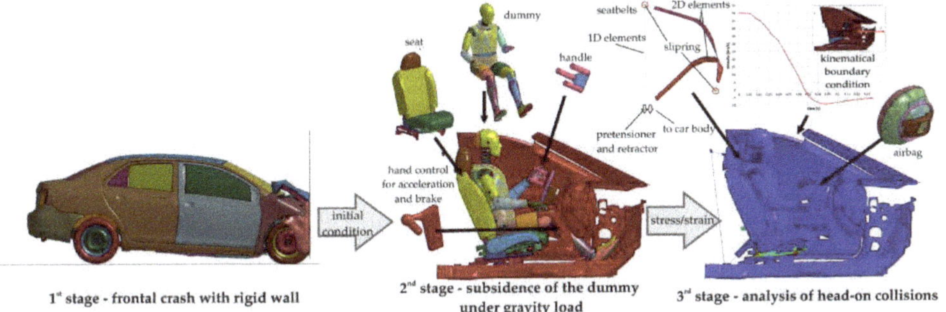


**Figure 1.** Numerical research strategy [28].

For numerical simulations carried out using the LS-Dyna system, two approaches to the realisation of this stage are popular. The former involves positioning the dummy in a roughly approximate position, positioning the belts and performing dynamic relaxation [18,29]. The advantage of this approach is its low numerical cost and rather low level of complexity. During the analysis, additional damping is added, which facilitates and shortens the subsidence process (oscillations of the manikin's position are dampened much faster). The dynamic relaxation is performed immediately before the actual numerical analysis, and the analysis time after the relaxation is 0.0 s, so there is no need to define appropriate shifts in numerical procedures defined later. Strains and deformations from dynamic relaxation are transferred automatically to the target analysis.

The second approach involves carrying out a full analysis of the subsidence process of the dummy (using an explicit or implicit time integration) [30,31]. In such analysis the dummy, under the influence of gravity, falls on the interior elements of the vehicle. The result is a file containing, for the time corresponding (termination time), the state of deformation, stresses and forces acting between the individual elements. Starting the collision analysis requires a full restart, which includes the procedure of loading the state from the end of the subsidence analysis (as an external file with the option stress initialisation), changing the initial (boundary change, etc.), boundary and inducement conditions by defining appropriate cards (preparing a new model file).

A simpler and less demanding method in terms of computer power is one that uses dynamic relaxation. However, it is limited by the problems with the safety belt retractor. During dynamic relaxation it is performed to a limited extent, which may result in a lack of adequate belt tension at the beginning of the final crash analysis. Therefore, the authors decided on an approach involving the performance of a full subsidence analysis using explicit time integration (stage 2 in Figure 1).

### 2.2. Numerical Model Description

In the numerical models developed by the authors, great emphasis has been placed on considering key elements of safety systems that can influence the behaviour of the driver's body during a frontal collision, while at the same time applying simplifications that do not significantly affect his behaviour. Therefore, on the basis of previously conducted research, it was decided to model only a section of the vehicle and give it the properties of a rigid body [28]. Deformable seat, steering wheel handle, airbag and seat belts were modelled inside the vehicle [28] (Figure 2).

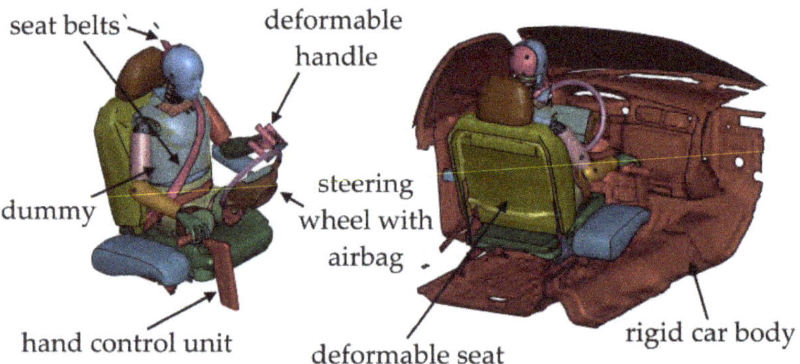


**Figure 2.** Numerical model used in simulations.

The seat belts were modelled on the basis of previously conducted experimental research [18,28,29]. Their arrangement was carried out using the seatbelt fitting procedure, available in the LS-Prepost pre-processor. Belts were modelled as 2D elements (for parts where their contact with the dummy is important) and as 1D elements for parts near attachment points (Figure 3). The lower belt is rigidly attached to the vehicle body on the right-hand side of the driver and to the ear connected to the seat base on the right. The function of the ear is performed by a special seatbelt slipring numerical element [30,31], thanks to which the shortening of one 1D element can be transformed into an extension of the other 1D element. The relation between the displacements of the ends of the two connected belts (Figure 3) is described using the following equation:

$$x_1 = x_2 + \Delta l_1 + \Delta l_2 \tag{1}$$

$$\Delta l_1 = (F_1 \cdot l_1)/(A_1 \cdot E_1) \tag{2}$$

$$\Delta l_2 = (F_2 \cdot l_2)/(A_2 \cdot E_2) \tag{3}$$

$$x_1 = x_2 + (F_1 \cdot l_1)/(A_1 \cdot E_1) + (F_2 \cdot l_2)/(A_2 \cdot E_2) \tag{4}$$

$$F_1 = F_2 + F_t = F_2 + F_1 \cdot \mu = F_2/(1 - \mu) \tag{5}$$

where: $x_1$, $x_2$—displacement of the ends of 1D seat belts elements, $\Delta l_1$, $\Delta l_1$—elongation and shortening of connected elements, $F_1$, $F_2$—tensioning force, $A_1$, $A_2$—cross section of elements, $E_1$, $E_2$—Young modulus of elements, $F_t$—friction force in slipring, $\mu$—coefficient of friction.

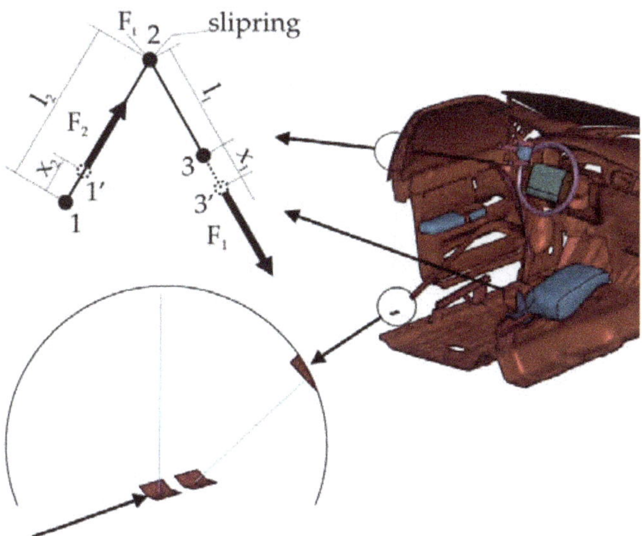


**Figure 3.** Numerical model of seatbelts.

If both belts (bottom and top) are made of the same material, the final formula is as follows.

$$x_1 = x_2 + ((F_2/(1 - \mu))\cdot l_1 + F_2\cdot l_2)/ (A\cdot E) = x_2 + (F_2\cdot(l_1/(1 - \mu) + l_2))/(A\cdot E) \qquad (6)$$

Thanks to the use of this type of numerical element, the "shortening" of the lower belt can be turned into the "lengthening" of the upper belt, i.e., it is possible to implement the rewinding of the belt through the assembly eye. In the developed numerical models, the upper belt starts and ends with the slipring elements. In the upper attachment, the belt changes direction and is connected to the attachment point in the lower part of the vehicle body, in which the elements representing the pretensioner and retractor are modelled. When modelling seat belts, the position of the attachment points should not be changed in relation to the points in the actual car, as this changes the length of the belt, which in turn affects its global deformation under the influence of force.

In numerical models, the pretensioner has been modelled using the SEATBELT_ RETRACTOR [30,31] type element that generates constant belt tension up to the pull-out force limit above which the retractor locks. The retractor is also locked when the pretensioner (SEATBELT_PRETENSIONER [30,31]) is activated, which retracts the belt until it reaches the user-defined belt tension limit. In these cases, the belt was activated by an acceleration sensor (SEATBELT_SENSOR) which registers the front acceleration of the vehicle. When the acceleration of 25.0 m/s$^2$ was exceeded, the sensor activated the pretensioner. The retractor and the pretensioner operate numerically similarly to the rewinder, with the difference that the shortening of the belt is not converted into the lengthening of another belt but is recorded as the retracted length of the belt.

### 2.3. Analysed Cases

During the research, three main groups of drivers were analysed: without disabilities (RD-reference driver), with disabilities requiring the use of a special steering wheel handle (H—group, HB—handle basic) and with a steering wheel handle and a manual gas and brake control unit (C—group, CB—control unit basic) (Figure 4). The use of specialist equipment may be required by various disabilities, such as paralysis or lack of a limb. Therefore, the HB and CB dummies have been modified so that it is possible to perform analyses for limbless dummies (Figure 5). The limb amputation, when compared to the paralysis, from the frontal impact analysis point of view primarily changes the position of the body's centre of gravity and reduces the areas of contact between the body and the vehicle interior. Figures 4 and 5 show the changes in the driver's mass (dm) and centre of gravity position for each of the analysed cases in the global coordinate system (dx, dy, dz).

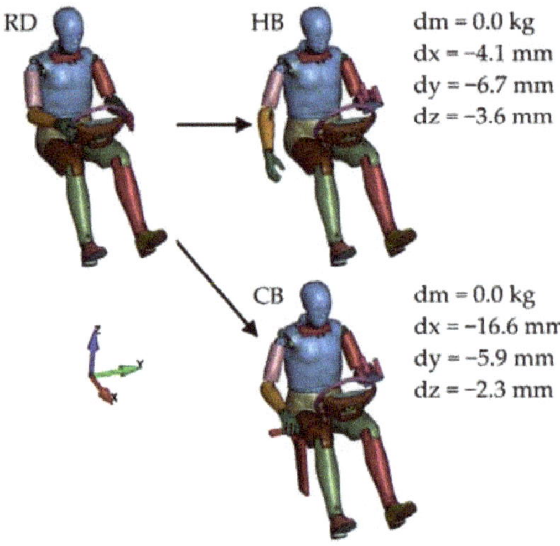


**Figure 4.** First stage of dummy model modifications.

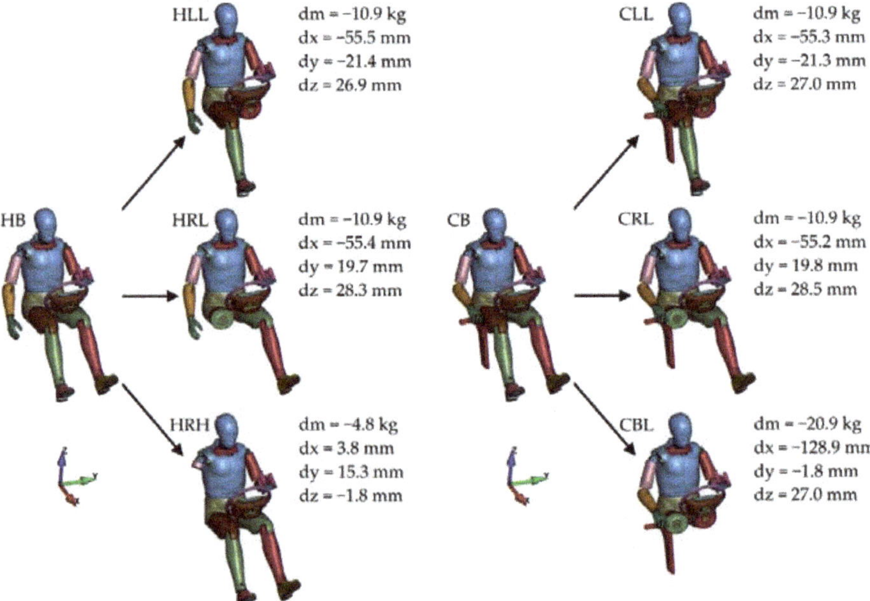


**Figure 5.** Second stage of dummy model modifications.

An individual subsidence analysis was conducted for each of the cases studied, preceded by positioning the dummy in an estimated target position. In each group, the initial setting was identical, so that only the impact of amputation of a given limb was examined.

## 3. Results

The results for the first and second stage, which covers the frontal impact of the whole car and subsidence, are presented in [28]. This study will only present the results for stage three.

During all analysed cases, characteristic time moments presented in Figure 6 can be distinguished. The subsidence stage ended at time t = 0.3 s and this is also the beginning of the analysis of the third stage. After approx. 10 ms, the airbag and belt pretensioner were activated. At t = 0.316 s the initial airbag opening is visible. After 36.0 ms, the airbag is close to full inflation, which pushes the upper limbs out of the steering wheel or handles. After 56.0 ms, the dummy's head encounters the airbag. At t = 0.380 s, the head reaches the maximum forward tilt, followed by the driver's rebound and a rearward movement to the seat. The contact between the driver's head and the airbag lasts until the time t = 0.400 s. At a time of approx. t = 0.432 s, the dummy hits the seat's backrest. This impact is asymmetrical to the seat.

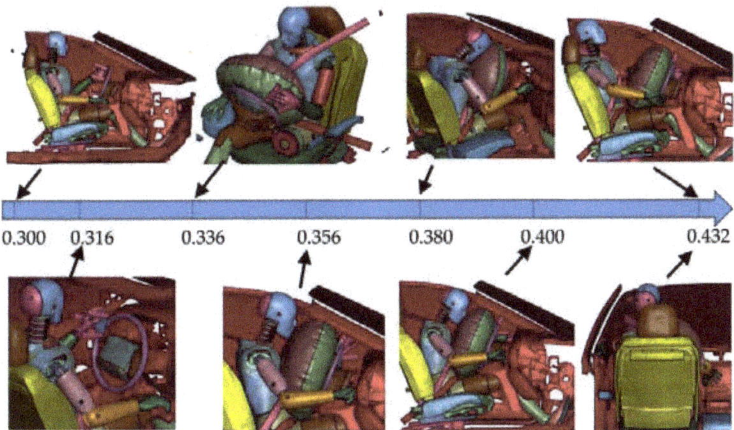


**Figure 6.** Characteristic time points of the simulations.

The maximum longitudinal displacement of the conventional centre of gravity of the dummy (i.e., corresponding to the centre of mass without amputation) is achieved at different times in each case (Figure 7). It is worth noting that at this moment, depending on the equipment used, the position of the dummy differs. In the RD case, the forearms and arms are pushed outwards while keeping the hands within the steering wheel. Knees hit the elements of the space under the steering wheel. In cases H, the left hand is pushed completely out of the steering wheel and the right hand hits the lower part of the central console. In cases C, the right hand strikes higher and in a different position. In cases H and C, the number of zones of contact between the dummy and the vehicle interior depends on the disability under consideration.

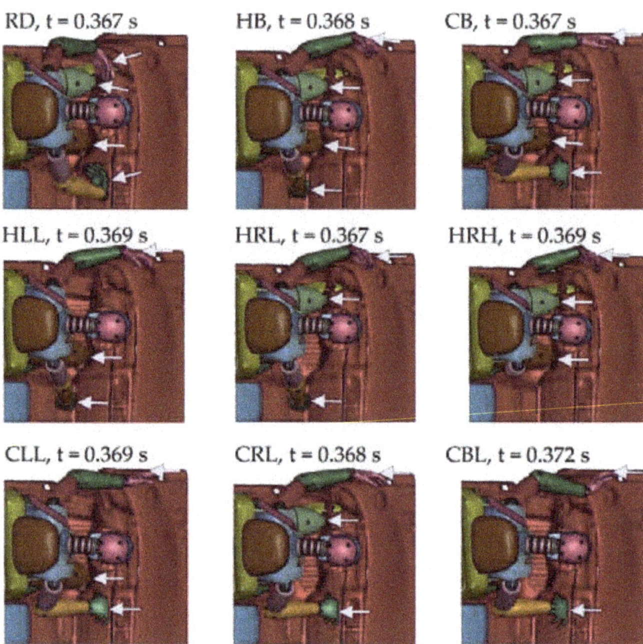



**Figure 7.** Positions of the dummy at the moments corresponding to the maximum longitudinal displacement.

The type of equipment used significantly affects the lateral displacements of the driver's conventional centre of gravity. In the RD case, the maximum displacement was about 10 mm and, importantly, in the final stage of impact the pelvis almost returned to its initial position (Figure 8). In the HB and CB cases, there is no return to the initial position and much greater maximum lateral displacement when the dummy hits the seat (approx. 15.5 mm and 18.0 mm respectively).

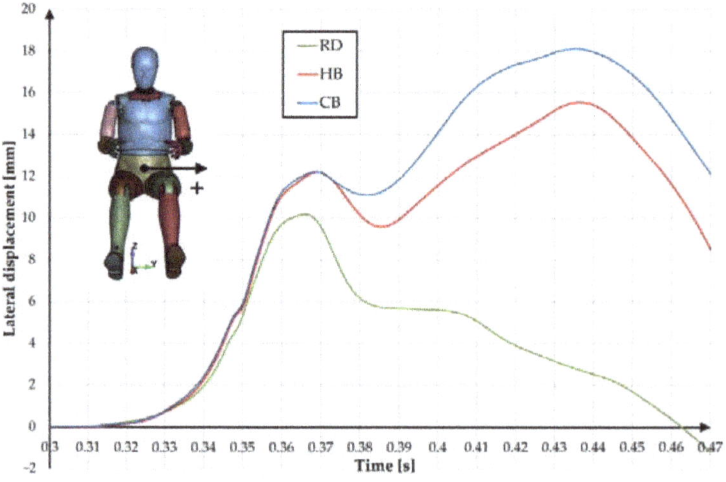



**Figure 8.** Lateral displacement of dummy H-point.

Analysing the H group, it can be seen that at the moment of the maximum longitudinal displacement of the dummy's centre of gravity in relation to the seat, the maximum lateral displacement occurs for HLL (Figure 9). In turn, when the dummy hits the seat's backrest, the maximum lateral displacement is for HRH (about 29 mm).

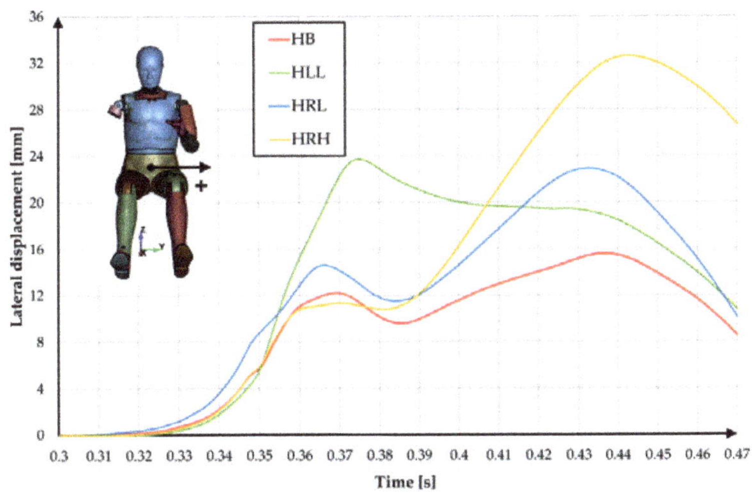

**Figure 9.** Lateral displacement of dummy H-point.

In group C the greatest lateral displacement can be observed for CLL, CRL and CBL cases (Figure 10).

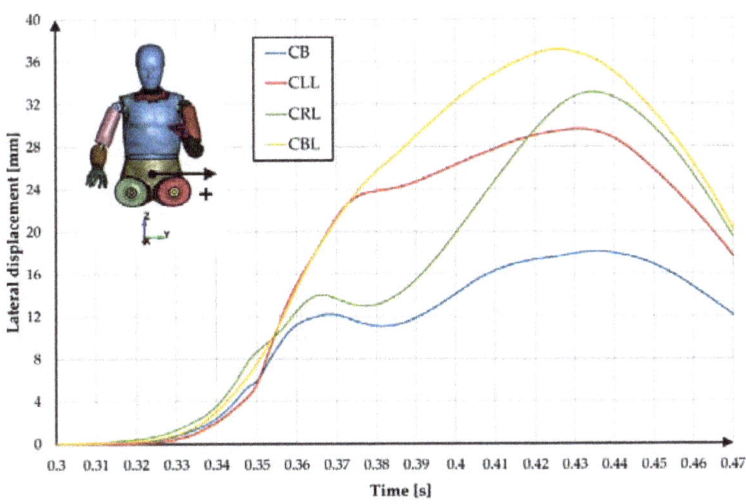

**Figure 10.** Lateral displacement of dummy H-point.

Lateral movements go with shoulder rotation. When analysing the influence of the equipment used for DP (Figure 11), there is no significant change. At the initial impact phase, in all cases, the right shoulder extends forwards more and, after rebounding from the airbag, the arms start to rotate in the opposite direction.

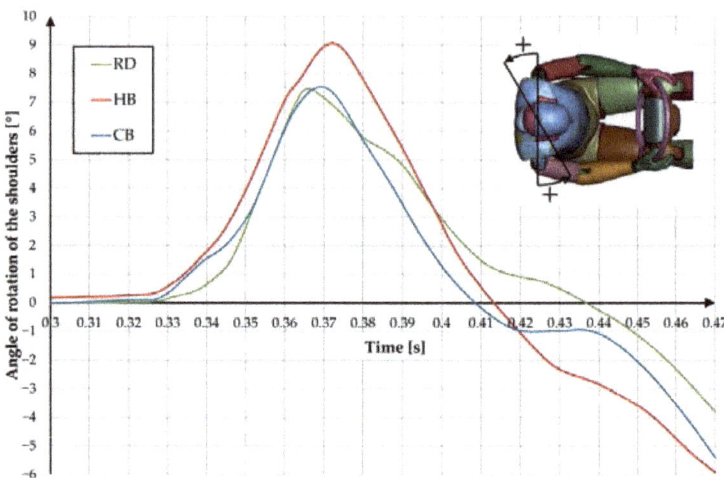

**Figure 11.** Angle of rotation of the shoulders.

Analysing group H, one can see the same type of changes, but with different levels of them (Figure 12). The highest rotation in the first impact phase is for HB and HRL cases (about 9°). In the second phase, for HLL and HRH (above 10°).

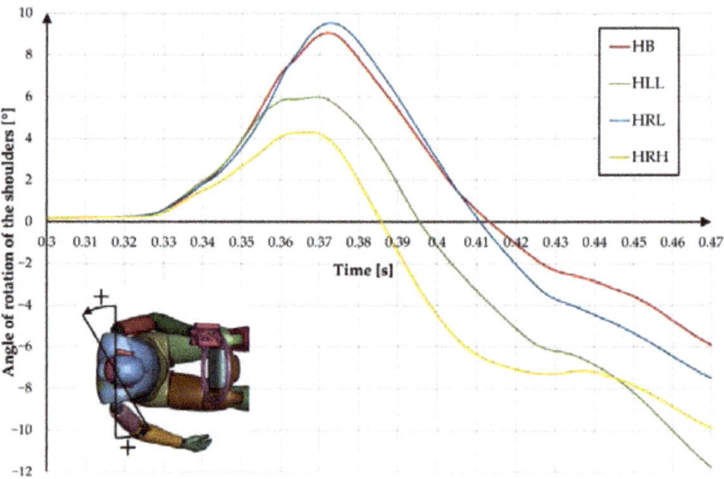

**Figure 12.** Angle of rotation of the shoulders.

Similarly, for group C (Figure 13). In the first phase, the highest rotation is for CB and CRL (about 7.5°) and in the second phase for CLL (over 12°) and CBL (over 15°).

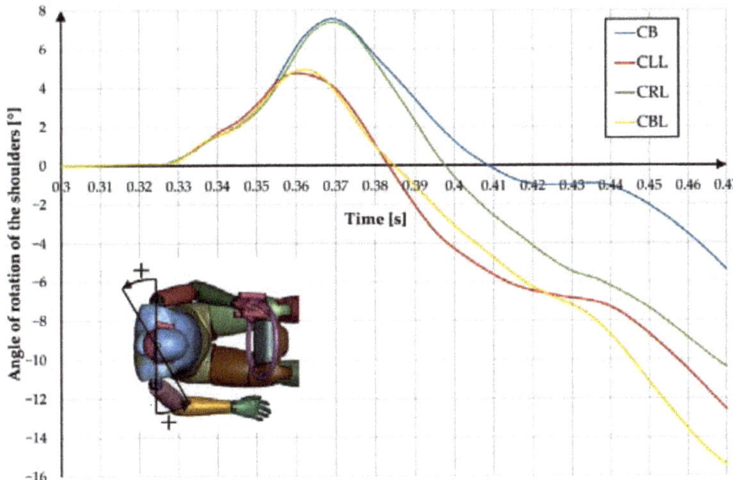

**Figure 13.** Angle of rotation of the shoulders.

The work also analysed the course of change of forces between the dummy and seat belts. In cases where the dummy had all the limbs, no significant changes between the individual runs are visible (Figure 14). The highest peak of strength occurs immediately before the head of the dummy (and thus part of the chest) contacts the airbag. After that time, the value drops by about 2 kN and is maintained until the maximum longitudinal displacement of the dummy in relation to the chair is reached.

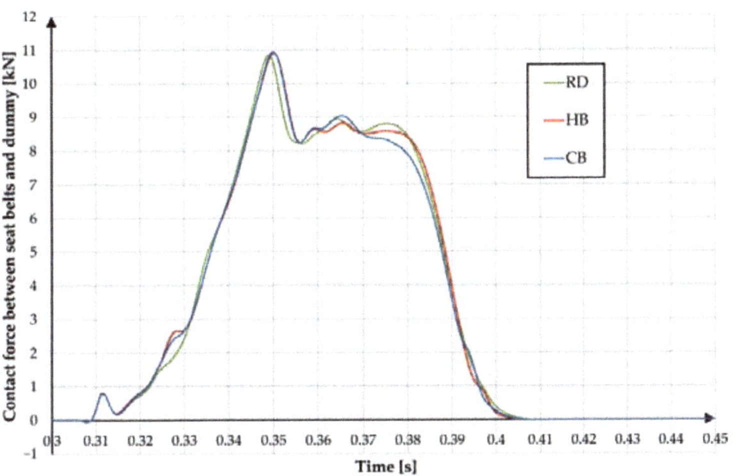

**Figure 14.** Comparison of contact force between the seat belts and the dummy.

A similar character of the course of force between the belts and the dummy can be observed for analyses from group H (Figure 15). Only in the HLL there is no decrease in force, it remains at a constant level until the time when the maximum movement of the longitudinal dummy in relation to the seat is reached.

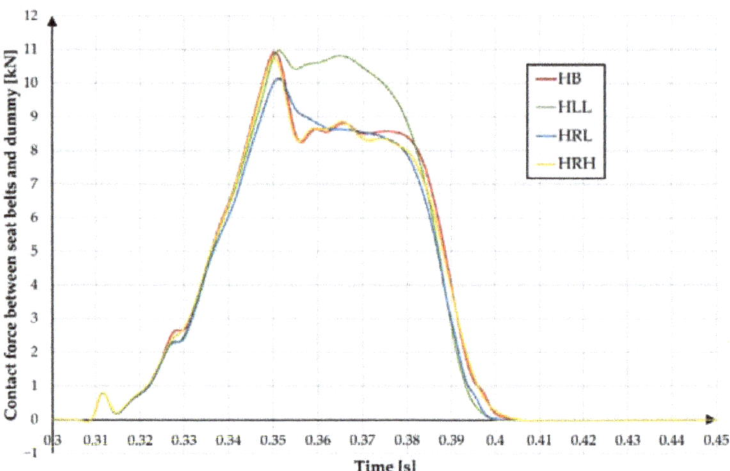

**Figure 15.** Comparison of contact force between the seat belts and the dummy.

In group C (Figure 16), no loss of strength after the dummy starts to contact the airbag is visible in two cases—CBL and CLL.

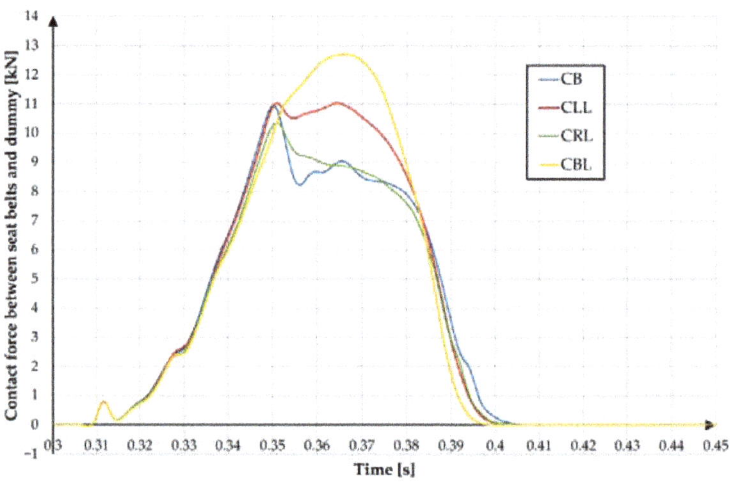

**Figure 16.** Comparison of contact force between the seat belts and the dummy.

The use of additional equipment for the DP, in addition to influencing lateral displacement, shoulder rotation and the course of force between the belts and the dummy, also affects skull injuries expressed as HIC (Head Injury Criterion) (Figure 17). Compared to RD, HB and CB have at least 6% lower HIC15 and 7% lower HIC36 values.

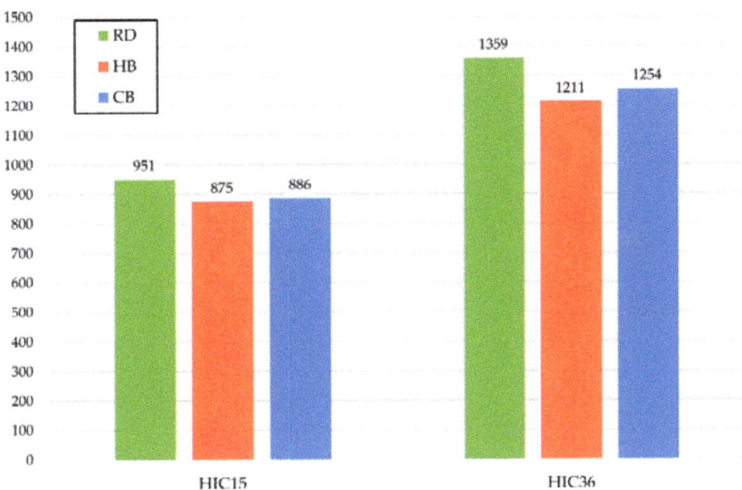

**Figure 17.** Head Injury Criterion (HIC) comparison.

In the H group (Figure 18), different HIC15 and HIC36 values were obtained for individual cases, but no significant differences were observed. The maximum reaches respectively 4% and 3%.

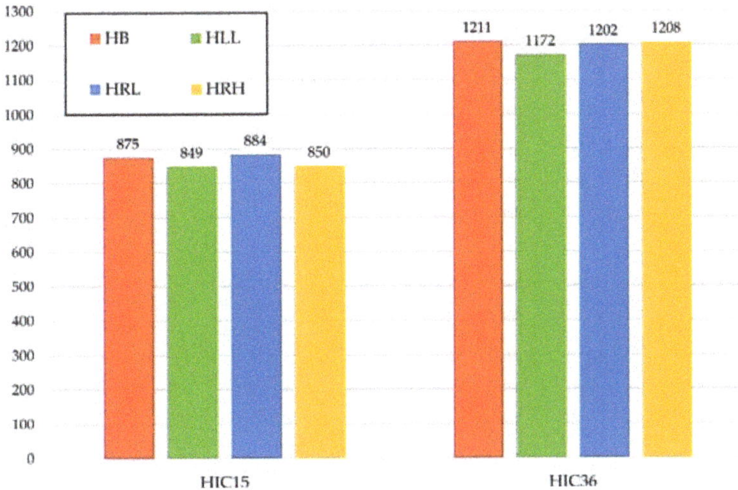

**Figure 18.** Head Injury Criterion (HIC) comparison.

In group C (Figure 19) for HIC15 the difference between the maximum and minimum result is about 4%. For HIC36 the difference increases less than 6%. The smallest value was obtained for CLL and the largest for CB.

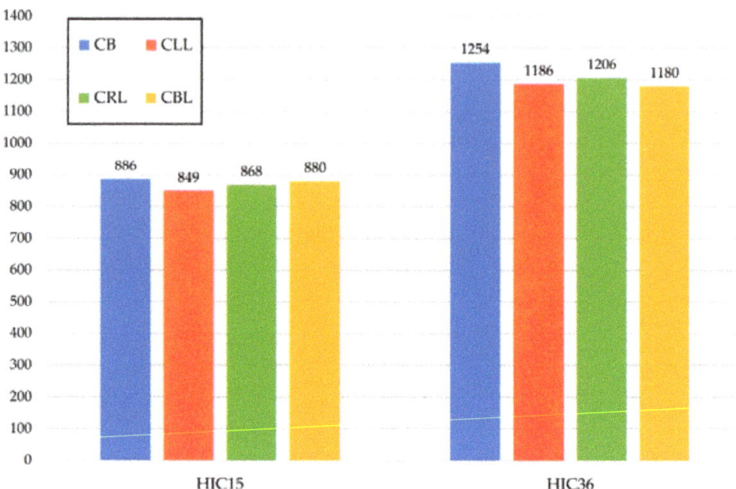

**Figure 19.** Head Injury Criterion (HIC) comparison.

## 4. Discussion

The main objective of the study is to examine the effect of the change in driver's centre of gravity due to various disabilities and the change in the points of support of the body due to the use of additional equipment on the driver's behaviour during a frontal collision. Due to the inability to carry out such studies through experimental tests, it was decided to employ a research methodology based on numerical analyses using the finite element method, with an explicit integration step. Thanks to FE studies, it was possible, among other things, to consider (in a virtual environment) the vulnerabilities of the car's interior structures, crucial from the point of view of the driver's behaviour, and of the dummy itself. A common approach to analyses of this type is to isolate a single phenomenon (with all the key elements) from the global system and to conduct studies on the influence of single parameters on the course of this phenomenon [19,21,32–34]. For frontal crash analyses aimed at assessing the driver's safety, this is most often a restriction of the numerical model to the area around the driver and modelling selected safety components such as seat belts [19,34] or an airbag. In this study the best possible reproduction of real objects and phenomena was sought, therefore a great deal of emphasis was placed on the use of the full characteristics of the car body delay during the collision, considering the deformation of the seat under the influence of the weight of the dummy, the friction between them, and the representation of the full operation of the seat belt system and the airbag. This was made possible by dividing the whole frontal impact into three phases—frontal impact of the car with a rigid barrier, subsidence of the dummy under the influence of gravity and the numerical analysis of the behaviour of the dummy inside an isolated part of the vehicle.

The authors were not able to carry out a validation test for the developed numerical models. Instead, the global model is built from smaller numerical models validated in independent tests [18,29]. In this study, a commercially approved numerical model of the dummy [35], an approved car model [36] and constitutive models of seat belts were used.

Frontal collisions, both full cars and selected vehicle parts, are very well documented in the literature. Nevertheless, the authors did not find any studies on the safety of disabled persons in the literature, so it is difficult to relate the results to other works.

In personal cars, the forces exerted on the body of a non-disabled driver during a frontal collision are asymmetrical. Excluding the asymmetrical positioning of the steering wheel in relation to the torso in many cars, asymmetry is mainly caused by the position of the legs and the operation of the seat belts. In the case of cars with automatic transmission, the right leg is usually set straight ahead or more inwards on the brake pedal during a collision. The left leg rests on a footrest that in many cars is positioned to the left and is closer to the seat than the fully depressed brake pedal. Seat belts, on the other hand, restrict the movement of the left arm, which naturally causes the rotation of the driver's shoulders visible in Figure 11. This rotation increases until the torso rests on the airbag and the driver rebounds from it. From now on, the arms start to rotate in the opposite direction. The lower limbs mentioned previously also play a major role in the first stage of the collision, as the driver moves longitudinally to the seat and hits the space under the steering wheel with his knees. This is considered by vehicle designers at the development stage.

The introduction of additional equipment on the steering wheel or the amputation of the driver's limb affects the process described above. When analysing the position of the dummy at the moment of maximum longitudinal displacement relative to the seat (Figure 7) it can be seen that the use of additional equipment on the steering wheel or between the seats affects the position of the driver's body immediately before the collision. The stiffness of these elements is so low that it does not generate significant movement resistance for the driver. The limb amputation, however, changes the number of support points of the driver, both immediately before the collision and during the loss of his kinetic energy. In addition, if one of the lower limbs is amputated, the latter is automatically more likely to suffer an injury as it will hit the components under the steering wheel with more force.

The changes in the position of the centre of gravity, as shown in Figures 4 and 5, clearly indicate that the sole use of a disabled person's equipment has a slight effect on the change in the driver's centre of gravity. Amputation, in turn, results in a significant reduction in the weight of the entire driver and thus in a significant shift in the driver's centre of gravity. On the one hand, the reduction in the driver's weight means less kinetic energy to reduce, on the other hand, a change in body position affects the operation of safety systems.

Figure 8 shows the lateral displacement of the driver's H-point, which shows that the use of a steering wheel handle or additional Hand Control Unit results in no return to the initial position of the driver's pelvis after a rebound from the airbag. This is particularly evident at time t = 0.43 s when the driver hits the seat's backrest. This is a very negative phenomenon, as it can result in missing the headrest and suffering a serious neck injury. This effect is even stronger in the case of amputation of any of the limbs (Figures 9 and 10).

However, the use of additional equipment does not have a significant influence on the rotation of the arms (Figure 11). The amputation, especially of the left leg, both legs or right hand, does influence the rotation (Figures 12 and 13). In the case of the amputation of the right hand, the mass that causes the rotation by its inertia decreases significantly and thus the rotation diminishes. A similar diminished rotation in the initial phase of movement can be observed in the case of the amputation of the left leg or both legs. This is because the left leg rests on a footrest that is closer to the seat than the brake pedal pressed to the end. In the event of a collision, the leg pushes against its entire surface, while the right leg slides on the brake pedal. Thus, in the case of drivers without disabilities, the left leg increases the rotation, and in the case of a disabled driver, the lack of the limb reduces the rotation.

Lack of support for the left leg means that the belts and the airbag must carry more load (Figures 15 and 16), which may result in increased chest injuries. The analyses carried out showed that the other factors examined did not have a significant impact on the course of force between the dummy and the belts. The modifications made also had no significant impact on HICs.

To sum up, the contribution of this paper is to present a three-stage scheme for conducting numerical analyses of the behaviour of people inside a car during a crash. The article justifies the importance of each of the stages and the use of the results of each of them in the analysis of the entire phenomenon. The paper also presents a method of seat belts modelling aimed at a faithful representation of their actual operation. Describes an approach using a combination of 2D elements (in area of contact with the driver) and 1D elements (which enable the operation of the retractor, pretensioner and slipring). A detailed diagram of the operation of the slipring elements is also presented. Based on the modelling performed, the paper presents the original results of analyses of the effect of changes in the position of the driver's centre of gravity caused by additional equipment for disabled people and amputations of the limbs on his behaviour during a frontal collision. Based on the results, it can be concluded that it is appropriate to carry out research aimed at assessing the safety of drivers using vehicles adapted to the needs of disabled people. The position of the driver immediately before the collision and the number of support points of the body affects his interaction with the safety systems and may affect his injuries.

Further work will be aimed at studying the impact of prostheses on the safety of a disabled driver.

**Author Contributions:** Conceptualization, K.S. and J.M.; methodology, K.S. and J.M.; software, K.S.; validation, K.S. and J.M.; formal analysis, K.S.; investigation, K.S.; resources, K.S.; data curation, K.S.; writing—original draft preparation, K.S.; writing—review and editing, K.S. and J.M.; visualization, K.S.; supervision, J.M.; project administration, K.S.; funding acquisition, J.M. All authors have read and agreed to the published version of the manuscript.

**Funding:** The article was written as part of the implementation of the university research grant supported by Military University of Technology (No. UGB 22-748/2020).

**Institutional Review Board Statement:** Not applicable.

**Informed Consent Statement:** Not applicable.

**Data Availability Statement:** Not applicable.

**Conflicts of Interest:** The authors declare no conflict of interest.

## References

1. WHO. Available online: https://www.who.int/features/factfiles/disability/en/ (accessed on 26 June 2020).
2. Palmera-suárez, R.; López-cuadrado, T.; Brockhaus, S. Severity of disability related to road traffic crashes in the Spanish adult population. *Accid. Anal. Prev.* **2016**, *91*, 36–42. [CrossRef] [PubMed]
3. Taylor, Z.; Józefowicz, I. Intra-urban daily mobility of disabled people for recreational and leisure purposes. *J. Transp. Geogr.* **2012**, *24*, 155–172. [CrossRef]
4. Solvoll, G.; Hanssen, T.S. User satisfaction with specialised transport for disabled in Norway. *J. Transp. Geogr.* **2017**, *62*, 1–7. [CrossRef]
5. Hassan, S.; Soltani, K.; Sham, M.; Awang, M. Accessibility for Disabled in Public Transportation Terminal. *Procedia Soc. Behav. Sci.* **2012**, *35*, 89–96. [CrossRef]
6. Chou, C.; Tsai, C.; Wong, C. A study on boarding facilities on wharves and ships for disabled and elderly passengers using public shipping transport. *J. Transp. Health* **2020**, *18*. [CrossRef]
7. Semeijn, J.; Gelderman, C.J.; Schijns, J.M.C.; Tiel, R. Van Disability and pro environmental behavior—An investigation of the determinants of purchasing environmentally friendly cars by disabled consumers. *Transp. Res. Part D* **2019**, *67*, 197–207. [CrossRef]
8. Suen, S.L. Adanced accessibility systems for elderly and disabled travellers in Canada. *IFAC Proc. Vol.* **1994**, *27*, 495–500. [CrossRef]
9. Monacelli, E.; Dupin, F.; Dumas, C.; Wagstaff, P. A review of the current situation and some future developments to aid disabled and senior drivers in France. *Irbm* **2009**, *30*, 234–239. [CrossRef]
10. Orsi, C.; Bertuccio, P.; Morandi, A.; Levi, F.; Bosetti, C.; La, C. Trends in motor vehicle crash mortality in Europe, 1980–2007. *Saf. Sci.* **2012**, *50*, 1009–1018. [CrossRef]
11. Koisaari, T.; Utriainen, R.; Kari, T.; Tervo, T. The most difficult at-fault fatal crashes to avoid with current active safety technology. *Accid. Anal. Prev.* **2020**, *135*. [CrossRef]
12. Zahabi, M.; Mohammed, A.; Razak, A.; Shortz, A.E.; Mehta, R.K.; Manser, M. Evaluating advanced driver-assistance system trainings using driver performance, attention allocation, and neural efficiency measures. *Appl. Ergon.* **2020**, *84*. [CrossRef] [PubMed]
13. Schneider, L.W.; Klinich, K.D.; Moore, J.L.; MacWilliams, J.B. Using in-depth investigations to identify transportation safety issues for wheelchair-seated occupants of motor vehicles. *Med. Eng. Phys.* **2010**, *32*, 237–247. [CrossRef]
14. Baranowski, P.; Damaziak, K.; Mazurkiewicz, L.; Muszynski, A.; Vangi, D. Analysis of mechanics of side impact test defined in UN / ECE Regulation 129. *Traffic Inj. Prev.* **2018**, *19*, 256–263. [CrossRef] [PubMed]
15. Klasztorny, M.; Nycz, D.B.; Szurgott, P. Modelling and simulation of crash tests of N2- W4-A category safety road barrier in horizontal concave arc. *Int. J. Crashworthiness* **2016**, *21*, 644–659. [CrossRef]
16. Ratajczak, M.; Frątczak, R.; Sławiński, G.; Niezgoda, T.; Będziński, R. Biomechanical analysis of head injury caused by a charge explosion under an armored vehicle. *Comput. Assist. Methods Eng. Sci.* **2017**, *24*, 3–15.

17. Arkusz, K.; Klekiel, T.; Sławiński, G.; Będziński, R. Influence of energy absorbers on Malgaigne fracture mechanism in lumbar-pelvic system under vertical impact load. *Comput. Methods Biomech. Biomed. Engin.* **2019**, *22*, 313–323. [CrossRef] [PubMed]
18. Mazurkiewicz, L.; Baranowski, P.; Karimi, H.R.; Damaziak, K.; Malachowski, J.; Muszynski, A.; Muszynski, A.; Robbersmyr, K.G.; Vangi, D. Improved child-resistant system for better side impact protection. *Int. J. Adv. Manuf. Technol.* **2018**. [CrossRef]
19. Joszko, K.; Wolański, W.; Burkacki, M.; Suchoń, S.; Zielonka, K.; Muszyński, A.; Gzik, M. Biomechanical analysis of injuries of rally driver with head supporting device. *Acta Bioeng. Biomech.* **2016**, *18*, 159–169. [CrossRef]
20. Kwasniewski, L.; Bojanowski, C.; Siervogel, J.; Wekezer, J.W.; Cichocki, K. International Journal of Impact Engineering Crash and safety assessment program for paratransit buses. *Int. J. Impact Eng.* **2009**, *36*, 235–242. [CrossRef]
21. Wilhelm, J.; Ptak, M.; Fernandes, F.A.O.; Kubicki, K.; Kwiatkowski, A.; Ratajczak, M.; Sawicki, M.; Szarek, D. Injury biomechanics of a child's head: Problems, challenges and possibilities with a new aHEAD finite element model. *Appl. Sci.* **2020**, *10*, 4467. [CrossRef]
22. Young, W.; Sobhani, A.; Lenné, M.G.; Sarvi, M. Simulation of safety: A review of the state of the art in road safety simulation modelling. *Accid. Anal. Prev.* **2014**, *66*, 89–103. [CrossRef] [PubMed]
23. Vangi, D. Impact severity assessment in vehicle accidents. *Int. J. Crashworthiness* **2014**, *19*, 576–587. [CrossRef]
24. Munyazikwiye, B.B.; Vysochinskiy, D.; Khadyko, M.; Robbersmyr, K. Prediction of Vehicle Crashworthiness Parameters Using Piecewise Lumped Parameters and Finite Element Models. *Designs* **2018**, *2*, 43. [CrossRef]
25. Munyazikwiye, B.B.; Karimi, H.R.; Robbersmyr, K.G. Application of Genetic Algorithm on Parameter Optimization of Three Vehicle Crash Scenarios. *IFAC-PapersOnLine* **2017**, *50*, 3697–3701. [CrossRef]
26. Vangi, D.; Begani, F.; Gulino, M.S.; Spitzhüttl, F. A vehicle model for crash stage simulation. *IFAC-PapersOnLine* **2018**, *51*, 837–842. [CrossRef]
27. Tang, L.; Zheng, J.; Hu, J. A numerical investigation of factors affecting lumbar spine injuries in frontal crashes. *Accid. Anal. Prev.* **2020**, *136*, 105400. [CrossRef] [PubMed]
28. Sybilski, K.; Małachowski, J. Sensitivity study on seat belt system key factors in terms of disabled driver behavior during frontal crash. *Acta Bioeng. Biomech.* **2019**, *21*, 169–180. [CrossRef]
29. Baranowski, P.; Bogusz, P.; Damaziak, K.; Malachowski, J.; Mazurkiewicz, L.; Muszynski, A. Analiza wpływu zastosowanego elementu energochłonnego mającego bezpośredni kontakt z głową dziecka w aspekcie minimalizacji obciążeń dynamicznych. *Logistyka* **2015**, *4*, 2355–2363.
30. LSTC. *LS-Dyna Keyword User's Manual, R11 ed.*; Livemore Software Technology Corporation (LSTC): Livermore, CA, USA, 2018.
31. Hallquist, J. *LS-Dyna: Theory Manual*; Livemore Software Technology Corporation (LSTC): Livermore, CA, USA, 2019.
32. Ptak, M. Method to assess and enhance vulnerable road user safety during impact loading. *Appl. Sci.* **2019**, *9*, 1000. [CrossRef]
33. Bose, D.; Crandall, J.R.; Untaroiu, C.D.; Maslen, E.H. Influence of pre-collision occupant parameters on injury outcome in a frontal collision. *Accid. Anal. Prev.* **2010**, *42*, 1398–1407. [CrossRef]
34. Xiao, S.; Yang, J.; Crandall, J.R. Investigation of chest injury mechanism caused by different seatbelt loads in frontal impact. *Acta Bioeng. Biomech.* **2017**, *19*, 53–62. [CrossRef] [PubMed]
35. Humanetic Innovative Solutions Inc. *Hybrid III 50th Dummy Dyna Model—Technical Report. Release Version 8.0.1*; Humanetics Innovative Solutions, Inc.: Farmington Hills, MI, USA, 2013.
36. Marzougui, D.; Samaha, R.R.; Cui, C.-D.; Opiela, K. *Extended Validation of the Finite Element Model for the 2010 Toyota Yaris Passenger Sedan*; The National Crash Analysis Center: Ashburn, VA, USA, 2012.

*Article*

# Modeling Head-On Collisions: The Problem of Identifying Collision Parameters

**Piotr Aleksandrowicz**

Department of Machine Operation and Transport, University of Science and Technology in Bydgoszcz, 85-796 Bydgoszcz, Poland; p.aleksandrowicz@utp.edu.pl

Received: 21 August 2020; Accepted: 4 September 2020; Published: 7 September 2020

**Abstract:** The analyses performed by the experts are crucial for the settlement of court disputes, and they have legal consequences for the parties to legal proceedings. The reliability of the simulation result is crucial. First, in article, an impact simulation was performed with the use of the program default data. Next, the impact parameters were identified from a crash test, and a simulation was presented. Due to the difficulties in obtaining the data identified, the experts usually take advantage of simplifications using only default data provided by the simulation program. This article includes the original conclusions on specific reasons of simplified collision modeling in Multi Body Systems (MBS) programs and provides specific directions of development of the V-SIM4 program used in the study to enhance the models applied. This manuscript indicates a direction for crash model development in MBS programs to consider a varied 3D body space zones stiffness related to the structure of the car body and the internal car elements instead of modeling the car body as a solid with an average stiffness. Such an approach would provide an alternative to Finite Element Method (FEM) convention modeling.

**Keywords:** collision modeling; mechanical parameters; contact detection

## 1. Introduction

One of the crucial problems accompanying the automotive sector development are road accidents resulting in human and material loss. Such events involve material loss and a loss of life. In 2018, in Poland, as many as 31,674 road accidents were reported, in which 2862 people were killed and 37,359 were injured. However, in the corresponding period, the total of 436,414 road collisions resulted only in material loss. The events classified as collisions of vehicles in motion, which accounted for 53.8% of all the accidents, were found to be most common. As many as 44.3% of all the people were killed in car crashes, and 58.6% were injured. There were 3104 head-on collisions, resulting in 510 deaths and 4676 people suffering injuries. In Poland, the death rate for every 100 accidents was found to be the highest in all the European Union member states in 2017 (8.6). The country which came right after Poland is Greece, with the rate of 6.7, whereas Germany had the lowest rate, at 1.1 [1], which suggests that Poland faces a serious road traffic safety problem to be urgently addressed.

Next to the human loss, one must also remove the material damage, repair the vehicles damaged in road accidents and collisions. Usually, only accidents involving material losses of road accidents and collisions are considered, whereas insurance companies are involved in compensation of both material and human loss related to the outcomes of accidents delayed in time. As such, they do not focus only on road accidents but also on road collisions and the subsequent costs of repairs. The costs of spare parts, painting, and man-hours for the post-accident vehicle repair are calculated from the car damage analysis.

In Europe, the USA, and Canada one must follow the requirements providing for the impacts of cars with a rigid barrier. Another study [2] discusses the vehicle superstructure designing method to develop the guidelines for the application of bumper beams and impact energy absorbers. The study also covered the head-on collision with rigid barriers. The low-speed collision requirements by the Insurance Institute for Highway Safety (IIHS) and Dunner tests, considered in this study, are demonstrated in Figure 1.

| | IIHS tests | | | | Dunner tests | |
|---|---|---|---|---|---|---|
| The test schema | Front impact | A front impact on the barrier | The rear-end impact | The rear-end impact | Front impact with an overlap of | The rear-end impact with overlap |
| Offset | 100% | | 100% | | 40% | 40% |
| Angle of inclination | 0° | 30° | 0° | 0° | 0° | 0° |
| Barrier | Not deformable | Not deformable | Fixed barrier | Fixed the column diameter 180 [mm] | Not deformable | Mobile deformable barrier with a mass of 1000 [kg] |

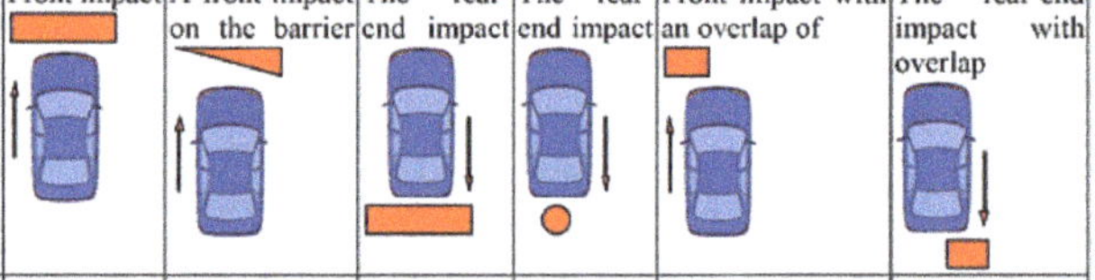

**Figure 1.** Requirements for low-speed collisions by the Insurance Institute for Highway Safety (IIHS) and Dunner tests.

For insurance companies, prior to the payment of damages for the car to be repaired, it is also very important to determine whether they are liable for damages or not. For that reason, in the process of insurance claim settlement, insurance companies usually use professional assessments provided by experts ordered by courts and public prosecutors to establish liability for the consequences of road accidents. Currently, the assessors use simulation programs to reconstruct road accidents, so manual calculations are hardly used. Adequate simulation results are of key importance as using incorrect simulation results leads to inadequate decisions in terms of liability for damages, which also triggers some legal effects for the parties to a car repair damages court case. The highest calculation accuracy is provided by programs with finite elements method modeling; the FEM convention, e.g., Abaqus, and LS-Dyna. Another study [3] uses LS-Dyna program for researching the vehicle structure stiffening elements during head-on collision. The test involved a Sport Utility Vehicle (SUV), very popular in China. The vehicle Finite Element (FE) model was initially validated against a full-frontal rigid barrier impact test by National Crash Analysis Center (NCAC). The study included also a head-on collision with a rigid barrier. A satisfactory compliance of the test and simulation deformations for each such collision was found (Figure 2).

**Figure 2.** Deformations of the car in test and simulation.

Other authors [4] report on using LS-Dyna program for the vehicle front protection system (VFPS). The study has been launched to address the problem of animals crashing into vehicles in Australia, causing damage to headlights, engine coolant radiators, and engines. The VFPS changes the vehicle crush characteristics during a head-on collision and affects the airbag activation characteristics. An inadequate VFPS structure results in a greater vehicle damage, hence resulting in higher repair costs and potential passenger injuries. Applying the FEM convention modeling program, a structure eliminating such defects has been developed (Figure 3).

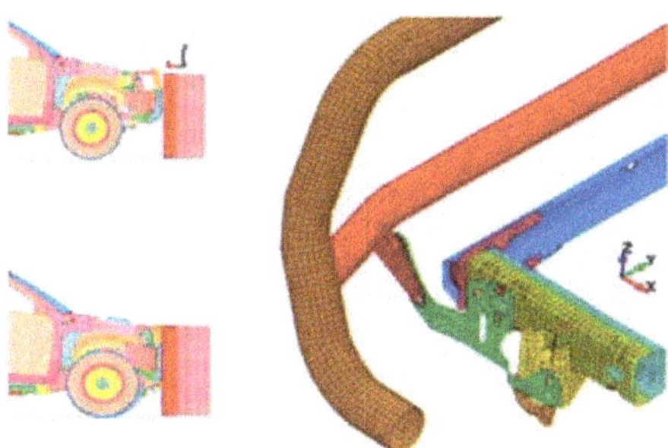

**Figure 3.** Over the bumper type front protection system (FPS) mount and its mating parts.

The FEM convention modeling programs are most commonly used in research facilities. However, they have not been commonly applied for the reconstruction of road accidents as geometric and material data must be introduced, which is discussed also by other authors [5,6]. The authors of this study have performed numerical research in the LS-Dyna environment. The results of that research was experimentally validated with crash tests using a full-size road infrastructure support structure dimension. The structure satisfied the Polish PN standard - PN-EN 12767 requirements: "Passive safety of road infrastructure support structures—Study requirements and methods". The study has demonstrated as a good compliance of the simulation with the experiment.

In practice, however, programs for modeling impacts and post-impact movement of vehicles in multi-body system convention (MBS) are used for analysis of vehicle collisions. They are not so precise as FEM programs; however, they offer significantly shorter calculation time and easier operation without the need to enter geometric and material constants. In consequence, modeling with the use of these programs is simplified. As practice shows, the users of these programs, who are commissioned to provide the court with assessment of road accidents, are not aware of it and make mistakes accepting incorrect simulation results as final. Programs for modeling in MBS convention include, for example, V-SIM, PC Crash, and Virtual Crash. Another study [7] reports on the tests of head-on collision with a rigid barrier at 13–95 km/h and full overlap, as compared with the experimental results. V-SIM, however, has not been verified with a full-frontal rigid barrier impact test. In other studies [8,9], contact parameters and other data were identified from crash tests and used in simulation. The articles present both the results of computer simulation of a passenger cars crash with a non-deforming barrier and a pole with V-SIM and the experimental results published by Allgemeiner Deutscher Automobil-Club and AUTOBILD. The results were received for the same initial conditions, which has facilitated evaluating the credibility of computer simulation. The V-SIM program can also be applied to verify damage using the SDC (Static, Dynamic, Characteristic) analysis method to prevent the payments of undue damages for vehicle repairs. It uses the results of geometrical parameters and the appearance of damage, as well as the results of a dynamic collation of the collision. A practical application of this method is described in another study [10], presenting research methods which facilitate an efficient verification of the vehicle collision to determine whether the crash actually occurred or not. It also presents an IT tool automating the decision-making process and supporting the expert's operations, verifying the impact with the SDC method.

The application of this program in Poland is common. Since it is so common to use the V-SIM program for the reconstruction of accidents and for insurance claim settlements, the problem of reliability of the simulation and the effect of the input data on the simulation results is addressed.

## 2. Materials and Methods

The V-SIM code is designed for the analysis of the impacts of mechanical vehicles (passenger cars, trucks, semi-trailer trucks, and semi-trailers and trailers). It also facilitates the analysis of the impact of vehicles with rigid barriers, such as a wall. The program models the movement of the vehicle, which is treated as a solid with 10 degrees of freedom, moving as a result of the external forces acting on it. The model involved the application of two systems of reference. The first is the global one, which describes the momentary positions of the simulation objects and the position of the movement environment elements designed, e.g., the terrain barriers. The coordinate axes of the reference system are marked $x$, $y$, $z$. The axis is oriented opposite to the force of gravity, and the beginning of the reference system is selected by the program operator. The second system of reference is related to the simulated vehicle, and its axes are marked $x'$, $y'$, $z'$. In that system, the external forces acting on the vehicle are determined. Axis $x'$ in that system is directed according to the movement of the vehicle, while axis $z'$ is directed vertically upwards the unladen vehicle, and its direction and sense result from the orthogonality and clockwise nature of the system. The positioning of the center of the vehicle mass is determined by radius vector $\vec{r}$. (Figure 4).

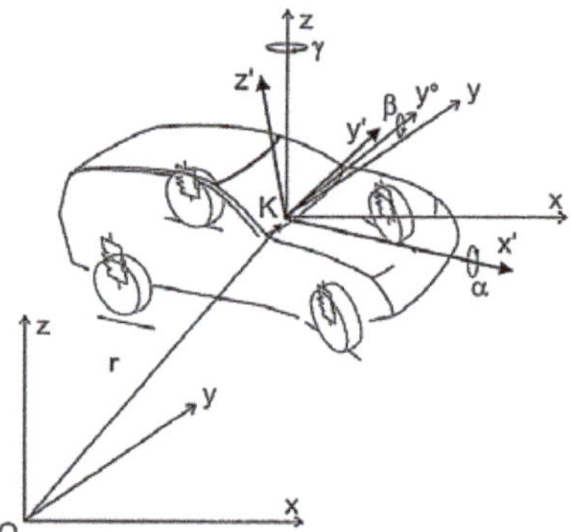

**Figure 4.** Model of vehicle with co-ordinate systems [11].

Equations of motion for the vehicle in the reference system related to the center of mass of the vehicle have been provided with the following formulas [11]:

$$\vec{F}' = \left( \ddot{\vec{r}}' + \dot{\vec{r}}' \times \vec{\omega}' \right) \cdot m, \tag{1}$$

$$\vec{M}' = \Theta_c \cdot \dot{\vec{\omega}}' + \vec{\omega}' \times \Theta_c \cdot \vec{\omega}', \tag{2}$$

where:

$\vec{r}$—radius vector,

$\vec{\omega}'$—rotational speed of the vehicle in the system connected to the vehicle, and

$\Theta_c$—tensor of the mass moment of inertia.

For the rotational movement, there has been assumed the tensor of mass moment of inertia with zero mass moments of inertia, besides the main moments, which has resulted in the system of scalar equations describing the rotational movement, presented below [11]:

$$\Theta_c = \begin{bmatrix} I_{x'} & 0 & 0 \\ 0 & I_{y'} & 0 \\ 0 & 0 & I_{z'} \end{bmatrix}, \tag{3}$$

$$I_{x'} \cdot \dot{\omega}_{x'} = M_{x'} - I_{z'} \cdot \omega_{y'} \cdot \omega_{z'} + I_{y'} \cdot \omega_{y'} \cdot \omega_{z'}, \tag{4}$$

$$I_{y'} \cdot \dot{\omega}_{y'} = M_{y'} + I_{z'} \cdot \omega_{x'} \cdot \omega_{z'} - I_{x'} \cdot \omega_{x'} \cdot \omega_{z'}, \tag{5}$$

$$I_{z'} \cdot \dot{\omega}_{z'} = M_{z'} - I_{y'} \cdot \omega_{x'} \cdot \omega_{z'} + I_{x'} \cdot \omega_{x'} \cdot \omega_{y'}, \tag{6}$$

where:

$I_{x'y'z'}$—the main moments of inertia of the vehicle,

$M_{x'y'z'}$—components of the moment of external forces acting on the vehicle in the system connected to the vehicle, and

$\omega_{x'y'z'}$—components of the rotational speed velocity of the vehicle against the axes selected.

The total force acting on the vehicle is the force of gravity, the forces of aerodynamics resistances, and the forces brought by the suspension of the vehicle wheels, which has been described with the following formula [11]:

$$\vec{F}' = \vec{F}'_g + \sum \vec{F}'_i + \vec{F}'_{ax} + \vec{F}'_{ay}. \tag{7}$$

The gravitational force has been determined by transforming from the global system of reference according to the following formula [11]:

$$\vec{F}_g = \begin{bmatrix} 0 \\ 0 \\ -m \cdot g \end{bmatrix} \xrightarrow[x \rightarrow x']{\vec{F}'_g}, \tag{8}$$

where:

m—mass of the vehicle, and

g—standard acceleration due to gravity.

The forces of aerodynamic resistance are determined irrespective of the frontal and side area of the vehicle from the speed of the vehicle, while considering wind speed $\vec{w}$ according to the formulas provided below [11]:

$$\vec{F}'_{ax} = \begin{bmatrix} \pm \rho/2 \cdot C_x \cdot A_x \cdot \left(\dot{r}_{x'} - w_{x'}\right)^2 \\ 0 \\ 0 \end{bmatrix}, \tag{9}$$

$$\vec{F}'_{ay} = \begin{bmatrix} 0 \\ \pm \rho/2 \cdot C_y \cdot A_y \cdot \left(\dot{r}_{y'} - w_{y'}\right)^2 \\ 0 \end{bmatrix}, \tag{10}$$

where:

$\rho$—air density,

$C_{x,y}$—coefficients of the frontal and side air resistance,

$A_{x,y}$—frontal and side areas of the vehicle,

$\dot{r}_{x'y'}$—linear velocities of the vehicle in the system connected with the vehicle, and

$w_{x,y}$—component wind speeds in the system connected with vehicle.

The vehicle suspension model also considers the independent vertical movement for each wheel of the vehicle. From the positioning of the vehicle and the local configuration of the road surface, deflection of the suspension $S_i$ is determined, whereas the normal force of the response of the suspension considering the elasticity with progressive characteristics and the damping values is calculated from the momentary deflection and the velocity of its changes according to the formula [11]:

$$F'_{zi}\left(S_i, \dot{S}_i\right) = \max\left(\max(C_{3i} \cdot S_i^3 + C_{1i} \cdot S_i + F'_{0i}, 0) + \begin{cases} D_{ci} \cdot \dot{S}_i & \dot{S}_i \geq 0 \\ D_{ri} \cdot \dot{S}_i & \dot{S}_i < 0 \end{cases}, 0\right), \tag{11}$$

where:

$F'_{zi}$—vertical component of the response of the vehicle suspension,
$S_i$—reduced value of deflection of the vehicle wheel suspension,
$C_{3i}$—reduced coefficient of progression of stiffness of the vehicle wheel suspension,
$C_{1i}$—reduced coefficient of stiffness of the vehicle wheel suspension,
$F'_{0i}$—normal force of the response of the vehicle wheel suspension,
$D_{ci}$—reduced damping coefficient for the compression of the vehicle wheel suspension, and
$D_{ri}$—reduced damping coefficient for the stretching of the vehicle wheel suspension.

The program for determining the interaction of the tires with the road surface used a nonlinear model developed at the Highway Safety Research Institute (HSRI), University of Michigan, by the Dugoff team and TM-Easy. The forces of the tire reactions are calculated in the system connected with the tangent point of the tire with the road surface, and the normal force of the tire is assumed as the normal force of the reaction of the suspension. The normal force of the tire, the rotational speed of the wheel, and the parameters of the road surface are used to determine the forces of the reaction of the tire in the other axes of the system. For each wheel of the vehicle, additional degree of freedom has also been considered, and rotational movement has been considered. The model considers the driving torque and the movement resistances according to the formula [11]:

$$I_i \cdot \dot{\omega}_i = M_{ni} - F''_{xi} \cdot R_{di} \pm (M_{hi} + M_{ei} + M_{ti} + M_{ri}), \tag{12}$$

where:

$I_i$—mass moment of inertia of the vehicle wheels,
$\dot{\omega}_i$—rotational speed of the vehicle wheels,
$M_{ni}$—reduced driving torque of the vehicle wheels,
$F''_{xi}$—component of the force of the response of the tire in the system connected with the tangent point of the vehicle tires with the road surface,
$R_{di}$—rolling radius of the vehicle wheels,
$M_{hi}$—break torque of the vehicle wheel brake,
$M_{ei}$—reduced engine brake torque which occurs on the vehicle wheels,
$M_{ti}$—road surface rolling resistance torque which occurs on the vehicle wheels, and
$M_{ri}$—vehicle-wheels own resistances torque.

The model of the brake system of the vehicle considers the corrector of the braking force of the rear axle for each load of the vehicle, and, optionally, it can consider also the effect of the anti-blocking system (ABS).

However, the driving torque of the engine is determined from its external characteristics from the following formula [11]:

$$M_e(\omega_e) = M_m - \frac{M_m - \frac{N_n}{\omega_n}}{(\omega_n - \omega_m)^2} \cdot (\omega_e - \omega_m)^2, \tag{13}$$

where:

$M_e$—driving torque of the vehicle engine,
$\omega_e$—rotational speed of the vehicle engine,
$M_m$—maximum torque of the vehicle engine,
$N_n$—maximum power of the vehicle engine,
$\omega_n$—speed for the maximum power of the vehicle engine, and
$\omega_m$—rotational speed for the maximum torque of the vehicle engine.

The operation of the steering system of the vehicle was analyzed using the kinematic model following the Ackermann steering principles, considering the susceptibility of the real-life system and the correction of shearing forces of the reaction of the tires of the wheels of the steered axle. The program can also model additional tasks for the vehicle and for the driver; the vehicle acceleration and a turn with the steering wheel, braking with the foundation and secondary brakes, using the clutch and blocking the wheel, a pressure drop in the tire, and the deformation of suspension.

A force-based model has been developed for an impact with the impact forces developing in a constant manner while two objects are in contact [12]. The impact of vehicles is detected with a 2D or 3D model developed for collision detection, depending on the choice of the operator. This choice, however, should be informed and adequate to the collision analyzed. It mostly applies to the impacts of vehicles the geometric compatibility of which is inconsistent, which has also been discussed in other studies [13]. Solving problems of collision modeling is also discussed in Reference [14–16].

An impact force model of the V-SIM program suggests a development of force $F$ in time during the impact compression, and then its collapsing, according to a certain function in the restitution phase. Such an approach assumes that all the external and mass forces, provided by the vehicle model and by the model of a wheel, occur during an impact [17,18]. Detailed methods for a vehicle impact modeling with a rigid barrier are used in another study and presented below.

Such modeling considers changes of force which appear during an impact and its influence on the vehicle movement, as well as the occurrence of two impact phases: the first is a compression phase (when the value of momentary $F$ force increases), whereas the second one is a restitution phase (when the value of momentary $F$ force decreases), and according to Newton's hypothesis, the above phases are associated with restitution coefficient $k$.

On the basis of the concept of force impulse for the compression and restitution phases, formulas for both phases are used [19]:

$$\vec{S}_k = \int_{t_0}^{t_k} \vec{F}(t)dt, \tag{14}$$

where:

$\vec{S}_k$—represents impulse of force for the compression phase,
$\cdot t_0$—represents the compression phase beginning time,
$\cdot t_k$—represents the compression phase end time, and
$\vec{F}(t)$—represents a momentary force of impact.

$$\vec{S}_r = \int_{t_k}^{t_r} \vec{F}(t)dt, \tag{15}$$

where:

$\vec{S}_r$—stands for impulse of the restitution phase, and
$t_r$—stands for restitution phase end time.

The total impulse of the impact force is presented as the sum of force impulses for the phases of compression and restitution, expressed using the formula [19]:

$$\vec{S} = \vec{S}_k + \vec{S}_r.$$ (16)

In turn, joining the values of both impulses of the force by applying the restitution coefficient is described by the formula [19]:

$$S_r = k \cdot S_k.$$ (17)

The applied impact model assumes that, during the impact, the value of temporary impact force $F$ is proportional to the volume of overlapping contours of the colliding simulation objects (Figure 5a) by the below formula [19]:

$$F(t) = c f(t),$$ (18)

where:

$c$—stands for stiffness coefficient of the impact area [N/m$^3$], and
$f(t)$—stands for the impact area volume [m$^3$].

After linearization of the course of action of forces that occur during the two earlier indicated phases, the impact force is presented with a diagram in Figure 5b. Modeling also includes the mean value of the force in the states of compression and restitution, expressing the coefficient of restitution by the following formula [19]:

$$k \cdot F_{kmid}\, t_k = F_{omid} \cdot t_o,$$ (19)

where:

$F_{kmid}$—stands for the mean value of force $F$ in the compression phase,
$F_{omid}$—stands for the mean value of impact force $F$ in the restitution phase,
$t_o$—stands for the restitution phase duration time, and
$t_k$—stands for the compression phase duration time.

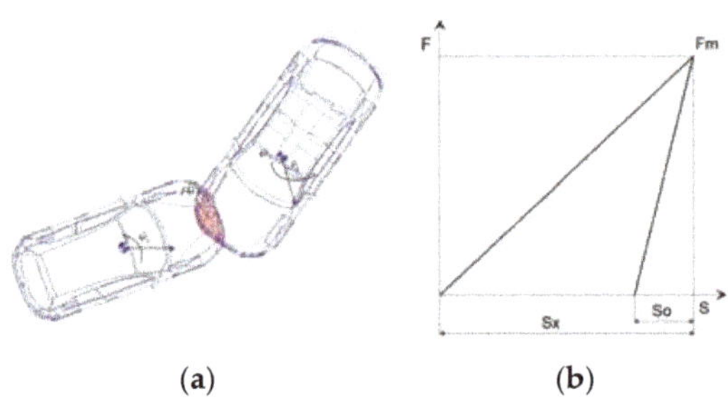


(a)     (b)

**Figure 5.** Impact area of vehicles (**a**,**b**) impact force $F$ in the compression and restitution phase: $S_k$—deformation depth in the compression phase, $S_o$—depth of the elastic disappearing deformation in the restitution phase, and $F_m$—maximum force value.

For the uniformly variable motion, the deformation depth has been described with the following formula [19]:

$$S_x = \frac{a \cdot t_x}{2},$$ (20)

where:

$S_x$—deformation depth, and
$a$—average deceleration.

The course of action of impact force $F$ during restitution is written by formula [19]:

$$F(t) = F_m\left(1 - \frac{1}{k^2}\left[1 - \frac{s(t)}{s_k}\right]\right).$$

(21)

For $k = 0 \rightarrow F(t) = 0$, where:

$F_m$—stands for maximal value of the impact force, and

$s_k$—stands for depth of deformation at the end of the compression phase.

It has been assumed that having been hit by a vehicle, the wall does not undergo deformation. Following that approach, the vector of momentary impact force $\vec{F}(t)$ is on the margin of a non-deforming barrier, in point $C$, found in the middle of the barrier length, in the area (Figure 6a), whereas, the position of this point for the vehicle is defined by radius vector.

Momentary impact force $F$ is represented by the following components: tangent and normal to the surface of the non-deforming area. In its application point, the direction of normal component is perpendicular to the surface on the non-deformable area. The unit vector which defines the direction of this component is marked (Figure 6b) $\vec{e}_{n_1}$, and it is defined as follows [19]:

$$\vec{F}_{n_1}(t) = F_n(t)\vec{e}_{n_1},$$

(22)

where:

$F_n(t)$ stands for the normal component value.

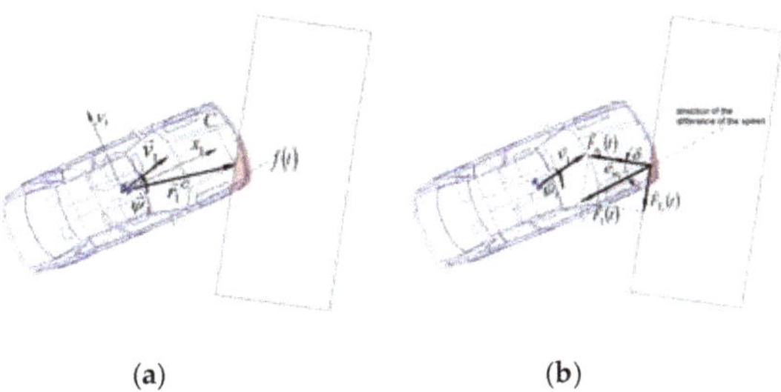

(a)        (b)

**Figure 6.** Diagram depicting the calculation method for an impact with a rigid barrier (a) and momentary force of impact with a rigid barrier (b).

In this model, friction force occurring in the place of impact is represented by tangential component $\vec{F}_T(t)$. Its value must be such that the direction of the speed difference is consistent with the direction of impact force resultant $\vec{F}_1(t)$. In Figure 6b, the angle between the direction of impact force component and the speed difference direction is marked as $\delta$. It was also accepted that the value of tangent component cannot exceed the product of normal component and friction coefficient $\mu$.

The value of tangent component is expressed with the formulas [19]:

$$F_{T_1} = F_{n_1}\tan\delta \leftrightarrow F_{n_1}\tan\delta \leq F_{n_1}\mu,$$

(23)

$$F_{T_1} = F_{n_1}\mu \leftrightarrow F_{n_1}\tan\delta > F_{n_1}\mu.$$

(24)

The compression phase finishes when the impact area $f(t)$ ceases to increase and the restitution phase begins. The value of normal component decreases from the maximal one down to zero, with the assumption of the above linear change in momentary impact force $F$ to be a function of time and restitution coefficient $k$.

## 3. Results

A crash test of Mercedes Benz C300 4Matic AWD from 2013, published by National Highway Traffic Safety Administration [20] was used to identify the impact parameters. The car tested was involved in a head-on collision with a rigid barrier (overlap 100%) with the speed of 56.3 km/h. The total mass of the vehicle was 1703 kg, its length: 4.581 m, its width: 1.770 m, its wheel base: 2.760 m, the front axle: 1.549 m and the rear axle: 1.552 m, the height above the ground: 0.552 m, the moments of inertia: roll 605 kgm$^3$, pitch: 2700 kgm$^3$, and yaw: 2771 kgm$^3$. The test was recorded from above and from the right and left sides of the vehicle, as well as from the bottom.

In practice, the experts and assessors have problems obtaining the experimental data to be used in the simulation calculation of a reconstructed road accident. To acquire the data, a research method involving measurements of the vehicle body position during an impact from the crash test film has been proposed. Crash tests are recorded with fast video cameras facilitating the separation of particular frames (images) of the moving vehicle, even every 1 ms. First, the assessor should, of course, establish precise technical data of the vehicle involved in the accident. A precise data determination is very important as, according to the author, it is the assessor's further task to find a crash test of a vehicle with structure similar to the one which was involved in the accident. Only then can they go on to separate all the single images from the video, preferably, every 1 ms. Next, they should adjust the scale of the vehicle images to its vector styling, which is in scale and depicts details of the car body. A division into images (frames) can be performed with programs designed for this kind of film processing, whereas precise scaling is possible with the use of the V-SIM program. For this purpose, a single image needs to be imported to the program, and the size of a vehicle image needs to be matched with its vector styling; those stylings, in turn, are imported from external databases. For the purpose of this study, the styling of the Mercedes available in the AutoView database was used. The vehicle scale tapes can be applied in crash tests, as well as technical data, including the width, length, and the axle base.

According to the research procedure proposed, after scaling, it is necessary to start the analysis with the image depicting the moment of contact consistent with simulation time being $t = 0$ ms. The results of the image scaling at the moment of contact corresponding to the beginning of simulation are presented in Figure 7a. Measuring the vehicle body position changes, it is necessary to consider the vehicle elements which do not undergo deformation in the picture; the analysis accepts the roof edge near the windscreen. No universal rule can be used here due to different kinds of impacts and different zones of body deformation. It is more difficult to scale the image of the undercarriage, and one must focus on a selected well-visible element. The rear part of the engine supporting frame, close to the gear box, was selected for analysis.

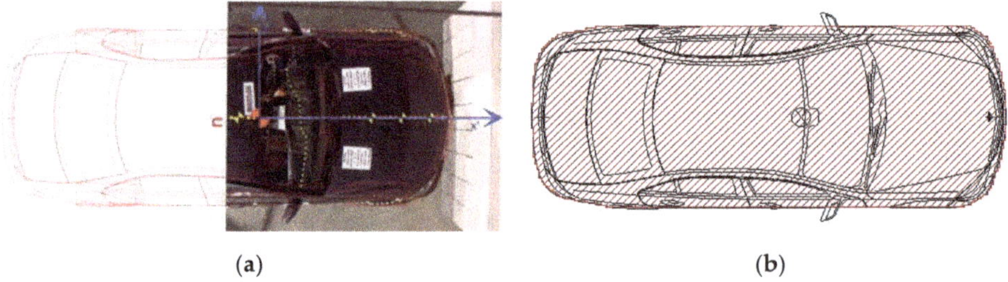

(a)        (b)

**Figure 7.** Scaled image of the vehicle with a coordinate system connected with the vehicle (**a**) and rigid area preview on its vector styling projection (**b**).

In Figure 7b, a rigid area preview is presented on a projection of a vector styling of a Mercedes. However, in this program, the changes in the vehicle body positions were measured in a system, the center of which is in the middle of the vehicle mass. Axis $x'$ of this system is directed in the forward-facing position of the car, axis $y'$ runs left clockwise, and axis $z'$ runs vertically, upward.

After scaling the images, according to the proposed research procedure, a simulation with default impact parameters, proposed by the V-SIM program, started. The results are presented in Figure 8a–c. After the measurements, an impact simulation was performed for the crash test parameters, and the results are presented in Figure 8d–f.

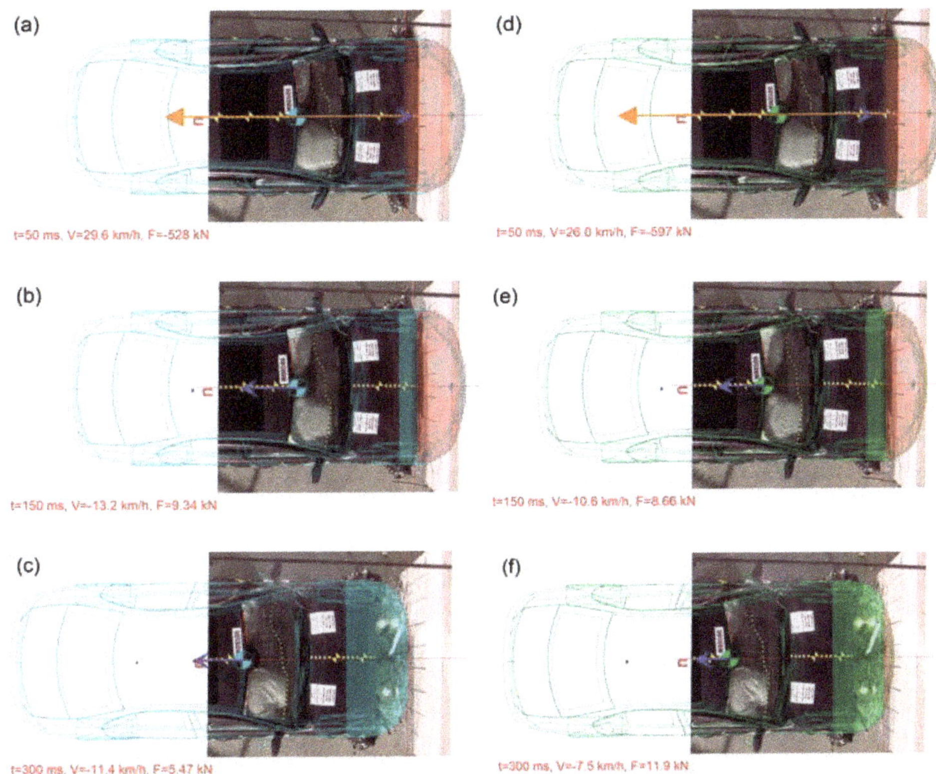


**Figure 8.** Comparison of car projections with recorded images for default data: $t = 50$ ms (**a**), $t = 150$ ms (**b**), $t = 300$ ms (**c**) and identified data: $t = 50$ ms (**d**), t = 150 ms (**e**), and t = 300 ms (**f**).

A comparison of the simulation results for the program default data with data identified from the crash test shows that the positions of vehicles were similar only in the initial phase. In time $t = 50$ ms, the stylings of the simulated vehicles are similar, and they overlap with the image of the real vehicle at the edge of the roof, near the windscreen (Figure 8a–d). However, over time, an error of the vehicle position develops for the program default data. In time $t = 150$ ms, the vehicle simulated for default data is definitely shifted forward in relation to the actual one; the edges of the roof near the windscreen do not overlap (Figure 8b), whereas, for a vehicle simulated for the data identified, entered into the program, a consistence of the position with the real vehicle was obtained (Figure 8e). In turn, in the case of post-impact movement, the simulated vehicle for default data is moved backward in relation to the real one, after the vehicle rebound from the obstacle, in time $t = 300$ (Figure 8c), and, for a vehicle simulated with the data identified, the consistence of its position with the real vehicle was found (Figure 8f).

It was also observed that during the impact, elements of different stiffness undergo deformation which affects the compression and restitution phases. The scaled images of the vehicle undercarriage were used to analyze the deformation of those elements (Figure 9a–f).

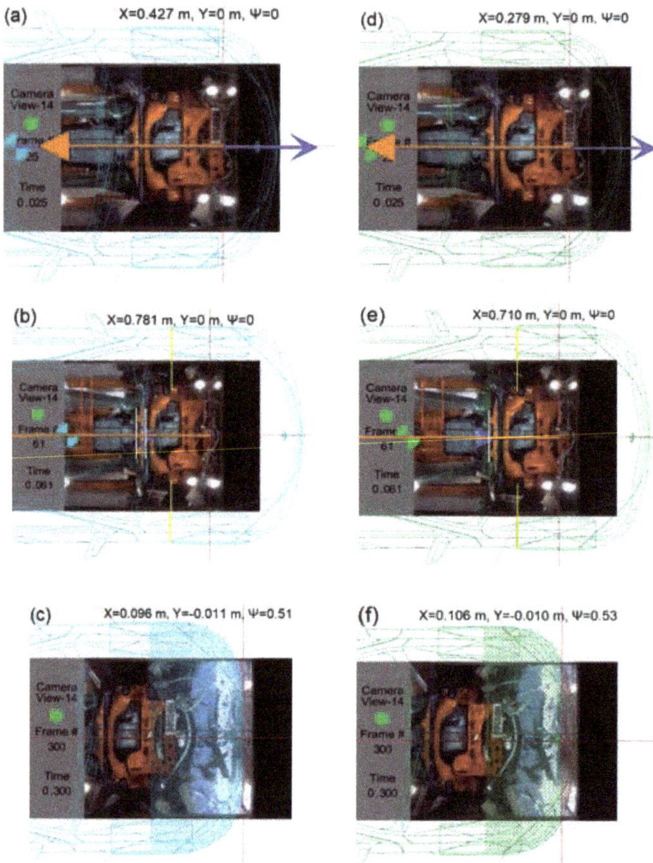


**Figure 9.** Comparison of the vehicle undercarriage projections with recorded images for default data: $t = 25$ ms (**a**), $t = 61$ ms (**b**), $t = 300$ ms (**c**), identified data: $t = 25$ ms (**d**), $t = 61$ ms (**e**), and $t = 300$ ms (**f**).

Initially, in the compression phase in time $t = 25$ ms, while the bumper lining and its cross beam are being deformed, interferences in the impact reconstruction simulation for default parameters are acceptable in comparison with the ones received for the data identified, as shown in Figure 9a,d. The impact consistence for accident reconstruction using the program lasted until time $t = 50$ ms, when the engine support frame and the engine had not been crushed yet. Further, in time $t = 61$ ms, it can be observed that the wheel suspension, engine supporting frame, and engine are crushed; however, the deformations do not increase any more, as shown in Figure 9b,e). The vehicle body stiffening occurred during the impact. The phenomenon is discussed in other studies [21,22], which shows the effect of the engine chamber elements (engine, drive block) on the stiffness and deformation of the vehicle body and discusses the simplifications accepted in linear characteristics of the body deformations during the impact.

The vehicle body zones show a different stiffness, which additionally changes during the impact due contact with those elements, susceptible to deformation, with a barrier. When, after deformation of the bumper lining, the rigid barrier hit the engine supporting frame, the stiffness of the vehicle body increased, which was not precisely reconstructed by the V-SIM program for default parameters. Therefore, in order to reconstruct the collision, it was necessary to identify the crash test parameters and to enter them for simulation. The increase in stiffness in the compression phase facilitated reducing the error in simulation reconstruction for the compression phase. A change in the restitution coefficient and the body stiffness increase in the restitution phase were due to the program-modeled stiffness increase by a crushed engine with its support frame. However, having entered the task to block the wheels, their braking was reconstructed, which made the vehicle pull back after the impact. A change in the plane of contact force $\Delta y'$ was also entered to ensure a better overlap of the simulated vehicle for crash test images. To show the differences in the course of the impact simulation for other V-SIM default data and identify them following the above methods, a crash test was used to develop a video [23].

The research method presented also facilitates determining the value of the coefficient of unitary body stiffness $k$ following the formula below [24]. In that approach, the measure of the body deformation is assumed to be its width, depth, and height. For the case considered, the parameters of the body deformation are presented in Figure 10a,b, whereas, obtained with those parameters, the value of unitary stiffness coefficient k is 564 243 N/m·m$^2$.

$$k = \frac{m \cdot v^2}{w \cdot h \cdot C^2}, \tag{25}$$

where:

$m$—gross vehicle mass,

$v$—velocity of the impact into the barrier, and

$w, h, C$—width, height, and depth of the vehicle undercarriage deformation.

Table 1 shows default parameters proposed in the V-SIM program and changed according to the data identified from the Mercedes crash test.

**Table 1.** Default and identified parameters.

| Parameter | Default Data | Identified Data |
| --- | --- | --- |
| Stiffness of the car body for compression phase | 764 kN/m$^3$ | 870 kN/m$^3$ |
| Stiffness of the car body for restitution phase | 764 kN/m$^3$ | 5000 kN/m$^3$ |
| Restitution coefficient | $0 \rightarrow 0.24$ | 0.19 |
| Front axle wheels blocking | Non-automatic | FR 35 ms FL 35 ms |
| Displacement of plane of contact force $\Delta y'$ | 0.000 m | −0.002 m |
| Adhesive friction coefficient | 0.9 | 0.9 |
| Slip friction coefficient | 0.8 | 0.8 |
| Rolling resistance coefficient | 0.015 | 0.015 |

The stiffness of the car body for compression phase was increased to decrease the depth of the penetration of the barrier into the car body. Right after the compression phase $t = 70$ ms, the car impact speed is already low, so the coefficient of restitution decreased, and stiffness for the restitution phase increased. During the restitution phase, stiffness is high, and some energy is released. In V-SIM, the car body stiffness for the phase of compression and restitution is the same, which is a simplification in the impact model applied. Once the program default parameters were changed into the ones identified in time $t = 70$ ms, the same depth deformation of the vehicle simulated in the top and side projections were also obtained, as compared with the actual car filmed (Figure 10).

The analysis has also demonstrated that for time t = 35 ms the front wheels of the Mercedes were blocked, which can be clearly seen in the crash test recording, showing the vehicle from the left and the right side, which is caused by a progressing deformation of the vehicle front, and the V-SIM program does not reconstruct it for default data. The problem of uncertainty which occurs in car crash reconstruction, with models of the same physical phenomenon providing different results, is described in another study [25,26]. In Reference [27], problems related to various types of car impacts and the impact of road surface condition on the reconstruction of object movement are discussed, while, in Reference [28,29], a discussion of problems of autonomous differential lock in a truck during movement in different terrain conditions is presented.

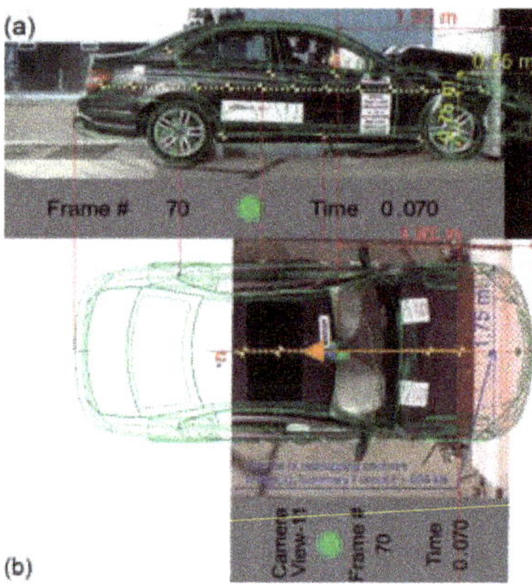


**Figure 10.** Test and simulation depth of deformations in the final compression phase t = 70 ms, side view (**a**) and top view (**b**).

To illustrate the above differences, time histories of changes: longitudinal component X, transverse component Y, yaw angle Ψ, and impact force F (Figure 11a–d) are presented, as well. In the simulation for default data, the maximum displacement of the vehicle body along X axle occurred in time *t* = 72 ms and was X = 0.767 m, whereas, for identified data, it was lower, X = 0.721 m and occurred earlier, in time *t* = 67 ms (Figure 11a). The differences in the position of simulated vehicles along Y axle were consistent with time t = 34 ms, after which the differences occurred (Figure 11b).

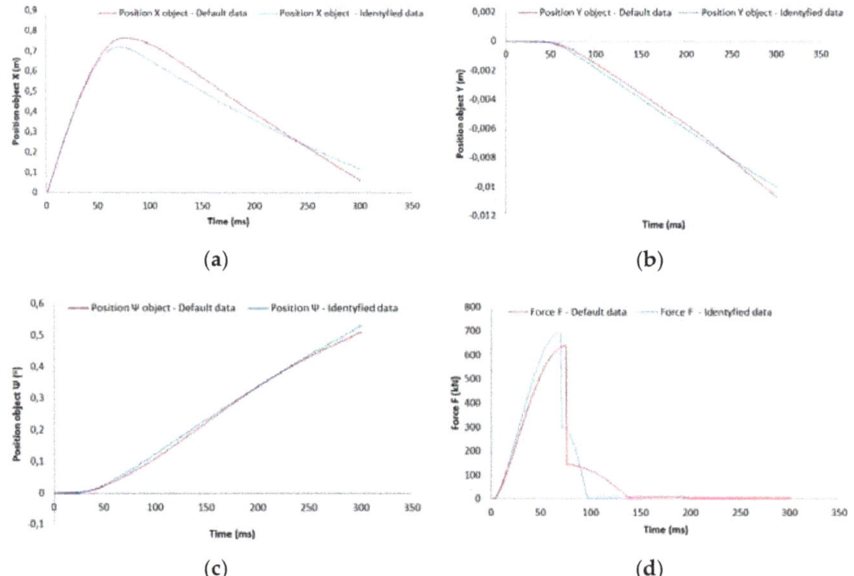

**Figure 11.** Time histories: longitudinal component X (**a**), transverse component Y (**b**), yaw orientation angle Ψ (**c**), and impact force F (**d**) recorded in V-SIM simulations.

The vehicle simulated for default data rotated faster than indicated in the crash test; hence, a correction of the plane of contact force $\Delta y'$, which is also reflected in the time histories of yaw angle $\Psi$ (Figure 11c).

In the compression phase, impact force F (Figure 11d) in the simulation for the data identified reaches the maximal value 696.86 kN in time t = 68 ms, next—298.62 kN, in time $t$ = 71 ms, and further drops down to zero. In turn, for the program default data, impact force F reaches, in the compression phase, a lower value 640.88 kN, and the maximum follows in time $t$ = 74 ms, then—144.39 kN in time $t$ = 75 ms, and decreases down to zero. The differences result from those phenomena related to an increase in the body stiffness while the engine and the engine supporting frame are being crushed.

## 4. Conclusions

The impact parameters identification method proposed involves the performance of the procedures: a test selection, a division of the film into single frames, the measurements and photograph scaling, the selection of the reference base for the overlap analysis for the test vehicle and the simulated vehicle, the top and bottom views, and a change in the impact parameters (stiffness in the compression and restitution phase, the coefficient of restitution, a change in the plane of contact force, and wheel blocking) for the best image matching. Having received the image overlap from the recorded test and vector styling of the simulated vehicle in time t of the collision, the impact parameters identified must be entered into the simulation of the incident reconstructed. The approach in that convention increases the possibilities of the impact reconstruction by the road accident experts and developing expertise for courts with no access to test impact forces data.

Further studies should, therefore, be performed towards the development of the crash models in MBS programs to facilitate considering a varied stiffness of the body zones in its 3D space connected with the vehicle body structure and its internal elements and not, as it has been the case so far, of the vehicle body, its solid with the average stiffness. A solution to this problem would provide an alternative to time-consuming modeling in the FEM convention.

The study shows that the parameters used for simulations in practice should be selected based on the actual accidents which have been filmed and on crash tests. The study results have demonstrated that, for the V-SIM default parameters, an accurate reconstruction of the course of the impact was not possible due to:

- a simplified model of the contact of the vehicle with a rigid barrier and the need to introduce a change in the contact force application point;
- the program not modeling the elements installed inside the vehicle, resulting in the stiffness changes during impact;
- inaccurate values of default parameters for a head-on collision with a rigid barrier; and
- not identifying the car body explorer deformations causing wheel blocking and the need for the operator to enter that task manually.

The expert witnesses applying programs for expertise development should consider the above modeling simplifications not to assume an inadequate simulation result as it has a negative effect for insurance companies and court decisions.

The study results also demonstrate that the vehicle impact model in V-SIM should be further developed to:

- apply the models which consider different stiffness of the car body particular zones or which divide the car body into separate blocks of different stiffness;
- apply automatic identification of wheel blocking in the function of the body depth deformations;
- introduce a possibility of 2D impact course visualization in the car side view and not, as it has been so far, the top view; and
- a substantial 3D collision course visualization enhancement.

To recapitulate, this research developed a methodology of identifying the crash parameters to be applied for various vehicles and various crash velocities, as well as various kinds of barriers and crashes. Thus, it is possible to provide the adequate, compliant with the real-life course of the crash, result of the simulation using the MBS programs, although, they currently use more simple models of the detection of collision and crash, as compared with the FEM programs. However, by offering a much shorter time of calculations and a high number of numerical vehicle models in their databases, they provide an interesting and developmental alternative for the FEM programs. The research performed also facilitated determining a further development of models in the MBS programs which should focus on developing the crash models to consider a varied stiffness of the body zones in its 3D space. Next to the development of the aspects of crash modeling presented in the article, the effects for practical applications have also been provided for the experts performing vehicle crash simulations, as well as the guidelines for optimizing the functionality and the calculations precision of the V-SIM program, in the future to be used by designers when working on the successive versions of that IT tool.

**Funding:** This research received no external funding.

**Conflicts of Interest:** The authors declare no conflict of interest.

## References

1. Road Accidents—Annual Reports. Available online: http://statystyka.policja.pl/st/ruch-drogowy (accessed on 14 June 2020).
2. Gulyaev, V.; Loginov, N.; Kozlov, A. Method of Designing the Superstructure of the Car Body Based on the Requirements of Low-Speed Collisions. In Proceedings of the 9th International Scientific Practical Conference on Innovative Technologies in Engineering, Yurga, Russia, 24–26 May 2018; Volume 1059, pp. 140–147. [CrossRef]
3. Guibing, L.; Jikuang, Y. Influence of Vehicle Front Structure on Compatibility of Passenger Car-to-SUV Frontal Crash. In Proceedings of the 3rd International Conference on Digital Manufacturing and Automation, Institute of Electrical and Electronics Engineers, Guilin, China, 31 July–2 August 2012; pp. 492–495. [CrossRef]
4. Thota, N.; Jayantha, A.; Epaarachchi, J.; Lau, K. CAE simulation based methodology for Airbag compliant vehicle front protection system development. *Int. J. Veh. Struct. Syst.* **2013**, *5*, 95–104. [CrossRef]
5. Stopel, M.; Skibicki, D.; Cichański, A. Determination of the Johnson-Cook damage parameter D4 by Charpy impact testing. *Mater. Test.* **2017**, *60*, 974–978. [CrossRef]
6. Stopel, M.; Skibicki, D.; Moćko, W. Method for determining the strain rate sensitivity factor for the Johnson-Cook model in Charpy tests. *Mater. Test.* **2017**, *59*, 965–973. [CrossRef]
7. Steffan, H.; Geigl, B.; Moser, A.; Hoschopf, H. *Comparison of 10 to 100 km/h Rigid Barrier Impact*; Paper Number 98-S3-P-12; Graz University of Technology, Institute for Mechanics: Graz, Austria, 1998.
8. Kostek, R.; Aleksandrowicz, P. Study of vehicle crashes into a rigid barrier. *Trans. Can. Soc. Mech. Eng.* **2019**, *44*, 335–343. [CrossRef]
9. Kostek, R.; Aleksandrowicz, P. Identification of the parameters of a vehicle crashing into a round pillar. *JTAM* **2020**, *58*, 233–245. [CrossRef]
10. Aleksandrowicz, P. Verifying a Truck Collision Applying the SDC Method. In Proceedings of the 58th International Conference of Machine Design Departments, Praque, Czech Republic, 6–8 September 2017; pp. 14–19.
11. Bułka, D.; Świder, P. *The Vehicle Model Used in the Computer Program V-SIM for Simulation of a Vehicle Move and Crashes*; Kielce University of Technology: Kielce, Poland, 2004; Volume 79, pp. 149–156.
12. Bułka, D. *V-SIM 4.0 Operating Manual*; CIBID: Cracow, Poland, 2019; pp. 149–153.
13. Aleksandrowicz, P. Selection of Collision Detection Model on the Basis of a Collision of Incompatible Vehicles. In Proceedings of the 24th International Conference Engineering Mechanics, Svratka, Czech Republic, 14–17 May 2018; pp. 21–24.
14. Zalewski, J. Selected problems of motor vehicle maintenance after side impact collision. *MATEC Web Conf.* **2018**, *182*, 01019. [CrossRef]

15. Smit, S.; Tomasch, E.; Kolk, H.; Plank, M.; Gugler, J.; Glaser, H. Evaluation of a momentum based impact model in frontal car collisions for the prospective assessment of ADAS. *Eur. Transp. Res. Rev.* **2019**, *11*, 2. [CrossRef]

16. Gidlewski, M.; Prochowski, L.; Jemioł, L.; Żardecki, D. The process of front-to-side collision of motor vehicles in terms of energy balance. *Nonlinear Dyn.* **2019**, *58*, 1877–1893. [CrossRef]

17. Rill, G. *Simulation von Kraftfahrzeugen*; Genehmigter Nachdruck; Vieweg-Verlag: Berlin, Germany, 2007.

18. Kruszelnicka, W.; Marczuk, A.; Kasner, R.; Bałdowska-Witos, P.; Piotrowska, K.; Flizikowski, J.; Tomporowski, A. Mechanical and processing properties of rice grains. *Sustainability* **2020**, *12*, 552. [CrossRef]

19. Świder, P.; Polański, A.; Grzegożek, W. Analysis of possibilities of using the method for impact force for road accident reconstruction purposes. In *Paragraph on the Road*; Institute of Court Expertise: Cracow, Poland, 2003; Volume 7, pp. 37–47.

20. A Crash Test of Mercedes Benz C300 4Matic from 2013. Available online: https://www.youtube.com/watch?v=Tvw_VLcUUwU&t=96s (accessed on 14 June 2020).

21. Prochowski, L.; Żuchowski, A. Dynamic loads of power unit during car impact. *KONES* **2006**, *13*, 467–475.

22. Żuchowski, A. The use of energy methods at the calculation of vehicle impact velocity. In *The Archives of Automotive Engineering*; Scientific Publishers of the Industrial Motorization Institute: Warsaw, Poland, 2015; Volume 68, pp. 86–111.

23. The Differences in the Course of the Impact Simulation for other V-SIM Default Data and Identified. Available online: https://youtu.be/hsLlxbLmcJM (accessed on 14 June 2020).

24. Prochowski, L.; Unarski, J.; Wach, W.; Wicher, J. *Automotive Vehicles. Basics of Road Traffic Accidents*; Transport and Communication Publishers: Warsaw, Poland, 2014; pp. 197–198.

25. Wach, W.; Unarski, J. Uncertainty of Calculation Results in Vehicle Collision Analysis. In Proceedings of the 4th European Academy of Forensic Science Conference, European Academy of Forensic Science Conference, Copenhagen, Denmark, 10–12 May 2006; pp. 181–188. [CrossRef]

26. Wach, W. Structural Reliability of the Reconstruction of Road Accidents. *Forensic Sci. Int.* **2013**, *228*, 83–93. [CrossRef] [PubMed]

27. Prentkovskis, O.; Sokolovskij, E.; Bartulis, V. Investigating traffic accidents: A collision of two motor vehicles. *Transport* **2010**, *25*, 105–115. [CrossRef]

28. Kučera, P.; Pištěk, V. Prototyping a system for truck differential lock control. *Sensors* **2019**, *19*, 3619. [CrossRef] [PubMed]

29. Kučera, P.; Pištěk, V. Testing of the mechatronic robotic system of the differential lock control on a truck. *Int. J. Adv. Robot. Syst.* **2017**, *14*, 1–7. [CrossRef]

 *applied sciences*

*Article*

# Reverse Engineering and Topology Optimization for Weight-Reduction of a Bell-Crank

**Toh Yen Pang * and Mohammad Fard**

School of Engineering, RMIT University, Bundoora Campus East, Bundoora, VIC 3083, Australia; mohammad.fard@rmit.edu.au

* Correspondence: tohyen.pang@rmit.edu.au; Tel.: +61-3-9925-6128

Received: 11 November 2020; Accepted: 29 November 2020; Published: 30 November 2020

**Abstract:** This paper describes a new design method that was developed to achieve an optimal design method for weight reduction of a bell crank, sourced from a Louis Christen Road Racing F1 Sidecar. The method involved reverse engineering to produce a 3D model of the mechanical part. The 3D bell crank model was converted to a finite element (FE) model to characterize the eigenvalues of vibration and responses to excitation using the Lanczos iteration method in Abaqus software. The bell crank part was also tested using a laser vibrometer to capture its natural frequencies and corresponding vibration mode shapes. The test results were used to validate the FE model, which was then analysed through a topology optimization process. The objective function was the weight and the optimization constraints were the stiffness and the strain energy of the structure. The optimized design was converted back to a 3D model and then fabricated to produce a physical prototype for design verification and validation by means of FE analysis and laboratory experiments and then compared with the original part. Results indicated that weight reduction was achieved while also increasing the natural frequency by 2%, reducing the maximum principal strain and maximum von Mises stress by 4% and 16.5%, respectively, for the optimized design when compared with the original design. The results showed that the proposed method is applicable and effective in topology optimization to obtain a lightweight (~3% weight saving) and structurally strong design.

**Keywords:** bell crank; topology optimization; natural frequency; reverse engineering; vibrometer; design; Abaqus

---

## 1. Introduction

A bell crank is a mechanical component that is primarily responsible for translating the motion of links through an angle. It acts as a link between a spring and a damper component at one end, and a pushrod/pull rod at the opposing end. The most common limitation associated with vehicle suspension in racing vehicles is the inability to mount the spring/damper mechanism vertically, compared with most ordinary cars. As a result, bell cranks are incorporated into the design of racing vehicles to translate the vertical motion of the wheel into horizontal motion to allow the suspension to be mounted transversely or longitudinally [1].

However, in racing vehicles, there are limitations on bell cranks that relate to their weight. When there is a limit in engine power capacity, a reduction in vehicle mass will improve the performance aspect of a racing vehicle known as a power-to-weight ratio. Hence, engineers are constantly trying to improve the performance of bell crank by reducing its weight, while maintaining its structural integrity. To achieve that, engineers have been using the structural topology optimization method to find an optimal material distribution in a structural design domain considering an objective function in presence of constraints [2–5].

However, only a limited number of studies in the literature examined topology optimization of bell cranks [4,6,7]. Fornace [6] used a topology optimization method to assess weight reduction of bell cranks for a Formula Society of Automotive Engineers (SAE) vehicle. The author considered reducing the mass of the rear bell crank component by at least 10% while maintaining the yield strength of the previous design. The author utilized destructive testing on three specimens under static loads of 445 N to validate their new design. Choudhury et al. [4] used a 'Fully Stressed Design' optimization procedure to achieve the mass and stress optimization of bell crank of Formula SAE vehicle from SRM University Chennai in India. The procedure allowed them to investigate and analyze the structural stress distribution of the bell crank in real-time conditions during the damping process and the spring actuation. The authors managed to produce an optimized computer aided design (CAD) bell crank model with a 22% and 20% of overall weight and stress reduction, respectively.

When reducing the mass (or volume) of a mechanical component such as bell cranks, engineers face challenges to maintain structural integrity. The structural integrity relevant to a mechanical component is its overall stiffness. An element that directly correlates with component stiffness is its natural frequency. The natural frequency of a certain vibration mode can be controlled in order to reduce the displacement or deformation of a particular point in the structure [8,9]. Thompson et al. [10] found that chassis stiffness is directly related to its flexibility when the stiffness of the chassis increases, the twist vibration decreases and the overall vehicle handling is improved by allowing the suspension components to control a larger percentage of the vehicle's kinematics.

No studies in the literature have utilized a non-destructive mode shape analysis to investigate the dynamic response of the bell crank. The benefit of using mode shape analysis is that it allows engineers to determine the dynamic stiffness of a structure. By changing the stiffness of the structure, engineers could minimize the amount of vibration, which is produced by the relative motion or deformation of the structure under applied loads [11]. In the case of the bell crank, a small relative motion in the suspension system will provide undesirable feedback to the operator in either lateral, longitudinal or cyclic loadings [12].

The purpose of this study is to propose a methodology that incorporates a reverse engineering technique and physical testing for accurately and efficiently assessing topology optimization problems.

## 2. Theory of Weight Reduction and Stiffness Matrices

### 2.1. Modelling Approach

In a static structural analysis, for an arbitrary structure that is subjected to external forces and under the prescribed boundary conditions, the FE matrix equation can be expressed as [13–15]:

$$[F] = [K]\{u\} \tag{1}$$

where $[F]$ the force vector, $[K]$ the stiffness matrix, and $\{u\}$ the displacement vector [16].

According to Guyan [13], the vector $\{u\}$ can be partition into boundary degrees of freedom (DOF) and internal DOF, and they are denoted with subscript of $b$ and $i$, respectively. Equation (1) is fully general and is applicable for two-dimensional and three-dimensional problems, with appropriate boundary conditions. The entire system can be partitioned to [14]:

$$\left\{ \begin{array}{c} F_i \\ F_b \end{array} \right\} = \left[ \begin{array}{cc} k_{ii} & k_{ib} \\ k_{bi} & k_{bb} \end{array} \right] \left\{ \begin{array}{c} u_i \\ u_b \end{array} \right\} \tag{2}$$

Once the global stiffness matrix $[K]$ is assembled, the unknown nodal displacement, $\{u_i\}$ and unknown reaction forces, $[F_b]$ can be obtained by solving Equation (2). From the first line of equation, we first solve for $\{u_i\}$, i.e., $u_i = k_{ii}^{-1}(F_i - k_{ib}u_b)$; subsequently, substitute $u_i$ into the second line of equations, and thus $F_b$ can be solved as $F_b = k_{bi}u_i + k_{bb}u_b$.

### 2.2. Topology Optimization: Structural Compliance

Once the nodal displacements and reaction forces are calculated from Equation (2), the topology optimization is to determine the optimal subset $\Omega^*$ of material distribution of a given design domain, $\Omega$. The generalized formulation [16,17] for the compliance of the design domain is given as:

$$C^{\Omega}(x_e) = F_i^T(x_e)\, u_i(x_e) - F_b^T(x_e)\, u_b \tag{3}$$

where $x_e$ is the vector of design variables.

Assuming a structure that is subjected to the prescribed external loads and under appropriate supports, the design domain is meshed with $N$ finite elements. To achieve a reduction in mass, the topology optimization statement is expressed as [5,16,18–20]:

$$\text{Find}: \rho = \{\rho_1,\ \rho_2,\ \cdots,\ \rho_N\}$$

$$\text{Minimize}: [F]_\rho^T\{u_\rho\}, \tag{4}$$

$$\left.\begin{array}{l}\text{Subject to}: \int_\Omega \rho(x_e)d\Omega = Vol(\Omega^*) \le V \\ \quad\quad\quad: 0 < \rho_{\min} \le \rho(x_e) \le 1 \end{array}\right\} \tag{5}$$

where $V$ is the total initial volume before optimization, and $\rho_{\min}$ is the lower bound of artificial material density, $\rho(x_e)$ of each element in the design domain, which can be determined:

$$(x_e) = \left\{\begin{array}{l} 1 \text{ if \textbf{solid}, } x_e \in \Omega_* \\ 0 \text{ if \textbf{void}, } x_e \notin \Omega_* \end{array}\right. \tag{6}$$

The displacement vector $u_\rho$, which is the function $\rho$ in Equation (4) can be obtained by solving:

$$[K(\rho(x_e))]\{u\} = [f] \tag{7}$$

where $[K]$ is a structural global stiffness matrix and is obtained by the assemblage of element stiffness matrix over the design domain. For a given isotropic material, the stiffness matrix of each element, which depends on its artificial material density $\rho(x_e)$ is given by the following Young's modulus:

$$E(x_e) = \rho(x_e)\overline{E} \tag{8}$$

where $\overline{E}$ is Young's modulus for the given isotropic material, which assumed to be the same for all elements. When in equilibrium state, the energy bilinear form (i.e., the internal virtual work done by the elastic body, within a given design domain $\Omega$) can be expressed as [18,19]:

$$\delta W(v,\ w) = \int_\Omega E(x_e)\varepsilon_{ij}(v)\varepsilon_{kl}(w)dV \tag{9}$$

where $E(x_e)$ the optimal stiffness tensor that attain the isotropic material properties, $V$ is the given total material volume, $v$ the internal work at equilibrium state, $w$ an arbitrary virtual displacement, and with linearized strains $\varepsilon_{ij}(v) = \frac{1}{2}\left(\frac{\partial u_i}{\partial x_j} + \frac{\partial u_j}{\partial x_i}\right)$ and linearized load as $l(v) = \int_\Omega f\delta u d\Omega + \int_{\Gamma_T} t\delta u dS$.

The first integral extends over the volume of the body and the second integral extends the surface of the body.

We know strain energy is stored within an elastic body when the body deformed under the external forces. The objective function in Equation (4) is to achieve the optimal design via minimize the structural strain energy, which can be expressed as:

Minimize: $C = \{u\}^T[K]u$

Subject to: $[K(\rho(x_e))]\{u\} = [f]$

### 2.3. Vibration and Eigenvalue Problems

The eigenvalues of vibration and responses to an excitation mechanism play an important role in the design process. Likewise, the efficiency and accuracy of the FE analysis that solves the eigenvalue problem plays a crucial role in the success of present topology optimization of continuum structures [21]. Consider the linearized equation of motion for a vibration excitation in a discrete undamped structure [15,22,23], in our case the bell crank, can be expressed in the form:

$$M\ddot{q} + Kq = F \tag{10}$$

with $M$ and $K$ the global mass and stiffness matrices of the structure, $F$ is the force vector, and $q$ is the vector of unknown displacement DOF.

The solution of Equation (10) results in an eigenvalue problem, with $\omega$ are the eigenvalues and $\phi^N$ the associated eigenvectors, which can be expressed as [23]:

$$\left(-\omega^2 M^{MN} + K^{MN}\right)\phi^N = 0 \tag{11}$$

where $M^{MN}$ is the mass matrix, $K^{MN}$ is the stiffness matrix, and $M$ and $N$ are the DOFs. The eigenvalues of Equation (11) are usually solved using the Lanczos or subspace iteration method.

## 3. Materials and Methods

### 3.1. Overview of the Process

The overall design optimization process of the proposed study is summarized in Figure 1. The process started with a physical bell crank, which went through non-destructive experimental tests and a reverse engineering process to create an equivalent finite element (FE) model. The experimental data were used to verify the FE model. The optimization was based on the generalized process of finding a solution to the objective functions, which were either to be maximized or minimized material distributions within a design domain [24]. The design domain was subjected to constraints (e.g., material layout within a given design space (volume constraint) or maximum stress values) to be satisfied.

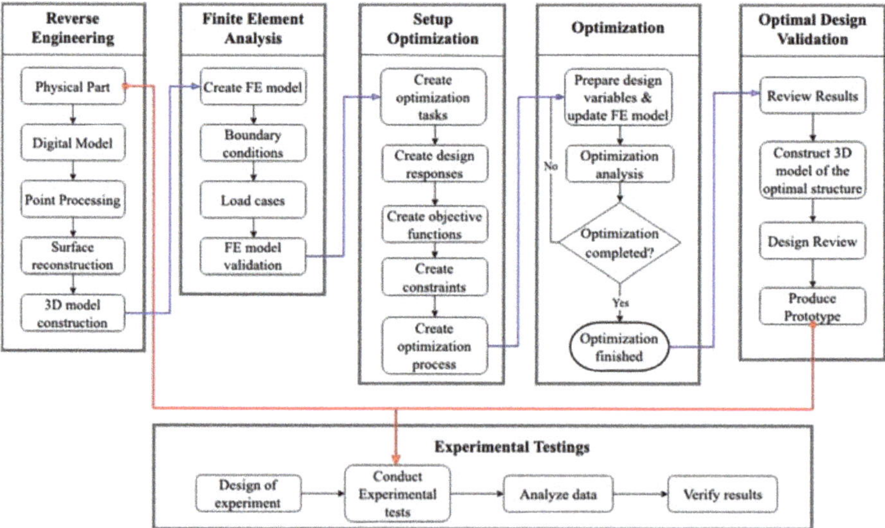

**Figure 1.** Flow chart depicting major stages of design optimization approach.

### 3.2. Physical Part of Reverse Engineering

The detail drawings and three-dimensional (3D) model for the bell crank were not available. Therefore, a reverse engineering technique known as 3D scanning was used to create a 3D model of the bell crank.

A bell cranks sourced from a Road Racing Sidecar and produced by Louis Christen Racing (LCR) in 1994, was used for the present study. The bell crank is made of anodized aluminum (Figure 2a) and its surface luster is relatively high, which resulted in a 'noisy' scan. Hence, the shiny surface required a coating that minimized the unwanted scanning noise. A product named *'PlastiDip'* in solid Matte White was used to coat the bell crank to achieve a matte finish surface. The coated bell crank was then placed on the rotating table and scan images were captured with FlexScan (Figure 2b). All scanned images were merged to create the shape of a 3D part model. In various locations, FlexScan was unable to duplicate the exact specifications and dimensions of the bell crank. Hence, a clean-up of the 3D model was necessary (highlighted in yellow (Figure 2c)). The surface of the material was cleaned using a *'defeature tool'* within 'Geomagic'. Holes were filled using the *'fill-up tool'*. Before the model was converted into an *'igs'* format, a *'mesh doctor'* tool was used to remove non-manifold edges, self-intersections, highly creased edges, spikes and small components. Various modifications of the 3D model were performed in the CATIA (V5R21) computer-aided design (CAD) program to assure the 3D model replicated the accurate dimensions and features of the actual bell crank. Figure 2d shows the revised and final 3D model that was then used for the validation and optimization tasks.

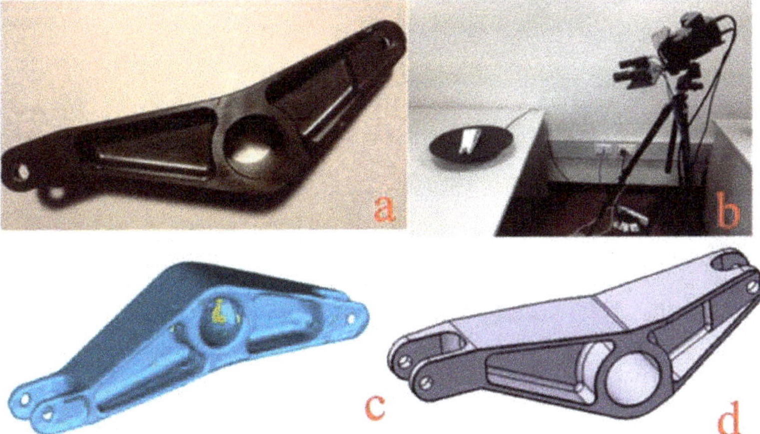

**Figure 2.** Bell crank scanning process: (**a**) Actual bell crank; (**b**) 3D scanning of the bell crank with FlexScan; (**c**) initial scanned 3D model; (**d**) final 3D model.

### 3.3. Experiment Method

Since only one bell crank part was available, a non-destructive method was adopted. The bell crank was tested using a laser vibrometer (PSV-400 Scanner, Polytec Inc., Irvine, CA, USA, coupled with the Polytec scanning program), which uses non-contact laser measurement [25]. The schematic diagram of the experimental equipment used for vibration testing is shown in Figure 3a. The bell crank was suspended in the air with strings, which has minimal damping characteristics, through the holes. The experiment's approach was aimed at extracting the modes of the frequency of the bell crank through applied sinusoidal vibration at five different points. The bell crank was excited by a modal shaker (model 2007E, miniature electrodynamic shaker, Cincinnati, OH, USA) as a sinusoidal force at predefined locations, indicated in Figure 3b. The tests were repeated five times at each location, so average eigenvalues for each location could be extracted. A mesh grid (x = 50, y = 20) was created over the top face of the bell crank, as this sizing of the grid gave the greatest coverage for the analysis (Figure 3c). Across the top face of the bell crank, the laser pointer recorded each separate point for a response. The laser vibrometer captures the natural frequencies and the corresponding vibration mode shapes of the bell crank in the required frequency band.

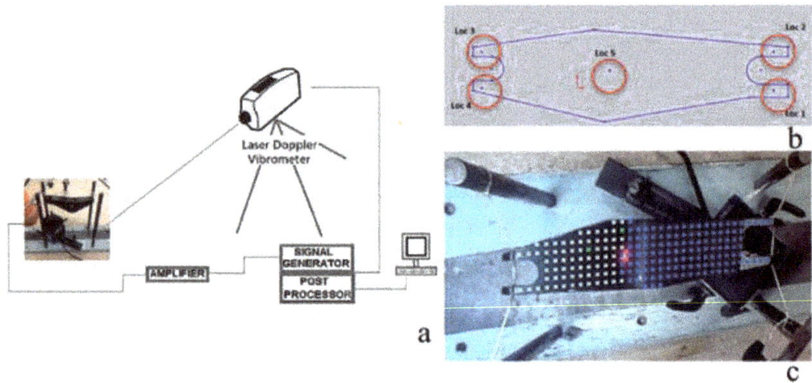


**Figure 3.** (**a**) Vibrometer test setup, (**b**) Locations for modal shaker attachment, (**c**) Scanning grid on top face of bell crank.

### 3.4. Bell Crank Model Validation

The 3D bell crank model was imported into the simulation program (Abaqus v6.14 [26]) for running a natural frequency analysis. Model geometry, material properties, element type, boundary conditions, step-to-request frequency modes, and a detailed description of the modelling are outlined in the following sub-sections.

### 3.4.1. Finite Element Mesh and Material Properties

The element size of the bell crank was set at 3 mm, and the 10-node quadratic tetrahedron element was chosen as the element type.

Figure 4 shows the meshed bell crank, which contains 20,827 elements and 35,803 nodes to the solid section. The mechanical properties of AW-6082-T6 aluminum alloy were assigned to the bell crank model [27,28]. The material properties used for the bell crank are as follows: Yield strength, $f_y$ = 310 MPa, density, $\rho$ = 2.7 × $10^{-9}$ kg/mm$^3$, Young's Modulus, E = 70 GPa and Poisson's ratio, $v = 0.33$.

**Figure 4.** Tetra-meshed of the 3D bell crank FE model.

### 3.4.2. Boundary Conditions

Boundary conditions were set up to replicate the best testing scenario previously conducted using a laser vibrometer. The bell crank was suspended by the string to avoid any external forces impeding the modal shaker; hence, the corresponding boundary conditions were imposed on the FE model. The two holes at either end of the FE model were restricted in both linear and rotational X, Y and Z directions. The eigenvalues of vibration and responses to excitation for the FE bell crank was determine using Equation (11) through the Lanczos iteration method in Abaqus software. The results obtained from the vibrometer testing were used for validation of the eigenvalue or natural frequency of the 3D bell crank FE model.

## 4. Bell Crank Optimization

### 4.1. Optimization Problem Statement

The optimization algorithms are based on published work [18,19,29]. Topology optimization for a continuum structure can be regarded as a material distribution problem, where the target is to find an optimized material distribution within the design domain, for the minimum compliance (or maximum global stiffness) according to Equation (3). Since $Stiffness \propto \frac{1}{Compliance}$, minimizing the compliance will maximize the stiffness.

In this study, the bell crank was discretized into finite elements and the Abaqus Topology Optimization Module (ATOM) was used to search for a minimum compliance design to minimize strain energy or compliance, based on the boundary conditions and forces being applied to the bell crank. The bell crank was discretized into finite elements. Assuming a constant $E_e$ for each element, and using Equations (4) to (7), the discrete form can be expressed as:

$$\text{Minimize}: \ C = \{F\}^T\{u\} \text{Subject to } [K \cdot E(x_e)]\{u\} = \{F\}$$

where $E(x_e) \in E_{ad}$, $K = \sum_{e=1}^{N} K_e(E(x_e))$, summing $e = 1, \dots, N$ elements. The objective function for this study was to minimize the overall displacements $\{u\}$, and a global measure of displacements is the strain energy, thereby minimizing the strain energy according to Equation (9).

$\{F\}^T$ represents the forces applied, $[K]$ is the global stiffness matrix, which depends on the stiffness of the individual elements $E_e$, whereas $E_{ad}$ is a set of admissible stiffness tensors for design problems. The entire design optimization process is summarized in Figure 5.

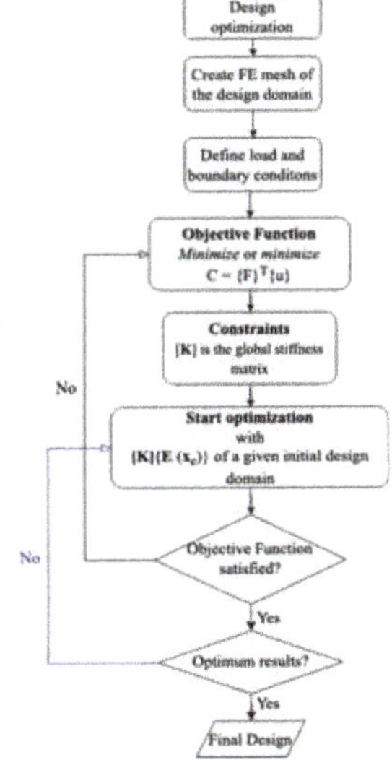


**Figure 5.** The design optimization process.

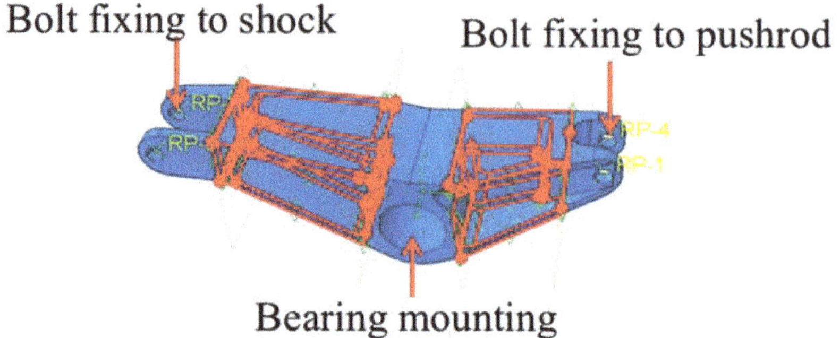



**Figure 6.** Design domains topology optimization (highlighted in red).

*4.2. Topology Optimization of the Bell Crank*

For the topology optimization analysis, only two design domains of the bell crank were considered to find an optimized material distribution. The regions were highlighted in red in Figure 6. The regions, which cannot be affected by the topology optimization, were located at the center bearing and the bolt holes on either end of the bell crank.

*4.3. Boundary Conditions*

During the optimization process, the bell crank model was subjected to two different boundary conditions. The first was applied to restrain all degrees of freedom at the center bearing hole. The second consists of point loads at four different locations. The loading forces were obtained from the Eibach Motorsport Catalogue 2014 [30], based on the 100-60-0080 spring, which is featured on the sidecar. A force value of 5299 N was provided as the block height of the spring and, therefore, will be considered as the worst-case loading scenario. Due to the pushrod and suspension orientations within the vehicle, these forces were divided into horizontal and vertical components. A horizontal force of 460 N and a vertical force of 2609 N were applied as a concentrated force at each reference point location indicated in Figure 7.

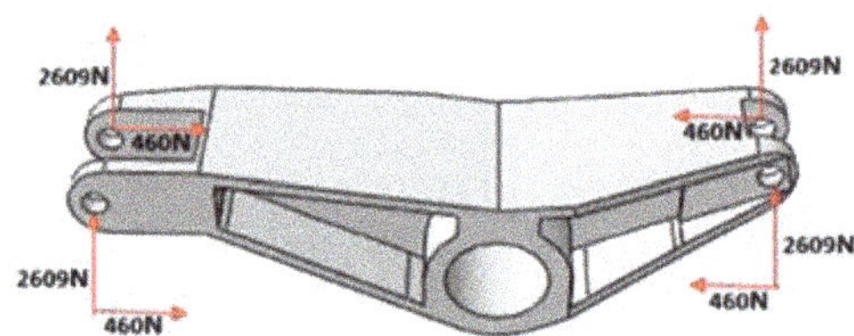



**Figure 7.** Applied boundary conditions for topology optimization.

## 5. Results and Discussion

*5.1. Experimental Testing Natural Frequency*

Values of natural frequency for the bell crank at the five tested locations are presented in Table 1. Note that an average natural frequency of 5046.75 Hz was obtained for all tested locations.

The value for the stiffness of the bell crank was 417.81 MN/m, which was solved by derivation of the natural frequency equation. The results obtained in the vibrometer testing were then used for validation of the 3D bell crank model before the topology optimization process.

Laser vibrometer measuring has high accuracy without causing any damage to the test object. However, displacements of a freely suspended test object during the acquisition process is a common source of minor error in the method [25]. However, in the measurement setup for this study, the bell crank was constrained by strings through the two side holes to ensure stability and to avoid any large displacement of the laser focal point during the data acquisition process. Hence, the properly constrained bell crank has enhanced the repeatability of the test results.

**Table 1.** Vibrometer Test Results.

| Location | Mean Peak Frequency (Hz) | Standard Deviation |
|---|---|---|
| 1 | 5061.25 | 1.30 |
| 2 | 5036.25 | 0.98 |
| 3 | 5045.00 | 1.15 |
| 4 | 5053.75 | 1.04 |
| 5 | 5037.50 | 0.64 |

*5.2. Finite Element Model Validation of Natural Frequency*

The accuracy of the created FE bell crank model was investigated through a proper process of validation to minimize any modelling errors, including the element and discretization errors (meshing), and the input errors from boundary and loading conditions [31]. Table 2 represents the results obtained from the natural frequency analysis of the FE bell crank model. The mode shapes from the FE model were analyzed to find the one that best represented a sinusoidal wave as observed in the physical test. It was found that, during the vibrometer testing, the first natural frequency was found in the fourth mode shape.

**Table 2.** Finite element bell crank modal analysis results.

| Mode | Frequency (Hz) |
|---|---|
| 1 | 1714.5 |
| 2 | 2560.7 |
| 3 | 2864.2 |
| 4 | 5285.7 |

A similar mode shape result was found in the FE bell crank model (Figure 8), which displayed a sinusoidal motion at the fourth mode with a resonant frequency of 5285.7 Hz.

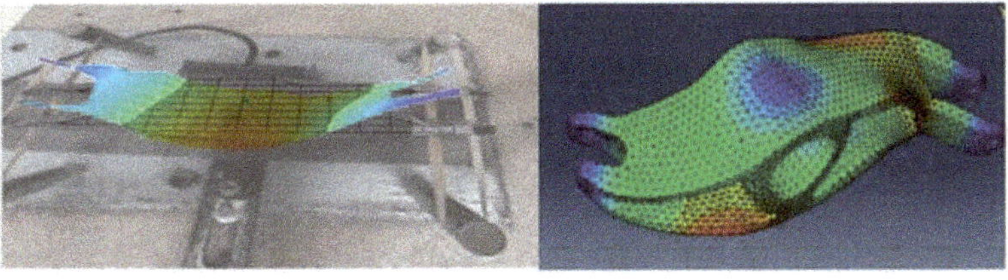


**Figure 8.** Fourth mode shapes of the physical bell crank and finite element model.

When analyzing both experimental and FE results, the difference in the natural frequency at the fourth mode of each case is below 5%; hence, the FE bell crank model achieved a satisfactory validation against the vibrometer experimental test. The validated FE bell crank model was then used for further topology optimization analysis.

*5.3. Topology Optimization*

A strain energy design response and a volume constraint were assigned in the optimization module. To improve the stiffness of the bell crank, the strain energy design response was assigned as the "minimizing strain energy". The stiffness of the structure was determined by its displacement corresponding to the assigned boundary conditions [32]. The volume constraint was assigned as it directly related to the overall mass of the bell crank model. In the analysis, a range of volume constraints, i.e., 95%, 90%, 85%, 80% and 75% of the original volume, was assigned to the FE model to achieve the optimal design.

Figure 9 shows the progressive removal and grouping of elements from the topology optimization analysis. At the volume constraint of 95%, the top surface and the two opposing arm locations of the bell crank (indicated by green contour) demonstrated a potential for element removal. Once the volume constraint was reduced to 90%, a significant material reduction was observed on the left arm region. From the 85% to 80% volume constraint results, significant reductions of material on both arm regions, as well as on the edges at the top of the bell crank, were achieved. The 75% volume constraint in the topology optimization featured the highest amount of material removal.

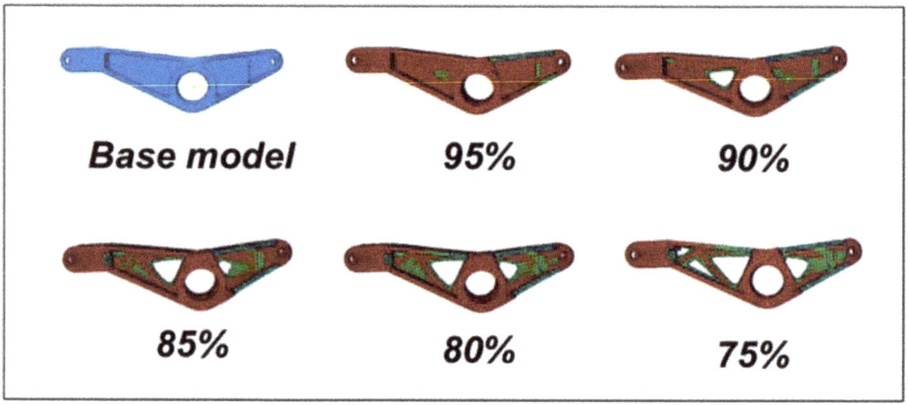


**Figure 9.** Topology optimization results with associated volume constraints, side view.

*5.4. Design Validation*

5.4.1. Geometry Interpretation and FEA

Topology optimization often produces complex contours in the geometry, which may be difficult for manufacturing and/or machining [8,33]. The results obtained from the iteration process of the topology optimization were used as guide geometries to determine where material should be withdrawn from the design domain. The guide geometries were interpreted toward the creation of optimal designs with suitable shapes and features for both machining and practical purposes. The guide geometries were exported from Abaqus into CATIA (V5R21), where their shapes and features were traced in the CAD environment. The optimized CAD design for the volume constraint models, i.e., 95%, 90%, 85%, 80%, and 75%, were then created.

To overcome unrealistic contour issues, the geometry was simplified to produce a practical optimized part. Sharp edges obtained through the topology optimization iterations were rounded with fillets, with a minimum radius of 3 mm for ease of machining. Profiles were smoothed out, due to the checkerboard contours resulting from the removal of elements. An example of the CAD design for the 80% volume constraint model is shown in Figure 10.

Re-analysis was carried out to assess the efficiency and structural integrity of the optimized bell crank design. The same 'static-general' and frequency-finite element tests were conducted, as examined on the original model, with identical set-ups of boundary and loading conditions. The maximum von Mises, stiffness, mass and maximum principal strain for the optimal design produced in the topology optimization with different volume constraints are shown in Figure 11.

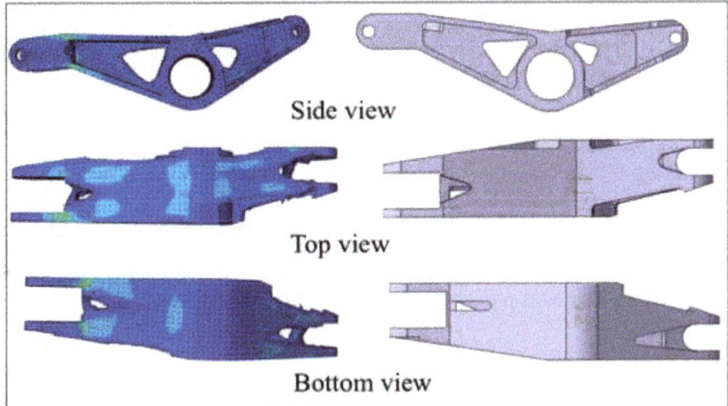



**Figure 10.** Revised CAD model for the 80% volume constraint. (**Left**) Optimization model, (**Right**) the CAD design free of irregular and complex shapes.

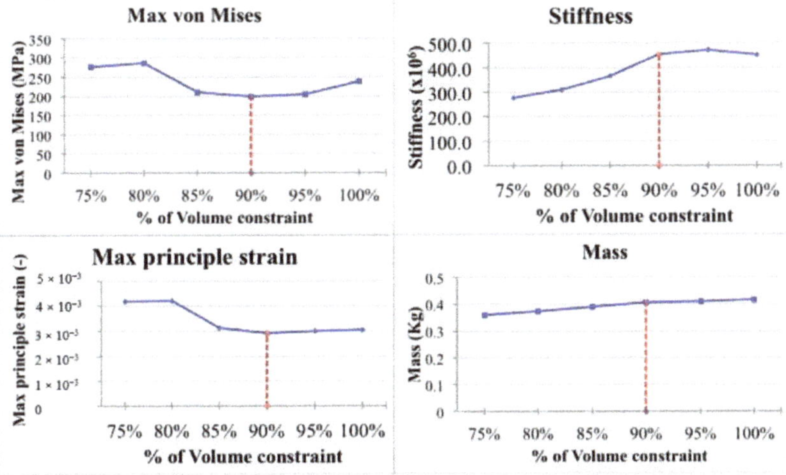



**Figure 11.** Summary of the structural integrity of the optimized bell crank CAD design.

The optimization results show that the 90% volume constraint model produced higher values of stiffness, with a corresponding natural frequency of 5334 Hz. It also featured a 16.5% reduction in maximum von Mises stress value, a 4.3% reduction in maximum principal strains, and a decrease in weight by 10 g from the original bell crank. The finite element result indicated that the highest stress concentration in the optimized model was 199.76 MPa. Based on the part that was machined out of Al6082-T6, the corresponding yield stress value was 310 MPa, which indicated a safety factor of 1.55 for the optimized bell crank. Hence, through the optimization iteration processes, the 90% volume constraint model was considered as the optimal design.

It is worth noting that the main technical challenges of converting the optimized FE analysis topological results in CAD design for further analysis and manufacturing. As highlighted previously by other investigators [29,33] and the current study, the initial topology optimization results still required manual process to eliminate holes, shape, irregular and complex edges, for example, see Figure 10. The topology optimization results were manually reconstructed to smoothen the shape and complex edges and to fill the missing pores to improve its manufacturability to the final design before the fabrication commencing.

### 5.4.2. Bell Crank Prototype Fabrication

After validating the FEA model, the prototype of the bell crank was fabricated for the next stage of physical testing and validation. A MU500VII-L 5-Axis Vertical Machining Station (Okuma, Charlotte, NC, USA) was used to machine the prototype. Due to cost limitations and the availability of the material, the optimized bell crank was machined out of Al6061 aluminum. The mechanical properties of Al6082-T6 and Al6061 are very similar, and the density, modulus of elasticity and Poisson's ratio all remain the same. The differences were in reductions in yield strength from 260 N/mm$^2$ to 240 N/mm$^2$ and tensile strength from 310 N/mm$^2$ to 260 N/mm$^2$. The changes in yield and tensile strength did not alter the result of the eigenvalue simulation and, therefore, did not alter the validation result from the laser vibrometer testing. The machined version of the optimized bell crank is illustrated in Figure 12.

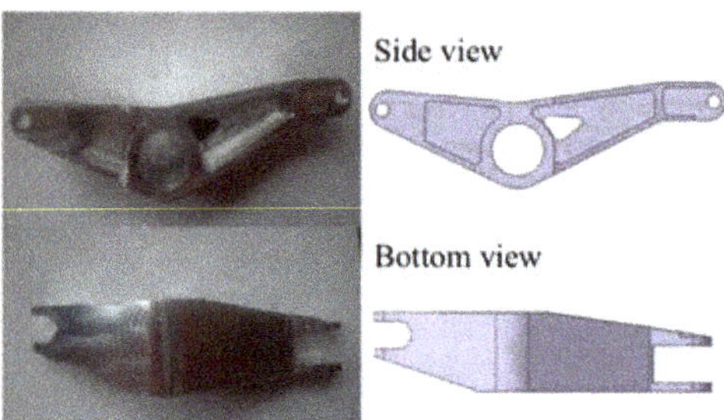



**Figure 12.** Prototype of optimal design and CAD comparison.

### 5.4.3. Experimental Validation of Prototype

A final validation after manufacturing of the optimized bell crank was carried out using the same laser vibrometer test. The original part was tested again, as well as the optimized part, to confirm that the optimization of the bell crank has been successful and the same testing conditions had been satisfied. Table 3 shows the natural frequency of the original bell crank compared with the optimized prototype. Overall, the optimized prototype showed an increase in natural frequency of 1.5% compared with the original bell crank. Since the difference was within the 5% limit, the optimization task and the machining can be deemed successful.

**Table 3.** Summary of laser Vibrometer testing of the original bell crank and optimized prototype.

| Location | Original Bell Crank, Frequency (Hz) | Standard Deviation | Optimized Prototype, Frequency (Hz) | Standard Deviation |
|---|---|---|---|---|
| Location 1 | 5055.5 | 1.45 | 5126.6 | 0.70 |
| Location 2 | 5037.5 | 0.70 | 5133.7 | 1.06 |
| Location 3 | 5056.9 | 1.48 | 5119.1 | 0.58 |
| Location 4 | 5087.8 | 1.15 | 5127.0 | 1.40 |
| Location 5 | 5029.6 | 1.41 | 5142.4 | 1.45 |
| Average | 5053.3 | | 5129.8 | |

The final mass of the optimized bell crank design achieved a weight saving of 2.6% (Table 4). Achieving a reduction in mass of the bell crank not only benefits the overall weight of the sidecar, but also reduces the unsprung mass of the suspension system and, therefore, improves the spring/damper unit efficiency [12]. With a careful design interpretation, the optimized design was able to reduce the density of elements within the design domain, while improving the overall stiffness of the bell crank.

Moreover, the optimized design achieved a higher value of natural frequency and lower von Mises stress value under the same static load conditions compared with the original part. The resulting topologically optimized bell crank was mounted in the LCF sidecar as shown in Figure 13.

**Table 4.** Comparison of original bell crank and optimized prototype.

| Parameters | Original Part | Prototype | % of Change |
|---|---|---|---|
| Natural Frequency (Hz) | 5053.3 | 5129.8 | +1.5 |
| Max von Mises (MPa) | 239.3 | 199.8 | −16.5 |
| Stiffness (k) (MPa) | 452.09 | 455 | +0.6 |
| Mass (kg) | 0.416 | 0.405 | −2.6 |
| Max strain (-) | $3.04 \times 10^{-3}$ | $2.91 \times 10^{-3}$ | −4.3 |

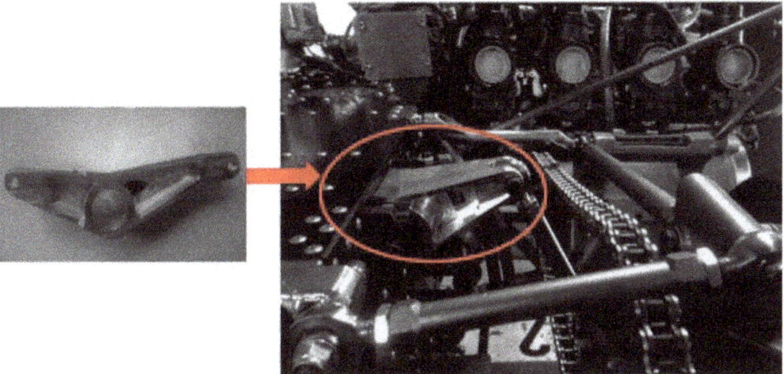

**Figure 13.** Optimized bell crank in the LCF sidecar.

The modern LCR F1 Sidecar weight is approximately 225 kg, including the engine weight (approximately 50 to 55 kg for current engine specifications). Considering an approximation of 30% to 40% for the sidecar's components (e.g., brake bell, steering, wheel hub and spindle, brake rotor etc.) and further work using the currently proposed methodology that incorporated the topology optimization approach, it was deemed possible that a 5 to 10 kg drop in net weight could be achieved. The methodology described in this paper can be used in mechanical, automotive and aerospace industries for weight reduction design optimization.

In producing a lightweight product, the present study also contributed to enhancing knowledge of how reverse engineering and non-destructive testing methods can be used effectively for structural topology optimization to solve complex designs without compromising performance and integrity.

The emerging of additive manufacturing, especially the 3D printing of metallic materials, will enable the possibilities to further optimize and manufacture bell crank with lattice structures [34], and hence, further work could be carried out for topological optimization of complex design, such as the bell crank and other mechanical components, as miniature lattice structures, which leads to even higher weight reduction.

## 6. Conclusions

In this paper, a design optimization methodology is described that incorporated: (1) a non-destructive measuring technique using a laser vibrometer; (2) a reverse engineering technique for 3D model construction; and (3) a structural topology optimization. The design optimization methodology was applied to a sidecar suspension bell crank to minimize its structural weight and then subjected to constraints including volume, strain energy and von Mises stress.

A 3D bell crank model was created using a reverse engineering technique and the corresponding FE bell crank model was validated using laser vibrometer test results. The difference between the experimental and FE results of the natural frequency at the fourth mode was less than 5%, which was deemed to be valid for further analysis.

At the beginning of the optimization process, the design domain for material removal was selected. Using the topology optimization approach, the geometry designs produced by the optimization algorithm were transformed into CAD models for the smoothing of the geometry to eliminate holes, irregular shapes and complex edges before fabricating the prototype for further testing. The results obtained from the design optimization methodology demonstrated its applicability and capability to produce an optimized bell crank, which increased its overall stiffness and natural frequency, while reducing its weight by 3%, maximum principal strain by 4.3% and maximum von Mises stress by 16.5%. This study demonstrated how reverse engineering and non-destructive testing methods can be used effectively for structural topology optimization to produce a suitable lightweight product. This is significant contribution to finding solution to complex designs without compromising structural performance and integrity.

Further work using the current methodology could be implemented in other sidecar components (e.g., brake bell, steering, wheel hub and spindle, brake rotor etc.), and additive manufacturing using 3D printing of lattice structures could achieve further net weight reduction.

**Author Contributions:** T.Y.P.: Conceptualization, methodology, software, writing—original draft preparation, M.F.: Methodology, writing, review and editing. All authors have read and agreed to the published version of the manuscript.

**Funding:** This research received no external funding.

**Acknowledgments:** The authors would like to acknowledge and give many thanks to Luka Grubic and Jamie Crass for their assistance on the project. The authors also acknowledged the contribution of the Laboratory Technician at RMIT Bundoora East Campus for their ongoing help and support in this project.

**Conflicts of Interest:** The authors declare no conflict of interest.

## References

1. Tony, A.; Marcel, E. Bell crank. In *A Dictionary of Mechanical Engineering*; Oxford University Press: Oxford, UK, 2013. [CrossRef]
2. Tsavdaridis, K.D.; Efthymiou, E.; Adugu, A.; Hughes, J.A.; Grekavicius, L. Application of structural topology optimisation in aluminium cross-sectional design. *Thin-Walled Struct.* **2019**, *139*, 372–388. [CrossRef]
3. Aulig, N.; Nutwell, E.; Menzel, S.; Detwiler, D. Preference-based topology optimization for vehicle concept design with concurrent static and crash load cases. *Struct. Multidisc. Optim.* **2018**, *57*, 251–266. [CrossRef]
4. Choudhury, P.; Suresh, N.; Panda, P. Shape Optimization Of A Suspension Bellcrank Using 3d Finite Element Methods. *Int. J. Eng. Res. Appl.* **2015**, *5*, 31–36.
5. Jaafer, A.A.; Al-Bazoon, M.; Dawood, A.O. Structural Topology Design Optimization Using the Binary Bat Algorithm. *Appl. Sci.* **2020**, *10*, 1481. [CrossRef]
6. Fornace, L.V. Weight Reduction Techniques Applied to Formula SAE Vehicle Design: An Investigation in Topology Optimization. Master's Thesis, University of California, San Diego, CA, USA, 2006.
7. Kim, B.-S.; Park, K. Kinematic Motion Analysis and Structural Analysis of Bellcrank Structures Using FEM. *JSAE* **2013**, *6*, 49–55. [CrossRef]
8. Lim, O.; Lee, J. Structural topology optimization for the natural frequency of a designated mode. *KSME Int. J.* **2000**, *14*, 306–313. [CrossRef]
9. Leader, M.K.; Chin, T.W.; Kennedy, G.J. High-Resolution Topology Optimization with Stress and Natural Frequency Constraints. *AIAA J.* **2019**, *57*, 3562–3578. [CrossRef]
10. Thompson, L.; Soni, P.; Raju, S.; Law, E. *The Effects of Chassis Flexibility on Roll Stiffness of a Winston Cup Race Car*; SAE Technical Paper 983051; Society of Automotive Engineers, Inc.: Warrendale, PA, USA, 1998.
11. Bonisoli, E.; Lisitano, D.; Dimauro, L.; Peroni, L. A Proposal of Dynamic Behaviour Design Based on Mode Shape Tracing: Numerical Application to a Motorbike Frame. In *Dynamic Substructures*; Linderholt, A., Allen, M., Mayes, R., Rixen, D., Eds.; Springer: Cham, Switzerland, 2020; Volume 4, pp. 149–158. [CrossRef]
12. Bolles, B. *Advanced Race Car Chassis Technology*; The Penguin Group (USA) Inc.: New York, NY, USA, 2010.
13. Guyan, R.J. Reduction of stiffness and mass matrices. *AIAA J.* **1965**, *3*, 380. [CrossRef]
14. Fard, M.; Yao, J.; Taube, R.; Davy, J.L. The Concept modeling method: An approach to optimize the structural dynamics of a vehicle body. *Proc. Inst. Mech. Eng. Part D J. Automob. Eng.* **2020**, *234*, 2923–2932. [CrossRef]

15. Donders, S.; Takahashi, Y.; Hadjit, R.; Van Langenhove, T.; Brughmans, M.; Van Genechten, B.; Desmet, W. A reduced beam and joint concept modeling approach to optimize global vehicle body dynamics. *Finite Elem. Anal. Des.* **2009**, *45*, 439–455. [CrossRef]

16. Niu, F.; Xu, S.; Cheng, G. A general formulation of structural topology optimization for maximizing structural stiffness. *Struct. Multidisc. Optim.* **2011**, *43*, 561–572. [CrossRef]

17. Asadpoure, A.; Valdevit, L. Topology optimization of lightweight periodic lattices under simultaneous compressive and shear stiffness constraints. *Int. J. Solids Struct.* **2015**, *60–61*, 1–16. [CrossRef]

18. Saleem, W.; Lu, H.; Yuqing, F. Topology Optimization- Problem Formulation and Pragmatic Outcomes by integration of TOSCA and CAE tools. In Proceedings of the World Congress on Engineering and Computer Science 2008, San Francisco, CA, USA, 22–24 October 2008.

19. Bendsøe, M.P.; Sigmund, O. *Topology Optimization Theory, Methods, and Applications*; Springer: Berlin/Heidelberg, Germany, 2004.

20. Lagaros, N.; Vasileiou, N.; Kazakis, G. A C# code for solving 3D topology optimization problems using SAP2000. *Optim. Eng.* **2018**, *20*, 1–35. [CrossRef]

21. Xia, Q.; Shi, T.; Wang, M.Y. A level set based shape and topology optimization method for maximizing the simple or repeated first eigenvalue of structure vibration. *Struct. Multidisc. Optim.* **2011**, *43*, 473–485. [CrossRef]

22. Spelsberg-Korspeter, G. Eigenvalue optimization against brake squeal: Symmetry, mathematical background and experiments. *J. Sound Vib.* **2012**, *331*, 4259–4268. [CrossRef]

23. Wu, B.; Yang, S.; Li, Z.; Zhong, H.; Chen, X. Computation of frequency responses and their sensitivities for undamped systems. *Eng. Struct.* **2019**, *182*, 416. [CrossRef]

24. Thomas, J.; Sahu, A.K.; Mahapatra, S.S. Simple Optimization Algorithm for Design of a Uniform Column. In *Advances in Applied Mechanical Engineering*; Voruganti, H., Kumar, K., Krishna, P., Jin, X., Eds.; Springer: Gateway East, Singapore, 2020; pp. 351–362. [CrossRef]

25. Trainotti, F.; Berninger, T.F.C.; Rixen, D.J. Using Laser Vibrometry for Precise FRF Measurements in Experimental Substructuring. In *Dynamic Substructures*; Linderholt, A., Allen, M., Mayes, R., Rixen, D., Eds.; Springer: Cham, Switzerland, 2020; Volume 4, pp. 1–11. [CrossRef]

26. *ABAQUS, Version 6.14*; Dassalt Systemés: Providence, RI, USA, 2014.

27. Huuki, J.; Hornborg, M.; Juntunen, J. Influence of Ultrasonic Burnishing Technique on Surface Quality and Change in the Dimensions of Metal Shafts. *J. Eng.* **2014**, *2014*, 8. [CrossRef]

28. Kim, D.Y.; Kang, W.J.; Cho, H.M.; Cheon, S.S. Analytical Evaluation of Al 6082-T6 Weld Zone Using Randomised Mixing Material Model. *Appl. Mech. Mater.* **2013**, *302*, 55. [CrossRef]

29. Manios, S.E.; Lagaros, N.D.; Nassiopoulos, E. Nested Topology Optimization Methodology for Designing Two-Wheel Chassis. *Front. Built Environ.* **2019**, *5*. [CrossRef]

30. Eibach. *Motorsport Catalog*; Eibach Springs, Inc.: Corona, CA, USA, 2014.

31. Prantil, V.C.a.; Papadopoulos, C.; Gessler, P.D. *Lying by Approximation: Truth about Finite Element Analysis*; Morgan & Claypool Life Sciences: San Rafael, CA, USA, 2013.

32. Liu, K.; Tovar, A. An efficient 3D topology optimization code written in Matlab. *Struct. Multidisc. Optim.* **2014**, *50*, 1175–1196. [CrossRef]

33. Zuo, K.-T.; Chen, L.-P.; Zhang, Y.-Q.; Yang, J. Manufacturing- and machining-based topology optimization. *Int. J. Adv. Manuf. Technol.* **2006**, *27*, 531–536. [CrossRef]

34. Terriault, P.; Brailovski, V. Modeling and simulation of large, conformal, porosity-graded and lightweight lattice structures made by additive manufacturing. *Finite Elem. Anal. Des.* **2018**, *138*, 1–11. [CrossRef]

**Publisher's Note:** MDPI stays neutral with regard to jurisdictional claims in published maps and institutional affiliations.

 *applied sciences*

*Article*

# Design and Structural Analysis of a Front Single-Sided Swingarm for an Electric Three-Wheel Motorcycle

**Polychronis Spanoudakis *, Evangelos Christenas and Nikolaos C. Tsourveloudis**

School of Production Engineering & Management, Technical University of Crete, 73100 Chania, Greece; echristenas@isc.tuc.gr (E.C.); nikost@dpem.tuc.gr (N.C.T.)
* Correspondence: hroniss@dpem.tuc.gr

Received: 31 July 2020; Accepted: 27 August 2020; Published: 1 September 2020

**Abstract:** This study focuses on the structural analysis of the front single-sided swingarm of a new three-wheel electric motorcycle, recently designed to meet the challenges of the vehicle electrification era. The primary target is to develop a swingarm capable of withstanding the forces applied during motorcycle's operation and, at the same time, to be as lightweight as possible. Different scenarios of force loadings are considered and emphasis is given to braking forces in emergency braking conditions where higher loads are applied to the front wheels of the vehicle. A dedicated Computer Aided Engineering (CAE) software is used for the structural evaluation of different swingarm designs, through a series of finite element analysis simulations. A topology optimization procedure is also implemented to assist the redesign effort and reduce the weight of the final design. Simulation results in the worst-case loading conditions, indicate strongly that the proposed structure is effective and promising for actual prototyping. A direct comparison of results for the initial and final swingarm design revealed that a 23.2% weight reduction was achieved.

**Keywords:** swingarm; single-sided; Finite Elements Analysis (FEA); three-wheel motorcycle; topology optimization

---

## 1. Introduction

Motorcycle technology advancements have been evident throughout the years, following or even exceeding the trends of automotive technologies. Accordingly, electric motorcycles are becoming a reality and they already have an important market share (over 30% in 2019) [1]. Several technology breakthroughs have been introduced up to now, providing numerous production vehicles and stimulating a huge effort towards their research and development by motorcycle manufacturers and startups. Electric motors, batteries and powertrains, are the key factors mainly researched by the industry. But electrification of powertrains also enables the redesign of several critical motorcycle parts.

One critical structural part of all motorcycles is the front suspension system that connects the chassis with the front wheel. Several alternative concepts of front suspension can be found, such as girder forks, leading link and hub center steering [2], introducing various advantages in motorcycle handling. The majority of the production motorcycles globally install telescopic forks, as front suspension systems, even though this entails specific disadvantages [3]. As the choice of the suspension system has major effects on the performance and handling of the motorcycle, alternative front suspension systems have been researched and have also been installed on production motorcycles [2,4]. Another issue of motorcycle design is the use of two or three wheels and specifically the setup of two wheels front and one rear. These kind of motorcycles can be considered as trikes or motorcycles, depending on the way they tilt as well as the lateral distance between the front wheels. Trikes are a different vehicle category, due to different handling and cornering. It is clear that up to now, two-wheel motorcycles dominate the market. However, in the last years different three-wheel concept vehicles have been introduced (Honda Neowing, Tokyo, Japan, Kawasaki J, Tokyo, Japan), production models are already on the road (Yamaha NIKEN, Shizuoka, Japan), and motorcycle magazine reviews have highlighted the advantages of two front wheels for handling, braking and safety feeling of the riders [5]. Linked to the above facts, extensive research on three-wheel motorcycles and their front suspension systems design, is needed and should be carefully evaluated. On this track, our research team have been focusing on the design and development of a new three-wheel electric motorcycle [6], with an innovative two-wheel front system arrangement using a single-sided swingarm for each front wheel and hub center steering. This motorcycle was developed to meet the challenges of future electric vehicles' and the development of the two-wheel front system arrangement was customized for installation on the specific vehicle. It must be noted, that the replacement of a gasoline engine by an electric powertrain in the same space, provides additional freedom in chassis design of future electric motorcycles, which makes it easier to install such an alternative front system. The work presented here is mainly focused on the structural analysis of the swingarm, which is a critical component of the front system.

Research on three-wheel motorcycles reported in the literature is limited. Related studies are mainly focused on dynamic modelling and stability analysis of these vehicles [7,8]. Simple and advanced dynamic models of three-wheel motorcycles are developed and specific modes of operation (capsize, weave and wobble) are recognized and evaluated compared to two-wheel vehicles through simulation. Other researchers are focused on the effects of passing over low friction coefficient surfaces, where the riders of three-wheel vehicles have indicated that one noteworthy characteristic is the vehicle's stability when cornering, over a low friction surface, like slippery or wet road. As indicated, under these conditions the decrease in lateral force on one front tire, passing over the low friction road surface, is compensated by the other front tire [9].

As mentioned, swingarm structural behavior is the main topic covered in this research. Based on related literature there are several publications focused on swingarm structural analysis, but they all explore single-sided [10–14] or double-sided [15–17] rear swingarms. Double-sided swingarms hold the rear wheel by both sides of its axle, while single-sided ones are a type of swingarm which lies along only one side of the rear wheel and hold it in one side, allowing it to be mounted like a car wheel. In all cases, a static analysis is considered, presenting stress and displacements based on different loading conditions. Focusing on the single-sided swingarms, in [10] the loads applied correspond to maximum traction and lateral bending forces, comparing an aluminum and Carbon Fiber Reinforced Plastic (CFRP) version. A multi objective optimization is performed for the redesign of the carbon fiber version targeting maximum stiffness and minimum weight. Bedeschi [14] presented a similar analysis, towards weight optimization, considering CFRP material. Smith [11,13] also used a static analysis based on torsional and vertical loads, trying to achieve higher torsional stiffness and minimum weight. Considering double-sided swingarms, a relative procedure is followed where structural analysis is based on braking and cornering [15], torsional loads [16], or extreme loading such as performing a wheelie [17]. Regarding material use, the majority is focused on aluminum alloys such as 7075-T6 [15], CFRP [18] and comparisons of aluminum and steel versions [16,17]. Based on the aforementioned literature, other valuable information was also gathered (Table 1), including: a) weight of swingarms developed, b) vehicle motor power and c) safety factor. It is evident that rear swingarm's weight depends on the motor power used on the motorcycle, as well as if it is single or double-sided. In Table 1, double- or single-sided swingarms are defined as (D) or (S), respectively. For motor power less than 40 KW, only double-sided can be found and their weight is 2.6–2.7 kg for aluminum and CFRP versions, respectively. On powertrains ranging between 130 and 185 KW, single-sided swingarms weight is 4.1–4.2 kg for CFRP and 5 kg for aluminum, while double-sided swingarms are 5.3–6.85 kg for different aluminum versions. The safety factor targeted by most researchers, even though data are limited, has values close to $N = 2$.

**Table 1.** Rear swingarm specifications in the literature.

| Reference | Double (D)/Single (S) Sided | Material | Weight (kg) | Safety Factor | Motor Power (KW) |
|---|---|---|---|---|---|
| Airoldi [10] | S | Aluminum A356.0T6 | 5 | | 129.7 |
| | | CFRP | 4.1 | | |
| Patiil [12] | S | Steel AISI 1018 | 9.26 | 2.07 | - |
| Bedeschi [14] | S | CFRP | 4.2 | - | 158.1 |
| Smith [11,13] | S | CFRP | 4.3 | - | - |
| Hasaan [15] | D | Aluminum 7075-T6 | 22.80 | 1.95 | 185 |
| Risitano [16] | D | Aluminum S2008 | 5.265 | | 138 |
| | | Aluminum MS 2008 | 6.85 | | |
| | | Aluminum BNG 2008 | 5.65 | | |
| Swathikrishnan [17] | D | AISI-4340 | 4.17 | 1.53–2.37 | 10 |
| | | Aluminum 6061-T6 | 2.6 | 2.34–2.39 | |
| Nigel O'Dea [18] | D | CFRP | 2.7 | - | 33 |

Braking and cornering forces are accounted for in most of the literature for rear swingarms. The magnitude of these forces is related to tire-ground friction and a simplification called "friction circle or ellipse" can be used to understand the maximum forces available for braking and cornering [2]. It assumes that the maximum tire friction force possible in any direction is a constant, which means that a tire can support a specific maximum force either for cornering, or braking, or any combination of steering and braking forces that result to the same maximum resultant force available, which is a simplification of reality. Other research findings exist, where more accurate tire models for combined braking and steering force components can be found, such as the Pajecka Magic formula [19] and BNP-MNC [20]. But according to the above and considering that, in our case, we are exploring the use of a front swingarm, the maximum loading scenario can be defined under emergency braking conditions in a longitudinal driving direction.

Basic theory of braking forces calculation can be found in various publications, including related books [21–23], as well as research focused on evaluation of the effects of braking loads applied on single-sided swingarms [12]. It must be noted though, that when brake forces are calculated, motorcycle tires and especially tire coefficient of friction are important as detailed in Section 2.3. Another important factor, directly linked to braking conditions, is the effect of deceleration. Braking deceleration of motorcycles is differentiated according to vehicle brake design and driver's ability to avoid skidding during braking [21]. Experimental brake tests have been conducted by different research teams measuring deceleration values, revealing the effect of the driver but also the differences that occur by using front, rear or both brakes simultaneously [24–26]. This part is exclusively presented in the following sections of this work, highlighting the importance of deceleration on braking calculations, in order to adequately define the loads applied on a swingarm.

To the best of our knowledge, there is no other analysis of a front single-sided swingarm in the relevant literature, as the majority of research is related to rear swingarms. Comparing the loads applied on a front or rear swingarm, several differences exist, such as: (a) the effect of torque during acceleration, which is not considered on a front swingarm, (b) motorcycle weight distribution loads, as also (c) higher braking forces applied on the front wheels. It can be easily understood, that weight and stiffness of rear swingarms cannot be directly compared to those of a front swingarm. Hence, the main contribution of this work is to present valuable results and insights, regarding stress and displacements calculated for the front single-sided swingarm design proposed. To cope with the forces applied during vehicle operation and at the same time to be lightweight, different designs are evaluated under different loading scenarios. Since braking conditions correspond to a major factor of forces applied, this is the worst-case loading scenario considered. For the evaluation procedure, a dedicated CAE (ANSA, Beta CAE Systems, Thessaloniki, Greece) software is used for a series of finite element analysis simulations. Based on preliminary results obtained, the redesign of the swingarm is discussed, targeting higher structural effectiveness and lower weight. For this purpose, a topology optimization procedure is carried out to assist and shape the new design. Finally, an evaluation of different swingarm versions under specific loading conditions is discussed, considering simulation results of stresses and displacements, which depend on design modifications.

## 2. Calculation of Forces on a Motorcycle

For the definition of the loadcases used in the simulations, the weight distribution and brake forces acting on the front swingarm must be calculated. In this section, a brief analysis of the relevant theory and experimental results found in the literature is presented.

### 2.1. Weight Distribution

Assuming a simple motorcycle model, the basic loads acting from weight distribution and braking can be determined as shown in Figure 1, where the motorcycle and the rider are modeled as a single rigid body. Supposing a flat road and ignoring the contribution from rolling resistance and aerodynamic forces, the loads on the vehicle's wheels are then calculated as shown below [21].

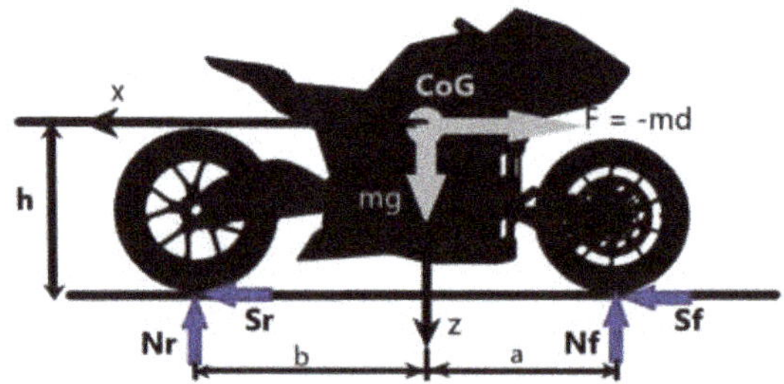


**Figure 1.** Simple motorcycle model of acting forces during braking.

In Figure 1, *a* represents the distance of front wheel from Center of Gravity (CoG), *b* is the distance of rear wheel from CoG and *h* measures the height of the CoG. Applying the sum of forces with respect to axes X and Z and sum of torques at the center of gravity, we obtain the following equations:

$$S_f + S_r = md \tag{1}$$

$$N_f + N_r = mg \tag{2}$$

$$N_r\,b - N_f\,a + \left(S_r + S_f\right)h = 0 \tag{3}$$

where, $S_f$ is the front braking force, $S_r$ is the rear braking force, $N_f$ is the reaction from the ground to the front wheel, $N_r$ the reaction from the ground to the rear wheel, $d$ represents the deceleration, $m$ the total mass and $m \times d$ the inertial force from the deceleration.

Rewriting the Equations (1)–(3) with respect to $N_f$ and $N_r$, results in:

$$N_f = mg\frac{b}{a+b} + md\frac{h}{a+b} \tag{4}$$

$$N_r = mg\frac{a}{a+b} - md\frac{h}{a+b} \tag{5}$$

Using the above equations, the weight distribution resulting in the loads acting on the front and rear wheels can be obtained.

*2.2. Braking Forces and Tire Friction*

As found in Equation (1), the total braking force ($S_{tot}$) is:

$$S_{tot} = S_f + S_r = md \tag{6}$$

A condition that should be avoided to maintain driver safety during braking is tire skidding, resulting in a loss of tire grip with the road [21]. Tire traction limit $D$ is based on tire characteristics and is related to the ratio $\mu$, which is the friction force (or otherwise the normalized braking force), effected by the braking force $S$ and the vertical load $N$. In order to maintain road grip during braking, this ratio should not exceed a maximum available value of $D$, for the front ($D_f$) and rear ($D_r$) tire traction, respectively:

$$\mu_f = \frac{S_f}{N_f} < D_f \tag{7}$$

$$\mu_r = \frac{S_r}{N_r} < D_r \tag{8}$$

As the total brake force ($S_{tot}$) increases, tire skidding may occur either in the front or in the rear tire depending on braking force distribution [21]. In order to examine this phenomenon, the ratio of braking balance ($\rho$) is introduced, linking the braking force on the rear wheel and the total braking force:

$$\rho = \frac{S_r}{S_f + S_r} = \frac{S_r}{S_{tot}} \tag{9}$$

By using Equations (1)–(3) and (7)–(9), normalized braking force $\mu$ may be expressed as a function of the deceleration $d$ and the braking balance $\rho$, as follows [21]:

$$\mu_f = (1 - \rho)\frac{(a+b)d}{ag + hd} \tag{10}$$

$$\mu_r = \rho\frac{(a+b)d}{ag - hd} \tag{11}$$

When $\rho = 0$ then $\mu_r = 0$ meaning only the front brakes are applied, while if $\rho = 1$ just the rear brakes are applied.

In order to calculate the brake ratio at the verge of skidding, the actual deceleration at this point must be known [22]. Assuming the tires (front and rear) have the same coefficient of friction, the total braking force at the point of skidding is calculated by Equation (2):

$$F_{skid} = (N_f + N_r)\mu = mg\mu \tag{12}$$

By equating (6) with (12), the expression for maximum deceleration is obtained:

$$F_{skid} = S_{tot} => d = -\mu g \tag{13}$$

The negative sign is an indication that the vehicle is in a state of deceleration.

In general, tire friction data are linked to road conditions (dry, wet, ice), as also on different road material types (such as asphalt, gravel etc.) [7,9,27]. Nevertheless, motorcycle tires are different than car tires and have different shapes of contact patch on the road [27,28]. Experimental results focused on motorcycle tire friction coefficient ($\mu$) calculation, indicate values close to $\mu = 0.8$ [22]. Experimental and theoretical results for car tires in contact with a road made from tarmac in dry conditions can be found in [23,27], presenting a range of values $0.7 < \mu < 1.0$. Research on motorcycle tires can also be found [22,27], indicating that for a dry road $\mu = 0.8$. Considering all the above information, for our calculations the reference coefficient of friction used for the front and rear tires is $\mu_f = \mu_r = 0.8$.

Therefore, based on Equation (13) and assuming $\mu = 0.8$ as a reference value, the maximum theoretical deceleration so that no tire skidding occurs is calculated at $d = 0.8$ g.

### 2.3. Urgent Braking in Real Conditions

The way motorcycle drivers use brakes in emergency situations affects motorcycle response and differentiates deceleration magnitude. For this purpose, several experimental tests have been conducted by different researchers, using professional or ordinary drivers, different sport or sport-touring motorcycle models and different braking conditions (using front, rear or both brakes) [24,29–36]. Relevant experiments on low-power motorcycles (100–150 cc) were conducted in [26], as well as in wet road conditions where decelerations are lower [25]. These tests reveal the range of decelerations achieved in reality (minimum–maximum) and can be compared to theory. A summary of these results is presented in Table 2 and at the last row the mean value for all minimum and all maximum decelerations is calculated. In this calculation, we have not accounted for the tests conducted in [26] due to the low power of motorcycles used in the tests. Also all values included are without the use of an anti-lock braking system (ABS).

**Table 2.** Braking deceleration rates in g reported in the literature (dry and wet road).

| | Deceleration (g) | | |
|---|---|---|---|
| **Reference** | **Front Brake Only** | **Rear Brake** | **Front and Rear Combined** |
| Bartlett [31] | 0.88–0.89 | 0.38–0.46 | 0.96 |
| Ecker [32] | - | - | 0.51–0.75 |
| Vavryn [33] | - | - | 0.67 |
| Bartlett, Baxter, Robar [34] | 0.44–0.76 | 0.31–0.43 | 0.59–0.89 |
| Anderson, Baxter, Robar [35] | 0.65 | 0.42 | 0.71 |
| Dunn, et al. [36] | 0.518–0.709 | 0.345–0.386 | 0.61–0.71 |
| Ariffin et al. [26] | 0.5 | 0.35 | 0.48–0.83 |
| **Min-Max decelerations** | **0.44–0.89** | **0.31–0.46** | **0.51–0.96** |
| **Mean values of min and max** | **0.62–0.75** | **0.34–0.42** | **0.57–0.78** |

Combined braking (front and rear brake use) is considered as the best way to brake in an emergency situation and this is represented in the results, where the maximum deceleration achieved is 0.96 g. Lower values are found for rear brake use (0.46 g), considered as the worst choice. Finally, using only front brakes results in maximum decelerations up to 0.89 g. It can be seen that the range of decelerations vary from test to test. For this reason, a mean value of all maximum and all minimum decelerations is calculated in the last row of Table 2 and is considered as a more adequate magnitude of reference. Based on that, when only front brakes are applied, the mean value of maximum deceleration is $d = 0.75$ g.

Comparing this experimental reference value (0.75 g) with the theoretical value of deceleration ($d = 0.8$ g) calculated in the previous Section 2.2, we observe that theoretical calculations are 5.8% higher than the experimental. Due to the small difference, we choose the worst case (theoretical deceleration value $d = 0.8$ g), as the reference value used for the maximum braking forces applied in calculations hereafter.

## 3. Testbed Vehicle Specifications and Loading Scenarios

The proposed swingarm design is installed on the front suspension system of a three-wheel electric motorcycle (Daedalus). Using two wheels at front, two identical single-sided swingarms are used to hold the wheels. The top and side view of this assembly is shown in Figure 2a,b, respectively. In order to calculate the forces acting on the testbed vehicle, the technical specifications and corresponding dimensions of CoG related to Figure 1 are presented in Table 3.

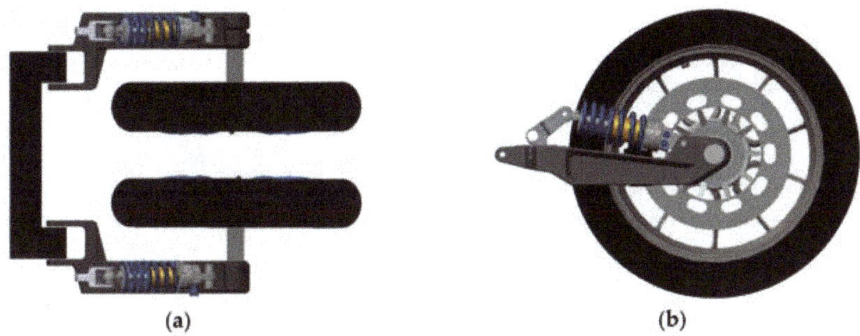



**Figure 2.** Front swingarm assembly: (**a**) top view; (**b**) side view.

**Table 3.** Testbed vehicle (Daedalus) technical specifications.

| Chassis | Aluminum |
| --- | --- |
| Motor | 100 KW-Brushless electric motor |
| Max motor torque | 230 Nm |
| Max motor RPM | 5200 RPM |
| Dimensions (L × W × H) | 2.1 × 0.8 × 1.2m |
| CoG Dimensions (a, b, h) | 795mm, 804mm, 541mm |
| Weight | 240 kg (excl. Driver) |

### 3.1. Front Swingarm Design and Materials

The initial design of the single-sided swingarm is presented in Figure 3, while the assembly of related components that are included in the simulation, is shown in Figure 4. As shown, at the back side the swingarm is connected to the chassis through an axle inserted in the spacers, while on the front it is connected to the wheel through the swingarm axle. The front axle is attached on the swingarm with two bearings and its other side is connected to the wheel using a hub center steering system.

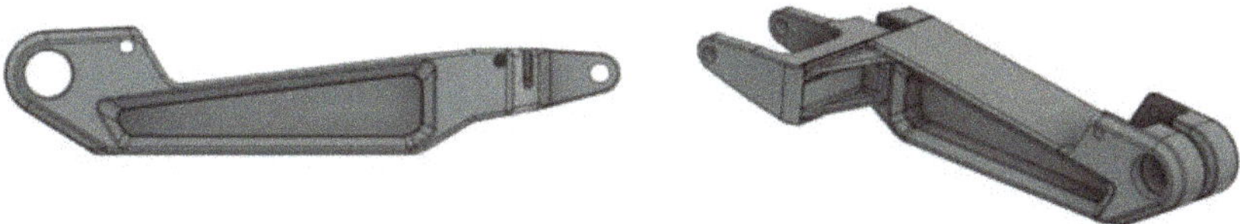

**Figure 3.** Front swingarm initial design.

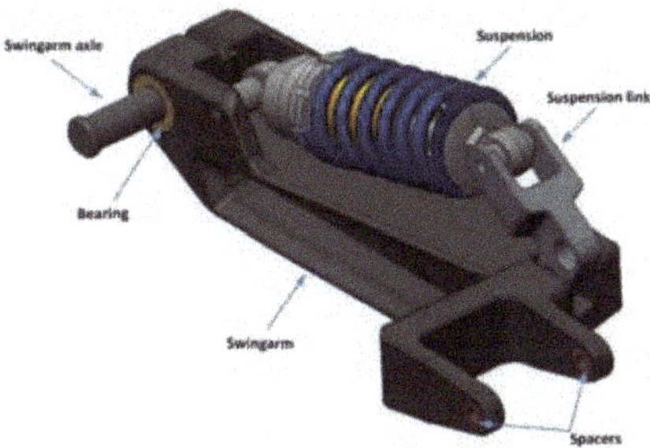



**Figure 4.** Front swingarm and related components assembly.

### 3.2. Weight Distribution Forces—First Loading Scenario

As explained in Section 2.1, the weight distribution forces on the wheels are determined by the position of the center of gravity (CoG) and the mass of the motorcycle including the driver, in our case 340 kg. By subtracting the unsprung mass of the wheels (21 kg), the total mass acting on the CoG is $m = 319$ kg. Weight distribution on the front and rear wheels is calculated using Equations (4) and (5), assuming deceleration $d = 0$, resulting in $N_f = 1572.9$ N and $N_r = 1555.9$ N. Force $N_f$ is distributed equally on the front two wheels, resulting in $N_{fleft} = N_{fright} = 786.45$ N. These forces correspond to the first loading scenario considered in the simulations.

### 3.3. Braking Forces—Second Loading Scenario

Using Equation (4) force $N_f$ can be calculated for different deceleration values ($d$). In addition, using Equation (10), the coefficient of friction ($\mu_f$) can be obtained, related to different braking distribution factors ($\rho$) and decelerations. The brake force acting on the front wheels ($S_f$) is then calculated based on $N_f$ and $\mu_f$. The same procedure is used to calculate the rear braking force $S_r$. The worst-case scenario of forces applied is considered when only the front brake is used ($\rho = 0$). Assuming a friction factor $\mu_f = 0.8$ and a maximum deceleration of d = 0.8g, results in $S_{fleft} = S_{fright} = S_f/2 = 1251.75$ N on each front wheel. The brake moment ($M_{br}$) generated by $S_f$, considering the front tire radius ($R = 0.32$ m), is $M_{br} = S_f \times R = 400.5$ Nm. The vertical load is also found equal to $N_{fleft} = N_{fright} = 1210$ N. These forces correspond to the second loading scenario considered in the simulations. Using Equation (13), the maximum deceleration that can be achieved in order to avoid tire skidding would be $d = 0.8$ g, assuming $\mu = 0.8$.

## 4. Modelling and Simulation

Finite elements modelling requires specific steps, depending on the pre-processor used, in order to prepare a CAD part for Finite Elements Analysis (FEA) simulation. This includes modelling of forces, constraints, connections of parts in the assembly, material specification and mesh generation. All these steps are detailed in the following paragraphs. A dedicated CAE software (ANSA, Beta CAE Systems, Thessaloniki, Greece) is used for this purpose [37], providing adequate results.

### 4.1. Geometry Mesh and Materials

The ANSA pre-processor is used for the development of the geometry mesh and volume elements are used for higher results accuracy (Figure 5). The mesh of the swingarm has 97,259 volume elements, from which 87,475 are Tetras and 9784 are Pyramids elements. The total mesh of all the components included in the swingarm assembly has 116,818 volume elements (Figure 6).

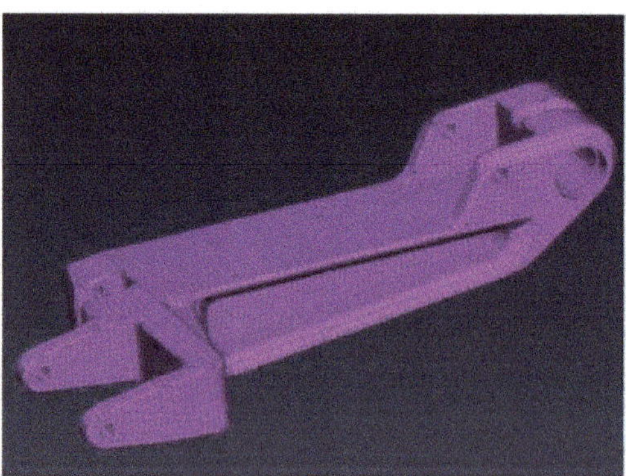

**Figure 5.** Mesh of the initial swingarm design.

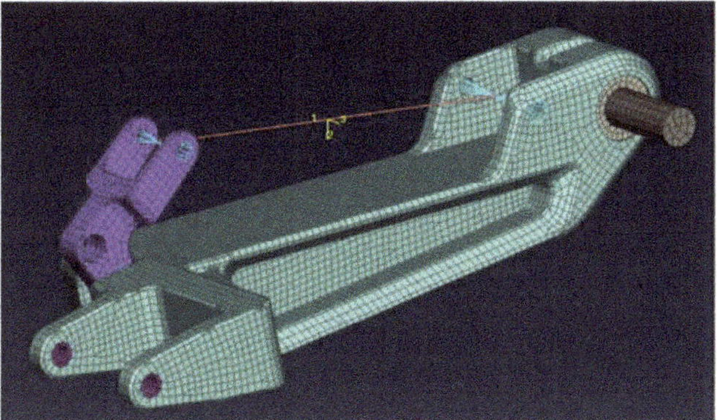

**Figure 6.** Mesh of components assembly for the swingarm design.

The modelling of material properties used for the preliminary simulations can be found in Table 4, where the required properties used by the pre-processor are: (a) Elastic modulus, (b) Poisson ratio, (c) Shear modulus, (d) Density and (e) Yield strength. The materials initially considered are: (a) aluminum 7075-T6 for the swingarm and the suspension link, (b) stainless steel AISI-304 for the swingarm axle and the bearings, (c) steel AISI-130 for the suspension, all modelled as isotropic materials. Suspension has been simplified to a steel round beam (shown with a line in Figure 6) in order to reduce mesh elements number and is assumed to be a rigid component.

**Table 4.** Materials properties used in preliminary simulations.

|  | Aluminum 7075-T6 | Stainless Steel AISI 304 | Steel AISI 4130 |
| --- | --- | --- | --- |
| Elastic modulus, E(GPa) | 71.7 | 200 | 205 |
| Poisson ratio, v | 0.33 | 0.29 | 0.29 |
| Shear modulus, G(GPa) | 26.9 | 86 | 80 |
| Density, $\rho$ ($\times 10^3$ kg/m$^3$) | 2.81 | 8 | 7.85 |
| Yield strength (MPa) | 503 | 215 | 435 |

### 4.2. Connections and Contacts

The connection of different components in the assembly is defined using bolts, related to suspension and suspension link, or contact functions for spacers and bearings. Figure 7a presents bolts connections placement and Figure 7b depicts the contact connections as modelled on the assembly.

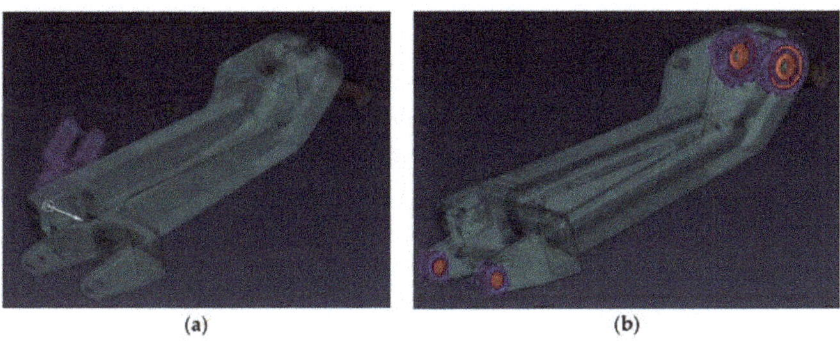

(a)          (b)

**Figure 7.** Connections modeled: (**a**) with bolts; (**b**) with contacts.

### 4.3. Forces and Constraints

Constraints are modelled as shown in Figure 8a, placed as fixtures on the rear part of the swingarm connected to the chassis. Suspension movement is also fixed using a constraint on the suspension link. The forces are applied considering the calculated values of the vertical force $N_f$ (first loading scenario) and a clockwise moment $M_{br}$ (used with $N_f$ for second loading scenario) placed on the end of the swingarm axle (Figure 8b).

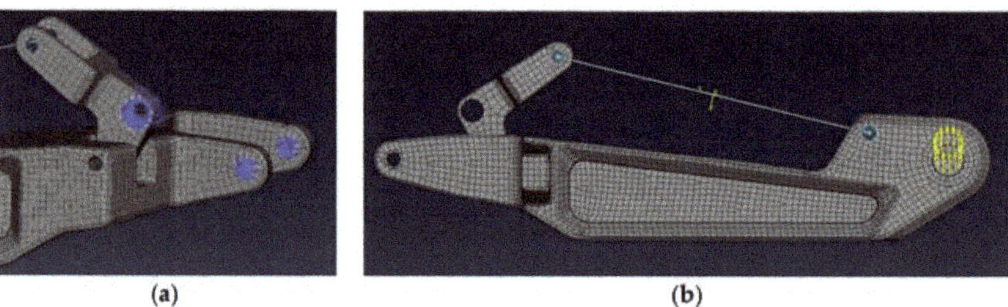

(a)         (b)

**Figure 8.** Finite element analysis model: (**a**) constraints; (**b**) loads.

The assumption that suspension is rigid and that no tire modelling is considered, is a simplification used to reduce computational time under the static analysis presented here. Stress and displacements calculated in this way are expected to be higher than in the case when the suspension was operating normally and tires could also compensate part of the loads. These assumptions help us on evaluating different swingarm designs, which is the main target here, but in order to obtain more precise results and achieve higher weight reduction, these modelling procedures should be applied.

### 4.4. Topology Optimization

A general optimization problem is identified firstly as a problem of finding the optimal topology and then as a problem of finding the optimal shape or finding the optimal cross sections [38]. Topology optimization is a mathematical process that aims to find the optimal distribution of the material of a construction, while satisfying its support conditions and loads. It is implemented by combining finite elements for analysis and mathematical programming techniques for solving [38,39]. ANSA's SOL200 was used to conduct topology optimization simulations in the present work [37]. The first important parameter defined is the design area, corresponding to the area of the part where the problem of topology optimization will be solved, while areas not included in the solution are called non-design areas [38]. In our case, the design area is the swingarm excluding the rear brackets connected to the chassis. The second parameter is the definition of the objective function and constraints. The objective function for this application corresponds to the minimization of the weighted deformation energy (min compliance). Using this objective function, the residual mass percentage and the static loading scenario are taken as constraints. The first restriction refers to the percentage of mass removed from the original mass of the swingarm.

## 5. Results

The linear static analysis of the front swingarm is based on the previously mentioned modelling of forces, materials, connections and constraints and the solver calculates the stresses (Von Misses) and the displacements. Two loading scenarios are conducted and discussed related to the forces applied. The first scenario includes the weight distribution of the forces (weight) resembling a vertical bending scenario, while the second investigates the effect of maximum braking forces. For the representation of the results, META post-processor was used. The first part of results shows the initial swingarm design under the two loading scenarios (Sections 5.1 and 5.2). In Section 5.3, details of the redesign process towards the final swingarm design are presented, based on topology optimization results and alternative designs evaluation. Finally, in Section 5.4 simulation results of the final swingarm design are shown and a detailed comparison to the initial design is discussed.

### 5.1. Initial Swingarm Design—First Loading Scenario

In this part, the force applied is $N_f = 786.45$ N, which corresponds to weight distribution on the front wheel (Section 3.2). Stress results and related displacements for the swingarm are shown in Figure 9.

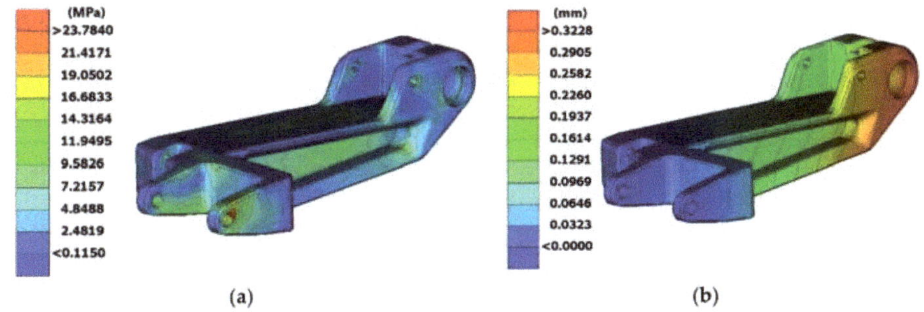

**Figure 9.** Stress (**a**) and displacement (**b**) results in 1st loading scenario.

A normal stress distribution is observed on the swingarm, where a maximum stress of 23.78 MPa occurs on one of the constrained points that connects it to the chassis. This result has a low magnitude that needs no further evaluation. The displacement results show a maximum value of 0.32 mm for the swingarm.

### 5.2. Initial Swingarm Design—Second Loading Scenario

Considering the case of braking, the forces applied are calculated in Section 3.3. As mentioned, the highest braking forces occur at deceleration of $d = 0.8$ g when using only the front brakes. According to this, the forces applied in the simulation are $N_f = 1210$ N and $M_{br} = 400.5$ Nm. The results of stresses and displacements obtained are presented in Figure 10.

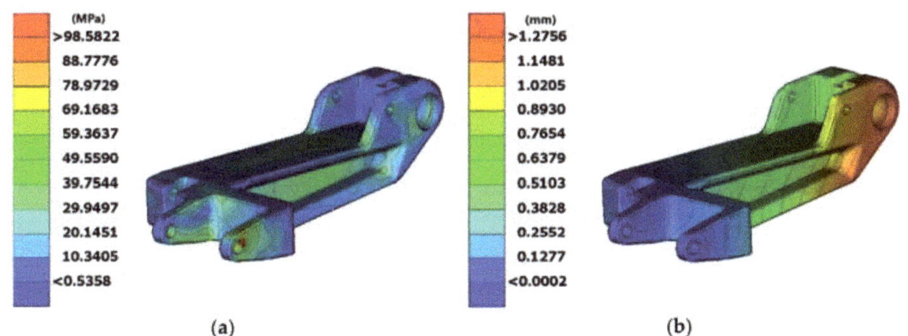

**Figure 10.** Stress (**a**) and displacement (**b**) results in 2nd loading scenario.

The maximum stress is 98.6 MPa placed on the right rear bracket, far lower than the material yield stress (503 MPa), corresponding to a safety factor $N = 5.1$. This yields that excess of material should be removed. Another important observation is that on the rest of the part stresses distribution are less than 60 MPa, while displacement results show a maximum value of 1.27 mm. According to these insights, certain modifications can be made on swingarm design in order to reduce weight.

Additionally, a change of material could be recommended in order to cut down raw material and production costs.

*5.3. Final Swingarm Design*

Considering the forces applied on the different loading scenarios, we mostly depend on results obtained for the worst-case scenario which is under emergency braking conditions. A swingarm redesign and material replacement is decided, targeting lower weight and production cost. It is clear that stress and displacements will be raised, but the objective is a safety factor close to $N = 2$ and at the same time to keep displacements as low as possible. A topology optimization procedure was used at first, so as to assist on redesign of the swingarm and obtain valuable results regarding specific areas of material removal (Figure 11). Modelling and simulation were conducted based on parameters referred in Section 4.4. The finite element model was solved using the second (worst) loadcase scenario for various topology optimization parameters and the new form obtained was again validated (solved) in a static analysis. In this way it was possible to fine tune certain parameters such as the percentage of mass removed from the swingarm, which was finally set to 40% reduction compared to the original mass. It must be noted that we also experimented with higher reduction percentages, but the results indicated: a) even more complex swingarm forms, which were difficult to manufacture with our production capabilities (CNC machining) and b) similar stresses and displacements. For these reasons, the specific percentage of mass reduction (40%) was chosen. As seen in Figure 11, material was mostly removed on the front part of the swingarm as well as in the middle, indicating that hollow parts should exist at these points. Most of the features of the form obtained through this procedure were incorporated in the new design.

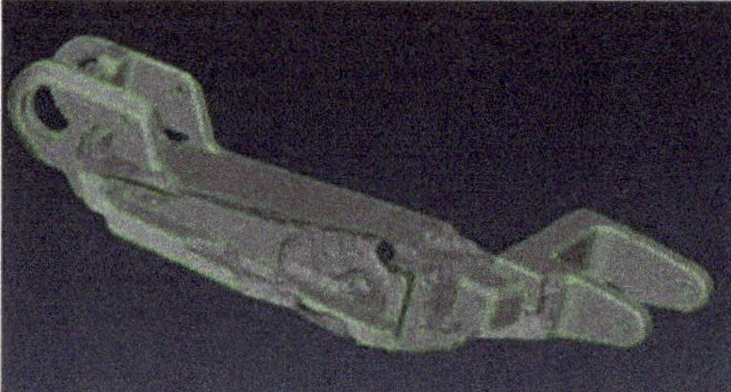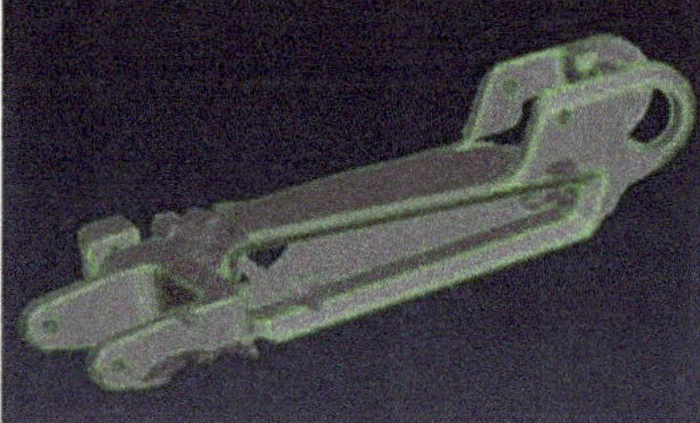

**Figure 11.** Topology optimization results for the swingarm.

Various designs were considered and evaluated and two new alternatives are shown in Figure 12a,b. Their analysis was based on loads of the first loading scenario and the material used was aluminum 7075-T6. The results obtained are presented in Figure 12c,d, respectively.

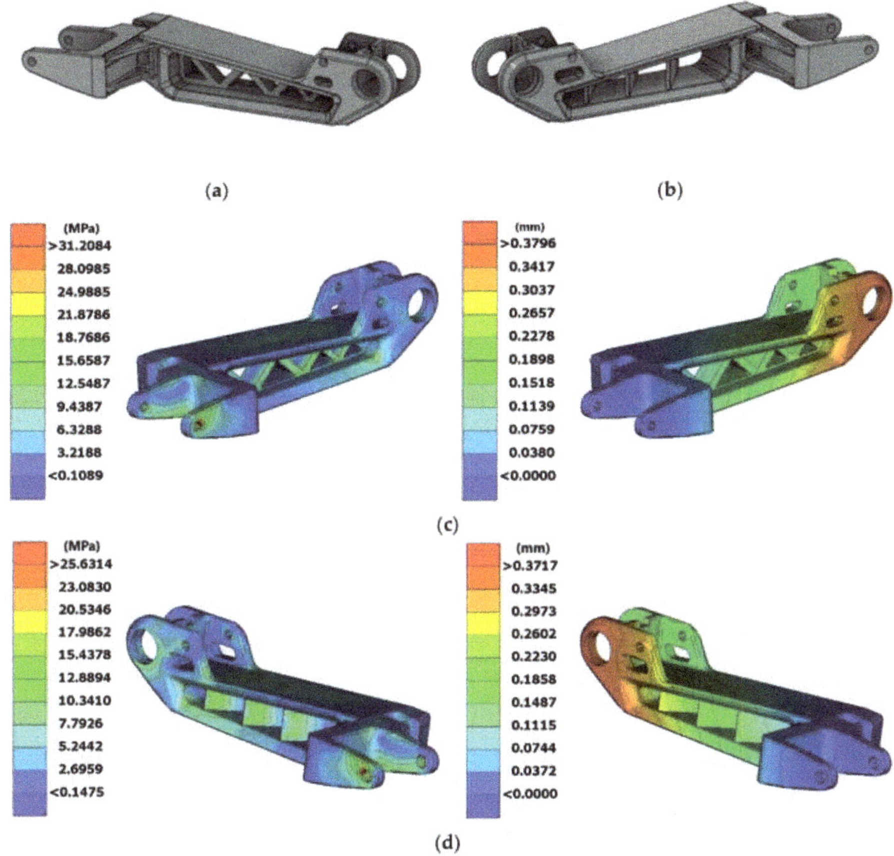

Figure 12. Different swingarm versions and results: (**a**) 1st alternative swingarm design; (**b**) 2nd alternative swingarm design; (**c**) results of stresses and displacements for 1st alternative swingarm; (**d**) results of stresses and displacements for 2nd alternative swingarm.

The first alternative design showed a weight reduction of 13% (4.94 kg), while the second alternative provided even higher weight reduction of 18.5% (4.62 kg). Considering the results of analysis, the first one presented higher maximum stress (37.2 MPa—first, 25.63 MPa—second) while both had almost the same displacements (0.38 mm—first, 0.37 mm—second). It is evident that the second design alternative was the type of design we should focus on for the final version. The final form was slightly changed, mostly affected by our production capabilities, where CNC machining manufacturing was chosen. According to the redesign procedure followed, the final swingarm design is presented in Figure 13.

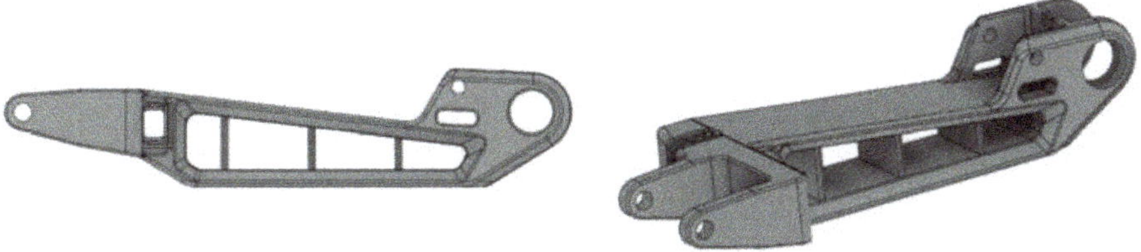

Figure 13. The final redesigned single-sided front swingarm.

As mentioned, low stress distribution and a high safety factor of the initial design gave us the flexibility to choose a new aluminum alloy (5083-H116) for the swingarm, targeting lower production cost. The modelling of the isotropic material properties used for the new swingarm assembly can be found in Table 5. Four different materials are used: (a) Aluminum 5083-H116 (swingarm and link), (b) Stainless steel 304 (suspension), (c) Stainless steel SS-A4-80 (swingarm axle), (d) Steel AISI4130 (bearings). The required properties used by the pre-processor are: (a) Elastic modulus, (b) Poisson ratio, (c) Shear modulus, (d) Density and (e) Yield strength.

**Table 5.** Materials properties used in the final modelling.

| | Aluminum 5083-H116 | Stainless Steel AISI 304 | Steel AISI 4130 | Stainless Steel A4-80 |
|---|---|---|---|---|
| Elastic modulus, E(GPa) | 71 | 200 | 205 | 193 |
| Poisson ratio, v | 0.33 | 0.29 | 0.29 | 0.29 |
| Shear modulus, G(GPa) | 26.4 | 86 | 80 | 86 |
| Density, $\rho$ ($\times 10^3$ kg/m$^3$) | 2.66 | 8 | 7.85 | 8 |
| Yield strength (MPa) | 228 | 215 | 435 | 600 |

*5.4. Final Swingarm Design Results*

An identical modelling process is followed for the new materials applied, as in the initial design. Only the worst-case scenario (second loading scenario) is used for the evaluation and comparison to the initial version. The forces applied are again $N_f$ = 1210 N and $M_{br}$ = 400.5 Nm and the results of stresses and displacements are presented in Figure 14.

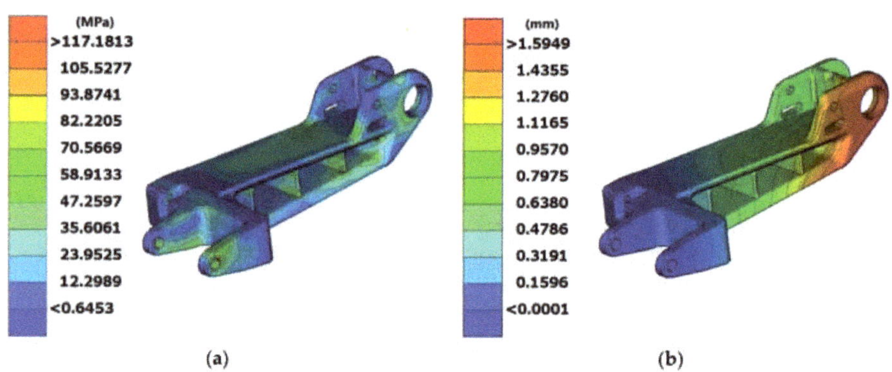


**Figure 14.** Stress (**a**) and displacement (**b**) results in 2nd loading scenario for the final swingarm design.

The maximum stress calculated on the swingarm has a magnitude of 117.2 MPa, resulting in a safety factor $N$ = 1.95 (yield strength is 228 MPa). As expected, maximum stress is once more observed on one of the constrained points that connects the swingarm to the chassis, as was the case in the simulation results of the first design. It must be also noted that stress distribution on the rest of the swingarm is below 70 MPa. The maximum displacement results in a value of 1.59 mm. Even though there are no comparative results for front swingarms on the literature, these displacements under heavy braking conditions are evaluated as acceptable. That means that no driving or handling problems would be noticed by the driver in braking conditions. A comparison of results from the simulations conducted for the initial and final swingarm design, can be found in Table 6, including maximum stresses, displacements, weight and safety factor.

**Table 6.** Results comparison of the initial and final swingarm design.

|  | Max Stress (MPa) | Max Displacement (mm) | Weight (kg) | Safety Factor |
|---|---|---|---|---|
| **Initial swingarm design** | 98.6 | 1.27 | 5.67 | 5.10 |
| **Final swingarm design** | 117.2 | 1.59 | 4.35 | 1.95 |
| **Difference** | +19% | +25% | −23.2% |  |

As shown, the maximum stresses are 19% higher in the final design, as also displacements are raised by 25%. These results were expected but at the same time do not raise any concerns, since they are lower than the material yield strength. Safety factor is reduced, which is normal for this kind of structure and ensures structural rigidity safety levels. The main target was of course to lower weight, which is accomplished, considering that a 23.2% reduction was achieved. Finally, the change of materials decided had minimum overall effects on our results, but on the other hand will help the most on reducing production costs.

The development procedure presented and decisions taken towards the final design, were based on finite element analysis simulations that were modeled based on our knowledge and experience. Due to the lack of relevant research on front single-sided swingarms, no real comparison can be made to similar research or tests in order to validate our results. One engineering parameter found from rear swingarm analysis (as referred to in Table 2 and used for validation in our case), is the safety factor. In rear swingarms it ranges from 1.53 to 2.39 with a mean value of 1.95, which is an important parameter that we successfully met on our new design. The next step would be to set up an experimental testing procedure on a custom test rig, which would provide additional strain and displacement data for the validation and tuning of our finite element model. We should also mention that since the suspension and tire effects are not considered in our model, the results of stresses and displacements are higher, providing additional certainty that we can further reduce the weight. According to the above, we are confident that the proposed design will provide the needed safety during driving and braking.

## 6. Discussion

The work presented is focused on the design and development of a front single-sided swingarm used on a new three-wheel electric motorcycle. To the best of our knowledge there is no literature for front single-sided swingarm analysis, since most of the research found is related to rear swingarms. Comparing a front and a rear swingarm, loading conditions are differentiated. One difference is the effect of loads from the motor through the chain on a rear swingarm and on the other hand higher braking forces applied on the front wheel of a motorcycle.

The main targets set for the development of this part, are structural safety and low weight. A dedicated CAE software was used for the modelling of loads, constraints and materials applied, in order to evaluate a front swingarm design through series of finite element analysis simulations. Results of stresses and displacements were calculated and presented. A review of braking conditions was also presented, in order to identify the braking forces applied and investigate their effects on the structural strength of the part under consideration. In the simulations conducted, two loading scenarios were performed. The first scenario included weight distribution forces, while the second scenario investigated the effect of braking forces, considered as the worst-case scenario. At first, an initial swingarm design was evaluated. In the worst-case scenario, the maximum stress calculated was 98.6 MPa, far lower than the yield stress (503 MPa), corresponding to a safety factor of $N = 5.1$. Displacement results showed a maximum value of 1.27 mm on this design. A swingarm redesign and material replacement was decided, targeting lower weight and production cost. A topology optimization procedure was used, so as to assist the redesign of the swingarm and obtain valuable results regarding specific areas of material removal. The final form was also affected considering our production capabilities, targeting manufacturing with CNC machining. In the final swingarm design, exactly the same modelling process was followed and only the worst-case scenario (second loading scenario—braking) was used in the simulations. A direct comparison of results for the initial and final swingarm design revealed that, the maximum stresses are 19% higher in the final design (117.2 MPa), as well as displacements were raised by 25% (1.59 mm). Safety factor was reduced to $N = 1.95$, which is normal for these type of structures (as related to rear swingarm literature) and ensures structural rigidity. The main target was of course to lower weight, which is accomplished since a 23.2% reduction was achieved, resulting in a weight of 4.35 kg.

The main contribution of this work was to present valuable results and insights, based on finite element analysis simulations, revealing stress and displacements that are calculated for various versions of a new front single-sided swingarm design. The results presented indicate strongly that the proposed structure is effective and promising for actual prototyping. For a future work, the modelling of suspension and tires could be applied, in order to obtain more refined results and target further weight optimization.

**Author Contributions:** Conceptualization, P.S.; Methodology, P.S. and E.C.; Software, E.C.; Validation, P.S. and N.C.T.; Formal Analysis, N.C.T.; Writing—Original Draft Preparation, P.S. and E.C.; Writing—Review and Editing, P.S., E.C. and N.C.T.; Visualization, E.C. and P.S.; Supervision, N.C.T. All authors have read and agreed to the published version of the manuscript.

**Funding:** This work has been partially funded by the TUC's internal project "TUC Eco Racing team".

**Acknowledgments:** The authors would like to acknowledge the DEADALUS/TUC team's members who contributed and helped in the development.

**Conflicts of Interest:** The authors declare no conflict of interest.

## References

1. Electric Motorcycles & Scooters Market Size by Product (Motorcycles, Scooters), by Battery (SLA, Li-ion), by Voltage (24V, 36V, 48V) Industry Analysis Report, Regional Outlook, Growth Potential, Price Trends, Competitive Market Share & Forecast, 2020–2026. Available online: https://www.gminsights.com/industry-analysis/electric-motorcycles-and-scooters-market (accessed on 20 July 2020).
2. Foale, T. *Motorcycle Handling and Chassis Design: The Art and Science*, 1st ed.; Tony Foale: Madrid, Spain, 2002.
3. Cossalter, V. *Motorcycle Dynamics*, 2nd ed.; Vittore Cossalter: Padova, Italy, 2006.
4. Mavroudakis, B.; Eberhard, P. Analysis of Alternative Front Suspension Systems for Motorcycles. *Veh. Syst. Dyn.* **2006**, *44*, 679–689. [CrossRef]
5. Cycle World. Available online: https://www.cycleworld.com/2019-yamaha-niken-gt-first-ride/ (accessed on 12 July 2020).
6. Deadalus Site. Available online: www.daedalus.tuc.gr (accessed on 27 July 2020).
7. Sponziello, A.; Frendo, F.; Guiggiani, M. Stability analysis of a three-wheeled motorcycle. *SAE Int. J. Eng.* **2009**, *1*, 1396–1401. [CrossRef]
8. Bartaloni, F. Multibody Analysis of a New Three Wheeled Vehicle. Ph.D. Thesis, University of Pisa, Pisa, Italy, 14 April 2008.
9. Terada, K.; Sano, T.; Watanabe, K.; Kaieda, T.; Takano, K. Investigation of the behavior of three-wheel vehicles when they pass over a low μ road surface. In Proceedings of the Small Engine Technology Conference, South Carolina, CA, USA, 15–17 November 2016; SAE International: Warrendale, PA, USA, 2016.
10. Airoldi, A.; Bertoli, S.; Lanzi, L.; Sirna, M.; Sala, G. Design of a motorcycle composite swing-arm by means of multi-objective optimisation. *Appl. Compos. Mater.* **2012**, *19*, 599–618. [CrossRef]
11. Bevan, I.S. Development of a Carbon Fibre Swingarm. Master's Thesis, University of the Witwatersrand, Johannesburg, South Africa, 2013.
12. Ketan, P.; Gaurav, R.; Rohan, M.; Akshay, S. Design and analysis of single sided swing arm for modified bike. *Int. Res. J. Eng. Technol.* **2019**, *6*, 876–879.
13. Smith, B.; Kienhöfer, F. A carbon fibre swingarm design case study. *J. S. Afr. Inst. Mech. Eng.* **2015**, *31*, 1–11.
14. Bedeschi, A. Carbon swingarm for the Ducati 1299 Superleggera. *JEC Compos. Mag.* **2018**, *55*, 35–36.
15. Syed, H.A.; Wajahath, A.R. Design of racing motorcycle swingarm with shape optimisation. *Int. J. Sci. Res. Dev.* **2018**, *6*, 179–183.

16. Risitano, G.; Scappaticci, L.; Grimaldi, C.; Mariani, F. Analysis of the structural behavior of racing motorcycle swingarms. In Proceedings of the SAE 2012 World Congress and Exhibition, Detroit, MI, USA, 24–26 April 2012; SAE International: Warrendale, PA, USA, 2012.

17. Swathikrishnan, S.; Singanapalli, P.; Prakash, A.S. Design and analysis of swingarm for performance electric motorcycle. *Int. J. Innov. Technol. Explor. Eng.* **2019**, *8*, 3032–3039.

18. O'Dea, N. Motorcycle swingarm redesigned in carbon composite. *Reinf. Plast.* **2011**, *55*, 38–41. [CrossRef]

19. Pacejka, H. *Tire and Vehicle Dynamics*, 3rd ed.; Elsevier Health Sciences: Amsterdam, The Netherlands, 2012.

20. Brach, R.; Brach, M. The tire-force ellipse (Friction Ellipse) and tire characteristics. In Proceedings of the SAE 2011 World Congress & Exhibition, Detroit, MI, USA, 12–14 April 2011; SAE International: Warrendale, PA, USA, 2011.

21. Cossalter, V.; Maggio, F.; Lot, R. On the braking behavior of motorcycles. *SAE Trans.* **2004**, *113*, 1274–1280.

22. Joakim, L. Development of Front Suspension for an Electric Two-Wheeled Amphibious Vehicle. Master's Thesis, KTH Royal Institute of Technology, Stockholm, Sweden, 7 April 2015.

23. Limpert, R. *Brake Design and Safety*, 3rd ed.; SAE International: Warrendale, PA, USA, 2011.

24. Rose, N.; Carter, N.; Neale, W.; Mckelvey, N. Braking and swerving capabilities of three-wheeled motorcycles. In Proceedings of the WCX SAE World Congress Experience, Detroit, MI, USA, 09–11 April 2019.

25. Ciepka, P. Effect of ABS and CBS on motorcycle braking deceleration on a wet road surface. *Probl. Forensic Sci.* **2016**, *107*, 557–568.

26. Ariffin, A.H.; Hamzah, A.; Solah, M.S.; Paiman, N.F.; Jawi, M.Z.; Isa, M.M.H. Comparative analysis of motorcycle braking performance in emergency situation. *J. Soc. Automot. Eng. Malays.* **2017**, *1*, 137–145.

27. Lambourn, R.F.; Wesley, A. *Comparison of Motorcycle and Car Tyre/Road Friction*; IHS: Berkshire, UK, 2010.

28. Cossalter, V.; Doria, A.; Lot, R.; Ruffo, N.; Salvador, M. Dynamic properties of motorcycle and scooter tires: Measurement and comparison. *Veh. Syst. Dyn.* **2003**, *39*, 329–352. [CrossRef]

29. Huertas-Leyva, P.; Savino, G.; Baldanzini, N.; Pierini, M. Loss of control prediction for motorcycles during emergency braking maneuvers using a supervised learning algorithm. *Appl. Sci.* **2020**, *10*, 1754. [CrossRef]

30. Cattabriga, S.; de Felice, A.; Sorrentino, S. Patter instability of racing motorcycles in straight braking manoeuvre. *Veh. Syst. Dyn.* **2019**. [CrossRef]

31. Motorcycle Braking and Skidmarks. Available online: http://mfes.com/motorcyclebraking.html (accessed on 20 July 2020).

32. Ecker, H.; Wassermann, J.; Hauer, G.; Ruspekhofer, R.; Grill, M. Braking deceleration of motorcycle riders. In Proceedings of the International Motorcycle Safety Conference, Orlando, FL, USA, 1–4 March 2001.

33. Vavryn, K.; Winkelbauer, M. Braking performance of experienced and novice motorcycle riders—Results of a field study. In Proceedings of the International Conference on Traffic & Transport Psychology, Nottingham, UK, 5–9 September 2004.

34. Bartlett, W.; Baxter, A.; Robar, N. Motorcycle braking tests: IPTM data through 2006. *Accid. Reconstr. J.* **2007**, *17*, 19–21.

35. Anderson, B.; Baxter, A.; Robar, N. Comparison of motorcycle braking system effectiveness. *SAE Tech. Pap.* **2010**. [CrossRef]

36. Dunn, A.L.; Dorohoff, M.; Bayan, F.; Cornetto, A.; Wahba, R.; Chuma, M.; Guenther, D.A.; Eiselstein, N. Analysis of motorcycle braking performance and associated braking marks. *SAE Tech. Pap.* **2012**. [CrossRef]

37. BETA CAE Systems, S.A. *ANSA Version 20.1 x User's Guide*; BETA CAE Systems S.A.: Thessaloniki, Greece, 2020.

38. Manios, S. Topology Optimization of a Two-Wheeled Electric Vehicle Frame. Master's Thesis, National Technical University of Athens, Athens, Greece, 2018.

39. Bendsoe, M.P.; Sigmund, O. *Topology Optimization: Theory, Methods and Applications*, 2nd ed.; Springer: Berlin, Germany, 2013; pp. 1–22.

 *applied sciences*

*Article*

# Bias Truck Tire Deformation Analysis with Finite Element Modeling

**Józef Pelc**

Faculty of Technical Sciences, University of Warmia and Mazury in Olsztyn, 10-719 Olsztyn, Poland; joseph@uwm.edu.pl

Received: 1 June 2020; Accepted: 22 June 2020; Published: 24 June 2020

**Abstract:** This paper presents a method for modeling of pneumatic bias tire axisymmetric deformation. A previously developed model of all-steel radial tire was expanded to include the non-linear stress–strain relationship for textile cord and its thermal shrinkage. Variable cord density and cord angle in the cord-rubber bias tire composite are the major challenges in pneumatic tire modeling. The variabilities result from the tire formation process, and they were taken into account in the model. Mechanical properties of the composite were described using a technique of orthotropic reinforcement overlaying onto isotropic rubber elements, treated as a hyperelastic incompressible material. Due to large displacements, the non-linear problem was solved using total Lagrangian formulation. The model uses MSC.Marc code with implemented user subroutines, allowing for the description of the tire specific properties. The efficiency of the model was verified in the simulation of mounting and inflation of an actual bias truck tire. The shrinkage negligence effect on cord forces and on displacements was examined. A method of investigating the influence of variation of cord angle in green body plies on tire apparent lateral stiffness was proposed. The created model is stabile, ensuring convergent solutions even with large deformations. Inflated tire sizes predicted by the model are consistent with the actual tire sizes. The distinguishing feature of the developed model from other ones is the exact determination of the cord angles in a vulcanized tire and the possibility of simulation with the tire mounting on the rim and with cord thermal shrinkage taken into account. The model may be an effective tool in bias tire design.

**Keywords:** bias tire; textile cord; shrinkage; rubber; inflation analysis; finite element modeling

## 1. Introduction

Nowadays, the vast majority of vehicles are equipped with radial tires, but they have not pushed bias tires out of the market. The latter are used mainly in all-terrain vehicles as well as agricultural and heavy-duty machines. Often, these are tires of large sizes. They are still utilized since they provide several benefits, including high sidewall resistance to mechanical damage, good performance on uneven ground surfaces, silent-running operation, good tread, self-cleaning properties, single-phase tire building connected with a simple design, and prices lower than radial tires. High demand for bias tires for vehicles is present in countries with a high percentage of unpaved roads [1]. The majority of leading tire producers design and manufacture bias tires (Goodyear: bias ply tire, Continental: crossply tire).

It must be noted that many experienced bias tire designers are leaving their occupation and retiring, and their successors need tools to aid their design work. Thus, actions aimed at improving the already existing tools and creating new tools facilitating bias tire design are well-founded.

Tire designing is an extremely difficult and complex task, arising from the fact that a pneumatic tire is a complex, heterogeneous, and anisotropic structure made of rubber, rubberized cord fabric (cord), and wire. The most important tire load-bearing members are bead wires made of wire coils, ensuring contact connection between the tire and the wheel rim, and body ply made of one or several cord layers, whose ends (edges) are fixed in beads through wrapping around bead wires. In a non-inflated state, a tire is quite a flaccid/flexible structure, and when inflated, it stiffens and may carry external loads. Thus, pneumatic tire mechanics is fully non-linear, i.e., physically, geometrically, and with respect to boundary conditions. Deformation analysis of tire and strength of its elements is always a major computational challenge.

The subject of pneumatic tire deformation modeling was discussed as early as the 1950s. Back then, Hofferberth [2] and Biderman [3] proposed a mathematical description of a tire's netting model in which rubber influence on tire deformation is disregarded. Even the simplest task of determining an inflated tire inner profile with the use of a simple netting model cannot be solved with an analytical approach. Differential equation describing this profile features no solution in elementary functions set [4]. Thus, in the past, in the tire designing process, there was the need to use atlases containing profiles for tires being in equilibrium [5]. Due to high labor consumption and the low accuracy of such actions, along with an increase in the availability of electronic digital machines, various algorithms were created to support this process with computers. In [6], based on the tire netting model, an automation method of determining bias and radial tire profiles was presented, and the entire set of computational tools, verified in the industry and successfully used in tire designing process, was described in [7].

Nowadays, calculation software based on the tire netting model is used in tire industry only for the initial tire profile design. Exact and reliable tire design verification is conducted with the use of the pneumatic tire finite element model. Since Dunn and Zorowski [8], being some of the pioneers, applied the finite element method (FEM) for pneumatic bias aircraft tire deformation modeling, many tire models were worked out based on this method. The complexity of the problem of aircraft tire deformation modeling is well presented in the paper [9]. Due to the radial tire application universality, many publications describing tire computational models dwell upon radial constructions, and papers on bias tires are scarce, e.g., [10–14].

In the paper [15], proprietary FEM software is presented for the analysis of pneumatic tire axisymmetric models aimed at supporting the design of all-steel radial tires. Many subroutines of this software have been included in the commercial MSC.Marc software (Ver. 2001, MSC.Software Corporation, Santa Ana, CA, USA, 2001) [16] as user subroutines. Robust solver and effective contact procedures, available in MSC.Marc, together with user subroutines created the possibility of advanced analyses of tire deformation. A technique based on overlaying finite elements characterizing reinforcement (cords) onto elements representing rubber [17], which ensured very stable numerical solutions, was utilized in the modeling [14,18].

This model was developed so that specific characteristics of textile cord used in bias tire building could be taken into account, i.e., non-linearity of the stress–strain relationship and thermal shrinkage. Such a model was used for the axisymmetric analysis of bias truck tire deformation caused by inflation pressure.

In the paper, specific characteristics of a cord-rubber tire composite resulting from the tire formation process are highlighted. The influence of thermal cord shrinkage onto the alteration of tire configuration when removed from the vulcanization mold was taken into account, and mounting on the rim was simulated. Friction contact between tire beads and wheel rim was considered. Profiles of a tire mounted on the rim and inflated as well as forces in body ply cords were determined. Displacements of two tire characteristic points and cord forces in the first body ply were compared with and without the effect of shrinkage. The possibility of analyzing the influence of cord angle in body plies arranged on the tire building drum onto the tire apparent lateral stiffness was presented.

## 2. Description of Bias Tire Model

For analysis of pneumatic bias tire mounting on the rim and its inflation, the axisymmetric finite element model was employed. The analysis was based on the tire meridional cross-section with a plane containing the tire axis of revolution, also referred to as the tire profile (Figure 1).

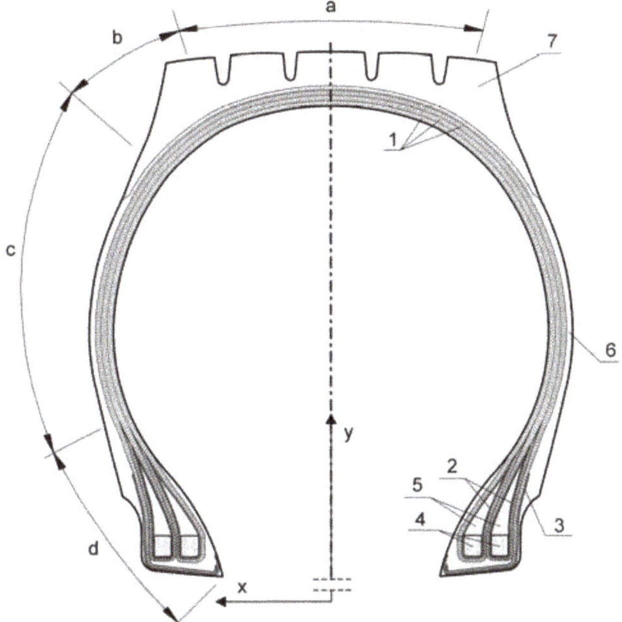

**Figure 1.** Meridional bias truck tire profile. a—crown, b—shoulder, c—sidewall, d—bead; 1—3 × ± $\theta$ doubled body plies, 2—flipper strips, 3—chafer, 4—bead wires, 5—bead fillers, 6—sidewall, 7—tread.

The tire profile is subject to discretization with axisymmetric finite elements. Each finite element node has two degrees of freedom—axial $x$ and radial $y$ displacement—but in each Gaussian point of the element, in the coordinate system $x$, $y$, $\varphi$ ($\varphi$—angular coordinate, measured around the tire rotation axis), there are three normal stresses and one shear stress. When analyzing tire deformation, large displacements occur, thus a non-linear mechanics problem was considered with total Lagrangian formulation [19]. Therefore, for description of deformation and stresses, the energetically conjugate Green-Lagrange strain tensor and the second Piola-Kirchhoff stress tensor were used. Their components are invariant against the rigid body motion. As an initial configuration, a tire profile in vulcanization mold was assumed immediately following the end of the vulcanization process, as, in this condition, the tire geometry is uniquely determined.

The cord-rubber composite used in pneumatic tire building is a specific one owing to two main reasons.

1.  The rubber ply, reinforced with cords, is characterized by a high degree of anisotropy. The composite matrix, i.e., rubber, is a hyperelastic material, and the cord has high tensile stiffness. By assuming the quotient of initial, tangent Young moduli for cord and rubber as composite anisotropy indicators, it amounts to ca. 640 for a nylon-rubber combination and ca. 6000 for a steel-rubber combination. For typical composites with a rigid matrix, this indicator amounts to 20–70 (glass/graphite-epoxy resin).

2.  In the tire forming stages, bead wires are moved towards the tire symmetry plane, and the body ply takes toroidal shape. As a result, the cord density and the cord angle (with exception of radial tire body ply) with the toroid parallel lines in a tire profile location become the function of the location distance to the tire rotation axis (Figure 2).

In the layer arranged on the tire building drum, the cord density $epdm_0$ (ends per decimeter) and the cord angles $\theta_0$ to drum surface meridians are constant. As a result of "lifting" of the layer from the building drum to the profile in the mold, the values of these two parameters vary along the profile of the cord ply. The lowest cord density *epdm* is in the tire crown area, and the highest is in the vicinity of bead wire. Similar variability can be observed for the cord angle $\theta$ in the layer of formed tire (Figure 3).

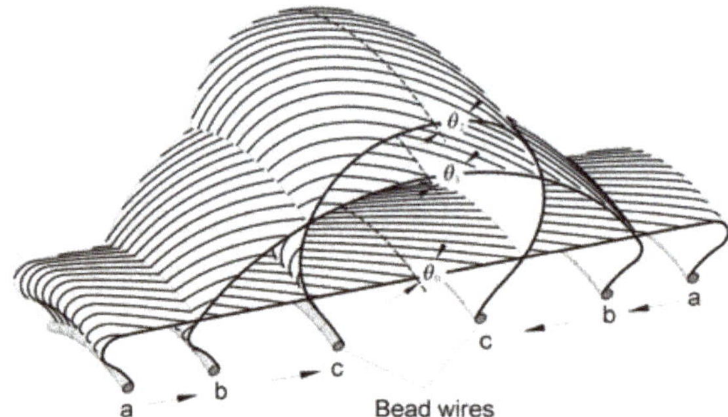



**Figure 2.** Changes in cords angle and density in a ply during tire forming stages: a—initial on the tire building drum, b—intermediate, c—final ($\theta_2 < \theta_1 < \theta_0$).

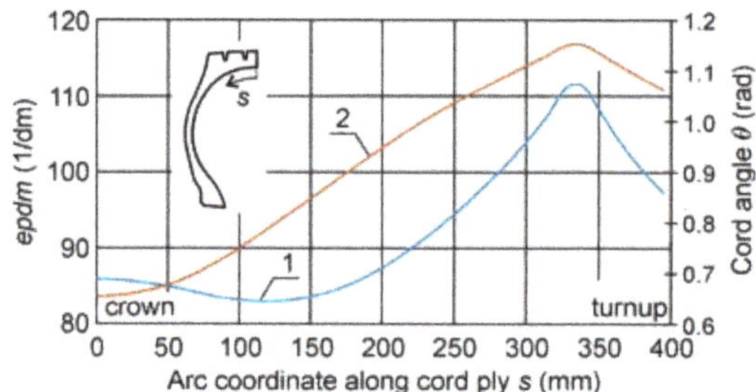



**Figure 3.** Cord parameter variations along body ply in a bias tire; 1—cord ends per decimeter (*epdm*), 2—cord angle $\theta$.

Equations for effective material constants of cord-rubber composite ply, considered as transversely isotropic, two-dimensional continuums, can be found in the literature, e.g., in [20]. However, due to the first characteristic feature given for such a composite, using material constants defined this way in the tire computational model results in lack of convergence of the solution that is searched for. Thus, in tire analysis, overlaying reinforcement elements onto rubber elements is commonly used. The elements have the same nodes and occupy the same geometrical space, but the materials have distinctly described stress–strain relationships. As in the paper [17], here, the determination of reinforcement material constants was adopted from the Halpin-Tsai equations, with the assumption of a very small value of Young modulus for rubber. The rubber elements were assigned the characteristics of a hyperelastic and incompressible neo-Hookean material. It was assumed that bead wire material is isotropic and linear elastic with high elasticity limit. The reinforcement area was discretized with quadrilateral, four-noded axisymmetric elements, and the rubber area was discretized with incompressible Herrmann elements with the same number of corner nodes. The latter had the fifth node with one degree of freedom, defining the hydrostatic pressure mean value.

The textile cord used for bias tire building is usually made of a polyamide (nylon). It is compliant in tension, and the curve of tensile stress–strain is non-linear (Figure 4, curve 1). The example graph shows mean strain values obtained experimentally for five samples of polyamide body ply cord with a measurement base length of 500 mm. Taking into account the non-linear characteristic of textile cord, the computational model accounted for the variability of its stiffness modulus in function of strain (Figure 4, curve 2). This curve was obtained by differentiation of the analytical form of curve 1.

The initial value of the stiffness modulus used in textile cord bias tire was 2232 MPa. Then, with the increase in strain, it dropped to 1272 MPa, and later it increased monotonically to almost 3565 MPa with a strain of 7.5%.

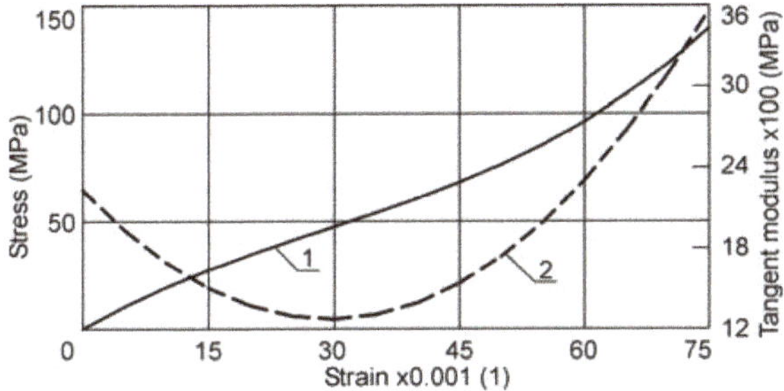


**Figure 4.** Tensile stress–strain for polyamide 1400dtex/3 cord; 1—tensile test, 2—tangential stiffness modulus.

Due to the negative coefficient of textile cord thermal expansion, when a tire is taken out of a mold, the cord is subject to shrinkage, and the tire changes its shape [21]. Therefore, before the mounting simulation, and with the use of a user subroutine, stress induced by cord shrinkage was applied, and a tire profile was obtained, which was shrunk compared with the form in the vulcanization mold. Once the tire was mounted on the rim in the simulation process, the tire loading was applied incrementally in the form of internal follower pressure, and subsequent system equilibriums were iteratively determined.

A major issue in bias tire computational model creation is caused by variable cord density and variable cord angle in a ply. The literature describes the use of reverse engineering methods to determine those parameters that are based on actual tire scanning [22–25]. Although this method requires laborious measurements, in the case of lack or inaccessibility of data for raw plies set on the tire building drum, it is the only way to obtain such parameters. For this paper, a more effective method of determining composite parameters in a vulcanized tire was used based on the reference to known and constant values of these parameters in ply set on the tire building drum (Table 1) and with the use of the so-called pantographing rule [2]. The cord-rubber ply stiffness matrix was then arranged with global coordinates $x$, $y$, $\varphi$, in which the FEM equation system was created and resolved. For this purpose, adequate transformations of the stiffness matrix were performed for the ply described in the system of its principal axes of orthotropy [26]. The calculations were executed with user subroutine coupled to the code of MSC.Marc. Similarly, to determine the forces in ply cords with the use of the user subroutine, the cord-rubber composite strain was calculated in the direction of the reinforcement cords.

**Table 1.** Geometric and mechanical properties of bias tire reinforcement.

| Reinforcement | $t$ (mm) | $A$ (mm$^2$) | $epdm_0$ (1/dm) | $\theta_0$ (rad) | $r_0$ (mm) | $E$ (GPa) | $N_r$ (N) |
|---|---|---|---|---|---|---|---|
| 5th and 6th body ply | 1.5 | 0.51 | 94 | ±1.03 | 338.2 ± t/2 | 2.232 | 299 |
| 3rd and 4th body ply | 1.5 | 0.51 | 94 | ±1.03 | 335.2 ± t/2 | 2.232 | 299 |
| 1st and 2nd body ply | 1.5 | 0.51 | 94 | ±1.03 | 332.2 ± t/2 | 2.232 | 299 |
| Flipper strip | 0.8 | 0.22 | 90 | 0.785 | 269.9 | 1.488 | 135 |
| Chafer strip | 0.9 | 0.30 | 40 | 0.698 | 269.9 | 0.992 | 90 |
| Bead wire | N/A | 110 | N/A | 0 | N/A | 206 | 134,000 |

$t$—layer thickness; $A$—cross-sectional area of the cord; $epdm_0$—number of cords per 1 dm (in the green layer laid on the tire building drum); $\theta_0$—cord angle orientation with respect to the parallel line of the drum; $r_0$—radius of the green layer on the drum; $E$—initial Young modulus of the cord/bead wire; $N_r$—cord/bead wire breaking force.

## 3. Results and Discussion

The described tire computational model was used to simulate the process of mounting on the rim and of inflating an actual 11.00-20 bias truck tire with nominal internal pressure of $p_n = 725$ kPa. The tire load-bearing members were: double bead wires, six nylon body plies (1400dtex/3), and reinforcement strips of bead fillers. In the area of beads that came into contact with rims, there were reinforcement plies known as chafers. All tire plies were made of textile cord. The ends of the first two body plies were fixed with the innermost bead wire, while the next two were held by the outer bead wire, and the ends of the two remaining ones—the outermost—reached the bead nose (Figure 1). Due to a large number of thin body plies, neighboring plies with angles of $\pm\,\theta$ were connected in the model, thus one finite element had the thickness of two joined plies. Effective material parameters for plies connected in such a way were calculated with a user subroutine. There were seven types of rubber compounds in the tire (cord layers rubber, tread, undertread, sidewall, bead filler, bead strip, protective strip). Their mechanical properties were described with a neo-Hookean material model, and the material constants necessary for calculations were determined based on tensile tests of adequate samples, as described in [18]. The $C_{10}$ parameter of the neo-Hookean material determined for these subsequent rubbers had the values (MPa) of 0.63, 0.603, 0.72, 0.484, 2.12, 1.23 and 1.83, respectively.

A starting point for tire deformation analysis was the discretized area of tire profile in a mold, right after the vulcanization process (Figure 5, item 1). The axial displacements of the nodes located in the tire profile plane of symmetry were blocked, whereas contact was defined between the outermost bead nodes and the rim. The mesh created for half the tire profile consisted of 600 finite elements and 421 nodes. All axisymmetric, isoparametric arbitrary quadrilateral elements used bilinear interpolation functions. The mesh consisted of 372 Herrmann elements with the properties of incompressible rubber materials, 208 orthotropic elements, and 20 isotropic elements. The orthotropic elements were overlaid on incompressible elements to model the properties of cord-rubber composite. Due to discretization error, the finite element mesh affected the accuracy of the obtained calculation results. In this case, the basic mesh was tested by generating a more dense one and comparing the results obtained using both meshes. Dense mesh was obtained by subdividing each element of the basic mesh into four elements. The results of the comparison were as follows: 4.0% and 2.3% for radial and axial displacement of points located at the tire crown and on the sidewall, respectively, and 3.2% for the maximum value of the cord tensile force. In each case, the values obtained for the basic mesh were smaller than for the dense one. The basic mesh was considered sufficient for the analyses carried out here.

When the rubber was being vulcanized, the tire polyamide cord was heated and showed a tendency to shrink, which was restrained in this state by a diaphragm pressing the tire to mold walls. Tensile forces were generated in cords. When the vulcanization process was finished and the tire was removed from the mold, the cords were able to shorten owing to the compliance of the surrounding rubber. The tire radically changed its shape (Figure 5, item 2), especially within its crown area, which moved radially towards tire rotation axis. It was assumed that, in the tire, the shrinkage of polyamide cords amounted to 5%. The shrunk tire profile was determined by applying self-equilibrated initial stresses in cords equivalent with this shrinkage. The calculation of cord shrinkage stress and its transformation from principal axes of orthotropy for cord-rubber ply into an $x$, $y$, $\varphi$ coordinate system was performed with the user subroutine.

The stage of mounting the tire on the rim was carried out in ten increments through concurrent displacement of the rim in the axial and the radial directions against the position, which was consistent with the actual rim dimensions of 8.0–20 assumed for this tire. At this stage, small values of internal pressure were also applied ($0.001p_n$ per increment) to provide for the correct setting of the bead on the rim. A non-deformable rim exerted contact forces on the bead. Forces tangential to the contact surface were calculated according to the Coulomb's law of friction. The rubber–steel friction coefficient was assumed at 0.6. The profile of a tire mounted on the rim is presented in Figure 5, item 3.

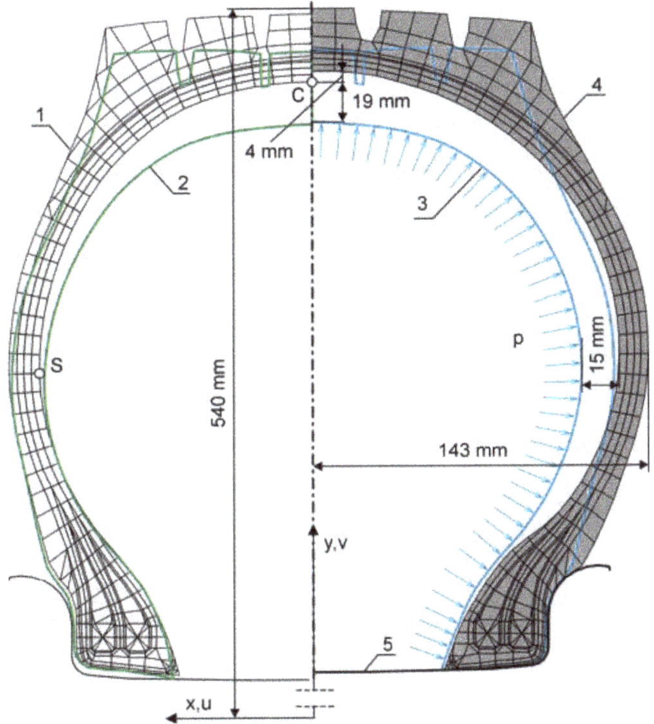

**Figure 5.** Bias truck tire; 1—profile in mold and finite element mesh, 2—outside the mold (shrunk), 3—profile after mounting on rim, 4—inflated up to $p_n$; 5—rim.

The simulation of tire inflation was divided into stages: (1) ten pressure increments, $0.01p_n$ each, when there was considerable tire profile alteration resulting from the "lifting" of a shrunk tire; (2) nine increments, $0.01p_n$ each, when the tire had already reached an adequate initial stiffness. The inflated tire profile is shown in Figure 5, item 4, together with the displacement of internal profile C and S characteristic points, i.e., the tire crown point and sidewall point. The shown profiles indicate the occurrence of large displacements in a tire subject to internal pressure. Axial displacement of the point located on sidewall was of the order of its thickness.

Displacements of the said C and S points but referenced to the tire profile in the mold amounted to 4 mm and 13 mm, respectively. The external diameter of the inflated tire and its width, as obtained from the model, amounted to 1080 mm and 286 mm, respectively. The diameter value was consistent with the catalog requirements for this tire, and the width difference was −1.7%. This shows that the tire profile in the mold was designed correctly, and the model correctly predicted the overall dimensions of the actual tire. It must be noted that this tire—once manufactured in a series—was correctly designed and fulfilled all requirements stipulated in the European Tyre and Rim Technical Organisation (ETRTO) documents.

In the case of tire inflation analysis and neglecting cord shrinkage, quite a mild fluctuation in body ply cord forces from the tire crown to the bead wire area was observed, which was consistent with the results obtained on the tire netting model (cf. [27] (p. 84, Figure 71a)). When considering the stresses generated by cord shrinkage, considerable variations in cord forces were observed. In particular, in the tire crown, there was almost a 60% increase in force value in the first and the second ply cords. It should be assumed that a certain influence on this increase was contributed by the first groove in the tire tread. The highest values of forces were obtained for innermost plies (first and second) cords, as this was where the internal pressure was applied, and these values decreased for plies located more outwards in the tire, which was consistent with the principles of deformable body mechanics.

Figure 6 presents the chart of body ply cord forces in the function of arc coordinate measured along the profile from the tire crown towards the beads.

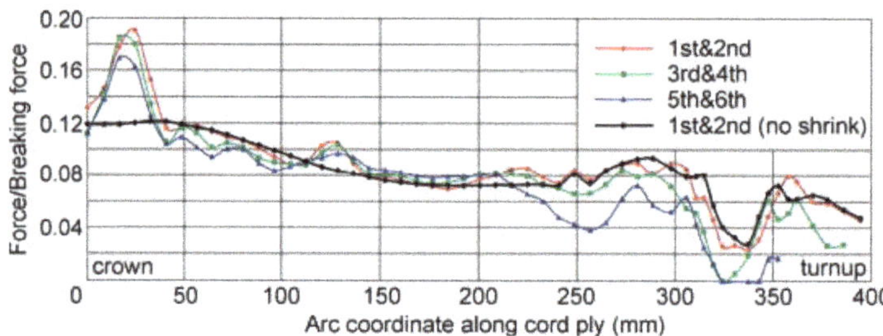


**Figure 6.** Cord tensile forces in all body plies and in two inner body plies with shrinkage neglected.

Although the impact of thermal shrinkage on the maximum force values in body ply cords was high, the final displacements of inner tire profile control points (C and S) were less dependent on whether the shrinkage was or was not taken into account (Figure 7). The differences were 15% and 3.5%, respectively. It must be stressed that the control point S was located at the sidewall and, as a result of tire inflation, it moved axially towards the symmetry plane. Finally, the tire width mounted on the rim and inflated was lower than its width in the vulcanization mold.

The stable finite element tire model makes it possible to obtain many usable results. Figure 8a presents a map of equivalent Cauchy stresses occurring only in purely rubber elements. The highest values of these stresses could be observed in the vertical bead–rim contact area. Large values also existed in the bead in the flipper strip ends zone and may have caused accelerated fatigue degradation of the material. In turn, the distribution of the maximum principal strain within the tire profile area (Figure 8b) can be important information in assessing the risk of tire components separation. Large deviations of the vectors from the course of layers indicated the occurrence of high values of shear strains and therefore the existence of high shear stresses in these zones of inflated tire.

For lateral tire stiffness analysis, a three-dimensional model should be created. Elaboration of such a model is quite a challenge both in terms of the modeling and the required intensity of calculations [26]. Thus, a method of determining apparent lateral stiffness with the use of an axisymmetric tire model was proposed. For this purpose, the deformable flat surface was moved towards the model along the y axis. The contact of this plane with the inflated tire tread produced radial displacement of tread nodes towards the tire rotation axis (Figure 9). Due to the axisymmetry of the tire model, the tire deformation was consistent with the result obtained by placing a tire in a cylinder of decreasing diameter. When the cylinder radius was reduced by an assumed value of 25 mm, it was displaced axially by 30 mm (Figure 9, item 4), and the resultant friction forces generated in tread nodes with the tread remaining with contact with the cylinder were being written. The graph for these forces for a few cord angle values $\theta_0$ in the function of tire tread axial displacement are presented in Figure 10, items 1–4. On this basis, the tire apparent lateral stiffness was determined as a derivative of the force with respect to tread axial displacement (Figure 10, items 1'–4'). The apparent lateral stiffness of the analyzed tire indicated the greatest sensitivity to angle $\theta_0$ variations for small tread axial displacement (up to 5 mm). By defining the correlation between the tire apparent lateral stiffness and the actual stiffness (e.g., obtained from measurements), it was possible to predict tire lateral stiffness based on an axisymmetric model, i.e., with low costs.

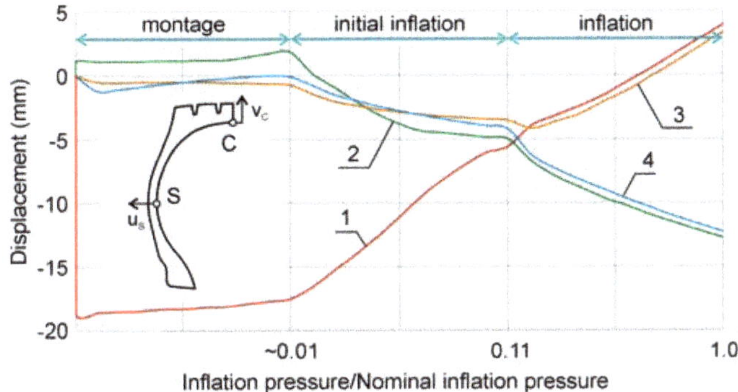


**Figure 7.** Displacement variation of C and S points located on inner profile of vulcanized tire in mold versus relative inflation pressure; 1—$v_C$, 2—$u_S$, and 3—$v_C$, 4—$u_S$ for shrinkage neglected.

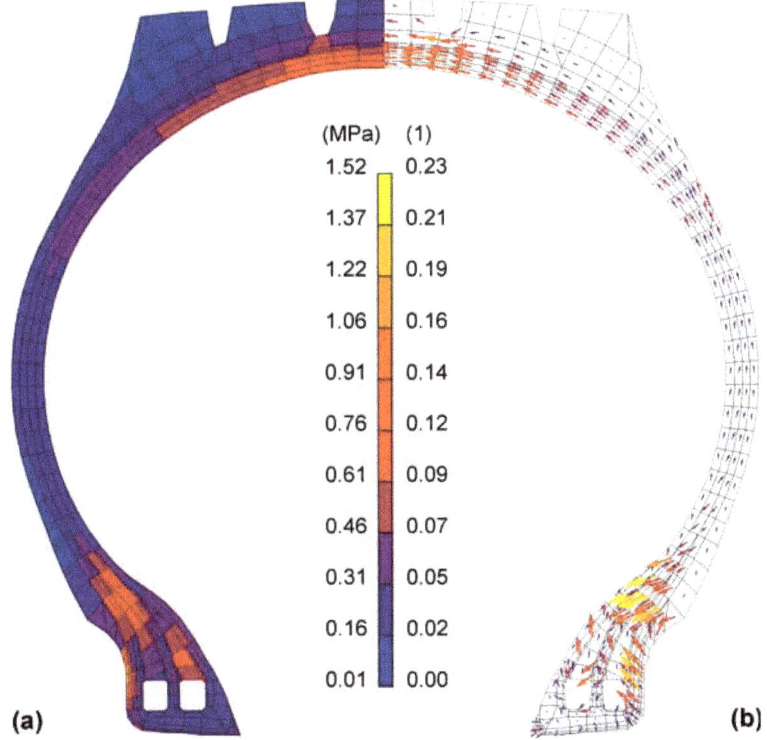


**Figure 8.** Inflated tire; (**a**) map of equivalent Cauchy stress in rubber, (**b**) distribution of the maximum principal strain in rubber.

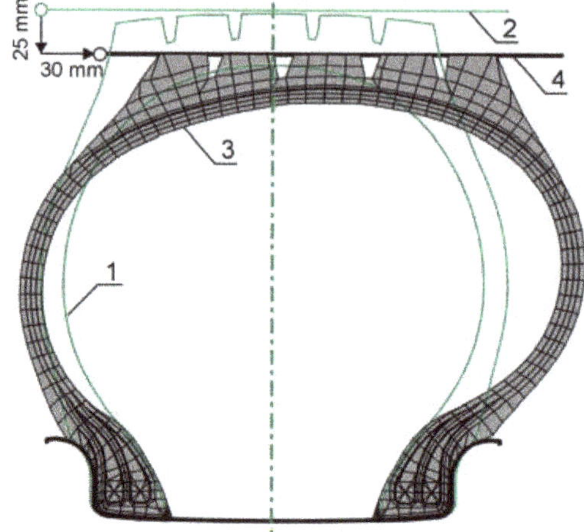

**Figure 9.** Procedure for determining of tire apparent lateral stiffness: 1—initial profile of inflated tire; 2—contacting plane at the initial position; 3—deformed profile; 4—contacting plane final position.

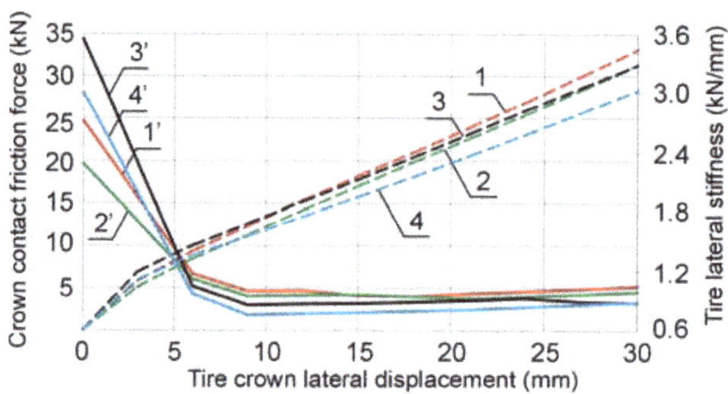

**Figure 10.** Crown contact friction force and tire apparent lateral stiffness (numbers with ') for some cord angles on drum $q_0$ (rad); 1—1.1, 2—1.06, 3—1.03, 4—1.00.

## 4. Conclusions

The cord-rubber tire composite plies, as opposed to typical composite materials, have specific characteristics resulting from the method of tire forming. Variable cord density and its angle must be taken into account in tire computational models. This causes a great difficulty in deformation analysis of such objects, as it requires a correct description of variations in these parameters along the tire profile and creating user subroutines, which execute the description within the computer code.

The textile cord, which is the most commonly used in the construction of pneumatic bias tires, features non-linear stress–strain characteristics. Thus, in the analysis of a tire, it is necessary to use tangential modulus varying together with the deformation. Moreover, due to the negative thermal expansion coefficient of such a cord, a tire removed from a vulcanization mold changes its shape under the influence of cord shrinkage forces. It is necessary to take the thermal shrinkage into account, as this has a considerable influence on force values in body ply cord when the tire is inflated. The low influence of the shrinkage on the final axial displacement of a point located on tire sidewall and a 15% influence on the displacement of the point placed at crown were observed.

The finite element model makes it possible to predict many tire parameters and greatly extends the possibilities of the classic netting model, taking into account not only the cord plies but also the rubber. Moreover, due to the consideration for the contact between the tire beads and the rim, it is possible to perform a realistic simulation of tire mounting on the rim. The model credibility may be evidenced by the fact that, in the case of an actual bias tire simulation, the overall sizes of an inflated tire obtained with the use of the model were practically consistent with the sizes on an actual tire.

The model enables, for example, determination of such a vulcanization mold shape, thus the ready tire after inflating attains the required sizes. The selection of mold shape is made in a virtual space at the design stage and thus at the stage of low costs. This does not require, as was the case with the previously utilized trial and error method, a time-consuming mold preparation to manufacture prototype tires. From the tire designer's point of view, the model provides considerable support in the difficult design work, making it possible to obtain valuable results, e.g., inflated tire profile, forces in tire load-bearing members and, thus, an evaluation of its strength. It may serve as a virtual prototyping tool.

The presented method for determination of tire apparent lateral stiffness may be used in a comparative analysis of different variants of the same tire model but with various design parameters (e.g., number of plies, *epdm*, cord angle, etc.).

It was shown that, when textile cord specific characteristics were taken into account in the previously created radial tire model, the modified and extended model was effective in bias tire deformation analysis.

**Funding:** This research received no external funding.

**Conflicts of Interest:** The authors declare no conflict of interest.

## References

1. JK Tyre & Industries Ltd., Truck/Bus Bias Tyres. Available online: https://www.jktyre.com/tbbtyre.aspx# (accessed on 27 May 2020).
2. Hofferberth, W. Zur Festigkeit des Luftreifens. *Kautsch. u. Gummi* **1956**, *9*, 225–231.
3. Biderman, V.L. *Raschet formy profilya i napryazheniy v elementakh pnevmaticheskoy shiny, nagruzhennoy vnutrennim davleniyem*; Goskhimizdat: Moscow, Russia, 1957.
4. Day, R.B.; Gehman, S.D. Theory for the meridian section of inflated cord tires. *Rubber Chem. Technol.* **1963**, *36*, 11–27. [CrossRef]
5. Biderman, V.L.; Bukhin, B.L.; Nikolaev, I.K.; Smetankina, R.N. *Atlas nomogramm ravnovesnykh konfiguratsiy pnevmaticheskikh shin*; Khimiya: Moskva, 1967.
6. Pelc, J.; Petz, E. Computer aid in inner tire profile design. *Polimery* **1988**, *33*, 381–383. [CrossRef]
7. Pelc, J.; Petz, E. Pneumatic Tyre Designing by CAD/CAE Technique. *Polimery* **1994**, *39*, 11–12. [CrossRef]
8. Dunn, S.E.; Zorowski, C.F. *A Study of Internal Stresses in Statically Deformed Pneumatic Tires*; U.S. National Bureau of Standards: Washington, DC, USA, 1970.
9. Behroozi, M.; Olatunbosun, O.A.; Ding, W. Finite element analysis of aircraft tyre—Effect of model complexity on tyre performance characteristics. *Mater. Des.* **2012**, *35*, 810–819. [CrossRef]
10. DeEskinazi, J.; Ridha, R.A. Finite element analysis of giant earthmover tires. *Rubber Chem. Technol.* **1982**, *55*, 1044–1054. [CrossRef]
11. Watanabe, Y.; Kaldjian, M.J. Modeling and analysis of bias-ply motocycle tires. *Comput. Struct.* **1983**, *17*, 653–658. [CrossRef]
12. Ghoreishy, M.H.R. Finite Element Analysis of a 6.45-14 Bias Tire under Contact Load. *Iran. Polym. J.* **2001**, *10*, 45–52.
13. Thein, C.K.; Tan, H.M.; Limn, C.H. Numerical modelling and experimental inflation validation of a bias two-wheel tire. *J. Eng. Sci. Technol.* **2016**, *11*, 70–81.
14. Pelc, J. Burst Test Simulation for bias Truck Tire. *Kautsch. Gummi Kunstst.* **2019**, *72*, 53–57.
15. Pelc, J. Numerical aspects of a pneumatic tyre model analysis. *Tech. Sci.* **2009**, *12*, 190–203. [CrossRef]
16. *Nonlinear Finite Element Analysis of Elastomers*; Whitepaper; MSC.Software Corporation: Santa Ana, CA, USA, 2010.
17. Pelc, J. Static three-dimensional modelling of pneumatic tyres using the technique of element overlaying. *Proc. Inst. Mech. Eng. Part D J. Automob. Eng.* **2002**, *216*, 709–716. [CrossRef]

18. Pelc, J. Simulation of Burst Test for all-Steel pneumatic Truck Tire. *KGK-Kautsch. Gummi Kunstst.* **2013**, *66*, 47–51.
19. Bathe, K.J. *Finite Element Procedures in Engineering Analysis*; Prentice-Hall: Englewood Cliffs, NJ, USA, 1982.
20. Halpin, J.C. *Effects of Environmental Factors on Composite Materials*; Technical Report; AFML-TR 67-423; Air Force Materials Lab: Wright-Patterson AFB, OH, USA, 1969.
21. Ridha, R.A. Analysis of tire mold design. *Tire Sci. Technol.* **1974**, *2*, 195–210. [CrossRef]
22. Małachowski, J. Numerical study of tires behaviour. *J. KONES* **2007**, *14*, 377–384.
23. Kondé, A.K.; Rosu, I.; Lebon, F.; Brardo, O.; Devésa, B. On the modeling of aircraft tire. *Aerosp. Sci. Technol.* **2013**, *27*, 67–75. [CrossRef]
24. Rosu, I.; Elias-Birembaux, H.L.; Lebon, F. Finite Element Modeling of an Aircraft Tire Rolling on a Steel Drum: Experimental Investigations and Numerical Simulations. *Appl. Sci.* **2018**, *8*, 593–609. [CrossRef]
25. Arif, N.; Rosu, I.; Elias-Birembaux, H.L.; Lebon, F. Characterization and Simulation of a Bush Plane Tire. *Lubricants* **2019**, *7*, 107. [CrossRef]
26. Pelc, J. Towards realistic simulation of deformations and stresses in pneumatic tyres. *Appl. Math. Modell.* **2007**, *31*, 530–540. [CrossRef]
27. Biederman, V.L.; Guslitzer, R.L.; Zakharov, S.P.; Seleznev, I.I. *Avtomobil'nye shiny (Automobile Tires)*; Goskhimizdat: Moscow, Russia, 1963.

 *applied sciences*

*Article*

# Automation of a Thin-Layer Load-Bearing Structure Design on the Example of High Altitude Long Endurance Unmanned Aerial Vehicle (HALE UAV)

**Wojciech Skarka * and Andrzej Jałowiecki**

Department of Fundamentals of Machinery Design, Silesian University of Technology, 44-100 Gliwice, Poland; andrzej.jalowiecki@polsl.pl
* Correspondence: wojciech.skarka@polsl.pl; Tel.: +48-32-237-14-97

**Abstract:** In the aerospace industry, thin-layer composites are increasingly used for load-bearing structures. When designing such composite structures, particular attention must be paid to the development of an appropriate geometric form of the structure to increase the structure's load capacity and reduce the possibility of a loss of stability and harmful aeroelastic phenomena. For this reason, the use of knowledge-based engineering support methods is complicated. Software was developed to propose and quickly evaluate a thin-layer load-bearing structure using generative modeling methods to facilitate development of the initial concept of an aerospace load-bearing structure. Finite Element Method (FEM) analysis verifies and improves such structures. The most important contributions of the paper are a methodology for automating the design of ultralight and highly flexible aircraft structures with the use of generative modelling, proposing and verifying the form of generative models for selected fragments of the structure, especially wings, and integration of the use of generative models for iterative improvement of structures using low- and middle-fidelity methods of numerical verification.

**Keywords:** HALE UAV; generative modelling; thin-layer composite structure

 check for **updates**

**Citation:** Skarka, W.; Jałowiecki, A. Automation of a Thin-Layer Load-Bearing Structure Design on the Example of High Altitude Long Endurance Unmanned Aerial Vehicle (HALE UAV). *Appl. Sci.* **2021**, *11*, 2645. https://doi.org/10.3390/app11062645

Academic Editor: Marek Krawczuk

Received: 7 February 2021
Accepted: 10 March 2021
Published: 16 March 2021

**Publisher's Note:** MDPI stays neutral with regard to jurisdictional claims in published maps and institutional affiliations.



## 1. Introduction

Composite materials are increasingly used in ultralight aerospace designs. Composite load-bearing structures allow design and manufacture unmanned aerial vehicles (UAVs) that weigh no more than a dozen kilograms and have wingspans exceeding 20 m. The strength of the composite materials is no longer a noticeable barrier. High-strength composite materials that significantly reduce weight are already widely available. However, other problems have not been solved so far. One of the most important is related to the loss of stability [1,2], which is the primary criterion for shaping such design structure. Among the other problems are the use of technologies to produce large-size structures from materials with thicknesses much less than 1 mm, and the challenge of designing large-size structures with extremely thin walls [3]. Forming a highly flexible structure, especially in aviation operating conditions—i.e., significant changes in geometry under varying load conditions—and aeroelastic phenomena associated with the operation of such an arrangement are issues that must be considered [4]. Structural shaping to meet the abovementioned requirements is becoming a fundamental issue. It is essential to analyze the spectrum of possible solutions at the concept and preliminary design stages in such conditions [5–9].

Time-consuming analyses and geometric modelling of structures make this process ineffective. Therefore, in such a particular case, using highly flexible thin-walled aircraft designs, knowledge-based engineering methods, and specific generative modelling was proposed [10,11]. This should allow the rapid generation of solutions and automate the support structure's modelling process based on simplified design criteria.

89

The generative model's operation is combined with a set of simplified verification methods based on Finite Element Method (FEM) used for a simplified structure model based on integrated load states. This method allows one to quickly evaluate individual solutions and choose a solution for further optimization analysis, which is carried out in the following design stages [10,11].

To solve the problem mentioned in the introduction, the authors propose a design methodology based on the use of the generative modelling technique, and the principles of forming highly flexible aircraft structures and thin-film composites with special attention focused on the FEM validation of the preliminary geometry [12,13].

The background of these issues is presented in Section 2 (state of the art) and the methodology proposed by the authors is described in Section 3—developed methodology. For the methodology verification, the example of a HALE UAV built by the authors was used, and verification details are presented in Section 4—use case. The article ends with conclusions and an assessment of the effectiveness of the methodology used.

## 2. State of the Art

In the design processes, strictly defined or even standardized activities are used, including the design of repeatable parts of the structure. A common practice is to make extensive use of these opportunities through Design Reuse (DR). Computer Aided Design (CAD) techniques make use of specific model solutions (CAD models) along with the principles of determining their features, which make the design process much easier [14].

The idea of using DR in the CAD systems environment gives rise to new possibilities, from simple parameterization to complex Generative Models (GMs). These methods can be applied to small parts, elements, and entire components, and products [10]. Such automation of constructing a geometric model in the CAD environment is also called Geometry-Based Design Automation (GBDA). It is not the only form of automation of the design process. Still, it stands out due to the widespread use of CAD systems in the engineering, automotive, aviation industries, etc. In individual CAD systems, tool-based forms of geometric modelling are quite different. These range from programming and scripting techniques, through extensive construction of various template complexity levels (high-level CAD templates), allowing for the integration of multiple forms of knowledge into the geometric model.

Automation itself accelerates the design process at some stage—e.g., it allows for faster generation of geometry and leads to the designer being able to study a much larger space of possible solutions simultaneously [1]. Such automation opens the way to integrating geometric modelling into optimization processes and, with complex multidisciplinary problems, to Multidisciplinary Design Optimization (MDO). However, it should be remembered that there is no definite methodology for the automation of design processes. It is also impossible to answer which of these works can be automated and with what techniques. This applies to the application of knowledge-based engineering methods. The proposed solutions are good practices rather than principles [15].

In a multidisciplinary approach, the geometric model plays a strategic role. So, it is possible to implement the MDO approach [16]:

(a)   without considering the geometric model (Figure 1);
(b)   with a geometric model implemented for visualization of results (Figure 2);
(c)   with a simplified geometric model in the form of the parametric grid used for optimization analyses (Figure 2);
(d)   with a detailed geometric model used in the optimization loop (Figure 2).

In both approaches, a and b, neither the geometric model nor the data that are used in the optimization process are saved in the form of a geometric model. There is no need to make such a model for the calculations themselves, which significantly speeds up the optimization loop time and reduces the labor and computational demand. The aforementioned approaches are used when such a model, by definition, is not needed in methods that use mathematical models of phenomena or semiempirical or statistical methods. If the model is built based on the geometric data used for the calculations, it is used by experts for visualization and comparative purposes only, which helps to interpret the results.

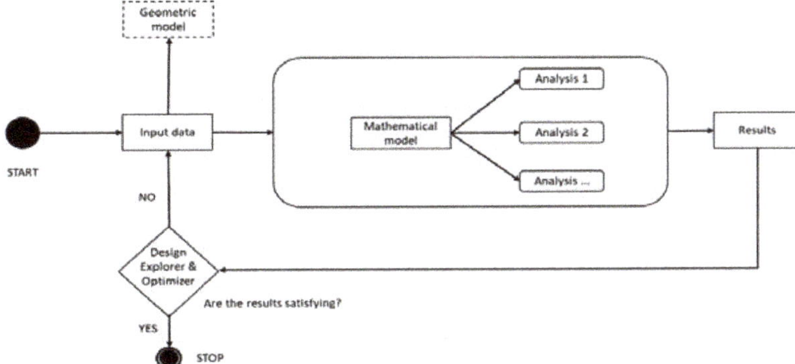

**Figure 1.** Multidisciplinary Design Optimization (MDO) framework without geometric model in the loop.

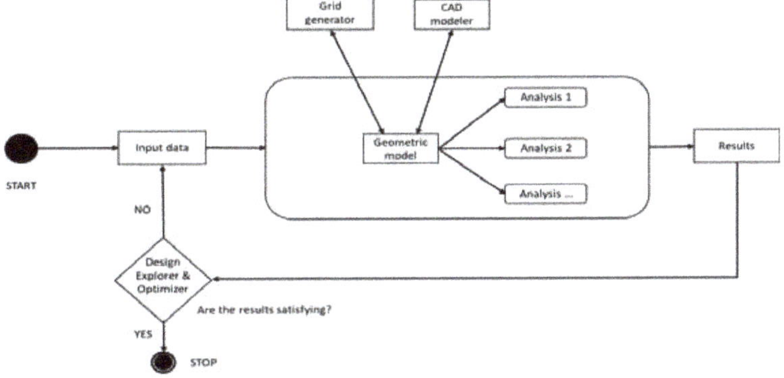

**Figure 2.** MDO framework with geometric model in the loop.

In the second group of the optimization approaches (c, d), we deal with a geometric model in the calculation process. The geometry used for calculations is taken from the model, or the geometry data used in the calculations are visualized through the visualization interface. Typically, these methods are used when some form of geometric structure model is required—e.g., the finite element method (FEM) or Vortex Lattice Method (VLM) or any calculation method using geometry input [11]. Depending on the calculation method and the form of the model used, the form of the model itself may be simplified or integrated into a complex CAD model. In both the first and second forms, generative models can be used to assist in the generation of geometry.

Especially for the latter and multidisciplinary problems, the optimization computation models can differ significantly from each other. Two types of Simultaneous Analysis and Design (SAND) and Nested Analysis and Design (NAND) models stand out from the several different possible configurations of multidisciplinary computation. With a SAND configuration, the disciplinary optimizer simultaneously determines the value of both design and state variables at a disciplinary level. In contrast, in the case of a NAND framework, the optimizer only determines the design variables, with state variables needing to be calculated at each iteration [17].

Automation of the design process in aviation applications and the application to the supporting structures of flying objects are the subject of intensive research. This is not only because of:

(1)     The need to organize the design process, increase its efficiency and record, and integrate knowledge and experience resources from other sources and previous projects.

(2)     Integration of these automation procedures in more extensive structural improvement and optimization procedures considering structural analysis—Computational Structural Mechanics (CSM). In terms of this approach, there are three main areas of application of the above type of analysis. The most characteristic examples include the consideration of topology optimization, the use of composite materials, and the definition of the structural layout.

(3)     Integration of load-bearing structure automation procedures in more extensive methods for improving and optimizing structures also considers other fields [6]. In terms of this approach, such relationships are most often investigated in aeroelastic analysis, aircraft performance, and costs or market demands.

Such integrated approaches are also called MDO (multidisciplinary design optimization) [7,16,18], Model-Based Design (MBD) [8,19], Model-Based Optimization (MBO) [20].

Usually, CSM methods are used as high-fidelity models in the detailed design phase and sometimes also as medium-fidelity models in the preliminary design stage on the simplified geometry. Although, in the conceptual design phase there is often a need to obtain alternative concepts load-bearing structures. On the other hand, relying on simplified geometry encounters a severe problem precisely because of the large amount of work involved in generating alternative concepts of support structures, which often exceed the time required for numerical analyses and their interpretation. Such a bottleneck is usually eliminated using methods of automating the process of generating a geometric model. It is necessary to use a model with a high degree of detail, considering the aspect of manufacturing technology, design dependencies, the engineering correctness of the model, and structural consistency. Generative modeling yields good results [10].

Generative modelling (GM) is one of the most popular knowledge-based engineering techniques used in the design field [8,15,21]. The main idea behind GM is to elaborate the CAD model, which will automatically, or at least semiautomatically, generate the model's geometry based on the design requirements and integrated knowledge [9,15,22]. The general schema of GM functioning is presented in Figure 3.

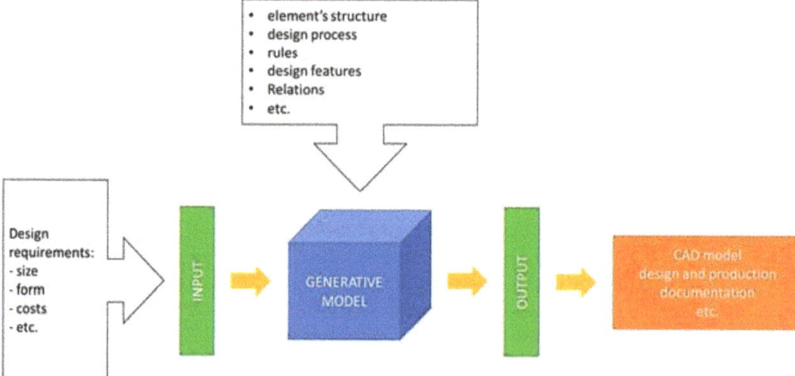


**Figure 3.** The general schema of the GM functioning [10,15,22].

The essential advantage of using the GM, which is one of knowledge-based engineering (KBE) methods of design automation, is the possibility to reduce the time a designer spends on some routine design tasks and put more effort into creative parts of the project, as presented in Figure 4 [15,22,23]. Thanks to this, it is possible to develop a new approach to some problems, which may be better and more efficient than those currently used. On the other side of the coin, development of the GM is time-consuming and usually requires the assistance of an expert in the field [15,21]. Moreover, the number of knowledge engineers who are essential during knowledge acquisition and structuralizing is limited [15,19,23]. However, the effort put into the GM development may bring more benefits in the long-term perspective, especially in the number of projects that can be realized in a shorter time and with less effort, as presented in Figure 4 [9,15,23].

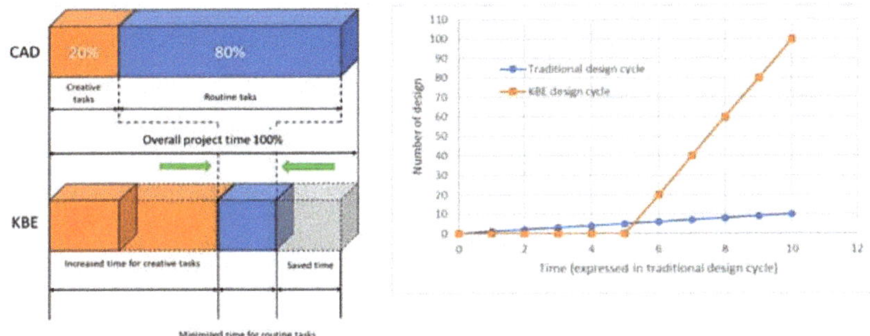


**Figure 4.** Benefits of the knowledge-based engineering (KBE) approach [10]. Copyright 2007 Elsevier.

### 3. Developed Methodology

The principal authors' purpose is to simplify the conceptual stage of the aircraft design as much as possible by using generative modelling techniques combined with FEM analysis. The result of these actions is the systematic approach presented in Figure 5 in a simple schema. Because the design process based on the GMs is slightly different from the traditional design method, the authors tried to adapt the generation procedure to specific requirements of the FEM validation.

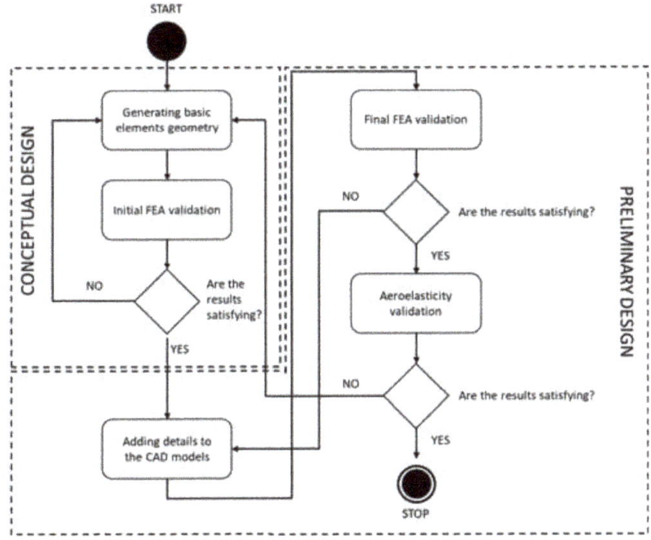


**Figure 5.** The general schema of the developed methodology.

At different stages of the project development, different approaches to defining material data were used. In the initial stages, the isotropic material model was used, and in the next stages the anisotropic model was used for composite materials. The latter was based on general manufacturer data in the initial stages and then the tests performed on samples with a given structure.

### 3.1. CAD Model Development

As we can see in the schema in Figure 5, the first stage of the process was the generation of the load-bearing structure's primary geometry. In this process, the series of the GMs were in use. Each component, such as a spar, a rib or a stringer, was designed individually with its own set of parameters.

The only common thing that connects all components is the geometrical input in the form of airfoil curves for the next sections of the wing. In the proposed approach, each section of a wing is designed independently, and as a section, we mean a part of a wing which contains the same type of an airfoil profile. Using the shared geometrical data makes any change in the geometrical input forces automatically rebuild all related elements without user attention.

The same approach was used in the stage of further development of the models. If the FEM validation results were satisfying, the user could add additional features related to the manufacturing method or specific requirements of the used material or technology.

In the presented approach, the GMs for additional features use existing geometry, from the previous stage of the design process, as the geometrical input. That creates a hierarchical feedback loop that forces automatically rebuilding all related elements—i.e., change of the airfoil causes revamp of the spar. Changes in the spar make changes in the spar features. The sequence of changes in this hierarchy is shown in Figure 6.

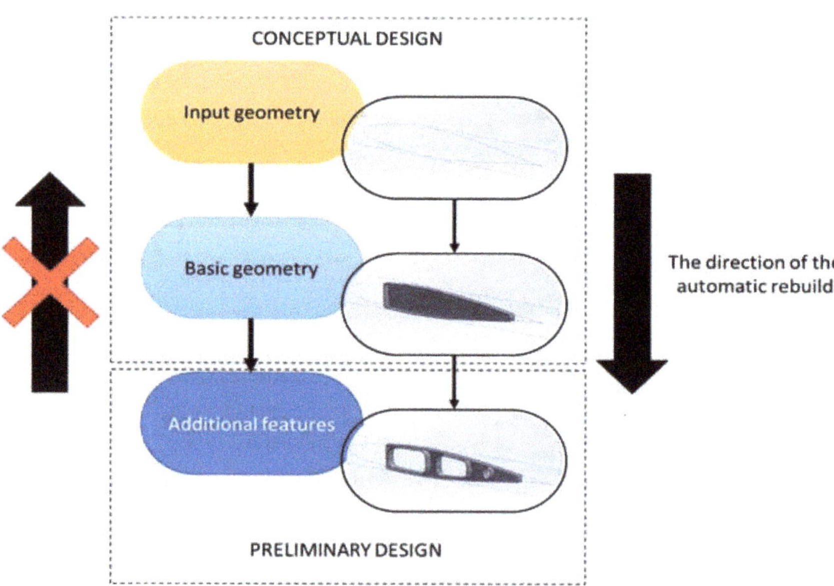


**Figure 6.** The hierarchical order of automatic rebuilding.

In the GM elaboration process it is essential to decompose the design object into smaller and simpler objects to generate. The authors prepared a series of hierarchical diagrams where all features were extracted and described by parameters, required geometrical inputs, rules, etc. In Figure 7, the main idea of the decomposition process is presented.

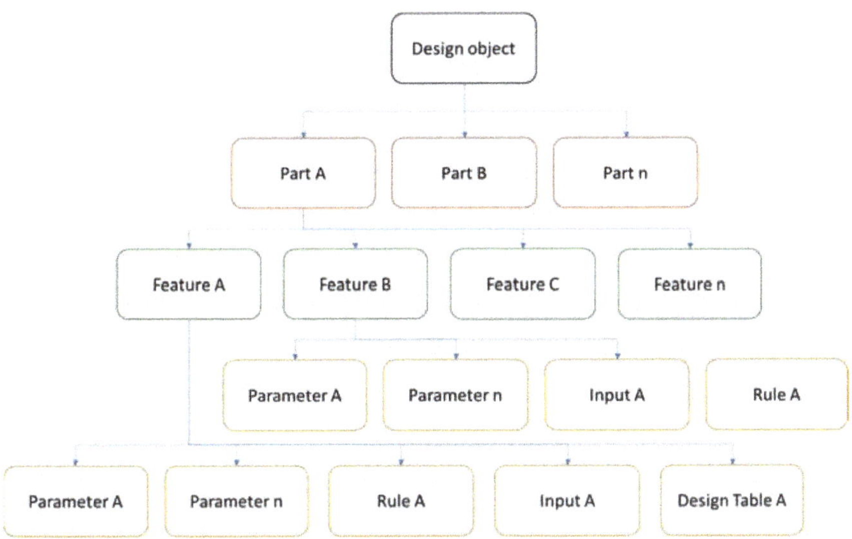



**Figure 7.** The idea of the decomposition process.

For all stages of model development, the numerical validation procedure was the same. The basic idea of the numerical validation process is presented in Figure 8. The proposed methodology is based on three standard stages: (1) preprocessing, where we worked with a CAD model to obtain the numerical representation of the validated model; (2) computation, which in our case consists of three stages: frequencies analysis, structural analysis, and buckling analysis; (3) postprocessing, where a user can decide if the model fulfils all required assumption or need some changes and to be revalidated based on the results.

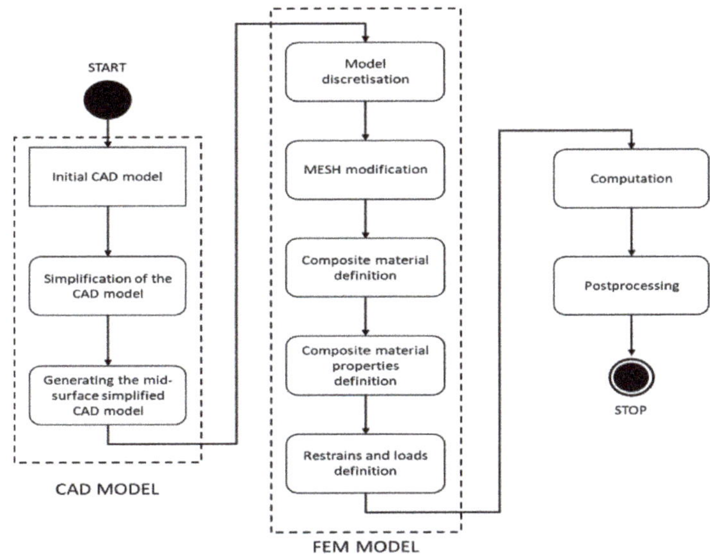



**Figure 8.** The methodology of the Finite Element Method (FEM) validation.

At different stages of the project development, a different approach was used to define material data. In the conceptual design, a very simplified approach is usually used—the material is defined as isotropic for analytical calculations and the calculations are to indicate the expected shape and cross-sections of the supporting structures. If FEM verification calculations are carried out at this stage, then simplified models and material data are also defined in this way. Material data were selected depending on the intended target material and many different options were explored. In preliminary design for composite elements, the preliminary internal structure of the reinforcement was selected and the material data were determined based on the general material data for the reinforcement class. Only in the next stages, and in particular during the production of the prototype, were samples made to fully identify the material data of the given composite structures; only such data give accurate results. Each subsequent stage of approximation brings the results closer to those expected in reality.

### 3.2. Numerical Model Preparation

The basis of Finite Element Analysis(FEA) analysis is the CAD model, and depending on the design stage, the model has different levels of detail. So, it was necessary to elaborate a process after which the input geometry for analysis will mostly be the same. This issue forced the authors to add additional steps that proceed with the central part of the study.

To unify the input geometry, the authors decided that the CAD model must be adequately prepared in the preprocessing stage. This preparation includes:

- geometry simplification;
- removing all unnecessary chamfers and filets as well as holes that will be smaller than the mesh size;
- removing all unrelated components such as fasteners, additional equipment, etc.

After the simplification process, the model must be divided into segments that correspond to the composite structure. An example of the simplification and segmentation process is shown in Figure 9.

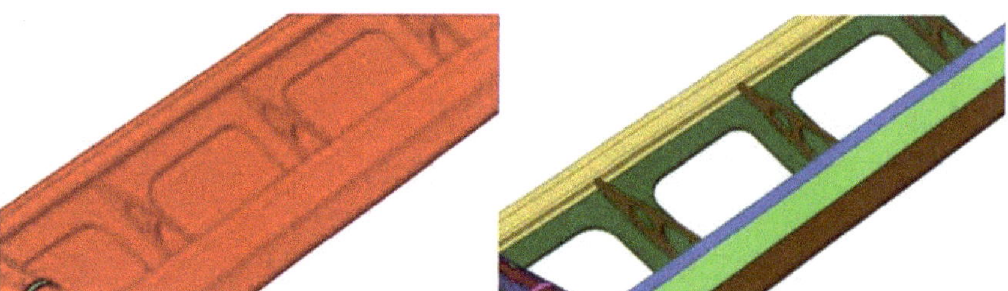

**Figure 9.** An example of the simplification and segmentation process. On the left is the model before the process, and the right is the model after the process.

In the segmentation process, each composite structure acquires a unique color code, which allows to quickly interpret the model material structure and define the material properties in the future steps of the analysis. The final phase of the geometry preparation is the extraction of the mid-surface for each component. This process provides the consistency geometry input for the loads and constraints definition in the later stages.

In general, composite structures can be calculated using a few different approaches. In the basic strategy, the layer modelling technique is used, that using 2D or 3D finite elements, is dominant. In the presented method, the authors decided to use only 2D shell elements. This decision is since in thin-layer composites, the thickness is the smallest dimension, and the ratio to the other dimensions is quite large.

The next step in the numerical model preparation process is a simplification of contact areas, especially where ribs or other elements are joined to outer structures. In case for a perpendicular connection, the authors reduced the connection to a single component which contains layers of both base laminate and rib laminate. An example of the contact area simplification is presented in Figure 10.

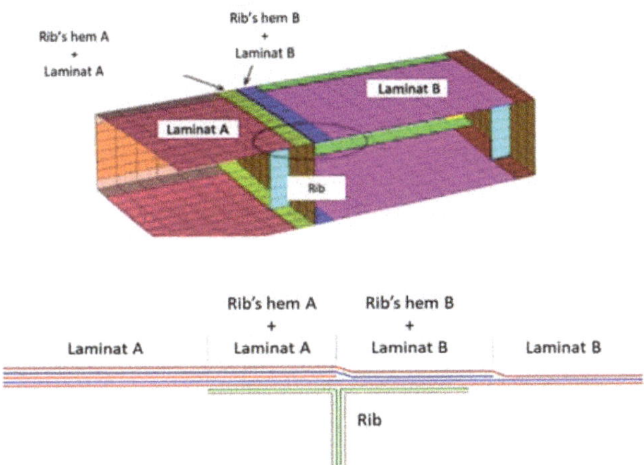



**Figure 10.** Example of connection reduction.

In case of a parallel connection, simple nods joining is enough and there is no need to create a new component, as in the case of perpendicular connections. After this stage, the discrete model was ready to apply loads and constraints.

*3.3. Loads and Constraints Definition*

Additionally, to unify the load application process, all forces and momentums were applied to the center of mass, laying on the cord line. Additionally, to discretize the continuous load, the authors decided to use loads in selected points—mostly the joints where ribs connect with a spar. The idea behind the load application is presented in Figure 11. The values of such distributed loads are calculated from the aerodynamic forces and mass loads, based on different flight scenarios.

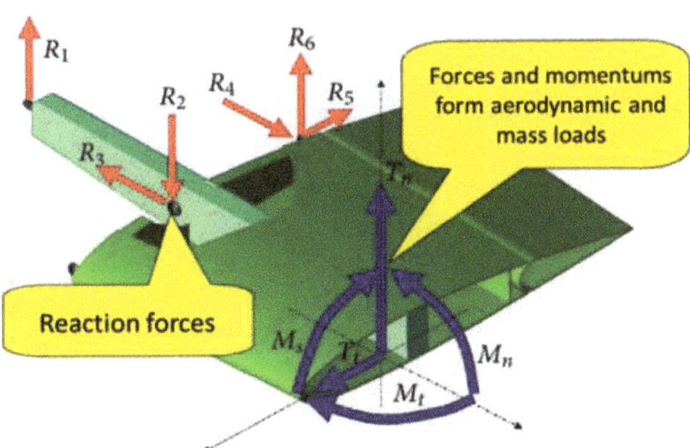



**Figure 11.** Loads and reactions in the wing section.

*3.4. Computation*

After the definition of the numerical model, the analysis can be computed. In the authors' approach, computation was realized in three stages. In the first stage, the first 50 natural frequencies were calculated. Knowledge about these frequencies is vital during the aeroelastic validation of the project. In this stage, the model was computed for both no constraints and with constraints applied.

In the second stage of the computation process, the standard structural analysis was performed. Based on the results of this stage, it was possible to evaluate the overall designed structure and detect areas where some problems were expected.

The last computation stage was devoted to performing buckling analysis. This type of research allows the verification of the stiffness of components and finds some potentially problematic elements.

Based on these results, the user can decide if the project needs to be redesign, what require only some parameters values change, thanks to the generative modelling, or to approve the project for the next stage of the designing/manufacturing process.

In the next part of the paper, the authors present how the presented methodology can be used in the use case of unique aircraft design.

## 4. Use Case

In this part of the paper, the authors present how the developed methodology can be applied in the real case—TWIN STRATOS UAV. TWIN STRATOS is part of the UAV family and was developed as part of the LEADER project implemented by an international consortium consisting of SkyTech eLab LLC, Silesian University of Technology, Universities of Warsaw, and Norwegian Research Center. The section presents a short description of the TWIN STRATOS UAV and further steps that have been made during the design and validation process.

The TWIN STRATOS project is a stratospheric unmanned aerial vehicle (UAV) family project designed for high atmospheric and stratospheric measurements of air pollution. The form of designed UAV family member is shown in Figure 12. The wingspan is 24 m with a 30 kg total payload. Thanks to the use of an electrical drive system and photovoltaic panels, the flight duration, in theory, is unlimited.

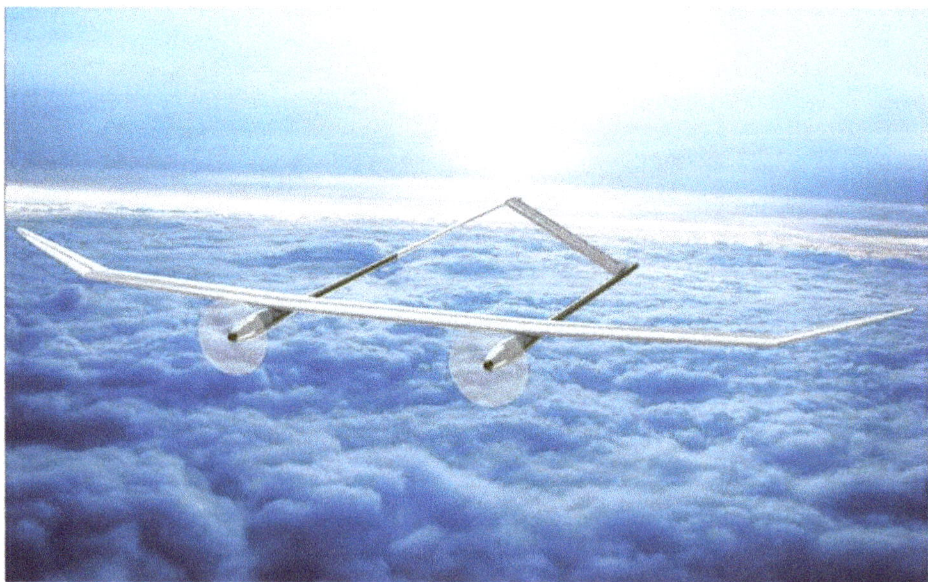

**Figure 12.** Visualization of the STRATOS unmanned aerial vehicle (UAV).

*4.1. Strain, Stresses and Buckling Analysis*

4.1.1. CAD Model Elaboration

Based on the project assumption, in the first step, the geometry of the UAV was designed. In this process, some load-bearing elements, i.e., spars, ribs, stinger, were intended as part of the generative model. To make this possible, the authors prepared a series of generative models for individual component types. Each model contained geometrical information about different designed elements' shape types, control rules, parameters, etc. As a main geometrical input, airfoil curve was used. This geometrical input was used because it is relatively easy to extract this type of curve along the wing. In the case of a fuselage as the geometrical input, a cross-section profile was used. Using the generative modelling technique allows quick changes to be made in the basic load-bearing structure if it is necessary, which makes the whole process of conceptual design much more comfortable and faster in comparison to the traditional design process. The prepared CAD model for numerical validation is presented in Figure 13.

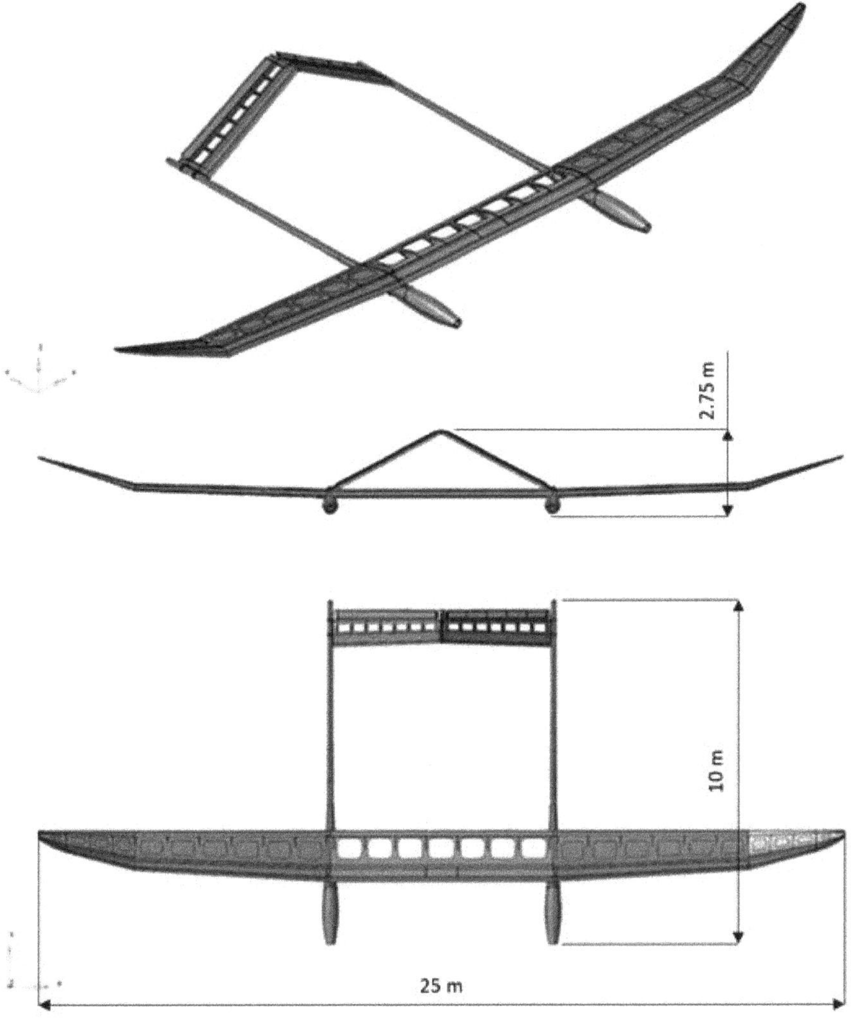


**Figure 13.** The CAD model prepared for the FEA analysis.

### 4.1.2. Numerical Model Preparation

After geometry development, the phase of the numerical validation of the project must be performed. As presented in the previous section, the delivered CAD geometry must be processed to be usable for the study. In Figure 14, general and detailed images presenting the TWIN STRATOS after the simplification and discretization process are shown. As can be seen, different components have different colors that correspond to the other composite structure, and which are related to the different material and structural properties. All preparation was related to the numerical model, the computation process, and the postprocessing of the results was conducted in Altair HyperWorks software.

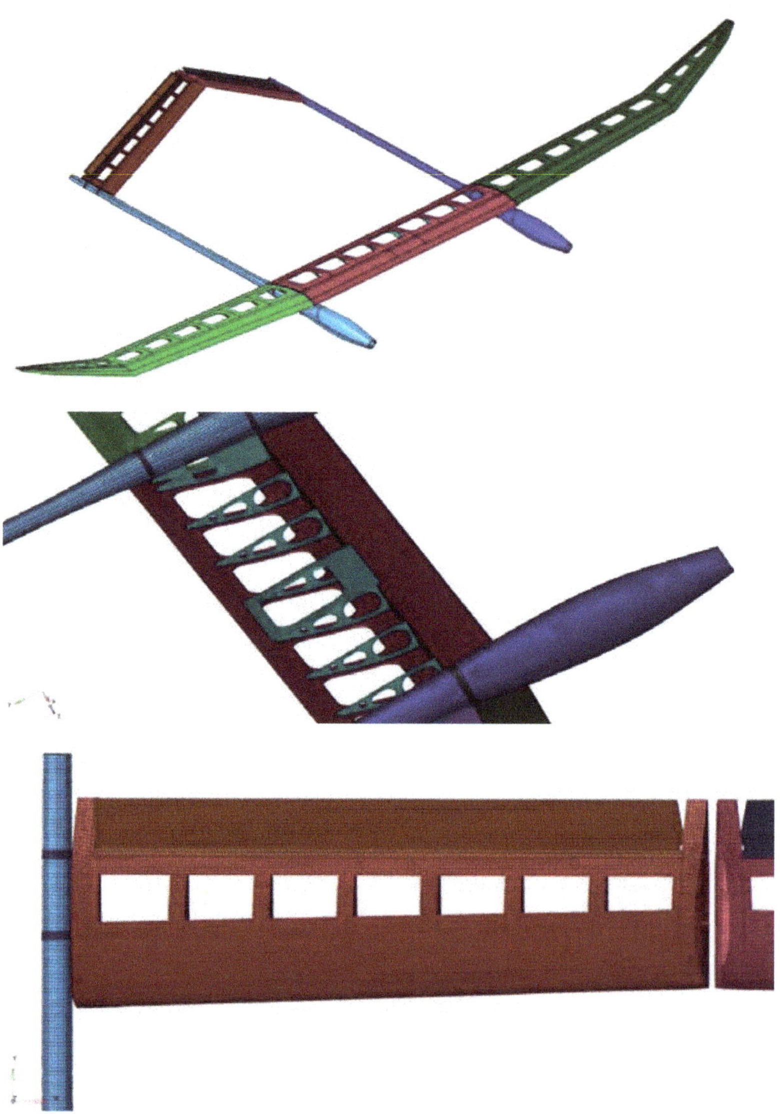

**Figure 14.** The simplified and discrete model of the STRATOS.

### 4.1.3. Material Properties

In Table 1, material properties that were used in the study definition are presented. For following approach for the calculations, general material data based on manufacturers' data were adopted.

**Table 1.** Anisotropic material properties used in the study definition.

| Property | | Carbon Fiber Composite | Honeycomb Expanded Core |
|---|---|---|---|
| Elastic modulus, MPa | $E_{cx}$ | 110,000 | 1700 |
| | $E_{cy}$ | 8500 | 1700 |
| | $E_{cz}$ | - | 576 |
| Shear modulus, MPa | $G_{cxy}$ | 4000 | 110 |
| | $G_{cyz}$ | - | 646 |
| | $G_{cxz}$ | - | 2500 |
| Allowable compressive stress, MPa | $X_c$ | 900 | - |
| | $Y_c$ | 170 | - |
| Allowable tensile stress, MPa | $X_t$ | 1500 | - |
| | $Y_t$ | 58 | - |
| Allowable shear stress, MPa | $S$ | 80 | - |
| Density kg/m$^3$ | | 1700 | 915 |

### 4.1.4. Loads and Constraints Definition

Preliminary calculations of loads were made due to the similarity with the use of software supporting glider analysis. Unfortunately, the analysis did not allow modelling the case of a double-hulled aircraft due to the different nature of the application. The better weight distribution of the two-hull system ensures that the actual loads will be lower. Load calculations were made for successive heights, flight speeds, angles of attack and flight conditions. In the final stage, a load envelope was developed for the subsequent sections of the wing. The envelope covered the maximum loads recorded in each section, although in adjacent sections they occurred in other flight conditions. In most cases, the maximum loads occur in flight at maximum flight altitudes with high speeds and gusts. Such an envelope was used in the preliminary strength calculations. At the next stages, the calculations were made based on proprietary software, considering the weight distribution system of the two-hull aircraft, and then using the XFLR5 system.

According to the presented method of load application, the authors applied properly calculated loads as well as constraints. An example of applied restraints added to fuselage mounting rings is presented in Figure 15.

### 4.1.5. Obtained Results

The prepared model was used to perform the computation procedure. In the first step, natural frequencies were calculated for both no constraints and with constraints. An example of the commutated natural frequency is shown in Figure 16. In the process, the first 50 frequencies were calculated for further aeroelastic analysis.

In the second step, standard structural analysis was performed. In this case, two analyses were conducted, the first one for the positive loads' envelope, and the second for the negative loads' envelope. As a result of this stage of the study, we obtained colored stress and displacement maps. Examples of the obtained results are presented in Figures 17 and 18.

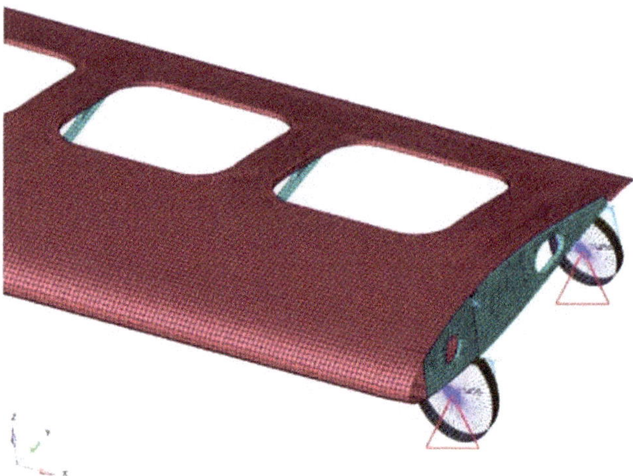

**Figure 15.** An example of applied restraints.

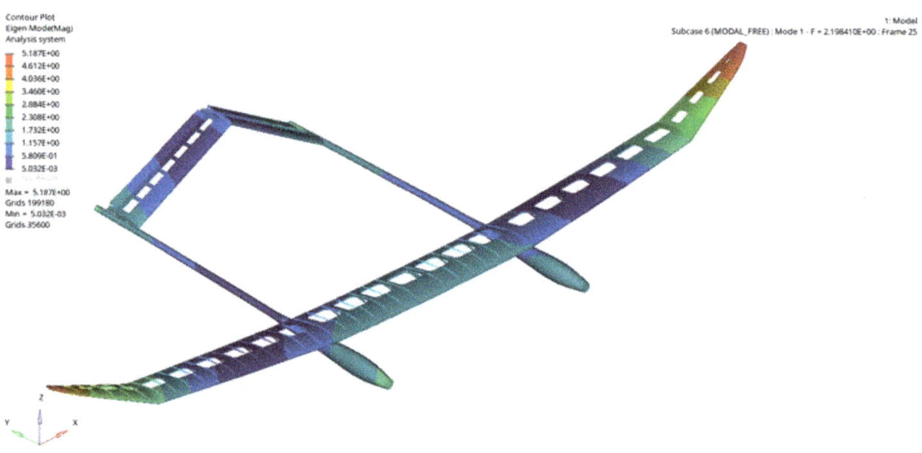

**Figure 16.** The first calculated natural frequency.

**Figure 17.** Global stress plot.

**Figure 18.** Global displacement plot.

The last part of the computation process is buckling analysis used for stiffens and stability evaluation. An example of the result of this analysis is presented in Figure 19.

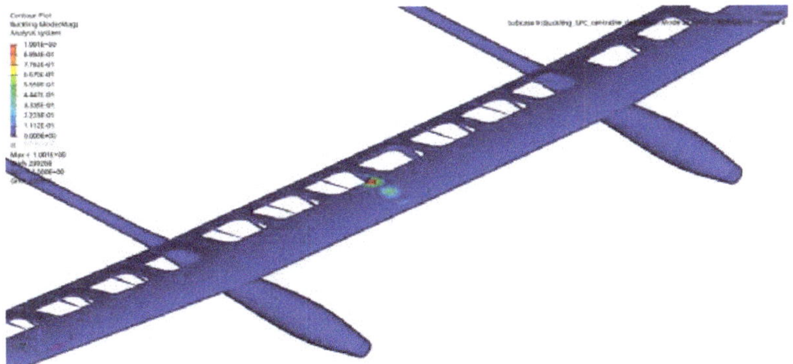

**Figure 19.** Example of stability lose in the central part of the UAV.

Based on the performed computations, it was possible to avoid some potential issues in the design. For example, the buckling analysis showed the area that needed to be revamped, because it is possible that in this area the UAV may lose stability and cause a crash. To address this issue, the authors, using generative modelling for honeycomb structures, changed the thickness to find out the proper value to fulfil the project's assumptions. In this case, the authors used the generative model as a tool to making models of the structure with different thickness. In the tests, the thickness of the filling layer of the sandwich structure was increased without changing the form of the filling. The study was intended to increase the viewer on the loss of stability in compressed structures. Finally, it finds out that in particularly endangered places 10 mm structure is an optimal value. The summary of the most important obtained results is presented in Table 2.

**Table 2.** Summary of the use case results.

| Parameter | Value | Unit |
| --- | --- | --- |
| Max von Mises stress | $1.197 \times 10^3$ | MPa |
| Max displacement | $4.118 \times 10^3$ | mm |
| 1st mode | 2.1984 | Hz |
| 10th mode | 11.817 | Hz |
| 50th mode | 47.078 | Hz |

*4.2. Aeroelastic Analysis with Flutter Verification*

The mid-fidelity tool ASWING was used to model aircraft aeroelasticity [24,25] ASWING couples interconnected nonlinear (specifically, Bernoulli–Euler) beam models with a general extended lifting line approach. The ASWING software enables the calculation of aircraft deflections, axial strains and shear stresses, aircraft aerodynamic properties, and aircraft stability derivatives and eigenvalues based on input data, including geometric, structural, and aerodynamic parameters airfoil sections along the aircraft wing and other surfaces. During analysis, aircraft stability derivatives and eigenvalues were used because the different outputs were identified using other methods. The model used as the input to ASWING was calculated to form the parametric CAD model (Figure 20); material data assumed for analysis and load distribution were identified using XFLR5 [26]. The data were integrated into excel file and subsequently transferred to a proper format which was used as the input to ASWING (Figure 21).

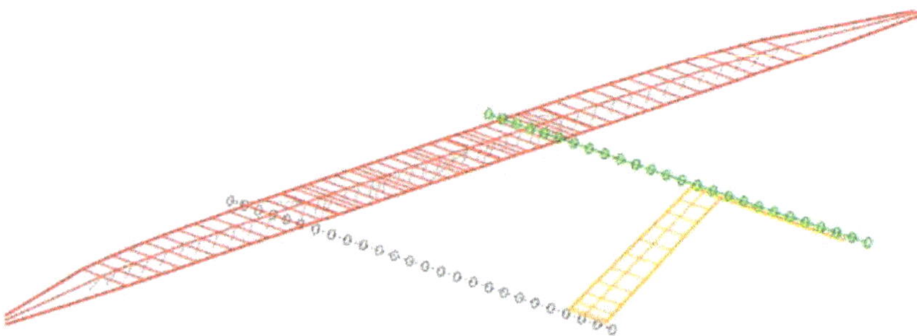

**Figure 20.** Geometric model used for ASWING analysis.

**Figure 21.** Data processing sequence for aeroelastic analysis using ASWING.

The flutter speed and actual flutter frequency are indicated by one of the eigenvalues crossing over from left to right of the imaginary axis. Flight-dynamic instabilities (e.g., spiral mode) will also have eigenvalues on the right side of the imaginary axis, but these will typically be at a near-zero frequency and will have completely different eigenmodes.

The results of the initial analyses indicated the risk of flutter at a speed of 20 m/s, (Figure 22), which is a speed exceeding the operating speed. A sensitivity analysis was performed during the flutter analysis. The influence of the position of the wing mass was investigated. The first simplified analysis assumed that the solar panels are positioned proportionally to the wing area. The model was then refined by adding mass points symbolizing solar panels at a distance of 0.4 of the wing chord from the leading edge. In the last iteration of calculations, the influence of shifting the mass of solar panels by 80 (mm) towards the leading edge was investigated. Sensitivity analysis showed the possibility of turning the critical flutter velocity in these cases to 25 and 35 m/s, respectively.

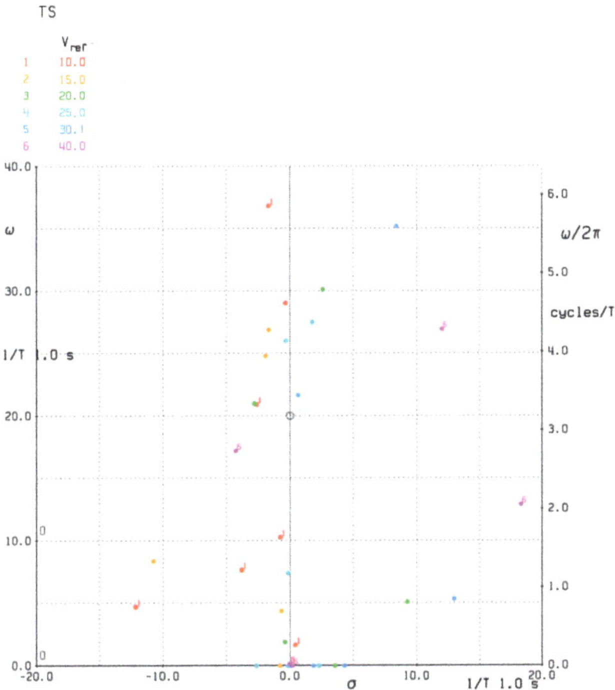


**Figure 22.** Eigenvalues of TWIN STRATOS UAV.

### 4.3. Comparative Analysis of Morphic Control

During the study, the main problem that should be solved is the considered design's aerodynamic performance. The airplane structure was reconstructed in XFLR5 software, as shown in Figure 23. The tests were similar to the aeroelasticity analyses (Figure 21) and carried out for various configurations of control systems [27].

The main aircraft wing was divided into several sections with differing chords and dihedrals. Different versions of roll control elements were taken into account. The following analysis compared three control options—classical with ailerons, tail control, wingtip twist morphing (Figure 24).

Figure 24 shows the distribution of the lifting force coefficient (Cl) in the given wing section as a function of the main wingspan and the tail wingspan measured from the symmetry plane for the roll operation for different angle of attack (ACA) (0°; 5,5°; 10°).

The computational simulation results confirmed the legitimacy of applying innovative solutions, both integrated in the tail and morphic wingtips. In both cases, the new control configurations became even 10% more efficient in the entire range of attack angles [27].

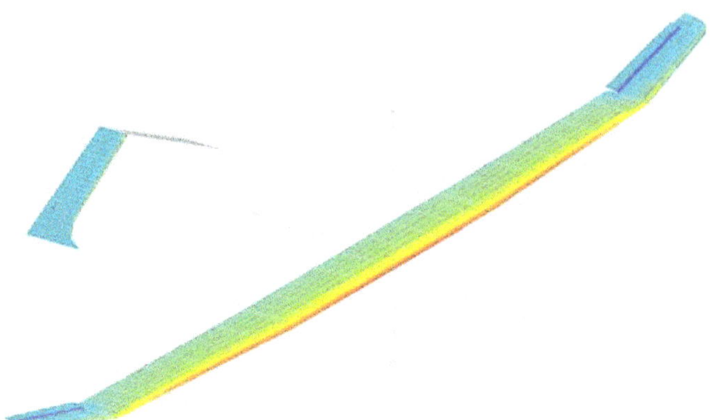

**Figure 23.** TWIN STRATOS UAV model in XFLR5 software.

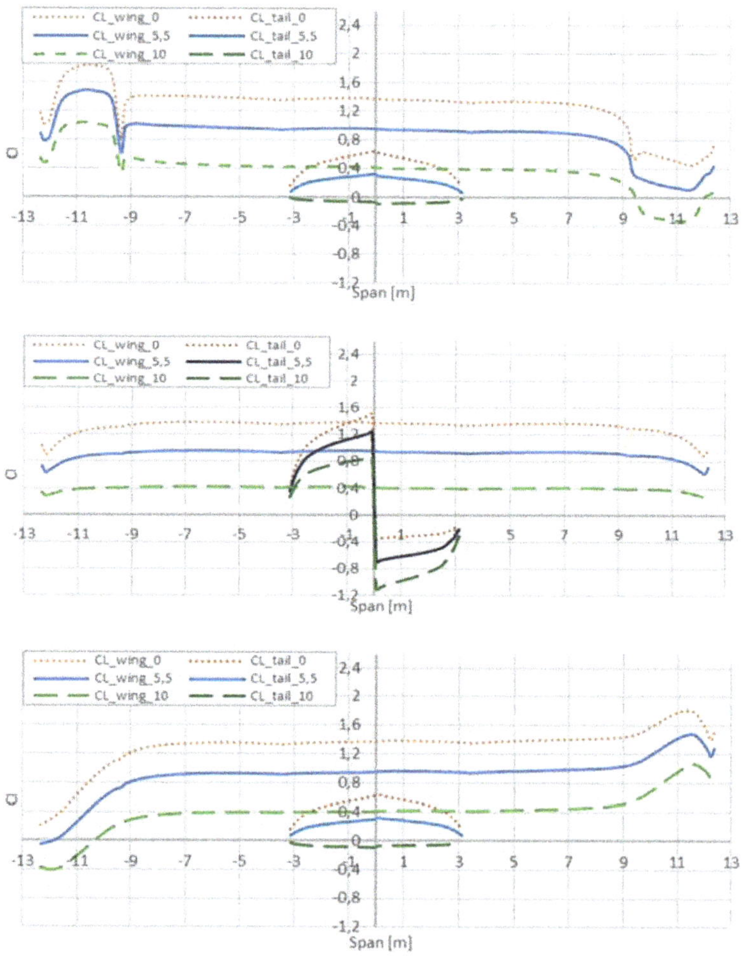

**Figure 24.** TWIN STRATOS UAV comparative analysis of roll operation for different steering arrangements. From above: ailerons, tail compliant trailing edge, wingtip twist morphing.

## 5. Conclusions

The design process is iterative, and the number of approximations in the case of a complex UAV object can be significant. Nowadays, CAD systems are commonly used to model the geometric form. In each of these approximations, it is necessary to generate a new geometric model, which is laborious and may lead to errors and inconsistencies. Automating the generation of a geometric model in the CAD system using generative modelling can significantly accelerate this process. Such generative models can be used both in the case of iterative improvement of the structure by designers and the application of optimization algorithms. The article describes improving a very unusual HALE UAV structure, where subsequent analyses are usually preceded by a labor-intensive rebuilding of the geometric model. Such an operation can be supported by using a generative model in the form of structured modules that enable structural modifications which are not understood in most low-fidelity verification methods of a parametric mesh but are in a multielement engineering model in the CAD system. The proposed solution does not eliminate the methods and models mentioned above—it perfectly complements them, as it allows for remodeling of a complex engineering structure and several engineering verifications that go far beyond general verification methods. In this way, the FEM analysis is performed not only on the general support structure but also on a highly verified design, including technological, assembly and general engineering relations.

The most important contributions of the paper are:

- A methodology for automating the design of ultralight and highly flexible aircraft structures with the use of generative modeling.
- Proposing and verifying the form of generative models for selected fragments of the structure, especially wings.
- Integration of the use of generative models for iterative improvement of structures using low- and middle-fidelity methods of numerical verification.

In future works, it is necessary to improve the design processes with the use of generative modelling, and in particular for further automation and integration with numerical verification methods and tools. Work will be carried out to extend the range of generative models to include less typical design solutions. Depending on needs, the generative models developed as part of the research will also be transferred to other CAD tools. Improvements of the design procedures for unusual classes of flying objects, such as HALE UAVs, are also required.

**Author Contributions:** Conceptualization, W.S. and A.J.; methodology, W.S. and A.J.; software, A.J.; validation, W.S. and A.J.; formal analysis, W.S. and A.J.; investigation, W.S. and A.J.; resources, A.J.; data curation, W.S. and A.J.; writing—original draft preparation, W.S. and A.J.; writing—review and editing, W.S. and A.J.; visualization, W.S. and A.J.; supervision, W.S.; project administration, W.S.; funding acquisition, W.S. All authors have read and agreed to the published version of the manuscript.

**Funding:** The APC was funded by the Rector's grant. Silesian University of Technology grant number 10/060/RGP18/0102.

**Institutional Review Board Statement:** Not applicable.

**Informed Consent Statement:** Not applicable.

**Data Availability Statement:** Restrictions apply to the availability of these data. Data was obtained from SkyTech eLab LLC and are available from the owners with the permission of SkyTech eLab LLC.

**Acknowledgments:** The authors would like to express their thanks to SkyTech eLab LLC for providing the data for research, moreover, we would like to thank the engineering staff and students who performed the calculations and analyses, in particular Mateusz Wąsik, Nikodem Ciomperlik and Michał Simon.

**Conflicts of Interest:** The authors declare no conflict of interest.

## References


1. Kumar, R. Investigation on buckling response of the aircraft's wing using finite-element method. *Aust. J. Mech. Eng.* **2020**, *18*, 122–131. [CrossRef]
2. Szklarek, K.; Gajewski, J. Optimisation of the Thin-Walled Composite Structures in Terms of Critical Buckling Force. *Materials* **2020**, *13*, 3881. [CrossRef] [PubMed]
3. Schor, P. Aerodynamic load of an aircraft with a highly elastic wing. *Acta Polytech.* **2017**, *57*, 272. [CrossRef]
4. Cesnik, C.; Senatore, P.; Su, W.; Atkins, E.; Shearer, C.; Pitchter, N. X-HALE: A Very Flexible UAV for Nonlinear Aeroelastic Tests. In Proceedings of the 51st AIAA/ASME/ASCE/AHS/ASC Structures, Structural Dynamics, and Materials Conference 18th AIAA/ASME/AHS Adaptive Structures Conference 12th, Orlando, FL, USA, 12–15 April 2010; American Institute of Aeronautics and Astronautics: Reston, VA, USA, 2010.
5. Alsahlani, A.; Rahulan, T. Conceptual and Preliminary Design Approach of A High Altitude, Long Endurance Solar-Powered UAV. In Proceedings of the CSE Annual PGR Symposium (CSE-PGSym 2017), Salford University, Salford, UK, 17 March 2017.
6. Amadori, K.; Jouannet, C.; Krus, P. Use of panel code modeling in a framework for aircraft concept optimisation. In Proceedings of the 11th AIAA/ISSMO Multidisciplinary Analysis and Optimization Conference, Portsmouth, VA, USA, 6–8 September 2006. [CrossRef]
7. Giesing, J.P.; Barthelemy, J.-F.M. A summary of industry MDO applications and needs. In Proceedings of the 7th AIAA/USAF/NASA/ISSMO Symposium on Multidisciplinary Analysis and Optimization, St. Louis, MO, USA, 2–4 September 1998; p. 4737.
8. Skarka, W. Model-Based design and optimisation of electric vehicle. In *Transdisciplinary Engineering Methods for Social Innovatio of Industry 4.0, Proceedings of the 25th ISPE International Conference on Transdisciplinary Engineering, Modena, Italy, 36 July 2018*; Peruzzini, M., Pellicciari, M., Bil, C., Stjepandić, J., msterdam, N.W., Eds.; IOS Press: Amsterdam, The Netherlands, 2018; pp. 566–575.
9. Papageorgiou, A.; Tarkian, M.; Amadori, K.; Ölvander, J. Multidisciplinary Design Optimization of Aerial Vehicles: A Review of Recent Advancements. *Int. J. Aerosp. Eng.* **2018**, *2018*, 1–21. [CrossRef]
10. Skarka, W. Application of MOKA Methodology in Generative Model Creation Using CATIA. *Eng. Appl. Artif. Intell.* **2007**, *20*, 677–690. [CrossRef]
11. Amadori, K.; Tarkian, M.; Ölvander, J.; Krus, P. Flexible and robust CAD models for design automation. *Adv. Eng. Inform.* **2012**, *26*, 180–195. [CrossRef]
12. Skarka, W.; Jałowiecki, A. Application of generative modeling methods to the design of thin layer composite aircraft structures. In Proceedings of the 7th International Conference Integrity-Reliability-Failure, IRF2020, Funchal, Portugal, 6–10 September 2020; Silva, J.F., Gomes, S.A., Eds.; INEGI-Instituto de Ciencia e Inovacao em Engenharia Mecanica e Gestao Industrial: Porto, Portugal, 2020; pp. 437–444.
13. Skarka, W.; Wąsik, M.; Skoberla, R.; Jałowiecki, A. Design automation of electric vehicle's aerodynamic parts. Transdisciplinary Engineering Methods for Social Innovation of Industry 4.0. In Proceedings of the 25th ISPE International Conference on Transdisciplinary Engineering, Modena, Italy, 36 July 2018; Margherita, P., Marcello, P., Cees, B., Josip, S., Nel, W., Eds.; IOS Press: Amsterdam, The Netherlands, 2018; pp. 556–565.
14. McMasters, J.H.; Cummings, R.M. Rethinking the Airplane Design Process—An Early 21st Century Perspective. *Aerosp. Sci. Meet. Exhib.* **2004**, 693. [CrossRef]
15. Verhagen, W.J.; Bermell-Garcia, P.; Van Dijk, R.E.; Curran, R. A critical review of Knowledge-Based Engineering: An identification of research challenges. *Adv. Eng. Inform.* **2021**, *26*, 5–15. [CrossRef]
16. Vadenbrande, J.H.; Grandine, T.A.; Hogan, T. The search for the perfect body: Shape control for multidisciplinary design optimisation. *Aerosp. Sci. Meet. Exhib.* **2006**, 928. [CrossRef]
17. Balling, R.J.; Sobieszczanski-Sobieski, J. Optimisation of Coupled Systems: A Critical Overview of Approaches. *NASA Contractor Rep.* **1994**, *34*, 6–7.
18. Wakayama, S.; Kroo, I. The Challenge and Promise of Blended-Wing-Body Optimization. In Proceedings of the 7th AIAA/USAF/NASA/ISSMO Symposium on Multidisciplinary Analysis and Optimization, St. Louis, MO, USA, 2–4 September 1998; pp. 239–250.
19. Tseng, M.M.; Jiao, J. Design for mass customisation. *Anal. CIRP* **1996**, *45*, 153–156. [CrossRef]
20. Targosz, M.; Skarka, W.; Przystałka, P. Model-Based Optimisation of Velocity Strategy for Lightweight Electric Racing Cars. *J. Adv. Transp.* **2018**, *2018*, 3614025. [CrossRef]
21. Niestrój, R.; Rogala, T.; Skarka, W. An Energy Consumption Model for Designing an AGV Energy Storage System with a PEMFC Stack. *Energies* **2020**, *13*, 3435. [CrossRef]
22. Yang, H.Z.; Chen, J.F.; Ma, N.; Wang, D.Y. Implementation of knowledge-based engineering methodology into ship structural design. *Comput. Aided Des.* **2012**, *44*, 196–202. [CrossRef]
23. Chandrasegaran, S.K.; Ramani, K.; Sriram, R.D.; Horváth, I.; Bernard, A.; Harik, R.F.; Gao, W. The evolution, challenges, and future of knowledge representation in product design systems. *Comput. Aided Des.* **2012**, *45*, 204–228. [CrossRef]
24. Drela, M. Technical Description—Steady Fromulation. 2015. Available online: https://web.mit.edu/drela/Public/web/aswing/asw_theory.pdf (accessed on 10 February 2021).

25. Drela, M. ASWING 5.97 User Guide. *MIT Aero & Astro*. 2012. Available online: http://web.mit.edu/drela/Public/web/aswing/aswing_doc.txt (accessed on 10 February 2021).
26. XFLR5. Available online: http://www.xflr5.tech/xflr5.htm (accessed on 10 February 2021).
27. Ciomperlik, N.; Skarka, W. Morphic arrangement of high flexibility and aspect ratio wing. In Proceedings of the 27th ISTE International Conference on Transdisciplinary Engineering, Warsaw, Poland, 1–10 July 2020.

*Article*

# Optimizing Roller Design to Improve Web Strain Uniformity in Roll-to-Roll Process

Yousung Kang [1], Yongho Jeon [1], Hongkyu Ji [2], Sin Kwon [3], Ga Eul Kim [1,3] and Moon G. Lee [1,*]

[1]  Department of Mechanical Engineering, Ajou University, Suwon-si 16499, Korea; gidalim89@ajou.ac.kr (Y.K.); princaps@ajou.ac.kr (Y.J.); autumn@kimm.re.kr (G.E.K.)
[2]  Global Technology Center, Samsung Electronics Co., Ltd., Suwon-si 16677, Korea; hongkyu.ji@samsung.com
[3]  Advanced Manufacturing System Research Division, Korea Institute Machinery and Materials (KIMM), Daejeon-si 34103, Korea; skwon@kimm.re.kr
*  Correspondence: moongulee@ajou.ac.kr; Tel.: +82-31-219-2338

Received: 17 September 2020; Accepted: 26 October 2020; Published: 27 October 2020

**Abstract:** In this work, we investigated the compensation of strain deviation in the machine direction of a web in the roll-to-roll process. As flexible devices have become popular, many researchers have begun to study roll-to-roll processes for the mass-production of flexible devices at low cost. In the continuous roll-to-roll process, an electronic circuit pattern is printed on the web while the web is transferring. Due to tension and Poisson's ratio, a non-uniform strain distribution can occur in the web. This strain distribution occurs mainly at the center of the web and causes a register error in the machine direction. In this work, we designed a roller to minimize the strain deviation. The design of the compensation roller was optimized using the design of experiments (DOE) methodology and analysis of variance (ANOVA), and the compensation performance was verified through experiments and simulations. According to the results of a comparative experiment conducted to confirm the correcting performance of the optimized roller, the strain deviation in the machine direction decreased by approximately 48% with the proposed roller compared to that of the conventional roller.

**Keywords:** web deformation; strain deviation; design of experiment; roll-to-roll process

## 1. Introduction

Flexible devices are part of the next generation of mobile technology and include foldable displays, wearable devices, and implant sensors. Their core component is a flexible printed circuit board (FPCB). To popularize flexible devices, FPCBs must be inexpensively and efficiently produced. However, conventional circuit board production methods (e.g., lithography and etching processes) cannot meet these requirements. Since the roll-to-roll process is suitable for the low-cost mass production of flexible substrates, researchers have attempted to develop the process for manufacturing flexible devices.

In the roll-to-roll process, electronic circuit patterns are printed onto flexible substrates which are transferred between unwinding and rewinding rollers. These flexible substrates are composed of a polymer film, which is called a "web". In the actual process, the two unwinding and rewinding rollers, along with many additional rollers, generate the electronic circuit patterns. The electronic circuit patterns are printed on the web, inspected, and packaged. The space between any two rollers in the process is called the "span". The direction along the transfer is "machine direction (MD)". The direction perpendicular to MD is "cross direction (CD)".

If the web has an uneven strain distribution in the span between these rollers, defects may form. In a given span, the two rollers at both ends exert tension on the web, generating a strain distribution in the web. To construct the electronic circuit pattern, multilayer printing is performed with several spans. A register error is a relative positional error between the patterns of the previous and next spans. A register error occurs if the strain generated on the web in each span is uneven. Further, an uneven strain can cause pattern errors because there is a contraction difference between the two patterns in the center. Therefore, this non-uniform strain distribution causes a register error in the machine direction (MD) and cross direction (CD) of the web [1]. If the strain deviation in the MD is reduced and the distribution becomes more uniform, this error is reduced.

Existing processes using rollers include rolling processes for metalworking and the rotary press printing process. In the rolling process, metal materials are reshaped to obtain a product. Thus, researchers have focused on plastic rather than elastic deformation [2,3]. In the rotary printing process, characters or images are printed on a paper substrate. Researchers have attempted to minimize the register error to improve print quality; however, micro-scale printing precision is not required [4–6].

The electronic circuit pattern created in the roll-to-roll process is finely detailed; therefore, micro-scale printing precision is important for multilayer printing. A register error due to a shape error or misalignment of the pattern creates defects in the electronic circuit. Thus, a great deal of research has focused on reducing register errors in order to manufacture high-performance electronic components. Hu, L. et al. measured the register error using a high-resolution microscope camera and studied motion-base calibration using the micro-movement of the print head [7]. Jung, H. et al. conducted and compensated for the register error by controlling the span length and tension using an active motion-based roller between the upstream and downstream printing rollers [8]. Kang, H. et al. compensated for the register error using the phase shift of the upstream roller and continuously compensating the register error of the downstream roller through PID control [9]. Chang, J. et al. reduced the register error by locally releasing tension on the web locally through a successive chucking system, clamping unit, and guide roller descending [10]. Kim, C. et al. reduced the strain on the web by controlling the tension using a dancer roller to reduce the web shrinkage that occurs during the drying process [11]. Lee, J. et al. used a register control algorithm to reduce the tension error [12]. Kim, S. et al. analyzed the deformation of the substrate in the ink transfer process using computational fluid dynamics (CFD) [13]. However, there are few studies that proactively correct the uneven strain distribution during the roller design stage prior to using a separate actuator or controlling the system. In the roll-to-roll process, the electronic circuit is printed on an arbitrary printing line in the CD. Therefore, the non-uniformity of the strain in the printing line is more important than the absolute value of the strain in the web. The strain is concentrated in the center of the CD and decreases toward both sides [14]. In addition, the strain deviation in the web is proportional to the web width and inversely proportional to the distance in the MD from the line where the roller and web touch. To reduce the register error, the strain deviation must not exceed a certain level within a single span; if the strain variation is great, the effective printing area is reduced, which limits productivity. In addition, the strain generated by tension during the process is recovered when the tension is disengaged after the process. This elastic change can also result in pattern error. If the strain distribution is uneven, the pattern is misaligned, resulting in an MD register error and poor print quality.

In this study, we investigated the strain distribution in the flexible web in the roll-to-roll process. We sought to equalize the strain distribution in the MD on the printing line in the CD. To maintain high print quality and productivity, we proposed a new roller design that compensates the non-uniformity of the strain. Moreover, we developed a finite element (FE) model of a single-span roller and web and evaluated the compensation performance of the roller. To maximize the compensation, we optimized the design parameters of the roller using the design of experiments (DOE) methodology and analysis of variance (ANOVA).

To minimize the strain deviation in the MD, we constructed the compensation roller with a convex roller core. The convex core was wrapped in a flexible material (rubber skin) such that the roller and web were always in contact. The outer diameter of the roller with the core and rubber is uniform. This roller configuration changes the contact stiffness along the CD on the web when the rollers and the web touch to compensate for the strain deviation.

To optimize the compensation performance, we varied the thickness gradient of the rubber skin surrounding the roller, the total diameter of the roller, and the material properties of the rubber skin. The thickness gradient is the thinnest minus the thickest diameters of the core. Additionally, we considered interactions between the design parameters. From this analysis, we found that the strain of the web was most uniform in the roller with a barrel-type core with a thickness gradient of approximately 8 mm. This roller compensated the uneven strain distribution by approximately 48% with respect to the conventional roller.

## 2. Strain of Web in Roll-to-Roll Process

The finite element analysis was used to analyze the strain distribution of the web in the roll-to-roll process. Through this analysis, it was confirmed that the MD strain deviation exists under uniform tension conditions. Additionally, the MD strain deviation with respect to the web width and the location of the printing line was analyzed. A series of experiments were conducted to verify the FEA results, and the MD strain deviation similar to the FEA results was measured. Therefore, we used the FE model to compensate the MD strain deviation.

### 2.1. Strain Analysis Model of Web

ANSYS Workbench R18.0 was used to conduct an FE analysis on the web strain caused by tension for a conventional roller in the roll-to-roll process. To analyze the strain of the web when tension is applied to a single span, the FE model consisted of a roller and a web. A schematic of the FE model is shown in Figure 1. The diameter and length of the roller were 60 and 350 mm, respectively, and the roller was composed of aluminum. The web was composed of a polyethylene terephthalate (PET) film with a width of 160 mm, a span length of 200 mm, and a thickness of 0.2 mm. The mechanical properties used in the analysis are shown in Table 1. PET films are viscoelastic materials; however, we considered them to be linear elastic materials because of the low strain in this case study. The element size of the web is 5 mm and the web was modeled as a uniform hexahedron mesh; the number of elements was 3324 and the number of nodes was 19,289. The element size and shape were predetermined after checking the FE model's convergency. In addition, boundary conditions such as the load, degrees of freedom, and friction conditions were set in the FE model. The tension (10 MPa) was applied at the load end along the transport direction of the web. At the fixed end of the web, which touched the roller, all degrees of freedom except the width direction of the web were constrained. The degree of freedom in the width direction of the web allowed for deformation due to Poisson's ratio. Moreover, the contact between the web and roller resulted in friction with a friction coefficient of 0.2.

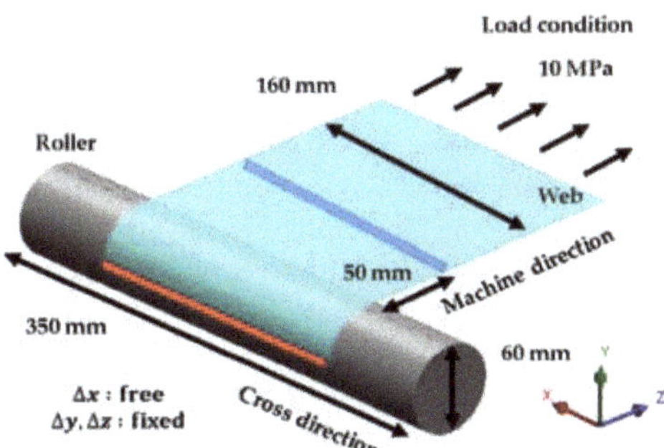


**Figure 1.** Schematic of finite element (FE) model: Red lines are fixed in the machine direction (MD); the blue line represents the target printing line.

**Table 1.** Mechanical properties in the finite element analysis (FEA).

|  | Young's Modulus | Mass Density | Poisson's Ratio |
| --- | --- | --- | --- |
|  | E (GPa) | $\rho$ (kg/m$^3$) | - |
| Roll (aluminum) | 71 | 2,770 | 0.33 |
| Web (PET) | 3.5 | 1,320 | 0.34 |

In the actual process, the transfer of the web is a dynamic process. In this study, however, we modeled the process with a constant transfer speed and under static conditions for simplicity. If the strain deviation is compensated under static conditions, it will also be compensated in the dynamic state.

### 2.2. Web Strain FEA Results: Conventional Roller

We used the FE model to analyze the strain distribution along the position of printing line or across the width of the web under uniform tension conditions. Figure 2 shows the relationship between the position along the printing line and strain deviation in the MD with respect to the distance from the contact line of the web (width of 160 mm) and roller when a tensile force of 10 MPa is applied to the web.

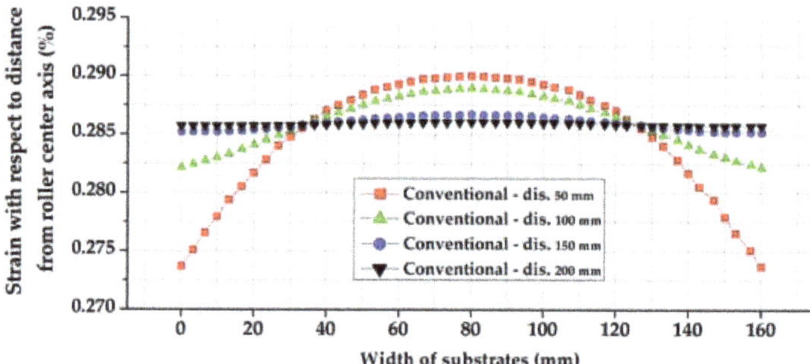


**Figure 2.** MD strain distribution along the printing line with respect to the change in distance between roller and line.

At a distance of 50 mm from the roller in the MD, the strain exhibited the most uneven distribution, while the most uniform strain was observed for a distance of 200 mm. The shorter the distance between the roller and the web in the MD, the greater the variation in strain distribution. Moreover, Figure 3 shows the deviation of the MD strain which occurs with changes in the web width for a distance 50 mm from the roller in the MD. When the film width was 160 mm, the strain deviation was minimal. The strain deviation increased as the web width increased. This strain deviation may cause a register error in the MD owing to the change in the tension in each span. In addition, when the tension is removed after the process, the elastic strain becomes zero and the positioning precision of the pattern is adversely affected. Therefore, we sought to design a roller to improve the precision of the printing pattern and to minimize the register error by compensating for the strain deviation in the MD.

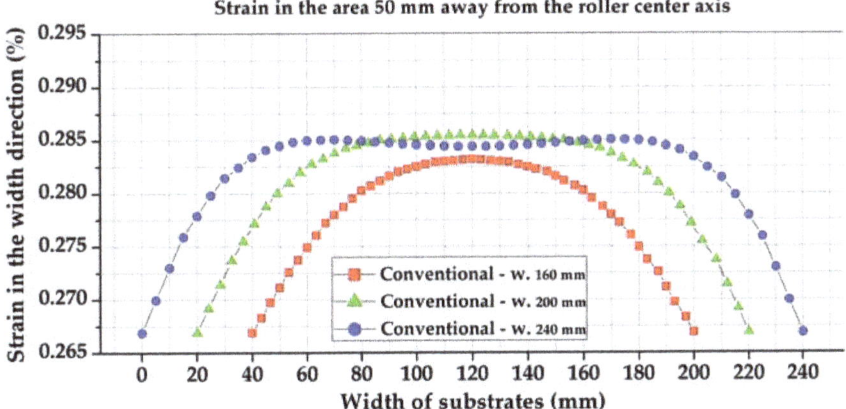



**Figure 3.** MD strain distribution along the printing line with respect to the web width.

### 2.3. Verification of FE Model for Analyzing Web Strain Deviation

Before using the FE model to design the roller, we verified the model. Figure 4 shows a schematic of the arrangement of the rollers and web of the experimental equipment. In general, the web is transferred by an unwinding roller and a rewinding roller. In a single span, one side uses a nip to simulate the fixed end, and the other side uses a dancer roller to simulate the load end. The tension acting on the web was measured with load cells (TS-100, Bongshin Loadcell Inc., Osan-si, Korea) at both ends of the dancer roll. The strain measurement was calculated from the relative displacement of reference marks on the web when tension was applied to the web, as shown in Figure 5. The mark displacement was measured with a vision camera (MV-BX30A, Crevis Inc., Yongin-si, Korea, pixel size = 4.65 μm).

To measure the relative displacement of the web, 18 marks were used—nine marks near the fixed end and nine marks near the load end. Each set of nine marks is aligned along the CD, parallel to each other. The marks were fabricated by using a $CO_2$ laser on the web. The marks were approximately $1 \times 1$ mm$^2$ and cross-shaped.

Figure 6 shows the relative displacement from the marks at the fixed end to the marks at the load end; the experimental results are similar to the results of the FEA. However, even if the transfer of the web was stopped during the experiment, displacement due to the inertia of the roller occurred; thus, it was impossible to stop it at the same position every time. Additionally, because the dancer roller used a hydraulic drive to add tension, it was difficult to compare quantitatively because there is a tension error for each test case. The experimental results were not symmetrical about the center because the position of the reference mark was not symmetrical.

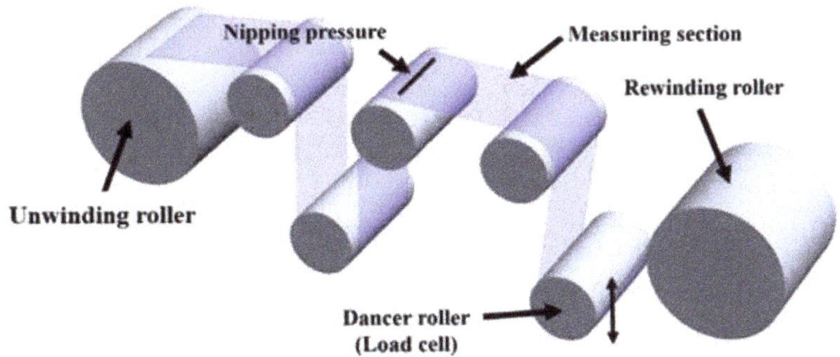



**Figure 4.** Schematic of equipment for web strain measurement.

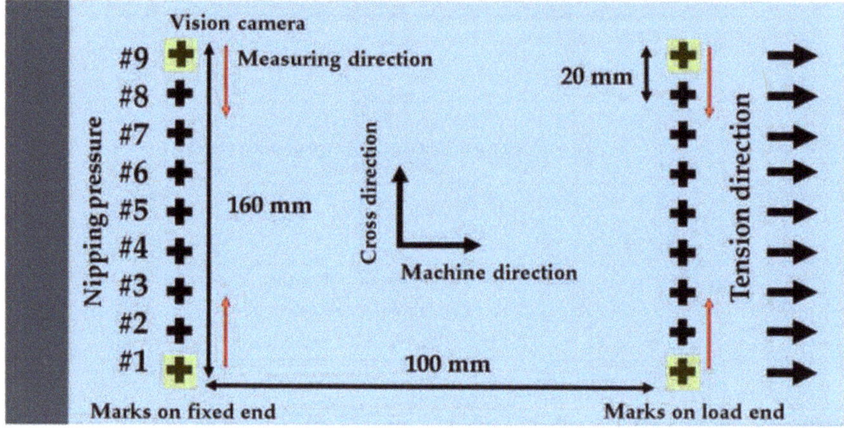



**Figure 5.** Schematic of the web strain measurement method (top view) with a conventional roller.

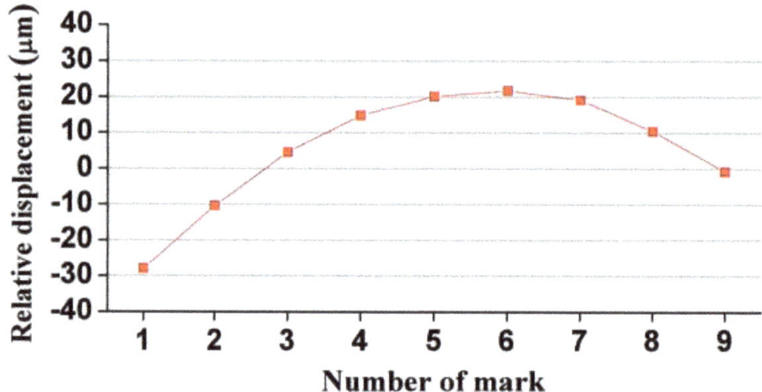



**Figure 6.** Strain measurement results along the cross direction (CD) of conventional roller.

Based on the results in Figure 6, we confirmed that the strain deviation occurred in the MD under uniform tension conditions; the strain in the MD was larger near the center of the CD and decreased toward both ends. Since non-idealities exist in reality, there are some errors; however, the FEA and the experiment largely agree. Thus, we used the FE model to design the compensation roller for minimizing the strain deviation in the MD.

## 3. Roller Design Optimization for Minimizing Strain Deviation

The compensation roller has a barrel-shaped core and is wrapped with an elastic rubber. This roller was named "barrel roller". In order to maximize the corrective performance of the barrel roller, we optimized the roller using DOE and ANOVA. After applying the optimized design factors to the FEA, we confirmed that the strain deviation was reduced.

### 3.1. Roller for Compensating Strain Deviation

According to the FEA and experimental results, the strain in the MD of the web was high in the center of the CD when a uniform tension was applied to the conventional roller. The strain deviation in the MD can be compensated by adjusting the contact stiffness between the roller and the film. If the contact stiffness decreases toward the roller end, the relatively low strain increases; thus, the strain deviation in the web is compensated.

To test this concept with an actual roller, a roller with a convex core was chosen and the roller was wrapped in an elastic material (rubber skin). Hence, the core is rigid and the skin is flexible. As shown in Figure 7, the contact stiffness between the web and roller is greatest in the center and decreases toward each side. The total diameter of the core and skin assembly is constant to prevent wrinkling of the web. Since the core of the roller is convex, the roller was named a "barrel roller".

We then used the FE model to optimize the design of the roller shown in Figure 7, focusing on the strain distribution in the MD 50 mm from the contact line. The goal of the optimization was to minimize the difference between the maximum and the minimum of the strain distribution in the MD along the CD.

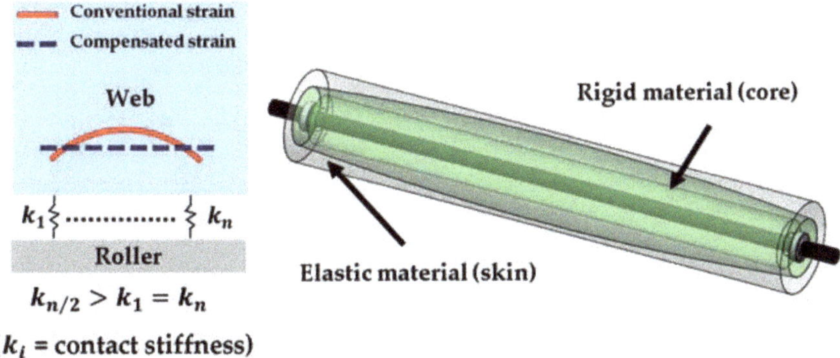


**Figure 7.** Compensation rollers for minimizing the non-uniform strain.

### 3.2. Design Factors and Their Interaction

The design of experiments (DOE) is a method of efficiently planning an experiment to obtain the maximum information while minimizing the number of experiments required [15]. DOE is used to analyze the factors that affect the output of the system. To minimize the MD strain deviation of the web under uniform tension conditions, design optimization was conducted based on DOE. Design factors were selected and the sensitivity analysis was performed according to the level change of design factors.

We varied the thickness gradient of the skin (A), the diameter of the entire roller including the core and skin (B), and the elasticity modulus (Young's modulus, C) of the skin. Considering the process with actual rollers, the minimum thickness of the skin was set to 3 mm.

Next, we investigated possible interactions between the selected design factors (i.e., whether one factor depends on another factor). Table 2 lists the initial values of the design factors used to study the interactions with a two-way factorial design. One of the three selected design factors was fixed, and the deviation of the strain in the MD of the web was analyzed as the other two factors were varied.

**Table 2.** Design factors and initial levels of each factor.

| Sort | | Factor | Level | | |
|---|---|---|---|---|---|
| | | | 1 | 2 | 3 |
| Control factor | A (mm) | Thickness gradient of skin | 3 | 6 | 9 |
| | B (mm) | Outer diameter of roller | 40 | 60 | 80 |
| | C (MPa) | Young's modulus of skin | 2.85 | 7.85 | 12.85 |

The results are shown in Figure 8. The graph shows that the curves representing the relationship between the thickness gradient of the skin (A) and the diameter of the entire roller (B) and the relationship between the diameter of the roller (B) and Young's modulus (C) of the skin intersect. However, the curves representing the relationship between the thickness deviation (A) of the skin and

Young's modulus (C) of the skin do not intersect. Hence, factor B is affected by the levels of factors A and C; these factors interact.

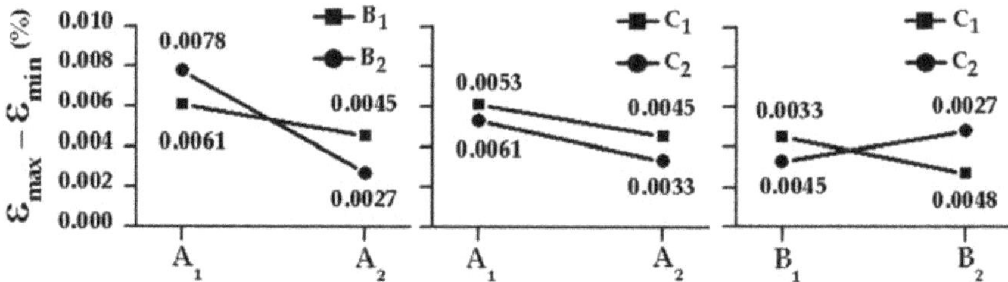


**Figure 8.** Interaction analysis for the two-way factorial design.

### 3.3. Design of Experiments (DOE) Considering Interactions

Since we confirmed that interactions between design factors occur, our DOE analysis considered these interactions. Table 3 presents an orthogonal array for the case in which the interactions between the design factors are considered. Factors A, B, and C are in columns 1, 2, and 5, and their interactions are indicated by AB, BC, and AC, respectively. In the DOE-based FEA, only columns 1, 2, and 5 with factors A, B, and C were considered. Then, in the process of selecting the optimum condition, the factors of the remaining columns were considered. The ABC factor, which is a higher-order interaction, was not considered and was pooled in the error term. Subsequently, a sensitivity analysis was conducted to determine whether changes in design factor values affected the strain deviation in the MD.

**Table 3.** Orthogonal array for interaction analysis.

| No. | A | B | AB | AB | C | AC | AC | BC | ABC | ABC | BC | ABC | ABC |
| | | | | | | | | Factor | | | | | |
| | 1 | 2 | 3 | 4 | 5 | 6 | 7 | 8 | 9 | 10 | 11 | 12 | 13 |
| | | | | | | Column | | | | | | | |
|---|---|---|---|---|---|---|---|---|---|---|---|---|---|
| 1 | 1 | 1 | 1 | 1 | 1 | 1 | 1 | 1 | 1 | 1 | 1 | 1 | 1 |
| 2 | 1 | 1 | 1 | 1 | 2 | 2 | 2 | 2 | 2 | 2 | 2 | 2 | 2 |
| 3 | 1 | 1 | 1 | 1 | 3 | 3 | 3 | 3 | 3 | 3 | 3 | 3 | 3 |
| 4 | 1 | 2 | 2 | 2 | 1 | 1 | 1 | 2 | 2 | 2 | 3 | 3 | 3 |
| 5 | 1 | 2 | 2 | 2 | 2 | 2 | 2 | 3 | 3 | 3 | 1 | 1 | 1 |
| 6 | 1 | 2 | 2 | 2 | 3 | 3 | 3 | 1 | 1 | 1 | 2 | 2 | 2 |
| 7 | 1 | 3 | 3 | 3 | 1 | 1 | 1 | 3 | 3 | 3 | 2 | 2 | 2 |
| 8 | 1 | 3 | 3 | 3 | 2 | 2 | 2 | 1 | 1 | 1 | 3 | 3 | 3 |
| 9 | 1 | 3 | 3 | 3 | 3 | 3 | 3 | 2 | 2 | 2 | 1 | 1 | 1 |
| 10 | 2 | 1 | 2 | 3 | 1 | 2 | 3 | 1 | 2 | 3 | 1 | 2 | 3 |
| 11 | 2 | 1 | 2 | 3 | 2 | 3 | 1 | 2 | 3 | 1 | 2 | 3 | 1 |
| 12 | 2 | 1 | 2 | 3 | 3 | 1 | 2 | 3 | 1 | 2 | 3 | 1 | 2 |
| 13 | 2 | 2 | 3 | 1 | 1 | 2 | 3 | 2 | 3 | 1 | 3 | 1 | 2 |
| 14 | 2 | 2 | 3 | 1 | 2 | 3 | 1 | 3 | 1 | 2 | 1 | 2 | 3 |
| 15 | 2 | 2 | 3 | 1 | 3 | 1 | 2 | 1 | 2 | 3 | 2 | 3 | 1 |
| 16 | 2 | 3 | 1 | 2 | 1 | 2 | 3 | 3 | 1 | 2 | 2 | 3 | 1 |
| 17 | 2 | 3 | 1 | 2 | 2 | 3 | 1 | 1 | 2 | 3 | 3 | 1 | 2 |
| 18 | 2 | 3 | 1 | 2 | 3 | 1 | 2 | 2 | 3 | 1 | 1 | 2 | 3 |
| 19 | 3 | 1 | 3 | 2 | 1 | 3 | 2 | 1 | 3 | 2 | 1 | 3 | 2 |
| 20 | 3 | 1 | 3 | 2 | 2 | 1 | 3 | 2 | 1 | 3 | 2 | 1 | 3 |
| 21 | 3 | 1 | 3 | 2 | 3 | 2 | 1 | 3 | 2 | 1 | 3 | 2 | 1 |
| 22 | 3 | 2 | 1 | 3 | 1 | 3 | 2 | 2 | 1 | 3 | 3 | 2 | 1 |
| 23 | 3 | 2 | 1 | 3 | 2 | 1 | 3 | 3 | 2 | 1 | 1 | 3 | 2 |
| 24 | 3 | 2 | 1 | 3 | 3 | 2 | 1 | 1 | 3 | 2 | 2 | 1 | 3 |
| 25 | 3 | 3 | 2 | 1 | 1 | 3 | 2 | 3 | 2 | 1 | 2 | 1 | 3 |
| 26 | 3 | 3 | 2 | 1 | 2 | 1 | 3 | 1 | 3 | 2 | 3 | 2 | 1 |
| 27 | 3 | 3 | 2 | 1 | 3 | 2 | 1 | 2 | 1 | 3 | 1 | 3 | 2 |

The initial factor levels in the sensitivity analysis were the factor levels in Table 2, and the results are shown in Figure 9. The objective function to minimize the strain deviation of MD was selected so that the difference between the maximum and minimum values of the strain distribution was to be the smallest. The same was applied to the DOE-based FEA and verification experiments. Accordingly, the strain deviation in the MD was minimal when the thickness gradient (A) of the skin was approximately 6 mm and the total diameter (B) of the roller was 40–60 mm. The lower the Young's modulus (C) of the rubber, the greater the reduction of the MD strain deviation.

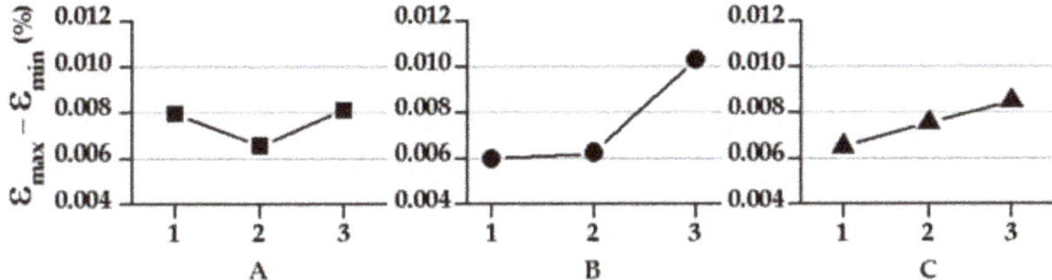

**Figure 9.** Sensitivity analysis considering interactions based on factor levels in Table 2.

Thus, the range of factor A was further sub-divided from 6 to 9 mm and the range of factor B was set to 40–60 mm for the FEA. The levels of the modified factors are shown in Table 4, and Table 3 shows the orthogonal array. Moreover, Figure 10 shows the results of the sensitivity analysis based on the modified factor levels in Table 4. For a skin thickness gradient (A) of 6 mm, roller diameter (B) of 60 mm, and low Young's modulus (C) of the skin, the strain deviation in the MD was minimized. In addition, each design factor influenced the reduction of the strain deviation in the MD.

**Table 4.** Modified factor levels for determining the global minimum.

|         | A (mm) | B (mm) | C (MPa) |
|---------|--------|--------|---------|
| Level 1 | 6      | 40     | 2.85    |
| Level 2 | 7      | 50     | 7.85    |
| Level 3 | 8      | 60     | 12.85   |

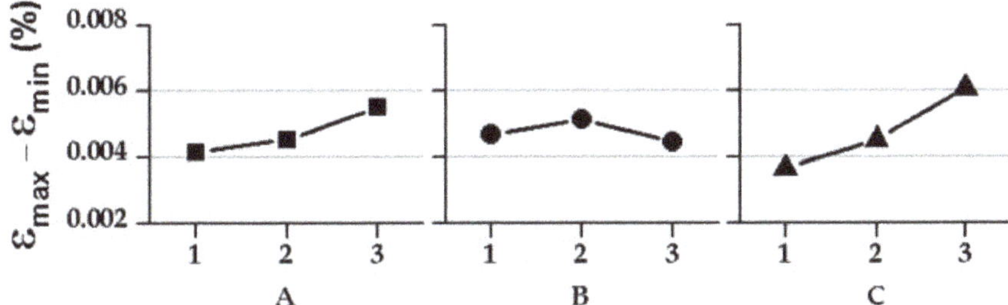

**Figure 10.** Sensitivity analysis considering interactions based on factor levels in Table 4.

Subsequently, we performed a quantitative analysis of variance (ANOVA) to select the optimal design. The ANOVA represents the variation in the characteristic values obtained as a result of the DOE as the total sum of squares [15]. This value is expressed as the total sum of squares for each factor, and the remainder is interpreted as an error variation. The variance of each factor is then compared to the variance of the error to determine how much influence each factor has on the characteristic values. The larger the F value in the analysis of variance, the greater the effect on the objective function. Table 5 shows the results of the ANOVA including the interactions between the factors; SS is the sum of squares, DF is the degrees of freedom, MS is the mean square, E is the error term, and T is the total sum of squares of the characteristic values obtained as a result of the DOE method. The F value was calculated by dividing the mean square value of each factor by the mean square value of the error (E).

**Table 5.** ANOVA results derived after performing FEA by applying the orthogonal array in Table 3 and the modified factor levels in Table 4.

| Factor | SS | DF | MS | F |
|---|---|---|---|---|
| A | $8.13 \times 10^{-10}$ | 2 | $4.07 \times 10^{-10}$ | 44.68 |
| B | $2.01 \times 10^{-10}$ | 2 | $1.00 \times 10^{-10}$ | 11.02 |
| C | $2.70 \times 10^{-9}$ | 2 | $1.35 \times 10^{-9}$ | 148.51 |
| AB | $2.07 \times 10^{-9}$ | 4 | $5.18 \times 10^{-10}$ | 56.94 |
| AC | $4.72 \times 10^{-12}$ | 4 | $1.18 \times 10^{-12}$ | 0.13 |
| BC | $1.91 \times 10^{-9}$ | 4 | $4.78 \times 10^{-10}$ | 52.57 |
| E | $7.28 \times 10^{-11}$ | 8 | $9.10 \times 10^{-12}$ | – |
| T | $7.78 \times 10^{-9}$ | 26 | – | – |

According to Table 5, the F value of the A $\times$ C factor was 0.13, which was less than the error value. Thus, the interaction between A and C had a small effect on the reduction of the strain deviation in the MD. The interaction between A and C was pooled in the error term in Table 6. Based on the results in Table 6, the modulus of elasticity of the skin had the greatest effect and the total diameter of the roller had the least effect.

**Table 6.** ANOVA results after pooling the A $\times$ C factor with an F value less than 1 into the error term.

| Factor | SS | DF | MS | F |
|---|---|---|---|---|
| A | $8.13 \times 10^{-10}$ | 2 | $4.07 \times 10^{-10}$ | 39.55 |
| B | $2.01 \times 10^{-10}$ | 2 | $1.00 \times 10^{-10}$ | 9.75 |
| C | $2.70 \times 10^{-9}$ | 2 | $1.35 \times 10^{-9}$ | 131.45 |
| AB | $2.07 \times 10^{-9}$ | 4 | $5.18 \times 10^{-10}$ | 50.40 |
| BC | $1.91 \times 10^{-9}$ | 4 | $4.78 \times 10^{-10}$ | 46.53 |
| E | $7.28 \times 10^{-11}$ | 12 | $1.03 \times 10^{-11}$ | - |
| T | $7.78 \times 10^{-9}$ | 26 | - | - |

*3.4. Determination of Optimized Design*

According to Table 6, the factor with the greatest influence on the reduction of the strain deviation in the MD was the elastic modulus of the skin (C); the total diameter of the roller (B) had the least effect. However, because these factors interact, the interaction must be considered when determining the optimal shape. Table 7 shows the results of the MD strain deviation for the factor levels in Table 4. The combinations with the least strain deviation in the MD are the $A_1B_1$ and $B_3C_1$ combinations. The results in Table 7 alone cannot be used to select the value of factor B because of the interaction. However, according to Table 6, factor C had the greatest influence on the strain deviation in the MD. Therefore, the $B_3C_1$ combination was chosen first. The level of factor B was fixed at $B_3$, and the level of A was determined. Among the combinations including $B_3$, $A_3B_3$ produced the smallest MD strain deviation.

**Table 7.** Two-way factorial table based on levels in Table 4.

| | $A_1$ | $A_2$ | $A_3$ | $C_1$ | $C_2$ | $C_3$ |
|---|---|---|---|---|---|---|
| $B_1$ | 0.0096 | 0.0110 | 0.0214 | 0.0154 | 0.0122 | 0.0144 |
| $B_2$ | 0.0142 | 0.0165 | 0.0154 | 0.0106 | 0.0152 | 0.0202 |
| $B_3$ | 0.0141 | 0.0134 | 0.0127 | 0.0069 | 0.0133 | 0.0200 |

Finally, the $A_3B_3C_1$ combination was selected as the optimal design and applied to the FE model to confirm the corrective effects on the strain deviation in the MD. Figure 11 compares the strain deviations in the MD of the conventional and barrel rollers. The total deformation of the web is shown in Figure 12. The proposed design is effective according to these results.

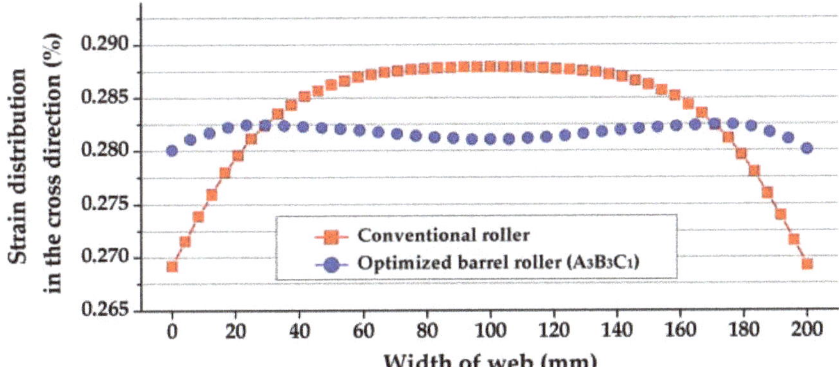

**Figure 11.** Comparison of strain distribution in the cross direction: Conventional versus optimized barrel roller.

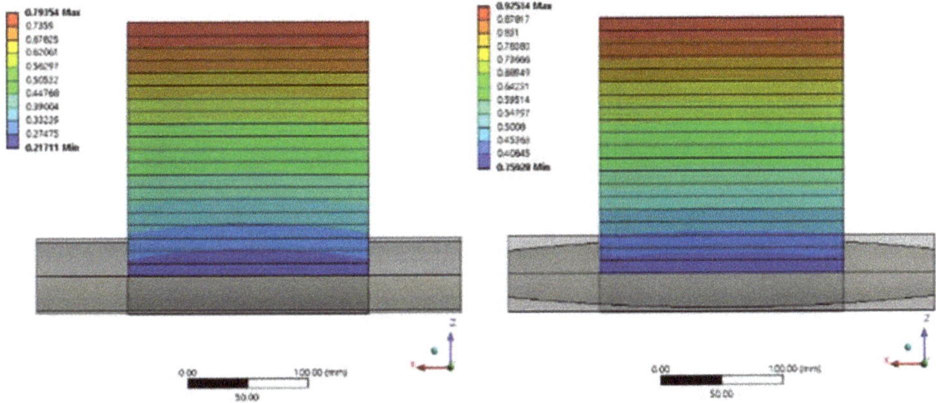

**Figure 12.** Comparison of total deformation results: Conventional versus optimized barrel roller.

## 4. Experiment

By manufacturing the optimized barrel roller, we were able to verify its compensation performance. For a span length of 200 mm, the average deviation was 8.54 μm for the conventional roller and 4.42 μm for the optimized barrel roller. This represents a compensation of approximately 48%. Under the same conditions as in the experiment, FEA produces a compensation performance of 85%.

### 4.1. Experiment for Verifying the Performance of Barrel Roller

We conducted an experiment to evaluate the compensation performance of the optimized barrel roller. The optimized barrel roller was a barrel-shaped $A_3B_3C_1$ combination; the thickness gradient of the skin was 8 mm, the total diameter of the roller was 60 mm, and the elastic modulus of the shell was 2.85 MPa (Figure 13).

**Figure 13.** Optimized barrel roller fabricated with final design parameters.

To evaluate the performance of the barrel roller, we conducted a comparative experiment by replacing the fixed-end roller with the barrel roller, as shown in Figure 14. The load-end roller was a conventional roller. Reference marks were also placed by a laser; the relative displacement of the laser marks was measured with a vision camera, and the strain was calculated. In this experiment, we used 28 marks—14 marks are close to the optimized barrel roller (fixed end) and 14 marks near the conventional roller (load end). The web used in the experiment was a PET film (Mitsubishi Inc. O321E188) with a width of 160 mm and a thickness of 188 μm. The experimental apparatus is shown in Figure 15.

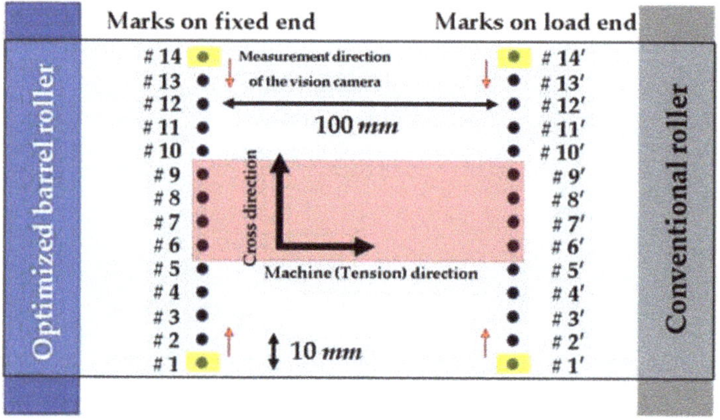


**Figure 14.** Schematic of web strain measurement (top view) with a barrel roller.

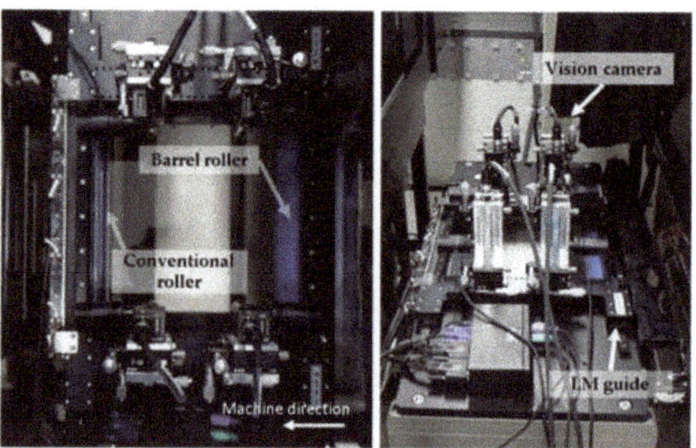


**Figure 15.** Experimental setup for strain measurement: linear motion (LM) guide and four vision cameras.

Tension in the film was engaged with dancer rollers and measured with load cells at both ends of the roller at the load end. In the FEA, the load condition was assumed to be 10 MPa; however, in the experiment, the maximum tension was limited to 5 MPa due to the load limit of the dancer roller.

Since we assumed that the deformation of the web and skin occurred only within the elastic region, a linear relationship between the strain and tension was assumed.

The vision cameras were positioned 50 and 150 mm away from the central axis of the roller and the web was transferred. When a reference mark entered the field of view of the vision camera, the transfer of the web was stopped and the position of the mark was measured. To measure the strain, a pair of reference marks at 100 mm intervals in the MD was pictured on the web under a load of approximately 3 MPa, and all reference positions were measured first. In the next step, the tension was increased to 5 MPa, and the positions of the marks were determined to calculate the strain. The procedure was repeated three times; marks #6 to #9 could not be measured due to mechanical interference between the vision cameras.

The experimental measurements of the strain deviation in the MD are shown in Figure 16 and Table 8. Figure 16 shows the relative displacements in the MD under tension condition at each reference mark. Table 8 compares the deviation in the relative displacements in the MD of each experiment. The mean deviations for the conventional and barrel rollers were 8.54 and 4.42 µm, respectively; thus, the deviation of barrel roller was 48% lower.

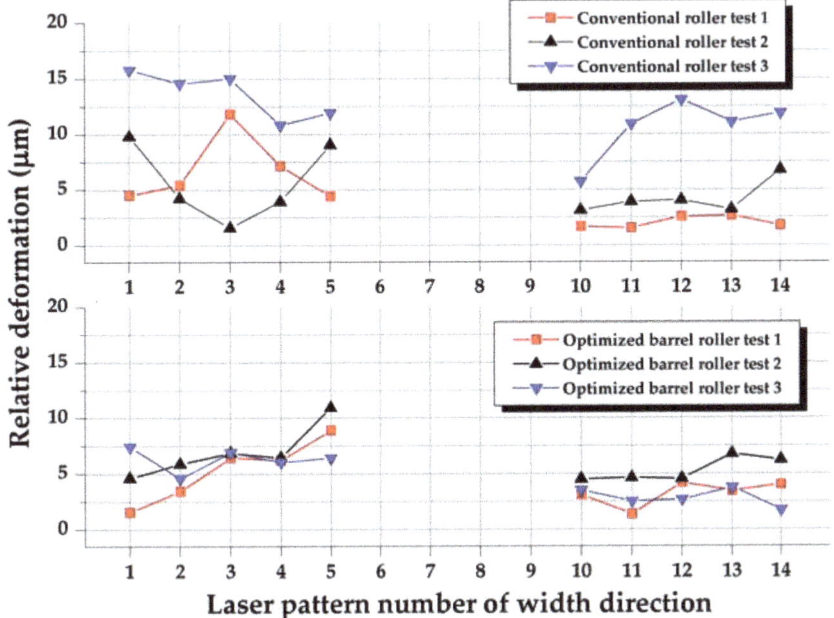


**Figure 16.** Comparison between relative displacements of conventional and optimized barrel rollers. The laser pattern number is an index in Figure 14. The spacing between the laser pattern numbers is 10 mm.

**Table 8.** Deviation of relative displacement in each experiment: Conventional versus optimized barrel roller.

| | Test No. | Deviation (µm) | Mean of Deviation (µm) |
|---|---|---|---|
| Conventional roller | 1 | 10.31 | |
| | 2 | 7.14 | 8.54 |
| | 3 | 8.18 | |
| Barrel roller | 1 | 5.38 | |
| | 2 | 3.79 | 4.42 |
| | 3 | 4.09 | |

*4.2. Comparison of Experimental and FEA Results*

In the DOE-based FEA, the tension was 10 MPa. However, the experiment was conducted under 5 MPa of tension due to the maximum load limitation of the dancer roller. In addition, reference markings for measuring the MD strain deviation were already engraved under a tension of 3 MPa. This is because the web is transferred to the patterning section. Therefore, in the verification experiment, the reference position of each mark was measured under 3 MPa of tension, and the strain was calculated by measuring the position of the mark after increasing the tension to 5 MPa. To compare the MD strain deviations of the experiments and FEA under the same conditions, an FE model with a tension of 5 MPa was constructed. In the 200 mm span, the printing lines were set at distances of 50 and 150 mm from the contact point of the roller and the web. The mechanical properties used in this FEA are shown in Table 9. Figure 17 shows the FEA results of the strain distribution in the MD for these conditions.

**Table 9.** Mechanical properties applied to FEA to simulate the MD strain deviation of the web under the same tension conditions as the experiment.

| | Young's Modulus | Mass Density | Poisson's Ratio |
|---|---|---|---|
| | E (GPa) | $\rho$ (kg/m$^3$) | - |
| Roll (aluminum) | 71 | 2,770 | 0.33 |
| Silicon rubber (skin) | 0.00285 | 1,200 | 0.5 |
| Web (PET) | 3.5 | 1,320 | 0.34 |

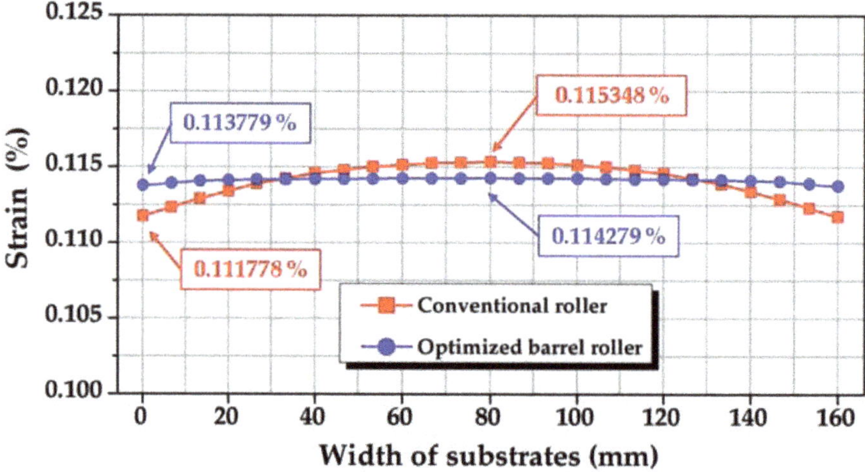


**Figure 17.** Comparison of FEA between the existing roller and barrel roller derived by calculating the MD strain distribution under 3 MPa of tension and then changing the tension to 5 MPa.

The strain deviations of the conventional and optimized barrel rollers were 0.00357% and 0.0005%, respectively. For the FEA model, the relative displacements were 7.14 and 1 μm, respectively. Based on these results, the compensation performance was approximately 85%.

There are a few explanations for the mismatch between the results of the experiment and the FEA. The reference marks were not placed exactly 50 mm from the central axis of the roller in the actual experiment. Additionally, uniform tension could not be applied with the dancer roller. Moreover, the reference marks may not have been symmetrically arranged around the center of the CD of the web, the laser mark array and the central axis of the rollers were misaligned, and the properties applied in the FEA were not perfectly accurate.

## 5. Conclusions

In this study, we analyzed the non-uniform strain distribution in a web in the roll-to-roll process. In addition, to mitigate the uneven strain distribution, we designed an optimized roller. According to the experimental results, strain uniformity is improved by approximately 48% with the proposed design compared to the conventional roller.

If the strain distribution in the transported web is non-uniform, a register error may occur in the machine direction (MD or transfer direction) of the web, reducing the shape accuracy or position precision of the print pattern. The distribution of the strain generated in a web being transferred under uniform tension is symmetrical along the width direction (CD) of the web with respect to the centerline of the web in the MD. Around this centerline, the strain is high and decreases toward both sides of the web. This deviation decreases as the web moves away from the roller and increases with the increasing web width. Improving this uniformity is important for productivity and precision.

In this study, we designed a novel roller to minimize the strain deviation in the MD. This roller consists of a rigid core and an elastic outer skin. The core is barrel-shaped and covered with the elastic skin. The assembly is a cylinder with a constant diameter. This configuration reduces the strain deviation by changing the contact stiffness between the web and the roller.

First, we used an FE model composed of a roller and web to simulate the strain due to tension in an actual process; based on this model, we optimized the roller design. For this purpose, we varied the thickness gradient of the skin, the total diameter of the roller, and the physical properties of the skin. Using DOE and ANOVA, we optimized the design to minimize the MD strain distribution at the printing line 50 mm from the roller central axis. The total diameter of the roller interacts with the thickness gradient; in addition, the mechanical properties of the skin interact with the diameter. The barrel roller with a thickness gradient of 8 mm, a roller diameter of 60 mm, and an elastic modulus of 2.85 MPa minimizes the strain distribution.

To compare the strain distributions of the conventional and optimized barrel rollers in the MD, we conducted a verification experiment. The rollers optimized with the FEA and DOE were employed, and the experiment was repeated three times for tensions of 3–5 MPa. The conventional and barrel rollers produce deviations of 8.54 and 4.42 μm, respectively. Thus, the compensation performance is improved by 48% by the barrel roller.

Our experiments had certain limitations; the center area could not be measured due to the mechanical interference between the vision cameras. This should be fixed with a jig in the future.

Despite the experimental limitations, the strain distribution is more uniform with the optimized than with the conventional roller. This is expected to improve the printing precision of fine patterns and to mitigate register errors in the multilayer printing process. The results of this study can promote the introduction of the roll-to-roll process into the industrial fabrication processes of FPCBs and displays.

**Author Contributions:** Conceptualization, funding acquisition, writing—review and editing, M.G.L.; writing—original draft, data curation, formal analysis, methodology, visualization, Y.K.; investigation, Y.K. and G.E.K.; supervision, project administration, M.G.L. and Y.J.; resources, S.K.; validation, M.G.L., Y.J., and H.J. All authors have read and agreed to the published version of the manuscript.

**Funding:** This study is based on work supported by the Ministry of Trade, Industry, and Energy (MOTIE, Korea) under the Industrial Technology Innovation Program No. 20000665. Development of an ecofriendly and highly durable surface treatment for super-omniphobic substrate on the large area over 4 m$^2$.



## References

1. Kang, H.; Lee, C.; Shin, K. A novel cross directional register modeling and feedforward control in multi-layer roll-to-roll printing. *J. Process Control* **2010**, *20*, 643–652. [CrossRef]
2. Yoon, J.S.; Kim, J.; Kang, B.S. Deformation analysis and shape prediction for sheet forming using flexibly reconfigurable roll forming. *J. Mater. Process. Technol.* **2016**, *233*, 192–205. [CrossRef]

3.  Bui, Q.V.; Ponthot, J.P. Numerical simulation of cold roll-forming processes. *J. Mater. Process. Technol.* **2008**, *202*, 275–282. [CrossRef]

4.  Lif, J.O. Hygro-viscoelastic stress analysis in paper web offset printing. *Finite Elem. Anal. Des.* **2006**, *4*, 341–366. [CrossRef]

5.  Parola, M.; Vuorinrn, S.; Linna, H.; Kaljunen, T.; Beletski, N. Modelling the web tension profile in a paper machine. In Proceedings of the Science of Papermaking, XII Fundamental Research Symposium, Oxford, UK, 19–21 September 2001; Volume 2, pp. 759–781.

6.  Parola, M.; Kaljunen, T.; Vuorinrn, S. New method for the analysis of the paper web performance on the press. In Proceedings of the 27th IARIGAI Research Conference, Graz, Austria, 10–13 September 2000; pp. 203–218.

7.  Hu, L.; Chen, W.; Liu, Z. An active register control strategy for multilayer R2R printed electronics based on microscope vision. In Proceedings of the 13th IEEE Conference on Industrial Electronics and Applications (ICIEA), Wuhan, China, 31 May–2 June 2018.

8.  Jung, H.; Nguyen, H.A.D.; Choi, J.; Yim, H.; Shin, K. High-precision register error control using active-motion-based roller in roll-to-roll gravure printing. *Jpn. J. Appl. Phys.* **2018**, *57*, 05GB04. [CrossRef]

9.  Kang, H.; Lee, C.; Shin, K. Modeling and compensation of the machine directional register in roll-to-roll printing. *Control Eng. Pract.* **2013**, *21*, 645–654. [CrossRef]

10.  Chang, J.; Lee, S.; Lee, K.B.; Lee, S.; Cho, Y.T.; Seo, J.; Lee, S.; Jo, G.; Lee, K.Y.; Kong, H.S.; et al. Overlay accuracy on a flexible web with a roll printing process based on a roll-to-roll system. *Rev. Sci. Instrum.* **2015**, *86*, 55108. [CrossRef] [PubMed]

11.  Kim, C.; Jeon, S.W.; Kim, C.H. Reduction of Linearly Varying Term of Register Errors Using a Dancer System in Roll-to-Roll Printing Equipment for Printed Electronics. *Int. J. Precis. Eng. Manuf.* **2019**, *20*, 1485–1493. [CrossRef]

12.  Lee, J.; Seong, J.; Park, J.; Park, S.; Lee, D.; Shin, K.H. Register control algorithm for high resolution multilayer printing in the roll-to-roll process. *Mech. Syst. Signal. Process.* **2015**, *60–61*, 706–714. [CrossRef]

13.  Kim, S.; Na, Y. Study on the web deformation in ink transfer process for R2R printing application. *Int. J. Precis. Eng. Manuf.* **2010**, *11*, 945–954. [CrossRef]

14.  Zillmann, B.; Wagner, M.F.X.; Schmaltz, S.; Schmidl, E.; Lampke, T.; Willner, K.; Halle, T. In-plane biaxial compression and tension testing of thin sheet materials. *Int. J. Solids. Struct.* **2015**, *66*, 111–120. [CrossRef]

15.  Montgomery, D.C. *Design and Analysis of Experiments*, 8th ed.; John Wiley & Sons, Ltd.: Chichester, UK, 2012; pp. 68–73, 183–192, 206–208.

**Publisher's Note:** MDPI stays neutral with regard to jurisdictional claims in published maps and institutional affiliations.

 *applied sciences*

*Article*

# Analysis of the Influence of High Peening Coverage on Almen Intensity and Residual Compressive Stress


Zhaorui Yang [1], Youngseog Lee [2], Shangwen He [1,*], Wenzhen Jia [1] and Jun Zhao [1]

[1]  School of Mechanics and Safety Engineering, Zhengzhou University, Zhengzhou 450001, China; zryang@zzu.edu.cn (Z.Y.); jiawenzhen1993@163.com (W.J.); zhaoj@zzu.edu.cn (J.Z.)

[2]  Department of Mechanical Engineering, Chung-Ang University, Seoul 06974, Korea; ysl@cau.ac.kr

*  Correspondence: hsw2013@zzu.edu.cn; Tel.: +86-1393-902-1862



Received: 1 November 2019; Accepted: 18 December 2019; Published: 21 December 2019



**Abstract:** The effectiveness of shot peening is mainly determined by the peening coverage. The peening coverage is required to be 100% for current technical standards of shot peening. With the increase of material strength, higher peening coverage is required in shot peening process. However, the influence of high peening coverage on Almen intensity and residual compressive stress is unclear, the difficulty mainly lies in the lack of quantitative description of peening coverage in finite element analysis. To analyze the influence of high peening coverage on Almen intensity and residual compressive stress, firstly an approximate quantitative description of peening coverage based on dent size, the distance of shots and shot numbers is proposed in this study. Based on this quantitative description of peening coverage, the arc height and residual stress of the Almen test are simulated with the finite element method. The simulation results of arc height and saturation curve agree well with that of the Almen test, by which the effectiveness of the quantitative description and FE simulation are proved. The further study indicates that in shot peening processes, the excessive peening coverage doesn't improve Almen intensity and residual compressive stress.

**Keywords:** shot peening; quantitative description of peening coverage; high peening coverage; Almen intensity; residual compressive stress


## 1. Introduction

Shot peening is a mechanical surface treatment that induces a surface layer of residual compressive stress (RCS). In the shot peening process, the surface of metal parts such as gears, springs, or turbine blades is sprayed by numerous small round metal balls and RCS is generated on the surface. RCS embedded in the surface plays a role in preventing the occurrence of cracks so that the fatigue strength of the pinched parts is improved [1–4].

Engineers in charge of shot peening usually to conduct the Almen test before shot peening is carried out [5]. In this test, a thin strip (dimensions of 76.2 mm × 18.9 mm with three commercially available thicknesses: 0.79, 1.29, and 2.39 mm) is peened with the same peening conditions as the actual parts of a given time period. The thin strip is called the "Almen strip" in the test. The Almen strip then bents since RCS induced by the peening leads the Almen strip to deflect from the peening direction [6]. The ejection of the balls is called impact, the balls are called shots. The maximum value of deflection of the Almen strip is called the arc height. In the Almen test, with the impact time increasing, the peening coverage increases and the arc height saturates. The value of the arc height at saturation of Almen strip is referred to as "Almen intensity".

In shot peening process, it is difficult to test the RCS. Therefore, Almen intensity has been used as the standard for shot peening, which determines the shot peening parameters. Guechichi et.al. [7] established a framework for predictive modeling of shot velocity. They found a relationship between Almen intensity and shot velocity. The effects of Almen intensity on microstructure and residual stress were investigated experimentally [8,9]. Ahmed et.al. [10] investigated the influence of the shot peening parameters on the hardness, residual stresses, and surface roughness. Jebahi et.al. [11] proposed a comprehensive methodology to simulate a real shot peening process with minimal computation effort. Atig et.al. [12] proposed a probabilistic methodology to evaluate the variability of the induced residual stress and Almen intensity of the shot peening parameters. Based beam bending principles, Divid et al. [13] quantify the relationship between the several parameters affecting measured Almen intensity. Cao et al. [14] proposed an analytic model to study the correlation of the Almen intensity with RCS, and compared the model-predictions with the measurements as a function of impact velocity, but a new problem was raised in assessing and interpreting the Almen intensity scale. Guagliano [15] performed a 3D finite element (FE) analysis to relate the Almen intensity with the terms of shot velocity, but didn't present a relation between the Almen intensity and RCS. Bhuvaraghan et al. [16] improved FEM-based approach that simulates the actual process of Almen strip peening by employing randomly located shots impacting on a part of Almen strip and the inclusion of strain-rate dependent target material properties. Bhuvaraghan et al. [17] used the discrete element method in combination with the finite element method to obtain reasonably accurate predictions of the residual stresses and plastic strains. Hu et. al. [18] examined the influences of the dimension and shape on the residual stress with 3D random impact FE model. In general, there have not been deeper studies about the influence of peening coverage on Almen intensity and RCS recently due to a lack of quantitative description of peening coverage.

During the strip test or general shot peening test, numerous shots are arbitrarily bumped onto the Almen strip or specimen surface in a random sequence and random locations [19–21]. Therefore, the peening coverage is difficult to define in FE simulation of shot peening. Miao et al. [22] proposed a 3-D finite element model that consists of some identical shots impacting randomly an aluminum target at normal or oblique incidence angles. The random coordinate of each shot that satisfied 100% peening coverage was generated from a MATLAB program combined with ANSYS program Design Language. However, Miao et al.'s model, the minimum distance between all the existing shots was set to be 0.75 mm, which is arbitrary. The radius of all shots was 0.5 mm and the maximum number of shots used for simulation was 96. Hence, Miao et al.'s model might apply to a unit cell-based 3-D finite element analysis where an impact region to be discretized into a fine mesh is very small, e.g. 2 mm × 2 mm × 1 mm. Majzoobi et. al. [23] and Klemenz et. al. [24] extended Kubler et. al.'s [25] work by developing a process map for random impact locations generation ensuring that no two shots occupies the same space mathematically, and applied it to simulate the Almen strip response under random multiple impacts. A slice of the Almen strip subjected to random multiple impacts was modeled. The radius of the shot used as 0.1778 mm. However, the deflected profile of the Almen strip computed using the FE model with random impact locations was not compared with the measured profile. Mylonas and Labeas [26] proposed a methodology that minimizes a total number of shots, and hence makes it possible practically to simulate the multiple shot impacts, which have statistics characteristics. They showed that predetermined shot patterns (locations and sequence of the multiple shots) on a reference area of 1 mm$^2$ could predict RCS distribution, surface roughness, and cold work by comparing the model-predictions with measurements. The previously mentioned methods demonstrate that the FE model based on randomly located impacts showed a new possibility to achieve a realistic model of shot peening. The relation of impact locations and sequence of shots has been investigated in these studies, which establishes the foundation for a quantitative description of peening coverage in this paper.

In all, the influence of high peening coverage on Almen intensity and RCS has not been studied in detail. The reason is that, in finite element analysis, it is difficult to describe the peening coverage quantitatively. In this study, the boundary conditions of the FE simulation of the Almen test were set up accord to reference [27]. By the definition of peening, peening coverage is the ratio of the area of the dents to the zone of the peening, a quantitative description of the peening coverage is proposed. The size of the peening zone is determined by the distance between two shots. After determining the size of the dent, the high peening coverage can be achieved by increasing the number of shots. This quantitative description peening coverages used to the simulation results of arc height and saturation curves are in accord with the Almen test results well. Therefore, the effectiveness of this quantitative description and FE simulation is proved. To analyze the influence of peening coverage on Almen intensity and residual compressive stress, further study is carried out. Through the discussion of the FE simulation and Almen test results, some new conclusions are obtained.

## 2. A Quantitative Description of the Peening Coverage

In this section, the peening coverage is described quantitatively by analyzing the effect of the distance between shots, dent size, and the number of shots on the peening coverage. Generally speaking, peening coverage is defined as the percentage of a surface area indented. For shot peening, the peening coverage is determined by three factors, (a) dent diameter, (b) the number of dents in each unit area of the peening surface, and (c) the time of peening. In this study, 'cycle' is used to describe the time of peening for FE simulation.

Before the peening coverage is described quantitatively, some assumptions should be put forward about locations and sequence of shots impact. Firstly, the locations and sequence of the multiple shots impacting is not generated randomly. Secondly, peening coverage depends on both dent size and the number of dents generated on Almen strip surface in the unit area, rather than the impact sequence. Thirdly, the locations of the multiple shots are defined in the same position and the shots impact one by one.

With assumptions above, the process of shots impacting is analyzed. Figure 1 shows the Schematic of dent size on Almen strip surface in 1-cycle, the second, fourth and fifth row of shots is omitted in Figure 1 for convenience. In Figure 1a, the shot balls are adjacent to each other in the row and impact Almen strip surface at the same time. The width of the shadow is defined as the size of each dent ($D_s$), and the distance between the center of the shots is defined as $C_d$. Figure 1b–d shows the process of dent expansion. In Figure 1d, peening coverage reaches 100%. This impact pattern, from the first to the sixth row of shots, is called one 'impact cycle'. The overlap of the dents is inevitable in the peening process, however, with the reasonable distance between shots, there is little overlap when full peening coverage is reached. Therefore, higher peening coverage could be got by the repetition of this cycle.

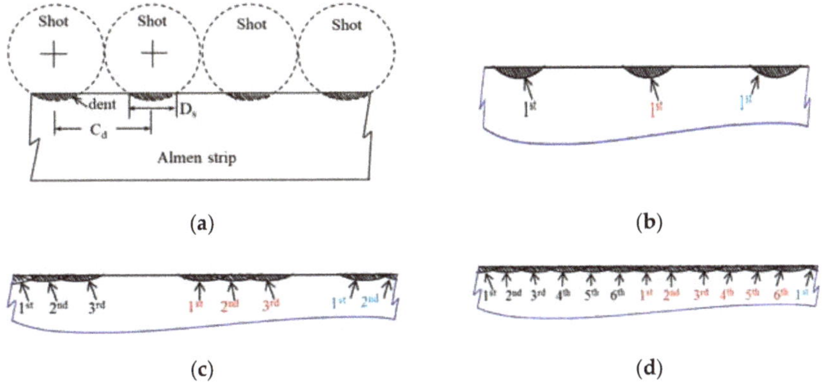


**Figure 1.** Schematic of dent size on the surface of Almen strip in 1-cycle (100% peening coverage). (**a**) the first row impacts the Almen strip surface, (**b**) the dent position of the first-row impact, (**c**) the dent position of the third-row impact, (**d**) the dent position of the 1-cycle.

With the analysis above, a quantitative description of peening coverage is proposed, and the peening coverage is computed by Equation (1).

$$R_o = \frac{D_s \cdot N_l - C_d}{N_l} \times \frac{1}{D_s} \times 100\%,\tag{1}$$

$$N_l = a \cdot \frac{C_d}{D_s},\tag{2}$$

where $R_o$ is the peening coverage, $D_s$ is dent size, $C_d$ is the distance between the centers of shots; and $N_l$ defined in Equation (2) characterizes the number of dents between the centers of the shot. As shown in Equation (2), the number of dents $N_l$ is determined by $C_d$, $D_s$, and the coefficient of proportionality ($a$). The coefficient of proportionality can be determined by the overlapping dents of the shot impacts [28]. The dent size is related to the size of the ball and the impact velocity. The horizontal distance between the centers of shots ($C_d$) is twice the radius of the shots. The effectiveness of the quantitative description of peening coverage will be proved in Section 3.

## 3. Comparison of Finite Element Simulation of Almen Test and Almen Test

Based on the quantitative description of peening coverage (Equation (1)), the finite element simulation of the Almen test is carried out. Meanwhile, the Almen test is also carried out. By comparing the results of the finite element simulation and Almen test, the effectiveness of the quantitative description of peening coverage and the FE simulation are proved.

### 3.1. FE Simulation of Almen Test

The commercial finite element program, ABAQUS®, which is suitable for analyzing the non-linear elastic-plastic deformation of metals subjected to multi-collision shot peening is introduced in this section [29]. The explicit time integration scheme is employed to solve the equation of motion for the shot-strip system. The element type used for the Almen strip is CPE4R (plane strain element). The shot bolls are treated as rigid bodies as their hardness is much greater than that of the strip.

### 3.1.1. Boundary Conditions

Figure 2 shows the initial mesh configuration and boundary conditions of the Almen strip when the modified strip holder [27] is employed, with which the boundary conditions can be greatly simplified. Traction free boundary condition is given to the surface of the Almen strip where shot balls are impacted, so is the opposite side of the Almen strip. U1 and U2 denote the displacement toward x-direction and y-direction, respectively. UR1, UR2, and UR3 stand for the rotation of the Almen strip on its x-direction and y-direction, and z-direction.

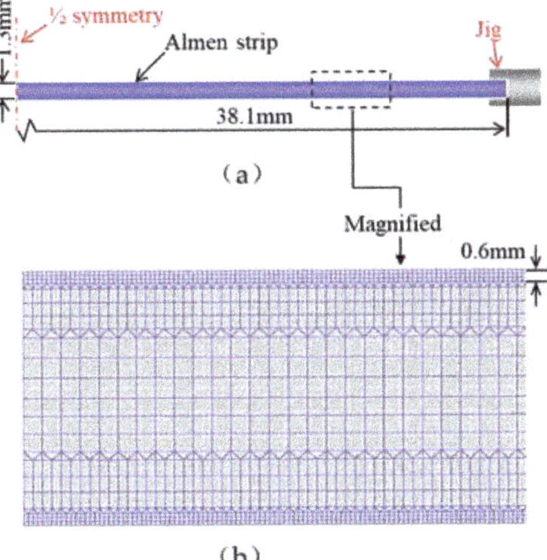



**Figure 2.** (a) Boundary condition of the Almen strip when a modified Almen strip holder is used; (b) Enlarged meshes at a part.

Enlarged meshes at a part are shown in Figure 2b. Since elastic-plastic deformation during shot peening remains in the thin layer of the Almen strip where shots are bombarding, the dense mesh is assigned in the thin layer zone and the coarse mesh is made in the middle zone of the Almen strip. Transition mesh is built between two zones.

During shot peening to avoid the stresses in the Almen strip oscillating around a mean value, the coefficient of material damping $\xi = 0.5$ is chosen [30]. The friction force of the relative motion of the shots and Almen strip surface is assumed to obey Coulomb's law. Meo and Frija [31,32] reported that the effect of the Coulomb friction coefficient on the residual stress and plastic strain is insignificant if it is in the range of 0.1–0.5. Coulomb friction coefficient $\mu = 0.2$ is used in this study. The temperature variation raised by impacting is neglected.

### 3.1.2. Material Model

In the previous studies [16,33–35], strain rate dependency of the Almen strip on the deformation behavior was assumed negligible even though high strain rates were engaged in the Almen strip during shot peening. However, the strain rate will have effects on the deformation profile, i.e. arc height, of the Almen strip. Hence constitutive equation which relates stress and strain to the current conditions of strain rate is necessary to describe the deformation behavior of the Almen strip under shot peening. In this study, the Johnson-Cook model is used for high strain rate deformation.

$$\sigma = \left[A + B(\bar{\varepsilon})^n\right]\left[1 + C\ln\left(\frac{\dot{\bar{\varepsilon}}}{\dot{\varepsilon}_0}\right)\right]\left[1 - \left(\frac{T - T_0}{T_m - T}\right)^m\right] \tag{3}$$

$A$, $B$, $C$, $m$ and $n$ are material constants to be determined from deformation tests. $\dot{\bar{\varepsilon}}_0$ and $T_0$ are reference values of strain rate and temperature, respectively. $\dot{\bar{\varepsilon}}$ and $T$ is the strain rate and temperature, respectively. $T_m$ is the melting temperature of the material.

The material parameters of the Johnson-Cook model used in this study were taken from the literature [16]. Figure 3 shows the strain rate dependent flow stress behavior of the Almen strip used in FE simulation.

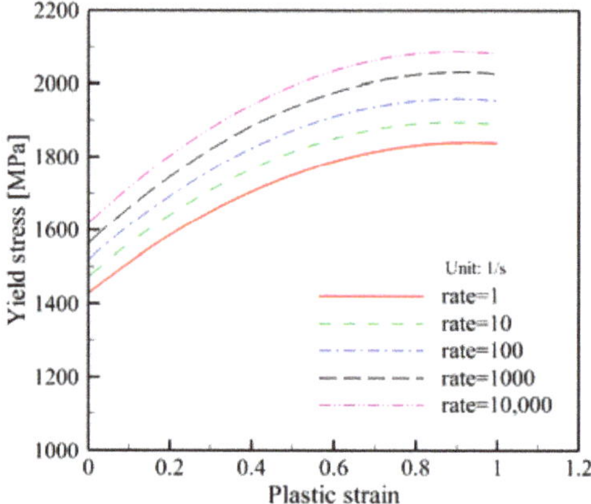

**Figure 3.** Variation of yield strength with strain rate.

### 3.1.3. Mesh Size

In finite element simulation of shot peening, the most challenging aspect is the evaluation of mesh size. In general, much finer elements have been preferred in the impacting area and coarser elements made in the area far from the impacting region. Since the size of the smallest element in explicit FE simulation determines run time, we cannot reduce thoughtlessly the element size. There are several studies [36–38] that assess the mesh size in FE simulation of the multiple impacts in terms of dimple (dent or indentation) diameter. The deformation of the mesh caused by the impact of shots is shown in Figure 4. The legend is the equivalent plastic strain (PEEQ) around the deformation area.

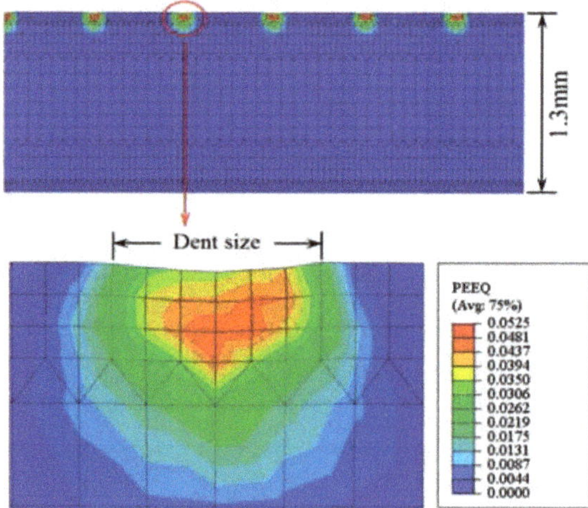

**Figure 4.** Schematic of dimple (dent or indentation) diameter on the Almen strip by a single shot.

Zimmerman et al. [35] chose an element size equal to 1/15th of dimple diameter produced by a single shot impact. The dimple diameter indicates the magnitude of the concave surface due to the single shot impact. Klemenza et al. [24] selected the size of elements that equals 1/10th of dimple diameter. In the Bagherifard et al.'s [37] works, to predict the generation of a nanostructure surface layer of material, the element size in the impact zone was selected to be 1/20th of dimple diameter.

In this study, where 2-D multiple impacts simulation is performed, the size of elements in the impact zone is chosen to be 0.02 mm, which is about 1/6th of dimple diameter.

### 3.1.4. Impact Sequence and Position

With the quantitative description of peening coverage, the impact sequence and position of shots for FE simulation could be determined. Figure 5 shows the schematic of the impact sequence and position in 1 impact cycle. The impact cycle is repeated to simulate 6 layers impacting irregularly on the Almen strip surface. In each layer, the distance between the two shots is 0.6 mm. To study the influence of high peening coverage on Almen intensity and RCS, higher peening coverage could be got by increasing the number of impact cycles. In the Almen strip test, impact velocity is 25, 30, and 35 m/s, respectively, dent size is 0.125, 0.14, and 0.16 mm in FE simulation.

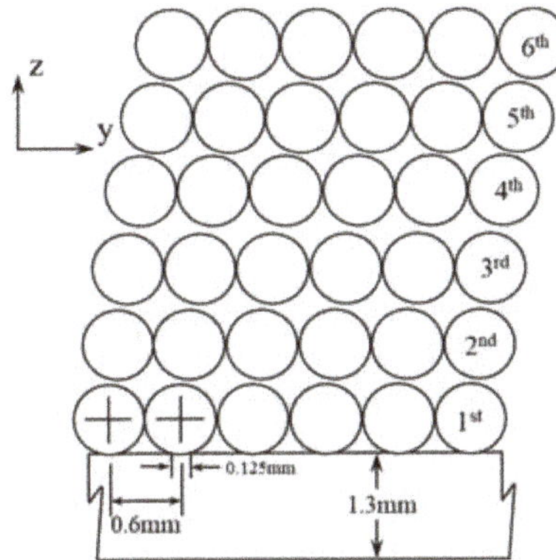


**Figure 5.** The impact sequence and position for 100% peening coverage at 1-cycle.

### 3.2. Almen Test

An air-compression type shot peening equipment is developed for this research. A cut wire round shot S230 (Diameter = 0.6 mm) is used for the shot materials in the Almen test. Mass flow rates of shots are approximately 2.5 kg/min. This equipment can propel shots with sizes up to 1/16 inch in diameter. The desired impact velocities are obtained by adjusting the air pressure and impact distance. A high-speed camera (Motion Xtra HG-LE, Redlake (USA)) is employed to measure the impact velocity (Figure 6). A number of frames used for measuring the impact velocity are 10,000 fps. When air pressure is 0.3, 0.5, 0.7 MPa, the impact velocity measured is 25, 30, and 35 m/s respectively. The distance from the nozzle to the surface of the Almen strip is 100 mm and the impact angle is 90°, which is typical for shot peening tests.

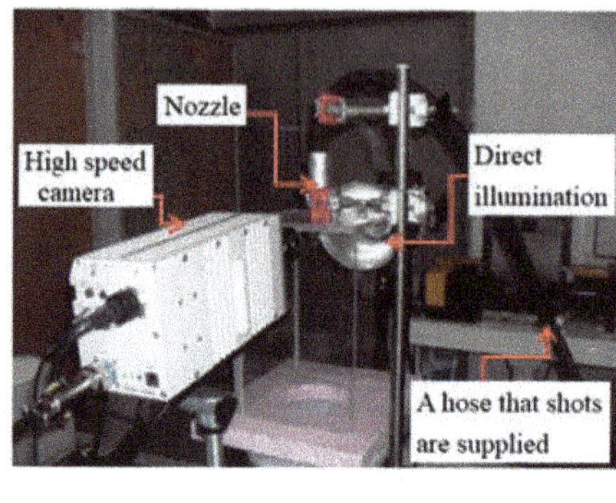


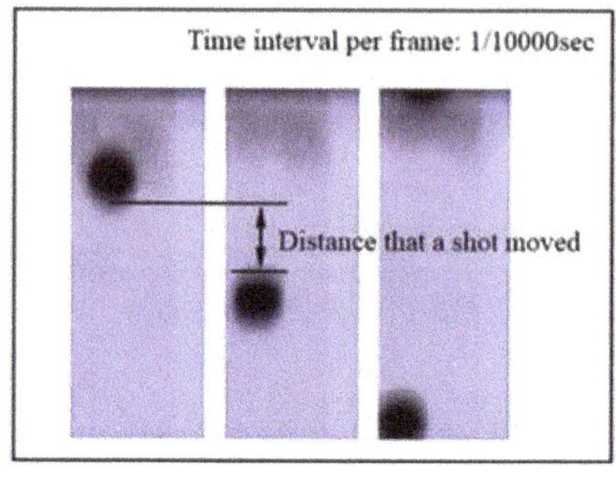


(**a**)                                                        (**b**)

**Figure 6.** Impact velocity is calculated by measuring the distance a shot traveled. (**a**) A high-speed camera is employed to measure the impact velocity, (**b**) measure the distance between the two balls to get the impact velocity.

SAE 1070 type-A Almen strip has been used [39]. The dimensions of the Almen strips are 76.2 (±0.38 mm) in length, 18.9mm (±0.064 mm) in width and 1.3 mm (±0.025 mm) in thickness. The material propert is listed of SAE 1070 steel in Table 1.

**Table 1.** Mechanical properties of SAE 1070 steel.

| Property | Value |
| --- | --- |
| Density | 7800 kg/m$^3$ |
| Poisson's Ratio | 0.29 |
| Young's Modulus | 205 GPa |
| Yield Strength | 1268 MPa |
| Ultimate Strength | 1422 MPa |
| Elongation | 8.2% |

In Almen test, the peening coverage of the Almen test is obtained by observation. Before peening, a penning area of the Almen strip surface is coated. After peening, the peened surfaces are observed with 10× to 30× magnification to verify the required level of peening coverage. Viewing the coating surfaces to determine the area of tracer removal, the all tracer of the Almen strip surface is removed at 5 s when the peening coverage reaches 100%. As the impact time is 10 s, the peening coverage is considered as 200%.

*3.3. Comparison of Results of FE Simulation and Almen Test*

The effectiveness of the quantitative description of peening coverage and the FE simulation is proved by comparing the Almen strip's deflection profile of the Almen test and FE simulation.

In the Almen test and FE simulation, the impact velocities are both 25, 30, and 35 m/s. In the Almen test, peening coverage reaches 100% when the impact time is 5 s, while in the FE simulation, the peening coverage is also 100%. The Almen strip deflection in the Almen test and FE simulation is shown in Figure 7.

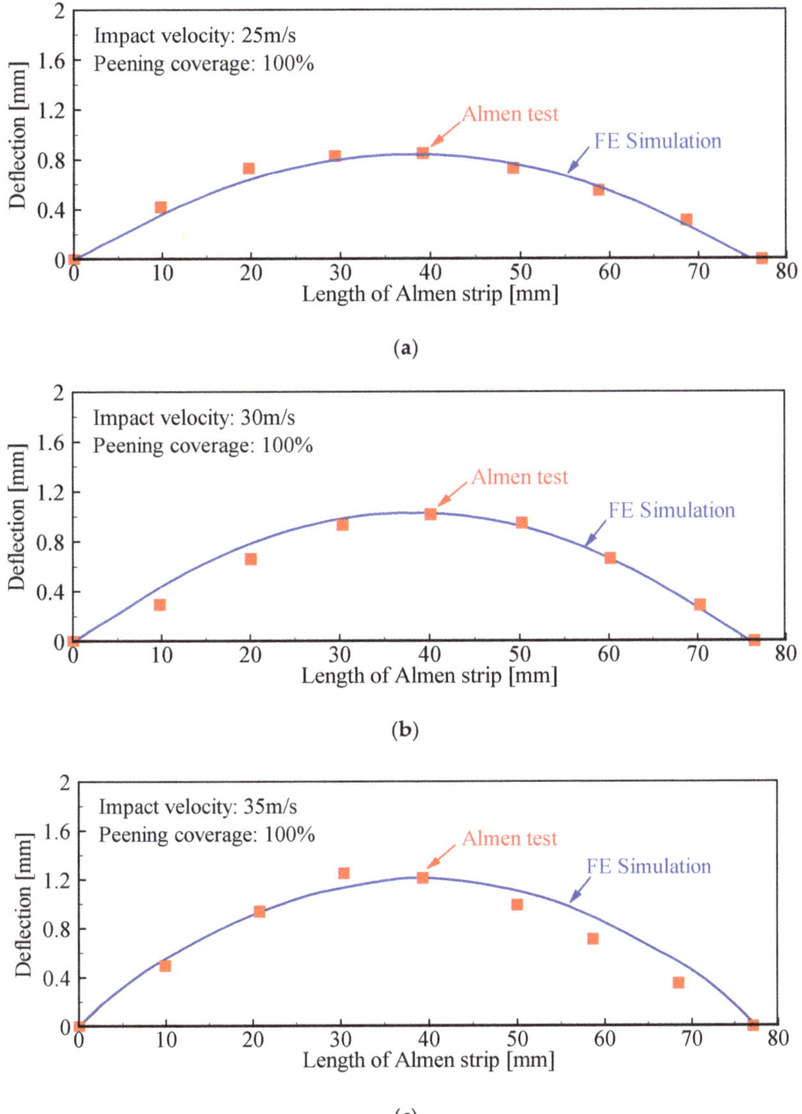



**Figure 7.** Variations of the Almen strip deflection profile at different impact velocities. (**a**) with impact velocity 25 m/s, (**b**) with impact velocity of 30 m/s, (**c**) with impact velocity of 35 m/s.

In Figure 7, The red dot points mark the data of the Almen test, while the blue line represents the result of the FE simulation. The simulation results agree well with that of the Almen test. Only small deviation arises near the center of the Almen strip, which may be due to the measurement error of the Almen test. The accuracy of arc height is very high at the center of the Almen strip.

The arc height (maximum value of the deflection) of the Almen strip is measured in the Almen test with different impact velocity of 25, 30, and 35 m/s, and the same impact velocity is used in the FE simulation, respectively. In Almen test the impact time is 5,10,20 and 40 s, the corresponding peening coverage is 100%, 200%, 400%, 800% respectively. The saturation curve of arc height in the Almen test and FE simulation is shown in Figure 8.

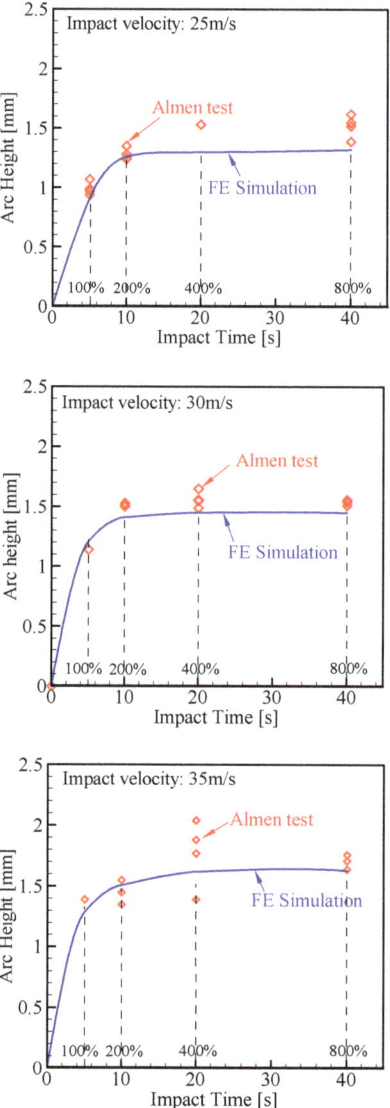

**Figure 8.** Variations of the saturation curves of the arc height at different impact velocities.

In Figure 8, the red dot points mark the arc height in the Almen test, while the blue line represents the arc height in the FE simulation. With the increase of peening coverage, the arc height both increases and then gradually saturate in Almen test and FE simulation. At the impact velocity of 30 m/s, the results of the arc height of the Almen test agree with that of FE simulation best, while at the impact velocity of 25 and 35 m/s, the simulation results basically agree with the Almen test results.

In all, with different conditions, the results of the Almen test and FE simulation are in good agreement, which indicates that the quantitative description of peening coverage and the FE simulation is effective.

## 4. Analysis of the Influence of High Peening Coverage on Almen Intensity and RCS

### 4.1. Influence of High Peening Coverage on Almen Intensity

In this section, the influence of the high peening coverage on the Almen intensity with three different dent sizes is studied.

The peening coverage increases from 200% to 800%. The dent size of three cases is 0.125, 0.14, and 0.16 mm. With three dent sizes, the variation of the Almen intensity with the peening coverage is shown in Figure 9, respectively. Figure 9a shown the variation of the Almen intensity at 0.125mm of dent size. Figure 9b,c shows the dent size at 0.14 and 0.16mm, respectively.

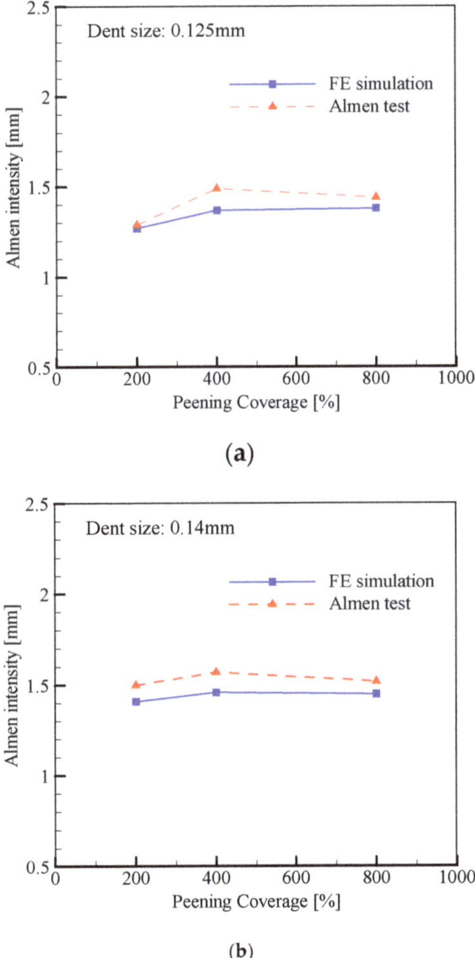


(a)



(b)

**Figure 9.** *Cont.*

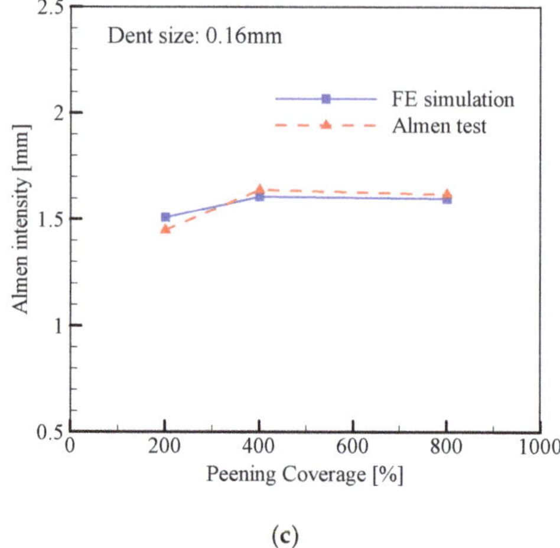

**(c)**

**Figure 9.** Variations of the Almen intensity with various peening coverage at different dent size. (**a**) 0.125mm, (**b**) 0.14mm, (**c**) 0.16mm.

From Figure 9, firstly, with dent size of 0.125, 0.14, and 0.16 mm, the FE simulation results agree well with that of the Almen test. Secondly, with dent size of 0.125, 0.14, and 0.16 mm, the Almen intensity all increases firstly and then decreases as the peening coverage increases, there is a maximum value of Almen intensity. In other words, it's not that the higher the peening coverage, the higher the Almen intensity, which is an important reference for the shot peening process in industry. The reason for the decrease of the Almen intensity with the peening coverage increasing might be the residual stress relaxation of the Almen strip surface.

*4.2. Influence of High Peening Coverage on RCS*

In this section, as the RCS at different depth is difficult to obtain in the Almen test, the distribution of residual compressive stress (RCS) at different depth with the increase of peening coverage is analyzed in FE simulation. Negative value of the RCS represents compression.

The peening coverage is 100%, 200%, 400% and 800%. The dent size of three cases is 0.125, 0.14, and 0.16 mm. With three dent sizes, the distribution of the RCS at different depth is shown in Figure 10 respectively. Figure 10a shown distribution of the RCS at 0.125mm of dent size. Figure 10b,c shows the dent size at 0.14 and 0.16mm, respectively.

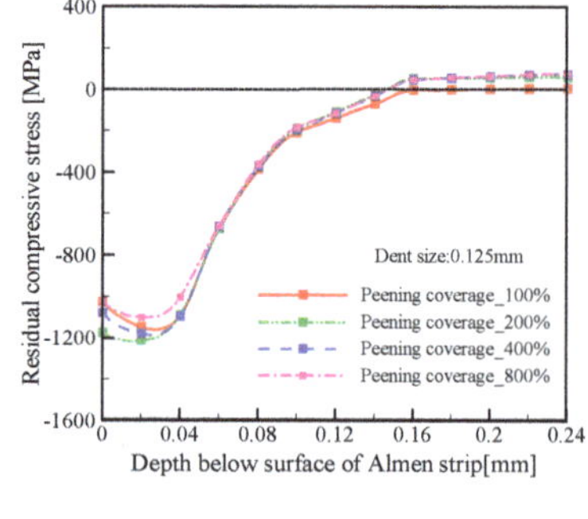

**(a)**

**Figure 10.** *Cont.*

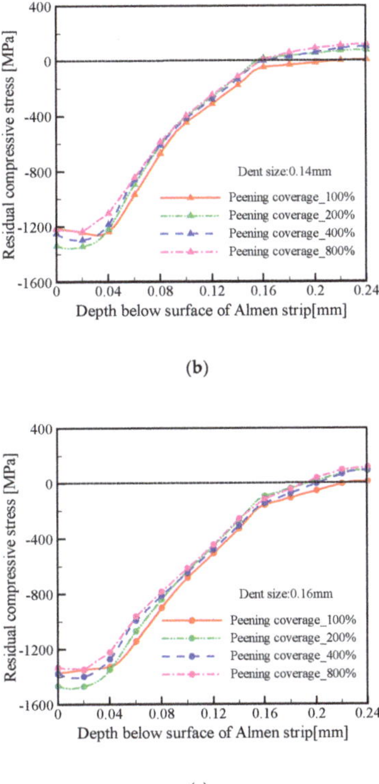

(b)

(c)

**Figure 10.** Residual compressive stress profiles for different impact velocity and peening coverage. (**a**) 0.125 mm, (**b**) 0.14 mm, (**c**) 0.16 mm.

From Figure 10, firstly, with dent size of 0.125, 0.14, and 0.16 mm, the absolute value of RCS firstly increases and then decreases with the depth increasing, there is a maximum absolute value of RCS around the depth of 0.02 mm. Secondly, with dent size of 0.125, 0.14, and 0.16 mm, at certain depth, the absolute value of RCS does not always increase with the peening coverage increasing.

With dent size of 0.125, 0.14, and 0.16 mm, a more detailed analysis of peening coverage and RCS is shown in Figure 11. Surface RCS is the RCS of the Almen strip surface, while the maximum RCS is the minimum value of RCS at different depth as the value of RCS is negative. Figure 11a shows the relation of the surface RCS and the peening coverage, while Figure 11b shows the relation of the maximum RCS and the peening coverage.

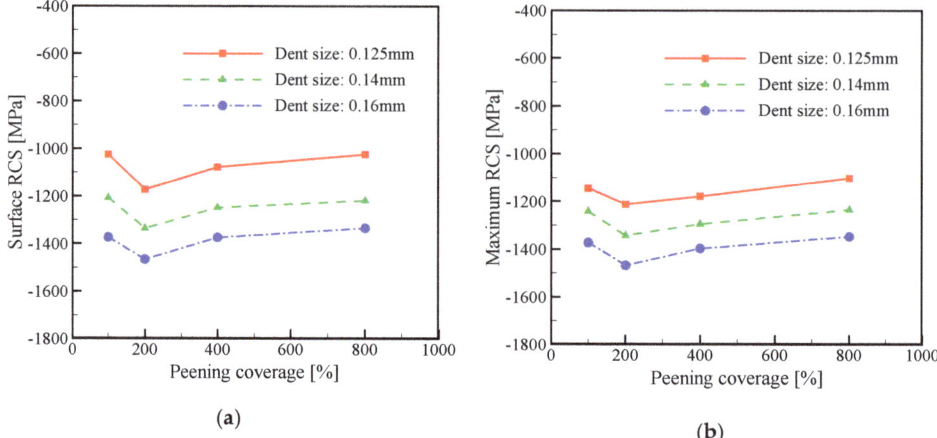

**Figure 11.** Residual compressive stress variation of different peening coverage. (**a**) surface residual compressive stress, (**b**) Maximum residual compressive stress.

From Figure 11, firstly, with dent size of 0.125, 0.14, and 0.16 mm, the absolute value of surface RCS and maximum RCS all increase firstly and then decrease as the peening coverage increases, there is a maximum absolute value of surface RCS and maximum RCS. Secondly, with certain peening coverage, the absolute value of surface RCS and maximum RCS increases as dent size increases.

## 5. Conclusions

To determine the influence of high peening coverage on Almen intensity and RCS, a quantitative description of the peening coverage has been developed in this paper firstly. Based on the quantitative description, the finite element simulation of the Almen test and Almen test are carried out. The following conclusions can be drawn:

1. With the same peening coverage, arc height of FE simulation and Almen test agrees well, which indicates that the quantitative description of peening coverage and the FE simulation is effective.
2. With dent size of 0.125, 0.14, and 0.16 mm, the Almen intensity all increases firstly and then decreases with the peening coverage increasing, there is a maximum value of Almen intensity, which is very important for shot peening process.
3. With dent size of 0.125, 0.14, and 0.16 mm, there is a maximum absolute value of surface RCS and maximum RCS with the peening coverage increasing. With certain peening coverage, the surface RCS and maximum RCS increases as dent size increases.

Before this study, it is difficult to predicate the influence of high peening coverage on the Almen intensity and RCS well. This study can supply an effective method to solve this problem.

**Author Contributions:** Z.Y. conceived the paper. Z.Y., Y.L. and S.H. wrote the paper and clarified the key methods and results. W.J. helped in the F.E. simulation. Z.Y., S.H. and J.Z. edited the manuscript. All authors have read and agreed to the published version of the manuscript.

**Funding:** The authors are grateful to the support by National Key R&D Program of China [Grants No. 2016YFE0125600] and the National Natural Science Foundation of China [Grants No. 51405452].

**Conflicts of Interest:** The authors declare no conflict of interest.

## References

1. Dimakos, K.; Mariotto, A.; Giacosa, F. Optimization of the fatigue resistance of nitinol stents through shot peening. *Procedia Struct. Integr.* **2016**, *2*, 1522–1529. [CrossRef]

2.  Gholami, M.; Altenberger, I.; Kuhn, H.A.; Wollmann, M.; Wagner, L. Effects of shot peening on the fatigue performance of ultrafine-grained CuNi3Si1Mg. *Surf. Eng.* **2017**, *33*, 706–714. [CrossRef]

3.  Segurado, E.; Belzunce, F.J.; Pariente, I.F. Effects of low intensity shot peening treatments applied with different types of shots on the fatigue performance of a high-strength steel. *Surf. Coat. Technol.* **2018**, *340*, 25–35. [CrossRef]

4.  Ferreira, N.; Antunes, P.; Ferreira, J.; Costa, J.; Capela, C. Effects of Shot-Peening and Stress Ratio on the Fatigue Crack Propagation of AL 7475-T7351 Specimens. *Appl. Sci.* **2018**, *8*, 375. [CrossRef]

5.  John, O.A.; Royal, O.M. Shot Blasting Test. U.S. Patent Office No. 440,987, 25 February 1944.

6.  Miao, H.Y.; Larose, S.; Perron, C.; Lévesque, M. An analytical approach to relate shot peening parameters to Almen intensity. *Surf. Coat. Technol.* **2010**, *205*, 2055–2066. [CrossRef]

7.  Guechichi, H.; Castex, L.; Benkhettab, M. An analytical model to relate shot peening Almen intensity to shot velocity. *Mech. Based Des. Struct.* **2013**, *41*, 79–99. [CrossRef]

8.  Unal, O.; Varol, R. Almen intensity effect on microstructure and mechanical properties of low carbon steel subjected to severe shot peening. *Appl. Surf. Sci.* **2014**, *290*, 40–47. [CrossRef]

9.  Unal, O. Optimization of shot peening parameters by response surface methodology. *Surf. Coat. Technol.* **2016**, *305*, 99–109. [CrossRef]

10. Ahmed, A.A.; Mhaede, M.; Basha, M.; Wollmann, M.; Wagner, L. The effect of shot peening parameters and hydroxyapatite coating on surface properties and corrosion behavior of medical grade AISI 316L stainless steel. *Surf. Coat. Technol.* **2015**, *280*, 347–358. [CrossRef]

11. Jebahi, M.; Gakwaya, A.; Lévesque, J.; Mechri, O.; Ba, K. Robust methodology to simulate real shot peening process using discrete-continuum coupling method. *Int. J. Mech. Sci.* **2016**, *107*, 21–33. [CrossRef]

12. Atig, A.; Sghaier, R.B.; Seddik, R.; Fathallah, R. Probabilistic methodology for predicting the dispersion of residual stresses and Almen intensity considering shot peening process uncertainties. *Int. J. Adv. Manuf. Tech.* **2018**, *94*, 2125–2136. [CrossRef]

13. David, K. Quantification of Shot Peening Intensity Rating. *Shot Peener Mag.* **2015**, *32*, 26–34.

14. Cao, W.; Fathallah, R.; Castex, L. Correlation of Almen arc height with residual stresses in shot peening process. *Mater. Sci. Technol.* **1995**, *11*, 967–973. [CrossRef]

15. Guagliano, M. Relating Almen intensity to residual stresses induced by shot peening: A numerical approach. *J. Mater. Process. Technol.* **2001**, *110*, 277–286. [CrossRef]

16. Bhuvaraghan, B.; Srinivasan, S.M.; Maffeo, B.; Prakash, O. Analytical solution for single and multiple impacts with strain-rate effects for shot peening. *Comp. Model. Eng. Sci.* **2010**, *57*, 137.

17. Bhuvaraghan, B.; Srinivasan, S.M.; Maffeo, B.; McCLain, R.D.; Potdar, Y.; Prakash, O. Shot peening simulation using discrete and finite element methods. *Adv. Eng. Softw.* **2010**, *41*, 1266–1276. [CrossRef]

18. Hu, D.Y.; Gao, Y.; Meng, F.C.; Song, J.; Wang, Y.F.; Ren, M.X.; Wang, R.Q. A unifying approach in simulating the shot peening process using a 3D random representative volume finite element model. *Chin. J. Aeronaut.* **2017**, *30*, 1592–1602. [CrossRef]

19. Marini, M.; Fontanari, V.; Bandini, M.; Benedetti, M. Surface layer modifications of micro-shot-peened Al-7075-T651: Experiments and stochastic numerical simulations. *Surf. Coat. Technol.* **2017**, *321*, 265–278. [CrossRef]

20. Tu, F.; Delbergue, D.; Klotz, T.; Bag, A.; Miao, H.; Bianchetti, C.; Brochu, M.; Bocher, P.; Levesque, M. Discrete element-periodic cell coupling model and investigations on shot stream expansion, Almen intensities and target materials. *Int. J. Mech. Sci.* **2018**, *145*, 353–366. [CrossRef]

21. Bagherifard, S.; Ghelichi, R.; Guagliano, M. On the shot peening surface coverage and its assessment by means of finite element simulation: A critical review and some original developments. *Appl. Surf. Sci.* **2012**, *259*, 186–194. [CrossRef]

22. Miao, H.Y.; Larose, S.; Perron, C.; Lévesque, M. On the potential applications of a 3D random finite element model for the simulation of shot peening. *Adv. Eng. Softw.* **2009**, *40*, 1023–1038. [CrossRef]

23. Majzoobi, G.H.; Azizi, R.; Nia, A.A. A three-dimensional simulation of shot peening process using multiple shot impacts. *J. Mater. Process. Technol.* **2005**, *164*, 1226–1234. [CrossRef]

24. Klemenz, M.; Schulze, V.; Rohr, I.; Löhe, D. Application of the FEM for the prediction of the surface layer characteristics after shot peening. *J. Mater. Process. Technol.* **2009**, *209*, 4093–4102. [CrossRef]

25. Kubler, R.F.; Berveiller, S.; Bouscaud, D.; Guiheux, R.; Patoor, E.; Puydt, Q. Shot peening of TRIP780 steel: Experimental analysis and numerical simulation. *J. Mater. Process. Technol.* **2019**, *270*, 182–194. [CrossRef]

26. Mylonas, G.I.; Labeas, G. Numerical modelling of shot peening process and corresponding products: Residual stress, surface roughness and cold work prediction. *Surf. Coat. Technol.* **2011**, *205*, 4480–4494. [CrossRef]
27. Yang, Z.R.; Park, J.S.; Lee, Y. A strip holding system for finite element simulation of Almen strip testing. *J. Mech. Sci. Technol.* **2014**, *28*, 2825–2830. [CrossRef]
28. David, K. The importance of work. *Shot Peen. Mag.* **2014**, *31*, 26–36.
29. Hibbit, H.D.; Karlsson, B.I.; Sorensen, E.P. *ABAQUS User Manual*; Version 6.12; Simulia: Providence, RI, USA, 2012.
30. Kim, T.; Lee, J.H.; Lee, H.; Cheong, S.K. An area-average approach to peening residual stress under multi-impacts using a three-dimensional symmetry-cell finite element model with plastic shots. *Mater. Des.* **2010**, *31*, 50–59. [CrossRef]
31. Meo, M.; Vignjevic, R. Finite element analysis of residual stress induced by shot peening process. *Adv. Eng. Softw.* **2003**, *34*, 569–575. [CrossRef]
32. Frija, M.; Hassine, T.; Fathallah, R.; Bouraoui, C.; Dogui, A.; de Génie Mécanique, L. Finite element modelling of shot peening process: Prediction of the compressive residual stresses, the plastic deformations and the surface integrity. *Mat. Sci. Eng. A* **2006**, *426*, 173–180. [CrossRef]
33. Hong, T.; Ooi, J.Y.; Shaw, B.A. A numerical study of the residual stress pattern from single shot impacting on a metallic component. *Adv. Eng. Softw.* **2008**, *39*, 743–756. [CrossRef]
34. Meguid, S.A.; Shagal, G.; Stranart, J.C. 3D FE analysis of peening of strain-rate sensitive materials using multiple impingement model. *Int. J. Impact Eng.* **2002**, *27*, 119–134. [CrossRef]
35. Meguid, S.A.; Maricic, L.A. Finite element modeling of shot peening residual stress relaxation in turbine disk assemblies. *J. Eng. Mater. Technol.* **2015**, *137*, 031003. [CrossRef]
36. Zimmermann, M.; Schulze, V.; Baron, H.U.; Löhe, D. A novel 3D finite element simulation model for the prediction of the residual stress state after shot peening. In Proceedings of the 10th International Conference on Shot Peening, Tokyo, Japan, 15–18 September 2008.
37. Bagherifard, S.; Ghelichi, R.; Guagliano, M. A numerical model of severe shot peening (SSP) to predict the generation of a nanostructured surface layer of material. *Surf. Coat. Technol.* **2010**, *204*, 4081–4090. [CrossRef]
38. Ullah, H.; Ullah, B.; Muhammad, R. Dynamic numerical simulation of plastic deformation and residual stress in shot peening of aluminum alloy. *Struct. Eng. Mech.* **2017**, *63*, 1–9.
39. SAE. *Test. Strip, Holder and Gage for Shot Peening*; Society of Automotive Engineers, Inc.: Warrendale, PA, USA, 2004; p. J442.

*Article*

# The Impact of the Strength of Roof Rocks on the Extent of the Zone with a High Risk of Spontaneous Coal Combustion for Fully Powered Longwalls Ventilated with the Y-Type System—A Case Study

**Magdalena Tutak [1,*] and Jarosław Brodny [2]**

[1]  Faculty of Mining, Safety Engineering and Industrial Automation, Silesian University of Technology, 44-100 Gliwice, Poland
[2]  Faculty of Organization and Management, Silesian University of Technology, 44-100 Gliwice, Poland; jaroslaw.brodny@polsl.pl
*  Correspondence: magdalena.tutak@polsl.pl; Tel.: +48-322-372-528

Received: 14 November 2019; Accepted: 3 December 2019; Published: 5 December 2019

**Abstract:** During the ventilation of longwalls in hard coal mines, part of the air stream migrates to the goaves with caving. These goaves constitute a space (void) filled with rocks following coal extraction. In the case where these goaves contain coal susceptible to spontaneous combustion, the flow of such an air stream through the goaves may lead to the formation of favourable conditions for coal oxidation, self-heating and spontaneous combustion. Such an area is referred to as the zone with a particularly high risk of spontaneous coal combustion (endogenous fires). The location and extent of this zone depend on many factors, with one of the most important being the permeability of the goaves which determines the tensile strength of the roof rocks forming the caving. This strength determines the propensity of these rocks to transform into the state of caving and the degree of tightness of the cave-in rubble (treated as a porous medium). The purpose of the present paper is to determine how the tensile strength of roof rocks influences the extent of the zone with a particularly high risk of spontaneous coal combustion (endogenous fires) in caving goaves of the longwalls ventilated with the Y-type system. To achieve this goal, model-based tests were conducted for a region of the longwall mined with caving and ventilated with the Y-type system. Critical air speed and oxygen concentration values in the caving goaves of this longwall were determined for the actual conditions of exploitation. These parameters define the risk zone of spontaneous coal combustion. The tests also helped to determine the extent of this zone, depending on the strength of the rocks forming the caving. The results obtained unequivocally indicate that the type of rocks forming the caving affects its permeability and the extent of the risk zone for spontaneous coal combustion. At the same time, the distribution of this zone is substantially different than in the case of other ventilation systems. The results obtained are of real practical significance for preventive measures to reduce fire risks. The effectiveness of these measures significantly improves the safety of mining exploitation.

**Keywords:** numerical modeling; finite volumne method; underground coal mine; endogenous fires; spontaneous combustion; longwall; ventilation system

---

## 1. Introduction

There are various types of rocks forming the hard coal seams in Poland. They include claystones, coal shales, mudstones, sandstones and, where the ongoing exploitation is divided into layers, also hard coal. All of these rocks have different strength properties [1,2].

So far, the strength properties of the rocks surrounding mine headings have been considered primarily in terms of their impact on the capacity to maintain the stability of mine headings [3–5].

When coal is exploited with caving, the analysis also encompasses their capacity to transform into the state of caving and to maintain the roof in a longwall heading (classification by Salustowicz [6]). The strength properties of these rocks were, therefore, mainly examined in terms of rock mass mechanics.

From the perspective of ensuring safe exploitation in longwalls, the impact of the strength of roof rocks should be considered not only in terms of maintaining the stability of the mine headings, but also in terms of the occurrence of ventilation hazards, and—more precisely—the risk of spontaneous coal combustion (endogenous fires) in the caving goaves of longwalls.

Spontaneous combustion of coal is a phenomenon that commonly occurs in hard coal mines [7–11]. It results from the self-ignition of coal following its self-heating in a mine heading or its immediate surrounding, e.g., in the goaves with caving.

Goaves with caving are a space formed after coal extraction, filled with rock rubble from the collapse of roof rocks hanging over the exploited seam. The degree to which this space is filled once the coal has been extracted depends on a number of factors. However, the most important of these is the type of roof rocks forming the goaves [12–14].

The parameter that is critical for the propensity of these rocks to transform into the state of caving is their tensile strength [15]. It determines the filling (tightness) degree of the cave-in rubble, and hence its porosity and permeability. Strong and solid roof rocks separating from the rock mass fill goaves with caving to a lesser extent than weak and brittle rocks. Therefore, goaves with caving filled with roof rocks of low tensile strength provide better filling of the space created after the mined coal (due to their lower permeability).

However, regardless of the type of rocks forming the goaves with caving, such goaves always include void spaces not filled with any rock material. These spaces, representing open contacts between chaotically arranged blocks of fractured roof rocks, form a particular type of a porous medium which allows for the flow of gases, including the air migrating from the longwalls.

The flow of air through goaves with caving results from its migration from the area of the longwall to which a stream of ventilation air is supplied. This stream is supplied to longwalls for their ventilation. There are several longwall ventilation systems, with the most common being the U-type ventilation system from the borders and the Y-type ventilation system [16] (Figure 1). Analysing both of these systems, it is possible to see a clear difference in the flow of air through the exploited headings.

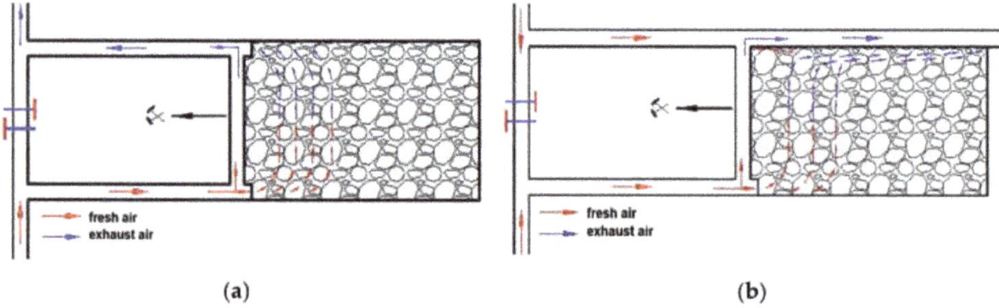


**Figure 1.** Diagram of the U-type (**a**) and Y-type ventilation systems (**b**).

The choice of a longwall ventilation system depends on a number of factors. The ventilation system must ensure proper chemical composition and temperature of the atmosphere [17–19]. One of the most important is the level of natural hazards, including gas-related risks, which occur in the region of active exploitation. In practice, there is no system that would be beneficial in the event of simultaneous occurrence of the self-combustion risk and the methane risk for the longwall under exploitation.

A system advantageous in the case of spontaneous combustion risk is less suitable for areas with methane hazard and vice versa. In longwalls with a high methane hazard which is typical of mines, it is increasingly common to use Y-type ventilation systems with air discharge along the goaves.

However, this system is characterised by a significantly higher migration of air to the goaves with caving compared to the U-type ventilation system. This offers more favourable conditions for the self-heating of coal left in the goaves, which, in turn, may cause an endogenous fire to arise.

The research conducted by Szlązak [20] revealed that the total migration of air to the goaves with caving depending on the type of roof rocks filling the goaves with caving may, in extreme cases, amount to as much as approximately 40% (for rocks with the highest value of tensile strength).

The air stream migrating to the goaves with caving poses a risk for a low-temperature process of coal oxidation to be initiated, which may lead to spontaneous combustion of the coal left in the goaves. The prerequisites for this process to be initiated, besides the presence of coal, include the flow of air with a specific speed and appropriate oxygen concentration. When these conditions are met, it is possible for the reaction of low-temperature coal oxidation to be initiated, during which heat is produced and then accumulated by the coal, thereby causing its temperature to rise. If these conditions continue for a specific period of time (the incubation time), spontaneous combustion of coal, i.e., an endogenous fire, may occur.

The most essential factor affecting the process of heat accumulation is the speed of the air stream flowing through the goaves with caving in the longwall. This speed depends on the type of roof rocks forming the caving (since they influence their sealing degree) as well as on the volumetric flow rate of the air supplied to the longwall. However, the issue of how the volumetric flow rate of the air supplied to the longwall impacts the risk of endogenous fires has been discussed in several publications [21–24].

Nevertheless, no conclusive range has been determined for the air speed flowing through the goaves that would contribute to the initiation and maintenance of the coal oxidation process.

In the paper by Cheng et al., this value was assumed to range from 0.004 to 0.0016 m/s [21]. Chumak et al. [22], on the other hand, assumed that the critical speed value ranges from 0.015 to 0.0017 m/s, whereas Szlązak reported that this value is between 0.015 and 0.0015 m/s [23]. On the other hand, Wang et al. [24] indicated that this value ranges from 0.001 to 0.02 m/s. The speed ranges indicated are quite extensive. Therefore, the present paper assumes that this speed ranges from 0.02 m/s to 0.0015 m/s.

In the case of the oxygen concentration in the air flowing through goaves with caving, it has been demonstrated that the lower limit for self-combustion of coal is 8%. The results of the tests carried out by Buchwald [25] indicate that no spontaneous combustion of coal occurs below this value due to the insufficient concentration of oxygen.

Coal oxidation in the goaves with caving may occur only in the area of the goaves which meets both of the above conditions, namely the presence of crushed coal susceptible to spontaneous combustion and the air flowing through the goaves at a specific speed and with a specific oxygen concentration. This area could be termed as the zone with a particularly high risk of endogenous fires. In this zone, the physical and chemical parameters of the air reach values conducive to the initiation of the oxidation process.

This was used as the basis for formulating the risk criterion for spontaneous coal combustion (endogenous fire) in the goaves with caving, which includes:

- the presence of fragmented coal left in the goaves with caving.
- the speed of the air flowing through the goaves with caving must range from 0.0015 to 0.02 m/s.
- the level of oxygen concentration in the air flowing through the goaves with caving should be higher than 8%.

Works involving determination of the distribution of physical and chemical parameters of the air flowing through the goaves with caving have already been published. However, they only concerned the determination of speed distributions and oxygen concentrations in the goaves of longwalls ventilated with the U-type system [26–35]. These papers failed to take into account the type of roof rocks forming the goaves with caving, which affect their permeability, and hence the possibility of air to migrate inside the goaves, and the distributions of parameters.

One of the first papers dedicated to a three-dimensional analysis of air flow through the goaves of a longwall with caving was by T. Ren and R. Balusu [24]. In this paper, a spatial model was used to present the results of numerical tests concerning the distribution of oxygen concentration in the goaves of a longwall with caving after inertisation. The subsequent works of the same authors [27,28] also present the distribution of oxygen concentration in the goaves with caving. However, the determination of this distribution served as an introduction to numerical tests related to various methods for supplying an inert gas both into longwall headings and through the holes drilled from the surface.

On the other hand, Esterhuizen and Karacan [29], using their own model for determining the permeability of goaves with caving, carried out numerical research and determined the speed of the air flowing through the goaves. They concluded that this speed reaches the highest value at the borderline of the goaves (at the starting line of the longwall and behind the longwall lining).

For their own model-based tests, Yuan and Smith [30,31] used the permeability model of the goaves with caving, created by Esterhuizen and Karacan [29]. These tests were related to the self-heating of coal left in the goaves with caving and to the determination of the temperature in those goaves. The tests were based on chemical reactions during which heat is released into the atmosphere upon contact of coal with oxygen. This served as the basis for determining the dependency between the oxidation rate and temperature, and oxygen concentration.

Another work which attempted to examine the flow of air through a three-dimensional model of goaves with caving was the one by Dai et al. [32]. It presented the distributions of the air speed in the goaves of a longwall with caving ventilated by means of the U-type system at two different flow heights, namely 1.5 m and 3.0 m from the floor of the exploited seam. In this work, the dangerous speed value of the air flowing through the goaves with caving, conducive to the self-heating of coal, was assumed to be equal to 0.004 m/s.

Tests on the flow of air through goaves with caving of a longwall, using a three-dimensional model, were also carried out by Xie et al., who presented the results of such tests in the paper [30]. They built a numerical model reflecting a real-world longwall and goaves with caving. The tests they conducted helped them to determine the distribution of air speed in the goaves with caving and the distribution of air pressure.

On the other hand, Shi et al., in the paper [34], presented the results of model-based tests on the distribution of oxygen concentrations in the goaves of a longwall with caving. They conducted these tests on a three-dimensional model which made it possible to determine the distribution of oxygen concentration in the goaves with the values from 8% to 18%. This concentration poses a risk that the oxidation process of the coal left in the goaves with caving could be initiated.

On the other hand, Brodny and Tutak [35] determined the impact of the volumetric flow rate of the air stream supplied to the longwall on the speed of the air filtrating through goaves and on the concentration of oxygen in this air.

Analysing the papers published to date, it is possible to conclude that none of them has determined how the type of the roof rocks forming the goaves with caving impact the formation of the zone with a particularly high risk of spontaneous coal combustion. This zone has also not been considered in terms of the Y-type ventilation system.

Therefore, the Authors conducted model-based tests whose purpose was to determine the impact of the type of roof rocks forming the goaves with caving on the formation of the zone with a particularly high risk of spontaneous combustion of the coal left in the goaves of longwalls ventilated with the Y-type system.

It is practically impossible to determine the zone with a particularly high risk of spontaneous coal combustion in real-world conditions because this zone is formed in an inaccessible area of the goaves. The attempts to measure the ventilation parameters in the goaves made to date have been unsuccessful in the majority of cases. For this reason, this zone was demarcated using model-based tests, which are successfully used for variant analyses of the processes related to ventilation of underground mine headings, as well as for analyses of emergency states occurring in these headings [36–38].

The tests were conducted for the actual layout of headings in one of the longwalls of a hard coal mine. The tests were based on the geometry of this longwall and the ventilation parameters registered during its exploitation. The tests (boreholes in the roof) also helped to define the strength parameters of the roof rocks forming the caving.

The main purpose of the works performed was to develop a methodology of model-based tests for spatial analysis of the ventilation phenomenon related to the identification of the area in the goaves with caving, where it is possible for spontaneous coal combustion, i.e., an endogenous fire, to occur.

In order to specify the impact of the type of roof rocks forming the goaves with caving on the location and extent of the zone with a particularly high risk of spontaneous coal combustion in the goaves, additional analyses were also conducted for five different tensile strengths of the rocks. The analysis was based on the geometry and ventilation parameters of the longwall in question. A total of six variants were considered for the tensile strength of roof rocks, and the dependency between this strength and the zone with a particularly high risk of spontaneous coal combustion in the goaves were determined.

## 2. Materials and Methods

### 2.1. The Porosity and Permeability of Goaves with Caving

One of the most important properties of roof rocks determining their ability to transform into caving is the tensile strength. This strength is the natural ability of the rock mass to resist stratification and caving of the roof rocks into the space (void) left after the mined coal as a result of vertical forces [39,40].

The value of the tensile strength of the roof rocks is determined by means of a down-hole penetrometer or the direct method—by stretching the sections of the vertical core of the borehole in the direction of the longitudinal axis of the borehole, and then it is determined from the following relationship:

$$R_{rri} = 0.8 \frac{F}{d^2} \tag{1}$$

where $R_{rri}$ is tensile strength of the rocks (Pa), $F$ is the applied axial load (N) and $d$ is core diameter (m$^2$).

This value depends on the type of roof rocks forming the caving. The maximum value of the tensile strength of rocks in Polish mines amounts to approximately 8 MPa [1,2]. In practice, however, such value is rare. Table 1 presents the types of roof rocks and the values of their tensile strength, as well as the characteristics of the roofs formed by these rocks.

**Table 1.** The types of roof rocks and the values of their tensile strength (own study based on [1,2]).

| Tensile Strength of Roof Rock, $R_{rrs}$, MPa | Description of Roof | Example of Rock |
|---|---|---|
| 0–0.5 | Roof falling immediately after unveiling | clay and sandy slates, coals |
| 0.5–1.5 | Falling roof (clay and sandy slates) and weakly self-supporting (coal) | clay and sandy slates, coals |
| 1.5–3.0 | Cracked roof, partially self-supporting and bearing, easily passing into a caving state | shales, sandy shale, coals |
| 3.0–4.5 | Self-supporting roof, it goes automatically into a caving state without sagging into goaves | coarse-grained sandstones |
| 4.5–6.0 | Supporting roof, without roof falls, hardly passing into caving state, sagging into goaves | sandstones (medium) |
| ➢ 6.0 | Strongly compacted roof, very difficult for passing into caving state | sandstones (hard) |

After such calculation of the tensile strength of roof rocks, it is possible to determine the permeability coefficient of goaves with caving, using the following equation [41]:

$$k(x) = \frac{\mu_g}{r_0 + ax^2} \ \text{for} \ 0 \leq x \leq 2/3 \cdot l \tag{2}$$

as well as the equation:

$$k(x) = \frac{\mu_g}{r_0 + a\left(\frac{4}{3}l - x\right)^2} \text{ for } 2/3 \cdot l \leq x \leq l \tag{3}$$

where $k(x)$ is permeability (m$^2$), $\mu_g$ is the coefficient of dynamic viscosity of air (Nsm$^{-2}$), $l$ the total length of the longitudinal longwalls (m), $r_0$ we determine from dependence $r_0 = \frac{\mu}{k_0}$ and $a$ we determine from dependence $a = 6 \cdot 10^9 R_{rrs}^{-1.74}$.

The value of the permeability coefficient of caving goaves $k_0$ behind the front of the longwall is determined from the following equation [41]:

$$k_0 = \frac{\mu_g}{6} \cdot 10^{-10} R_{rrs}^{1,44} \tag{4}$$

The porosity of goaves varies on the basis of "O-zone theory" [42]. The porosity distribution along the strike direction in the middle of the working face goaves is determined from the following equation [41]:

$$n_x = 0.2e^{-0.0223x} + 0.1 \tag{5}$$

where $n_x$ is the porosity distribution along the middle line of working face of longwall in goaves (%); $x$ is the x position of the goaves (m).

The porosity of goaves in dip distribution can be determined from the following equation [43]:

$$n_y = e^{-0.015y} + 1 \text{ for } 0 < y < \frac{L}{2} \tag{6}$$

$$n_y = e^{-0.015(L-y)} + 1 \text{ for } \frac{L}{2} < y < L \tag{7}$$

where $n_y$ is the porosity distribution along dip direction (%), $y$ is the y position of the goaves (m); $L$ is the length of the working face of longwall (m).

*2.2. Methods*

The objective of the tests conducted was to determine the impact of the type of roof rocks forming the goaves with caving on the extent of the zone with a particularly high risk of spontaneous combustion of coal in the goaves of a longwall ventilated with the Y-type system.

The analyses were conducted for a spatial model representing a real-world longwall along with longwall headings and goaves with caving, making use of Computational Fluid Dynamics (CFD). The Authors' experiences and the results obtained by other researchers indicate that this method may be used successfully for such analyses of the phenomena related with the flow of gases and the transfer of mass and heat [44].

The analyses were carried out by means of the ANSYS Fluent 18.2 commercial software. This software uses the finite volume method (FVM) for discretisation of the geometric model. The methodology for conducting tests by means of this programme involves development of a geometric model, a discrete model and a mathematical model of the phenomenon in question, as well as adoption of boundary conditions, performance of calculations and analysis of the results obtained. The most important stages of the methodology for the tests conducted are briefly discussed in the subsequent chapters of the article.

In the case at hand, this methodology is also supplemented with tests in real-world (actual) conditions. This is because the results of these tests serve as the basis for developing a geometric model for the region under analysis and for adopting the boundary conditions. The process of analysing the results of model-based tests also involves their verification with reference to real-world conditions.

### 2.2.1. Mathematical Models

The flow of air stream through the longwall and the longwall headings is deemed to be of a turbulent nature, while the flow of air stream through the goaves with caving of a laminar nature.

The mathematical mapping of a model of a longwall region and the goaves with caving takes the form of a set of equations which describe the aforementioned types of flows. These equations describe the flow of the mixture of air and mining gases released from the rock mass and generated as a result of the ongoing mining operations. Examining the three-dimensional flow of the air stream through region under analysis, encompassing the flow through the longwall, longwall headings and goaves with caving, one must consider the analytical models describing a turbulent and laminar flow towards the component parts of the Cartesian x, y and z coordinate system, in the particular calculation domains of the model. Modelling the flow of a multi-component mixture also requires solving additional equations for the transportation of the mixture components.

#### Basic Flow Equations

The flow of the air stream mixture is described by means of constitutive equations, which include the equations of mass, momentum and energy conservation and species transport equation. Conservation equations for mass, momentum, and species can be expressed as [45]:

$$\frac{\partial \rho}{\partial t} + \nabla \cdot \rho \boldsymbol{v} = 0 \tag{8}$$

$$\frac{\partial}{\partial t}(\rho \boldsymbol{v}) + \nabla \cdot \rho \boldsymbol{v} \boldsymbol{v} = -\nabla p \cdot \nabla \tau + \rho \boldsymbol{g} \tag{9}$$

$$\frac{\partial}{\partial t}(\rho c_p T) + \nabla \cdot (\rho c_p \boldsymbol{v} T) = \nabla \cdot \left(k_{eff} + \frac{c_p \mu_t}{Pr_t}\right) \nabla T \tag{10}$$

$$\frac{\partial}{\partial t}(\rho \omega_i) + \nabla \cdot (\rho \omega_i \boldsymbol{U}) = \nabla \cdot \left(\rho D_{i,eff} + \frac{\mu_t}{Sc_t}\right) \nabla \omega_i \tag{11}$$

where: $\rho$ is the gas density (kg/m$^3$), $v$ is the gas velocity (m/s), $p$ is pressure (Pa), $\tau$ is the viscous stress tensor (Pa), $g$ is gravity acceleration (m·s$^{-2}$), $c_p$ is the specific heat of the gas, $k_{eff}$ is the effective gas thermal conductivity, $T$ is the temperature (K), $\omega_i$ is the mass fraction of species $i$ (N$_2$, O$_2$ and CH$_4$), $\mu_t$ is turbulent viscosity (Pa·s), $D_{i,eff}$ is the effective diffusivity of species $i$ (m$^2$/s), $Sc_t$ is the turbulent Schmidt number (0.7) and $Pr_t$ is the turbulent Prandtl number.

#### Turbulence Model

The stream of the air and methane mixture flowing through the longwall and longwall headings, as well as through the initial section of the goaves with caving, is of a turbulent nature.

The analyses related to the flow of gas mixtures, including air, use the Reynolds number as the criterion specifying the type of flow, whose critical value makes it possible to determine the critical state of the flow separating the area of static laminar flow from the turbulent flow.

Therefore, by taking a time average of the Navier–Stokes equations, the Reynolds-averaged Navier–Stokes (RANS) equations have the following form [46]:

$$\frac{\partial \rho}{\partial t} + \frac{\partial (\rho v_i)}{\partial x_i} = 0 \tag{12}$$

$$\frac{\partial (\rho v_i)}{\partial t} + \frac{\partial (\rho v_i v_j)}{\partial x_j} = -\frac{\partial}{\partial x_i} + \frac{\partial}{\partial x_j}\left[\mu\left(\frac{\partial v_i}{\partial x_j} + \frac{\partial v_j}{\partial x_i} - \frac{2}{3}\delta_{ij}\frac{\partial v_l}{\partial u_l}\right)\right] + \frac{\partial}{\partial x_j}\left(-\overline{\rho v_i' v_j'}\right) \tag{13}$$

As can be seen from Equation (11), a new variable, the Reynolds stress $\overline{\rho v_i' v_j'}$ is introduced to the equations and it must be solved to achieve the closure of the equations. Two approaches are adopted to calculate the Reynolds stress, i.e., the Reynolds stress models (RSM) and the Boussinesq hypothesis. The Reynolds stresses are related to the mean velocity gradients in the Boussinesq hypothesis [46]:

$$-\overline{\rho v_i' v_j'} = \mu_t \left( \frac{\partial v_i}{\partial x_j} + \frac{\partial v_j}{\partial x_i} \right) - \frac{2}{3}\left( \rho k + \mu_t \frac{\partial v}{\partial x_k} \right)\delta_{ij} \tag{14}$$

In the turbulence model $k - \varepsilon$, in the standard variation, the basic Navier–Stokes equation has been transformed into the Reynolds averaged equation. This equation includes an additional term in the form of the Reynolds stress tensor. Due to this term, the set of equations is not closed. To close the set of equations, it is necessary to introduce additional differential equations, which include the equation of kinetic turbulent energy and the equation of kinetic turbulent energy dissipation in the following form [46]:

$$\rho \frac{\partial k}{\partial t} + \frac{\partial}{\partial x_i}(\rho k v_i) = \frac{\partial}{\partial x_j}\left[ (\mu + \frac{\mu_t}{\sigma_k}) \frac{\partial k}{\partial x_j} \right] + G_k + G_b - \rho\varepsilon - Y_M + S_k \tag{15}$$

$$\rho \frac{\partial \varepsilon}{\partial t} + \frac{\partial}{\partial xi}(\rho\varepsilon v_i) = \frac{\partial}{\partial x_j}\left[ (\mu + \frac{\mu_t}{\sigma_\varepsilon}) \frac{\partial \varepsilon}{\partial x_j} \right] + C_{1\varepsilon}\frac{\varepsilon}{k}(G_k + C_{3\varepsilon}G_b) - C_{2\varepsilon}\rho \frac{\varepsilon^2}{k} + S_\varepsilon \tag{16}$$

where: $C_{1\varepsilon}$, $C_{2\varepsilon\rho}$, $C_{3\varepsilon}$ are constans, $\sigma_k$, $\sigma_\varepsilon$ are turbulent Prandtl numbers for $k$ and $\varepsilon$, $G_b$ is the generation of turbulence kinetic energy due to buoyancy, $G_k$ is the generation of turbulence kinetic energy due to the mean velocity gradients, $Y_M$ is contribution of the fluctuating dilatation in compressible turbulence to the overall dissipation rate, $S_k$, $S_\varepsilon$ are user-defined source terms.

In the Ansys Fluent software, the porous medium (the goaves with caving) is represented as a fluid characterised by two additional parameters. These parameters include porosity and the permeability coefficient. In order for the tests to take into account the porous medium through which the flow occurs, it is required to consider the source term $S_i$ in the equation of momentum preservation. The additional source term assumes the following form [46]:

$$S_i = -\left( \sum_{j=1}^{3} K_{ij}\mu u_i + \sum_{j=1}^{3} C_{ij}\frac{1}{2}\rho v_i^2 \right) \tag{17}$$

where $S_i$ is the pressure loss items defined by Darcy's law and $C_2$ is the inertial resistance factor.

### 2.3. Problem Statement and Boundary Conditions

The basis stage of the tests conducted was to develop a numerical model for the real-world (actual) region of the longwall under analysis. This area covers the goaves with caving, the longwall and the longwall headings. The actual tensile strength of the roof rocks forming the caving in this longwall amounted to 3.06 MPa. Additionally, tests were also conducted for the following strength values of these rocks: 2, 4, 5, 6 and 7 MPa.

The geometric parameters of the entire region under analysis were taken into consideration during the development of the model. The geometric model of the longwall under analysis, ventilated with the Y-type system along with the equivalence conditions adopted, is presented in Figure 2.

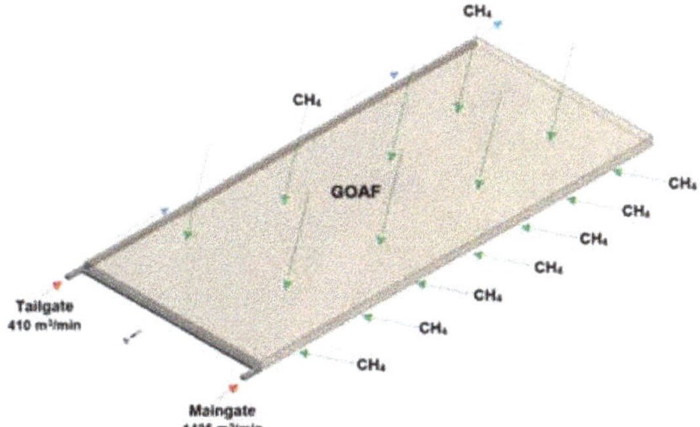



**Figure 2.** The geometric model with the assumed equivalence conditions for the longwall under analysis ventilated with the Y-type system.

The geometric model of the longwall area investigated takes into account the following:

- a section of the longwall gallery with a length of 20.0 m and a cross sectional area of $A = 15.0$ m$^2$.
- a tailgate maintained along the goaves with caving, with a length 525.0 m.
- a longwall with a height of 3.0 m and a length of 220.0 m, and an inclination of 0°.
- a section of goaves with caving with a length 500.0 m (constituting 2/3 of the longwall panel length).

In order to examine the air flow through the caving goaves of longwalls, it was also necessary to determine the vertical extent of this flow. Generally, it is assumed that the zone of air flow through the goaves with caving is equal to three to four times the thickness of the exploited seam (layer) [36].

The height of this flow is determined by the extent of the full caving and is equal to three times the thickness of the exploited layer. Taking into consideration the fact that the air also flows in the space (void) of the mined longwall, we obtain a flow zone measuring four times the thickness of the seam. On the other hand, taking into account the settlement of the basic roof by a value of $0.5 \div 0.6$ of the seam thickness, it was assumed that the height of the air flow in the goaves amounts to 3.5 times the thickness of the exploited seam (layer).

Such geometrical models were subjected to the process of discretization. The selection of the right size of the numerical mesh elements was preceded by an analysis of its sensitivity to the calculation results obtained. Based on the analysis, it was concluded that, for model-based tests of air flow through goaves with caving, one may adopt a structural numerical mesh with the size of cubic elements equal to 0.05 m × 0.05 m × 0.05 m for a longwall and longwall headings, as well as a structural numerical mesh consisting of cuboidal elements (type of mesh: hexahedron) measuring 0.025 m × 0.025 m × 0.025 m for goaves with caving. Smaller elements significantly extend the time of calculations, without making any changes in the results obtained.

The "inlet" and "outlet" boundary conditions were defined in longwall galleries. The "inlet" boundary condition was set in the distance of 20.0 m from the longwall, in the maingate as well as in the tailgate. It was assumed that the length of longwall galleries amounting to 20.0 m would allow for full development of the speed profile for the air stream supplied to the longwall.

The volumetric flow rate of the air supplied to the longwall under analysis was equal to 14,250 m$^3$/min. The volumetric flow rate of the air supplied through the tailgate amounted to 410 m$^3$/min.

The "outlet" boundary condition (pressure-outlet) was defined in the tailgate (this reflects the actual condition in the longwall region).

For longwall-related interactions of the flow, standard functions and zero values were adopted for the flow speed in "wall-type" conditions (longwalls treated as the sidewalls of headings), whose surface roughness corresponded to the height of 0.1 m, and their temperature (treated as the temperature of the surrounding rock mass) amounted to 40 °C. The temperature of the air stream at the inlet to the headings was 23 °C. The oxygen concentration in the air stream at the inlet to the headings was assumed to be equal to 20.8% (the value registered by automatic gasometry sensors, oxygen metres).

The analysed systems of headings took into consideration the flow of methane in the goaves with caving, which was equal to 8 m$^3$/min (according to actual measurements).

The computational domain consisted of two parts, with one mapping the longwall and longwall headings and the other the goaves with caving. In the domain reflecting the goaves with caving, a definition was provided for a change in the permeability coefficient of the goaves with caving as a function of distance from the longwall front, by means of the created user definition function (*UDF*).

The geometrical models failed to incorporate the machinery and devices forming the equipment of longwall headings.

The simplifications adopted in the models developed, in relation to the real-world (actual) objects, arise out of their sizes and constitute a certain compromise between calculation precision and the time of finding a solution.

The ANSYS Fluent 18.2 software was employed for all numerical simulations. The pressure–velocity coupling and scheme-coupled algorithm, the second-order upwind discretization method and the algebraic multigrid method were used to solve the equation.

Such models, along with the adopted conditions of uniqueness, were subjected to numerical analysis.

The each calculation required approximately 1500–1800 iterations, with a convergence tolerance of $10^{-6}$ for all variables (as per the "Fluent Theory Guide" support documentation).

## 3. Results and Discussion

The analyses conducted helped to determine a series of physical and chemical parameters of the air stream and methane flowing through the region under investigation.

In order to illustrate the processes related to this flow, Figure 3 shows the trajectories of the mixture of air and methane flowing through the caving goaves of the longwall ventilated with the Y-type system with the air being discharged along the goaves. A preliminary analysis of the distribution obtained was enough to demonstrate its great difference from the distributions concerning, for instance, the U-type ventilation system.

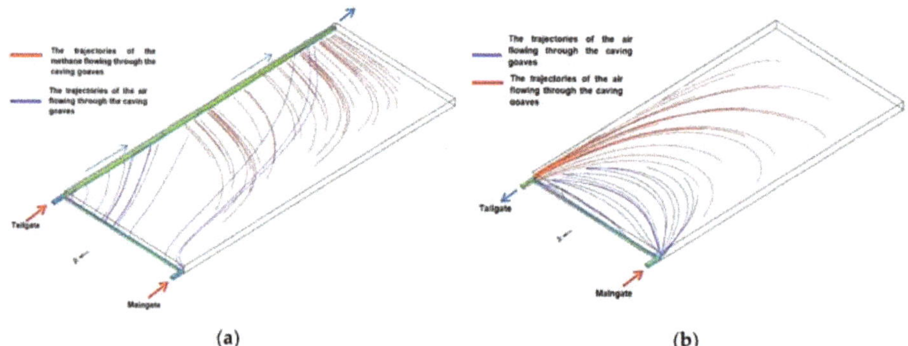


**Figure 3.** The trajectories of the mixture of air and methane flowing through the goaves with caving for Y-type ventilation system (**a**) and U-type ventilation system (**b**).

Figures 4–15 present the results of the analysis for the goaves with caving formed by roof rocks whose tensile strength is equal to 3.06 MPa (as is the case in a real-world system).

Figures 4–6 present, respectively, the distributions of the air speed and the dangerous speed due to the risk of endogenous fires, as well as the oxygen concentration levels in the goaves with caving at a distance of 0.5 m from the floor of the exploited seam.

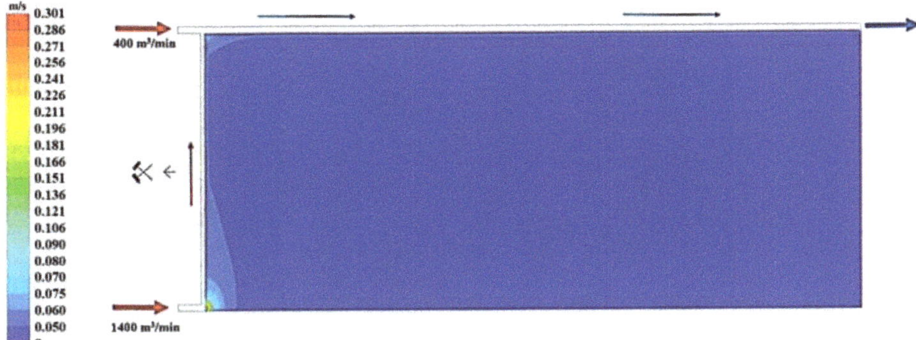


**Figure 4.** The distribution of the air speed flowing through the goaves with caving at a distance of 0.5 m from the floor of the exploited seam.

**Figure 5.** The distribution of the air speed within the range from 0.02 m/s to 0.0015 m/s flowing through the goaves with caving at a distance of 0.5 m from the floor of the exploited seam.

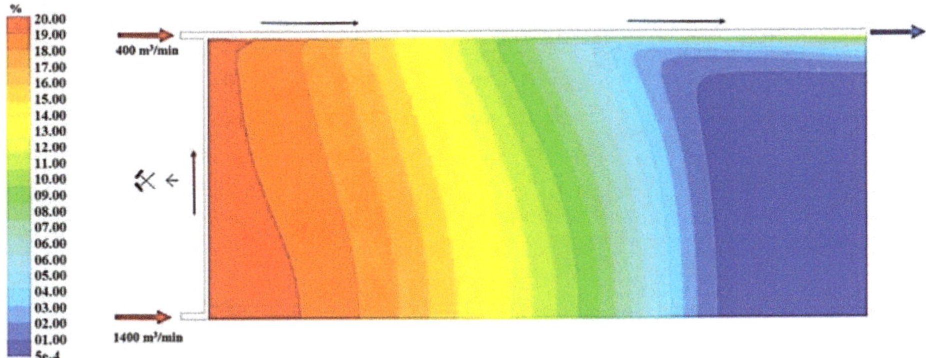


**Figure 6.** The distribution of oxygen concentration in the air flowing through the goaves with caving at a distance of 0.5 m from the floor of the exploited seam.

Figures 7–9 present, respectively, the distributions of the air speed and the dangerous speed due to the risk of endogenous fires, as well as the oxygen concentration levels in the goaves with caving at a distance of 2.0 m from the floor of the exploited seam.

**Figure 7.** The distribution of the air speed flowing through the goaves with caving at a distance of 2.0 m from the floor of the exploited seam.

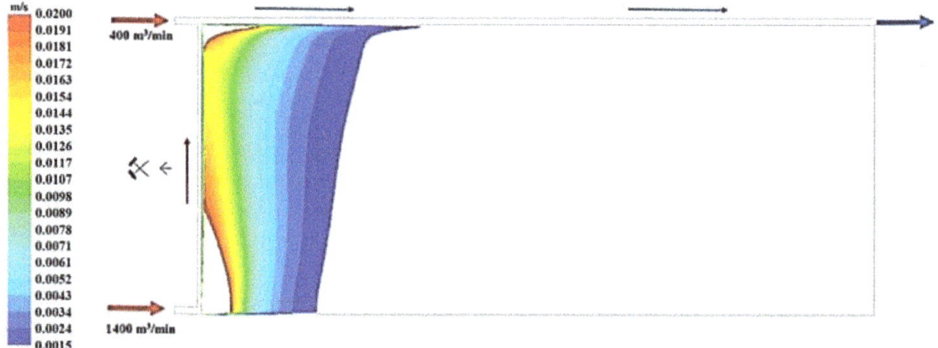

**Figure 8.** The distribution of the air speed within the range from 0.02 m/s to 0.0015 m/s flowing through the goaves with caving at a distance of 2.0 m from the floor of the exploited seam.

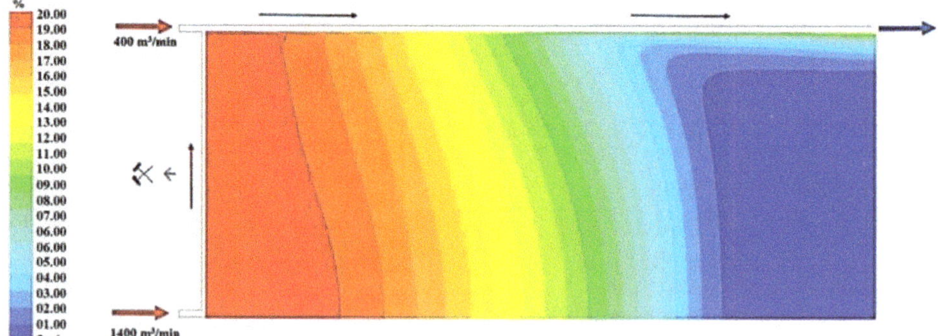

**Figure 9.** The distribution of oxygen concentration in the air flowing through the goaves with caving at a distance of 2.0 m from the floor of the exploited seam.

Figures 10–12 present, respectively, the distributions of the air speed and the dangerous speed due to the risk of endogenous fires, as well as the oxygen concentration levels in the goaves with caving at a distance of 7.0 m from the floor of the exploited seam.

**Figure 10.** The distribution of the air speed flowing through the goaves with caving at a distance of 7.0 m from the floor of the exploited seam.

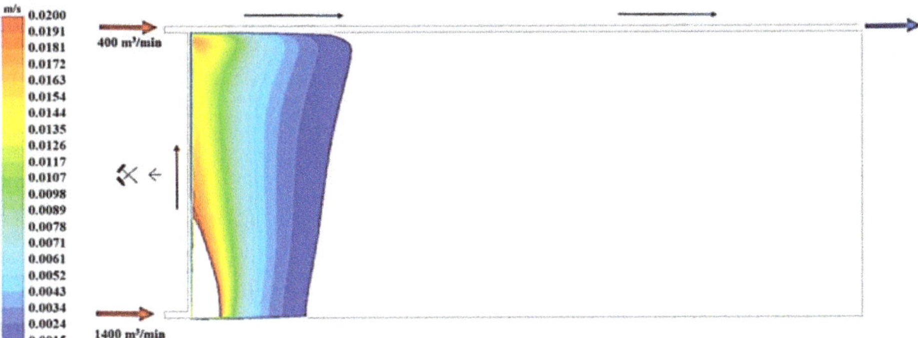

**Figure 11.** The distribution of the air speed within the range from 0.02 m/s to 0.0015 m/s flowing through the goaves with caving at a distance of 7.0 m from the floor of the exploited seam.

**Figure 12.** The distribution of oxygen concentration in the air flowing through the goaves with caving at a distance of 7.0 m from the floor of the exploited seam.

Figures 13–15 present, respectively, the distributions of the air speed and the dangerous speed due to the risk of endogenous fires, as well as the oxygen concentration levels in the goaves with caving at a distance of 10.5 m from the floor of the exploited seam.

**Figure 13.** The distribution of the air speed flowing through the goaves with caving at a distance of 10.5 m from the floor of the exploited seam.

**Figure 14.** The distribution of the air speed within the range from 0.02 m/s to 0.0015 m/s flowing through the goaves with caving at a distance of 10.5 m from the floor of the exploited seam.

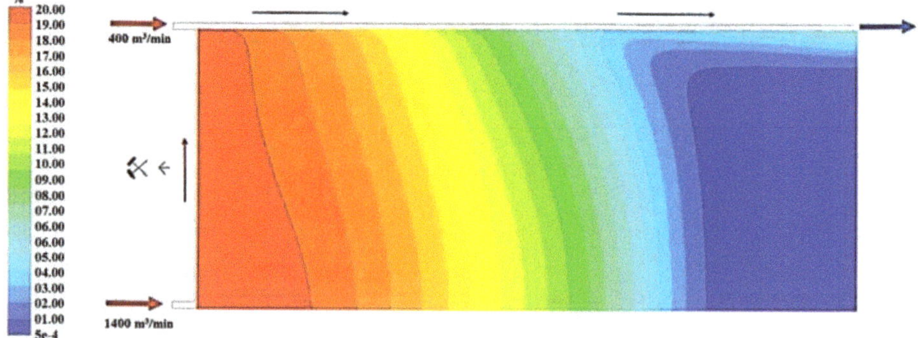

**Figure 15.** The distribution of oxygen concentration in the air flowing through the goaves with caving at a distance of 10.5 m from the floor of the exploited seam.

Based the results obtained, it may be concluded that the air speed value decreases along with an increase in the distance from the floor of the exploited seam. The air flowing through the goaves with caving reaches the highest speed value, irrespective of the flow height in the goaves, behind the caving line from the inlet side to the longwall, as well as along the tailgate maintained at the goaves. The highest speed value occurs at the flow height of 2.0 m from the floor of the exploited seam in the bottom corner of the longwall, and amounts to 0.36 m/s. The distribution of oxygen concentration for this ventilation system is also different than in the case of the U-type system [47]. It is clearly visible that the air stream moving along with tailgate leads to an increase in this concentration along this route.

Figure 16 presents the distribution of air speed values in the goaves with caving as a function of distance from the longwall front for the actual values of the tensile strength of roof rocks amounting to 3.06 MPa. Red horizontal lines were used to mark the range of dangerous speed values due to the risk of endogenous fires in these goaves.

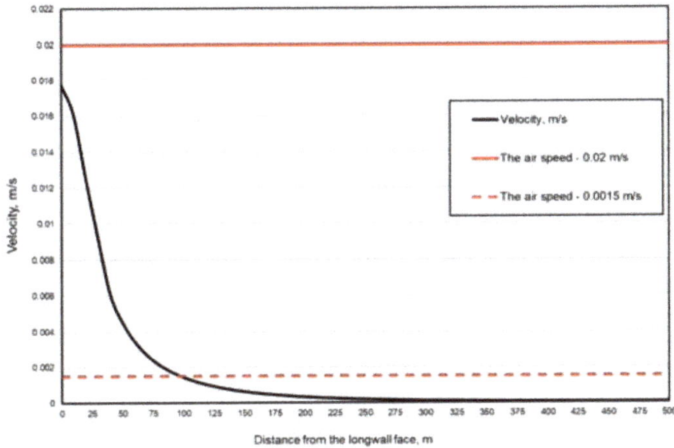



**Figure 16.** The distribution of air speed values in the goaves with caving as a function of distance from the longwall front for the actual values of the tensile strength of roof rocks amounting to 3.06 MPa.

Based on the determined speed characteristics, it can be concluded that the air speed in goaves with caving decreases along with the increasing distance from the longwall front.

At a distance of up to 96.0 m from the caving line into the depths of the goaves, the speed of the flowing air reaches the critical value due to the risk of endogenous fires, i.e., the value from 0.0015 m/s to 0.02 m/s. After exceeding the distance of 96.0 m from the longwall front, the speed of the air flowing through goaves with caving reaches a value lower than 0.0015 m/s.

Figure 17 presents the distribution of oxygen concentration in the air flowing through the goaves with caving as a function of distance from the longwall front.

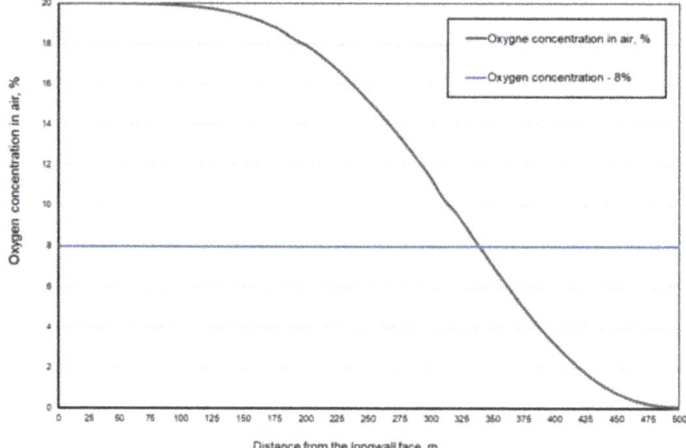

**Figure 17.** The distribution of oxygen concentration in the goaves with caving as a function of distance from the longwall front for the actual values of the tensile strength of roof rocks amounting to 3.06 MPa.

Based on the determined speed characteristics, it can be concluded that the concentration of oxygen in goaves with caving decreases along with the increasing distance from the longwall front.

At a distance of up to 335.0 m from the caving line inside the goaves, the oxygen concentration in the air flowing through the goaves with caving falls within the critical range due to the risk of endogenous fires, i.e., reaches a value higher than or equal to 8%.

The speed characteristics determined for the air flowing through the goaves with caving and for the oxygen concentration in this air served as the basis for demarcating the zone with a particularly high risk of spontaneous coal combustion (in which both of these conditions are met) (Figure 18).

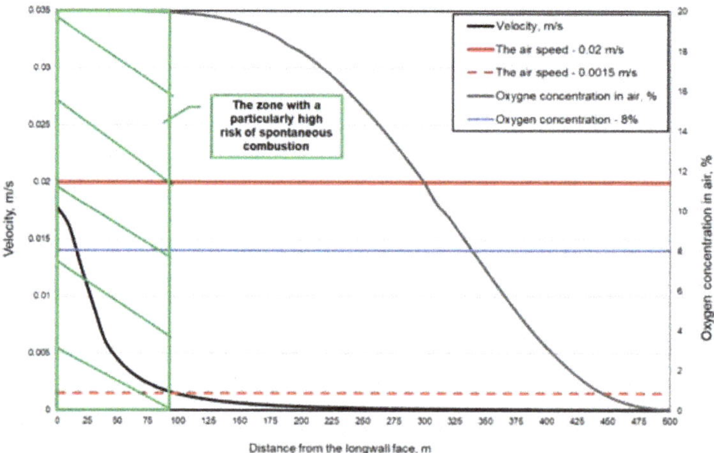

**Figure 18.** The zone with a particularly high risk of endogenous fires in the goaves formed by rocks with tensile strength equal to 3.06 MPa.

Based on the tests and the results obtained, it was concluded that the zone with a particularly high risk of spontaneous combustion, for the longwall ventilated with the Y-type system, is formed immediately behind the longwall front, and reaches 100.0 m inside the goaves.

Therefore, it can be assumed that the goaves with caving formed by roof rocks whose tensile strength amounts to 3.06 MPa have no cooling zone in which the air (behind the powered roof support along the entire length of the longwall) reaches a flow value higher than 0.02 m/s. This value was exceeded only in the upper and bottom corner of the longwall at the flow height of up to 8.0 m from the floor of the exploited seam.

Behind the zone with a particularly high risk of spontaneous combustion, at a distance of more than 100.0 m from the longwall front to approximately 345.0 m, there forms a zone with insufficient air speed, yet with sufficient oxygen concentration in the air, in terms of the risk of spontaneous coal combustion.

In order to determine the impact of the strength of the rocks forming the caving on the extent of the zone with a particularly high risk of spontaneous combustion for longwalls ventilated with the Y-type system with the air being discharged along the goaves and supplied along the tailgate, additional tests were conducted for different values of this strength (2, 4, 5, 6 and 7 MPa).

Figures 19–30 present the results of the analysis for the goaves with caving formed by roof rocks whose tensile strength is equal to 3.06 MPa (as is the case in a real-world system).

Figures 19–21 present, respectively, the distributions of the air speed and the dangerous speed due to the risk of endogenous fires, as well as the oxygen concentration levels in the goaves with caving at a distance of 0.5 m from the floor of the exploited seam.

**Figure 19.** The distribution of the air speed flowing through the goaves with caving at a distance of 0.5 m from the floor of the exploited seam.

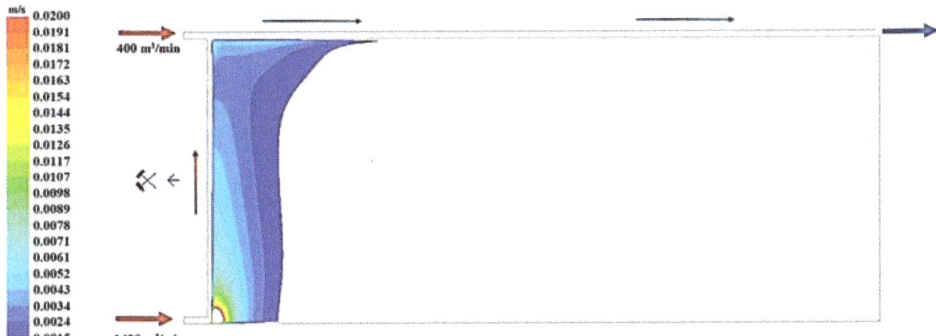


**Figure 20.** The distribution of the air speed within the range from 0.02 m/s to 0.0015 m/s flowing through the goaves with caving at a distance of 0.5 m from the floor of the exploited seam.

**Figure 21.** The distribution of oxygen concentration in the air flowing through the goaves with caving at a distance of 0.5 m from the floor of the exploited seam.

**Figure 22.** The distribution of the air speed flowing through the goaves with caving at a distance of 2.0 m from the floor of the exploited seam.

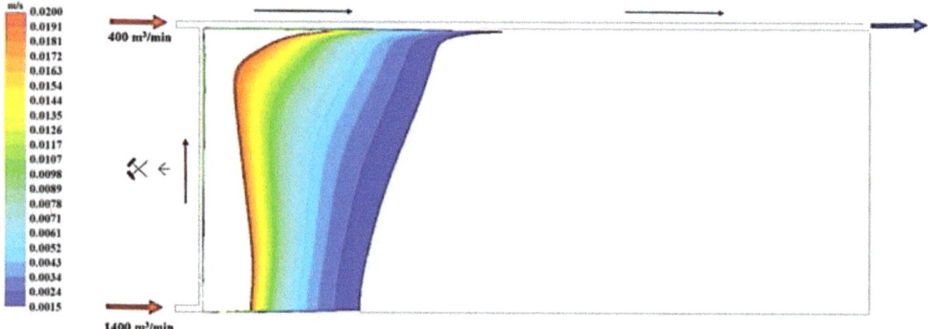

**Figure 23.** The distribution of the air speed within the range from 0.02 m/s to 0.0015 m/s flowing through the goaves with caving at a distance of 2.0 m from the floor of the exploited seam.

**Figure 24.** The distribution of oxygen concentration in the air flowing through the goaves with caving at a distance of 2.0 m from the floor of the exploited seam.

**Figure 25.** The distribution of the air speed flowing through the goaves with caving at a distance of 7.0 m from the floor of the exploited seam.

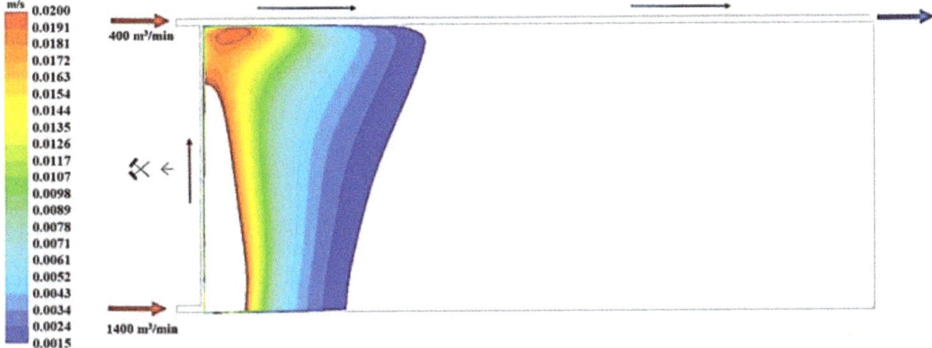

**Figure 26.** The distribution of the air speed within the range from 0.02 m/s to 0.0015 m/s flowing through the goaves with caving at a distance of 7.0 m from the floor of the exploited seam.

**Figure 27.** The distribution of oxygen concentration in the air flowing through the goaves with caving at a distance of 7.0 m from the floor of the exploited seam.

**Figure 28.** The distribution of the air speed flowing through the goaves with caving at a distance of 10.5 m from the floor of the exploited seam.

**Figure 29.** The distribution of the air speed within the range from 0.02 m/s to 0.0015 m/s flowing through the goaves with caving at a distance of 10.5 m from the floor of the exploited seam.

**Figure 30.** The distribution of oxygen concentration in the air flowing through the goaves with caving at a distance of 10.5 m from the floor of the exploited seam.

Figure 31 presents the distribution of air speed values in the goaves with caving as a function of distance from the longwall front for the actual values of the tensile strength of roof rocks amounting to 6.0 MPa. Red horizontal lines were used to mark the range of dangerous speed values due to the risk of endogenous fires in these goaves.

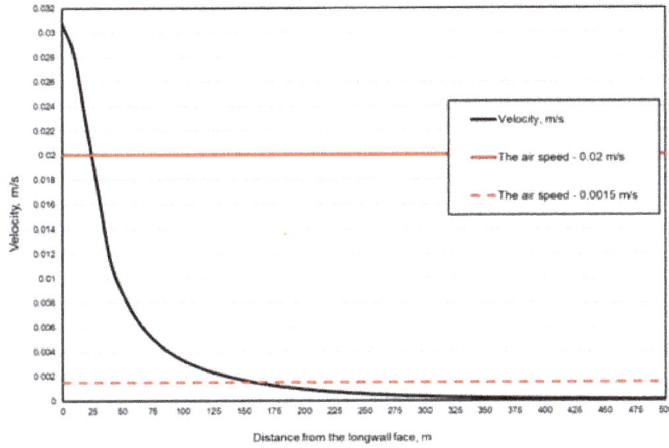



**Figure 31.** The distribution of air speed values in the goaves with caving as a function of distance from the longwall front for the actual values of the tensile strength of roof rocks amounting to 6.0 MPa.

Based the results obtained, it may be concluded that the air speed value decreases along with an increase in the distance from the floor of the exploited seam.

At a distance of 25.0 m behind the longwall front to 160.0 m inside the goaves, the speed of the flowing air reaches the critical value due to the spontaneous combustion risk, i.e., the value from 0.0015 m/s to 0.02 m/s. After exceeding the distance of 160.0 m from the longwall front, the speed of the air flowing through goaves with caving reaches a value lower than 0.0015 m/s.

Figure 32 presents the distribution of oxygen concentration in the air flowing through the goaves with caving as a function of distance from the longwall front.

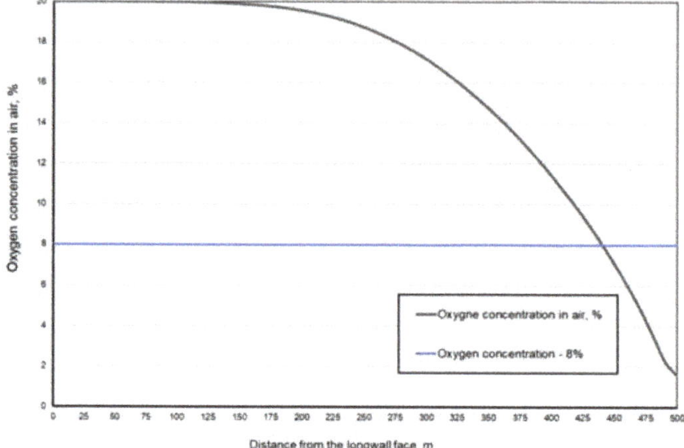

**Figure 32.** The distribution of oxygen concentration in the goaves with caving as a function of distance from the longwall front for the actual values of the tensile strength of roof rocks amounting to 6.0 MPa.

Based on the determined speed characteristics, it can be concluded that the concentration of oxygen in goaves with caving decreases along with the increasing distance from the longwall front.

At a distance of up to 440.0 m from the caving line inside the goaves, the oxygen concentration in the air flowing through the goaves with caving falls within the critical range due to the risk of endogenous fires, i.e., reaches a value higher than or equal to 8%.

The speed characteristics determined for the air flowing through the goaves with caving and for the oxygen concentration in this air served as the basis for demarcating the zone with a particularly high risk of spontaneous coal combustion (in which both of these conditions are met) (Figure 33).

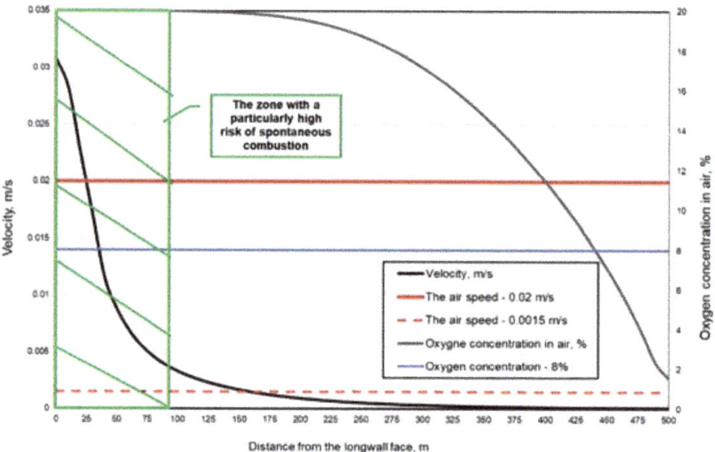

**Figure 33.** The zone with a particularly high risk of endogenous fires in the goaves formed by rocks with tensile strength equal to 6.0 MPa.

Based on the tests and the results obtained, it was concluded that the zone with a particularly high risk of spontaneous combustion is formed immediately behind the longwall front, and reaches 160.0 m inside the goaves.

In the goaves with caving formed by roof rocks whose tensile strength amounts to 6.0 MPa, the cooling zone in which this air (behind the powered roof support along the entire length of the longwall) reaches a flow value higher than 0.02 m/s occurs at a distance of up to 25.0 m from the longwall front.

Behind the zone of a particularly high risk of spontaneous combustion, at a distance of over 160.0 m from the longwall front to approximately 460.0 m, there forms a zone with insufficient air speed, yet with sufficient oxygen concentration in the air (due to the risk of spontaneous combustion).

The tests conducted made it possible to demarcate the zone with a particularly high risk of spontaneous combustion in the goaves with caving formed by rocks whose tensile strength was equal to 2, 3.06, 4, 5, 6 and 7 MPa. Table 2 summarises the extents of these zones for the actual conditions of the longwall in question as well as for the additional ones obtained from the analyses conducted.

**Table 2.** The zone with a particularly high risk of spontaneous combustion in the goaves with caving formed by rocks whose tensile strength was equal to 2, 3.06, 4, 5, 6 and 7 MPa.

| Tensile Strength of Roof Rock, $R_{rrs}$, MPa | Critical Air Velocity Zone, m | Critical Oxygen Concentration Zone, m | The Zone with a Particularly High Risk of Spontaneous Combustion, m |
|---|---|---|---|
| 2.00 | 0–70.0 m | 0–260.0 m | 0–70.0 m |
| 3.06 | 0–96.0 m | 0–335.0 m | 0–96.0 m |
| 4.00 | 8.0–115.0 m | 0–370.0 m | 8.0–115.0 m |
| 5.00 | 18.0–138.0 m | 0–427.0 m | 18.0–138.0 m |
| 6.00 | 28.0–160.0 m | 0–440.0 m | 28.0–160.0 m |
| 7.00 | 37.0–184.0 m | 0–460.0 m | 37.0–184.0 m |

The results obtained unambiguously indicate that the type of roof rocks (defined by their tensile strength) has a significant impact on the value of the air speed flowing through the goaves with caving, and on the value of oxygen concentration in this air. The more resistant the rocks, the greater the extent of the zones in which the air speed and oxygen concentration reach critical values due to the risk of spontaneous coal combustion.

The greater extent of the zone with a particularly high risk of endogenous fires in the caving goaves of the longwall ventilated with the Y-type system (compared, for example, to the U-type system) arises out of the necessity to maintain a tailgate along the goaves, through which the air can flow. Part of the ventilation air stream flowing through this tailgate migrates to the goaves through the sidewalls, thereby increasing the extent of this zone in the goaves.

## 4. Conclusions

Spontaneous combustion of coal is a highly dangerous phenomenon that occurs during mining exploitation. It leads to major economic losses for mining enterprises and poses a threat to the working crew. The products of coal combustion are also highly damaging to the natural environment. This is because they penetrate into the mining atmosphere along with the ventilation stream, and then to the surface into the natural environment through the ventilation system. As a result, it is necessary to undertake various steps to reduce the risk of these phenomena in mines.

The method developed and presented in the paper for determining the zone with a particularly high risk of spontaneous coal combustion, in this case for longwalls ventilated with the Y-type system, serves this purpose. The determination of such a zone may serve as the basis for taking effective preventive measures. They involve choosing the exploitation speed, isolating this zone, introducing inert gases, sealing the goaves, etc. For these measures to be effective, it is necessary to identify the sites (areas) where spontaneous coal combustion may occur.

It must also be stressed that mining exploitation also generates other hazards that determine, for example, the application of different ventilation systems. Generally speaking, the Y-type ventilation system under analysis is highly favourable in the case of methane threats, and slightly less favourable for the threat related to spontaneous coal combustion. Therefore, the tests conducted are of a particular significance in the process of limiting this threat.

The method developed, thanks to the application of advanced spatial models and the use of actual measurement data from the analysed region, allows for early identification of areas in which spontaneous coal combustion may occur.

The results obtained clearly indicate that the demarcated risk zones of spontaneous coal combustion are significantly higher for this ventilation system than for the U-type system.

The comprehensive analysis also indicates that the type of roof rocks forming the caving has a significant impact on the size and location of the zone with a particularly high risk of spontaneous coal combustion. The different tensile strength of these rocks leads to changes in the porosity and permeability of the caving, which in turn has a significant impact on the ventilation parameters of the air flowing through the caving. No such tests have been conducted so far, and the results obtained indicate the significant changes in the location and extent of the risk zone along with the changing values of this strength.

The results obtained also enhance knowledge about the ventilation of underground exploitation regions and should become an important source of information for the ventilation service teams in mines. In particular this concerns the essential differences in this process for the U-type and Y-type ventilation systems, as unequivocally indicated by the results obtained.

These results also demonstrate the great impact exerted on the ventilation process in mine headings by the goaves with caving, which—due to their porosity—must be taken into consideration in this process.

The methodology developed and presented in the paper is of a universal nature and may be successfully applied to multivariate analyses of the spontaneous combustion hazard, as well as on the mining landfill sites.

The authors hope that the results obtained and the methodology developed will find broader application for the support of preventive measures in terms of limiting the risk of spontaneous coal combustion.

**Author Contributions:** Conceptualization, M.T. and J.B.; methodology, J.B. and M.T.; software, M.T.; validation, J.B. and M.T.; formal analysis, J.B. and M.T.; investigation, J.B. and M.T.; resources, J.B. and M.T.; data curation, J.B. and M.T.; writing of the original draft preparation, J.B. and M.T.; writing of review and editing, J.B. and M.T.; visualization, M.T.; supervision, J.B.; project administration, J.B.; funding acquisition, M.T. and J.B.

**Funding:** This research was funded by SUBB, grant numbers 06/030/BKM_18/0041, 06/030/BKM_19/0048.

**Conflicts of Interest:** The authors declare no conflict of interest.

## References

1. Bukowska, M. Post-critical mechanical properties of sedimentary rocks in the Upper Silesian Coal Basin (Poland). *Arch. Min. Sci.* **2015**, *60*, 517–534. [CrossRef]
2. Bukowska, M. Mechanical properties of carboniferous rocks in the Upper Silesian Coal Basin under uniaxial and triaxial compression tests. *J. Min. Sci.* **2005**, *4*, 129–133. [CrossRef]
3. Hebblewhite, B.K.; Lu, T. Geomechanical behaviour of laminated, weak coal mine roof strata and the implications for a ground reinforcement strategy. *Int. J. Rock Mech. Min. Sci.* **2004**, *41*, 147–157. [CrossRef]
4. Majcherczyk, T.; Małkowski, P.; Niedbalski, Z. Analysis of maintaining the stability of dog headings in the long-term. *Prz. Gór.* **2015**, *1*, 53–60.
5. Prusek, S.; Masny, W. Analysis of damage to underground workings and their support caused by dynamic phenomena. *J. Min. Sci.* **2015**, *51*, 63–72. [CrossRef]
6. Salustowicz, A. *Zarys Mechaniki Górotworu*; Śląsk: Katowice, Poland, 1968.
7. Vijay, R.; Singh, K. Spontaneous heating and fire in coal mines. *Procedia Eng.* **2013**, *62*, 78–90.
8. Ren, W.-x.; Kang, Z.-h.; Wang, D.-m. Causes of spontaneous combustion of coal and its prevention Technology in the tunnel fall of ground of extra-thick coal seam. *Procedia Eng.* **2011**, *26*, 717–724.

9. Yang, S.Q.; Zhang, R.W.; Di, Z.Q. Distribution Regularity of Spontaneous Combustion "Three-Zone" in Goaf of Fully-Mechanized Coal Faces. *J. Chin. Univ. Min. Technol.* **2000**, *29*, 93–96.

10. Cygankiewicz, J.; Dudzińska, A.; Żyła, M. Examination of sorption and desorption of hydrogen on several samples of polish hard coals. *Adsorption* **2012**, *18*, 189–198. [CrossRef]

11. Dudzińska, A.; Cygankiewicz, J. Analysis of adsorption tests of gases emitted in the coal self-heating process. *Fuel Process. Technol.* **2015**, *137*, 109–116. [CrossRef]

12. Rezaei, M. Long-term stability analysis of goaf area in longwall mining using minimum potential energy theory. *J. Min. Environ.* **2018**, *9*, 169–182.

13. Majdi, A.; Hassani, F.P.; Yousef Nasiri, M. Prediction of the height of destressed zone above the mined panel roof in longwall coal mining. *J. Min. Environ.* **2012**, *9*, 62–72. [CrossRef]

14. Palchik, V. Formation of fractured zones in overburden due to longwall mining. *J. Environ. Geol.* **2003**, *44*, 28–38. [CrossRef]

15. Kłeczek, Z. *Geomechanika Górnicza*; Śląskie Wydawnictwo Techniczne: Katowice, Poland, 1994.

16. Tutak, M. Analysis of ventilation methods for mining longwalls in Polish hard coal mines. In *Mining-Prospects, Threats. Air Conditioning, Aerological Hazards*; Plewa, F., Badura, H.P.A., Eds.; NOVA: Gliwice, Poland, 2014.

17. Khattri, S.K.; Log, T.; Kraaijeveld, A. Tunnel Fire Dynamics as a Function of Longitudinal Ventilation Air Oxygen Content. *Sustainability* **2019**, *11*, 203. [CrossRef]

18. Krause, E.; Smoliński, A. Analysis and Assessment of Parameters Shaping Methane Hazard in Longwall Areas. *J. Sustain. Min.* **2013**, *12*, 13–19. [CrossRef]

19. Roghanchi, P.; Kocsis, K.; Sunkpal, M. Sensitivity analysis of the effect of airflow velocity on the thermal comfort in underground mines. *J. Sustain. Min.* **2016**, *15*, 175–180. [CrossRef]

20. Szlązak, J. Ocena zagrożenia pożarowego w zrobach ścian zawałowych na podstawie intensywności ich przewietrzania. *Arch. Górnictwa* **1990**, *35*, 339–345.

21. Cheng, J.; Li, S.; Zhang, F.; Zhao, C.; Yang, S.; Ghosh, A. CFD modelling of ventilation optimization for improving mine safety in longwall working faces. *J. Loss Prev. Process Ind.* **2016**, *40*, 285–297. [CrossRef]

22. Chumak, A.S.; Pashkovsky, P.S.; Yaremchuk, M.A. Prevention of spontaneous fires by directed nitrogen supply. In Proceedings of the 7th International Mine Ventilation Congress, Krakow, Poland, 17–22 June 2001.

23. Szlązak, J.; Szlązak, N. Numerical and mine research on airflow in goaf of longwalls. *Kwart. Gór. Geoinż.* **2004**, *3*, 59–78.

24. Wang, Y.; Zhang, X.; Sugai, Y.; Sasaki, K. A Study on Preventing Spontaneous Combustion of Residual Coal in a Coal Mine Goaf. *J. Geol. Res.* **2015**, *2015*, 712349. [CrossRef]

25. Buchwald, P.; Jaskólski, Z.; Kajdasz, Z. Determining the Impact of Model Gas Mixtures on the Directions of Fire and Methane Prevention. In *Materials of the Mining Aerology School*; Polish Academy of Sciences Kraków: Zakopane, Poland, 2002; p. 399. Available online: http://old.e-teberia.pl/bibliografia.php?a=showarticle&ArticleID=6006 (accessed on 15 November 2019).

26. Ren, T.X.; Balusu, R. CFD modeling of goaf gas migration to improve the control of spontaneous cobbustion in longwalls. In *Coal Operators' Conference*; Aziz, N., Ed.; University of Wollongong & the Australian Institute of Mining and Metallurgy: Wollongong, Australia, 2005; pp. 259–264.

27. Ren, T.X.; Balusu, R. Proactive goaf inertisation for controlling longwall goaf heatings. *Procedia Earth Planet. Sci.* **2009**, *1*, 309–315. [CrossRef]

28. Ren, T.X.; Balusu, R. The use of CFD modelling as a tool for solving mining health and safety problems. In *Proc. 2010 Coal Operators' Conference*; Aziz, N., Ed.; University of Wollongong & the Australian Institute of Mining and Metallurgy: Wollongong, Australia, 2010; pp. 339–349.

29. Esterhuizen, G.S.; Karacan, C.O. A Methodology for Determining Gob Permeability Distributions and Its Application to Reservoir Modeling of Coal Mine Longwalls. In Proceedings of the SME Annual Meeting, Denver, CO, USA, 23–25 February 2007; Available online: https://stacks.cdc.gov/view/cdc/8533 (accessed on 10 September 2019).

30. Yuan, L.; Smith, A.C. Numerical study on effects of coal propertis on spontanous heating in longwall gob areas. *Fuel* **2008**, *87*, 3409–3419. [CrossRef]

31. Yuan, L.; Smith, A.C. CFD modeling of spontanous heating in a large-scale coal chamber. *J. Loss Prev. Process Ind.* **2009**, *22*, 426–433. [CrossRef]

32. Dai, G.; Zhang, S.; Tang, M. Determination of spontaneous combustion "three zones" in goaf of no. 713 fully mechanized longwall of qinan coal mine. *AGH J. Min. Geoeng.* **2012**, *36*, 99–113.

33. Xie, Z.; Jin, C. Numerical simulation of air leakage in goaf of fully mechanized face with high dip and hard roof. *Comput. Model. N. Technol.* **2013**, *17*, 36–43.

34. Shi, G.Q.; Liu, M.; Wang, Y.M.; Wang, W.Z.; Wang, D.M. Computational Fluid Dynamics Symulation of Oxygen Seepage in Coal Mine Goaf With Gas Drainage. *Math. Probl. Eng.* **2015**, *2015*, 723764. [CrossRef]

35. Brodny, J.; Tutak, M. The impact of the flow volume ventilation to the location of the special hazard spontaneous fire zone in goaf with caving of operating longwalls. In Proceedings of the 16th International Multidisciplinary Scientific GeoConference, SGEM, Albena, Bulgaria, 30 June–6 July 2016; Volume 2, pp. 897–904. [CrossRef]

36. Mishra, D.P.; Kumar, P.; Panigrahi, D.C. Dispersion of methane in tailgate of a retreating longwall mine: A computational fluid dynamics study. *Environ. Earth Sci.* **2016**, *75*, 475. [CrossRef]

37. Tutak, M.; Brodny, J. Analysis of Influence of Goaf Sealing from Tailgate on the Methane Concentration at the Outlet from the Longwall. *IOP Conf. Ser. Earth Environ. Sci.* **2017**, *95*, 042025. [CrossRef]

38. Brodny, J.; Tutak, M.; John, A. The application of model-based tests for analysing the consequences of methane combustion in a mine heading ventilated through a forcing air duct. *Mechanika* **2019**, *3*, 204–209.

39. Hobbs, D.W. The tensile strength of rocks. *Int. J. Rock Mech. Min. Sci.* **1964**, *1*, 385–396. [CrossRef]

40. Prusek, S.; Rajwa, S.; Wrana, A.; Krzmień, A. Assessment of roof fall risk in longwall coal mines. *Int. J. Min. Reclam. Environ.* **2017**, *31*, 558–574. [CrossRef]

41. Szlązak, J. The determination of a co-efficient of longwall gob permeability. *Arch. Min. Sci.* **2001**, *46*, 451–468.

42. Xu, Q.; Yang, S.Q.; Wang, C.; Chu, T.X.; Ma, W.; Huang, J. Numerical simulation of gas flow law in stope under stereo gas drainage. *J. Min. Saf. Eng.* **2010**, *27*, 62–66.

43. Li, Z.X.; Wu, Q.; Xiao, Y.N. Numerical simulation of coupling mechanism of coal spontaneous combustion and gas effusion in goaf. *J. Chin. Univ. Min. Technol.* **2008**, *37*, 38–42.

44. Veersteg, K.K.; Malalasekera, W. An Introduction to Computational Fluid Dynamics. In *The Finite Volume Method*; Pearson Education: London, UK, 2007.

45. Kurnia, J.; Sasmito, A.; Mujumdar, A. Simulation of a novel intermittent ventilation system for underground mines. *Tunn. Undergr. Space Technol.* **2014**, *42*, 206–215. [CrossRef]

46. ANSYS. *Ansys Theory Guide*; ANSYS, Inc.: Canonsburg, PA, USA, 2011.

47. Tutak, M.; Brodny, J. Impact of type of the roof rocks on location and range of endogenous fires particular hazard zone by in goaf with caving. *E3S Web Conf.* **2018**, *29*, 00005. [CrossRef]

*Article*

# Finite Element Study of the Effect of Internal Cracks on Surface Profile Change Due to Low Loading of Turbine Blade

**Junji Sakamoto** [1,*]**, Naoya Tada** [1]**, Takeshi Uemori** [1] **and Hayato Kuniyasu** [2]

[1]  Graduate School of Natural Science and Technology, Okayama University, 3-1-1, Tsushimanaka, Kita-ku, Okayama 700-8530, Japan; tada@okayama-u.ac.jp (N.T.); uemori@okayama-u.ac.jp (T.U.)
[2]  Shimadzu Corporation, 1 Nishinokyo Kuwabara-cho, Nakagyo-ku, Kyoto 604-8511, Japan; plut6fkl@s.okayama-u.ac.jp
*  Correspondence: sakamoto-junji@okayama-u.ac.jp; Tel.: +81-86-251-8031

Received: 13 June 2020; Accepted: 14 July 2020; Published: 16 July 2020

**Abstract:** Turbine blades for thermal power plants are exposed to severe environments, making it necessary to ensure safety against damage, such as crack formation. A previous method detected internal cracks by applying a small load to a target member. Changes in the surface properties of the material were detected before and after the load using a digital holographic microscope and a digital height correlation method. In this study, this technique was applied in combination with finite element analysis using a 2D and 3D model simulating the turbine blades. Analysis clarified that the change in the surface properties under a small load varied according to the presence or absence of a crack, and elucidated the strain distribution that caused the difference in the change. In addition, analyses of the 2D model considering the material anisotropy and thermal barrier coating were conducted. The difference in the change in the surface properties and strain distribution according to the presence or absence of cracks was elucidated. The difference in the change in the top surface height distribution of the materials with and without a crack was directly proportional to the crack length. As the value was large with respect to the vertical resolution of 0.2 nm of the digital holographic microscope, the change could be detected by the microscope.

**Keywords:** nondestructive inspection; crack detection; low loading; surface profile; turbine blade; finite element analysis

## 1. Introduction

Thermal power generation using gas turbines is expected to expand in the long term as a clean and economical power generation method. Gas turbines are a type of internal combustion engine. They are thermal engines that obtain power from the combustion of gases by expanding high-temperature gas obtained by burning petroleum, natural gas, or other fuels to rotate the turbine. At present, the improvement in the thermal efficiency of various high-temperature equipment in power plants is required to further improve performance. In order to achieve high efficiency, it is essential to increase the temperature of the combustion gas and the turbine inlet temperature. It is therefore necessary to design the turbine blades for high temperatures. A cooling passage can be formed inside the turbine blade, which is then cooled by passing a cooling medium, such as air or steam, through this cooling passage, thereby ensuring the sufficient heat resistance of the blade. This technology contributes the most to such requirements for high temperature resistance. A typical blade cooling structure is the pin fin cooling system [1,2]. This system performs convection cooling, impingement cooling, and film cooling inside the moving blade, and enables the cooling air to flow out of the blade. In addition, the mainstream method of lowering the surface temperature of the turbine blade is applied in thermal barrier coatings (TBCs), which not only lowers the surface temperature but also improves thermal fatigue characteristics. Therefore, ceramics are coated on the gas turbine surface using the electron-beam physical vapor deposition (EB-PVD) method proposed by Bose et al. [3]. However, there are concerns about damage and deterioration due to an increase in the thermal load applied to the TBC, and the improvement of durability is an issue. Therefore, in recent years, in addition to TBC, transpiration cooling of the turbine has been considered as an effective method for improving the durability of the turbine blades by using a porous material to form them. Arai [4] developed a porous ceramic coating (P-TBC) through which cooling air could pass. P-TBC has a thermal conductivity as low as 50% of a conventional TBC, and it has been shown that its adhesion strength is almost the same.

During the operation of the power plant, the turbine blades are subjected to high centrifugal loads due to the rotation of the rotor, flexural loads due to the working fluid, and vibration loads. These occur under the severe condition of a high temperature, hence cracks may develop in the cooling passage. Because of this reason, power plants regularly conduct non-destructive inspections, such as ultrasonic inspection tests and radiation transmission tests based on Japanese industrial standards [5] to detect turbine defects. Additionally, the cooling structure is checked, and the remaining life of the turbine is determined. Yoshioka et al. [6] proposed a prediction technique based on metallographic image analysis and a life assessment technique using damage trend analysis. These were based on statistical parameters of information from regular inspections of gas turbines where surface damage and material deterioration were prominent. In order to extend the inspection interval and increase the operation rate by accurately diagnosing the remaining life of the turbine, it is necessary to improve the precision of the inspection method to detect cracks generated in turbine blades in power plants.

To meet this requirement, one of the authors has proposed a new flaw detection method using digital holographic microscopy (DHM). This method can quickly obtain a nanometer-order height distribution over a wide area of 1 mm$^2$ on the material surface and in an atmospheric environment [7–9]. Digital holographic microscopes can acquire 3D images of a sample surface in real time with a high resolution by using holograms. The vertical resolution of these is excellent at 0.2 nm, and the fine behavior of the material surface can be captured. In this method, a small load is applied to the target member, and the internal defect is detected based on the displacement of the member surface before and after the application of the load. This can be realized by combining DHM and digital height correlation method (DHCM), which identifies the same region before and after deformation, by referring to the patterns of the ultra-fine irregularities on the material surface. In this study, finite element (FE) analysis was performed on the new method using a model simulating a turbine blade to investigate the possibility of detecting cracks. Two-dimensional models were used to investigate the possibility of detecting cracks in cases where anisotropic materials or coating-treated materials are used for the turbine blade. A 3D model was used to investigate the possibility of detecting cracks by each trace line with DHM. Sections 2 and 3 describe the FE analysis conditions and analysis results, respectively.

## 2. Finite Element Analysis

FE analysis was performed using three types of 2D models and a 3D model for the turbine blade. The nonlinear FE program ADINA 9.3.4 (ADINA R&D, Inc., Watertown, MA, USA) was used for the FE analysis.

Figure 1 shows the form and dimensions of the 2D analysis model. Figure 1a shows Model 2D and Model 2D-Anisotropy, and Figure 1b shows Model 2D-TBC. Table 1 shows the material properties of the four models. As shown in Table 1, the material properties of Model 2D and Model 2D-Anisotropy are different. These were determined based on the material properties of the Ni-based superalloy NCF625 (isotropic material) and a Ni-based directionally solidified (DS) alloy (anisotropic material), respectively. As these Ni-based superalloys have excellent high-temperature strength, they are widely used as materials for gas turbine blades [10–12]. The Ni-based DS alloy is a material with a controlled crystal orientation and grain growth direction (hence anisotropic [12]), and has high strength at high temperatures. Therefore, in order to investigate the effect of anisotropy on the deformation behavior of the turbine blade, analysis was performed using Model 2D-Anisotropy. For Model 2D-TBC, the authors considered a heat-resistant coating (bond coat layer and top coat layer). The material properties of the bond coat and top coat were determined with reference to the ceramic coating [13]. The materials of the bond coat and the top coat were CoNiCrAlY alloy and yttria-stabilized zirconia, respectively. In this study, we focused on the materials that may be used in the actual turbine blade. A plane strain condition was used for the 2D models. The actual turbine blade is large in the depth direction. Therefore, the plane strain condition can be more suitable than the plane stress condition. Figure 2 shows the form and dimensions of the 3D analysis model. The material properties of the 3D model were determined based on the Ni-based superalloy NCF625 (isotropic material). The load values in the case of the 2D and 3D models are 1916 N. The load range is determined by the following conditions. The lower limit of the load should provide a large enough displacement difference for DHM to be distinguishable, and the higher limit of the load should limit the elastic deformation regime. If a several millimeter-sized crack can be detected in the actual turbine blade, the crack detection method can be practically useful. Based on the ideal, the size and depth of the crack were determined.

**Table 1.** Material properties of the four types of models.

| Model | Young's Modulus, $E$ [GPa] | Poisson's Ratio, $v$ |
|---|---|---|
| Model 2D | 207 | 0.3 |
| Model 2D-Anisotropy | $E_x = 125.5, E_z = 168.7, G_{xy} = 133.9, G_{zy} = 64.5$ | $v_{xz} = 0.333, v_{zx} = 0.447$ |
| Model 2D-Thermal barrier coating (TBC) | 207 (base material), 200 (bond coat), 40 (top coat) | 0.3 (base material), 0.3 (bond coat), 0.3 (top coat) |
| Model 3D | 207 | 0.3 |

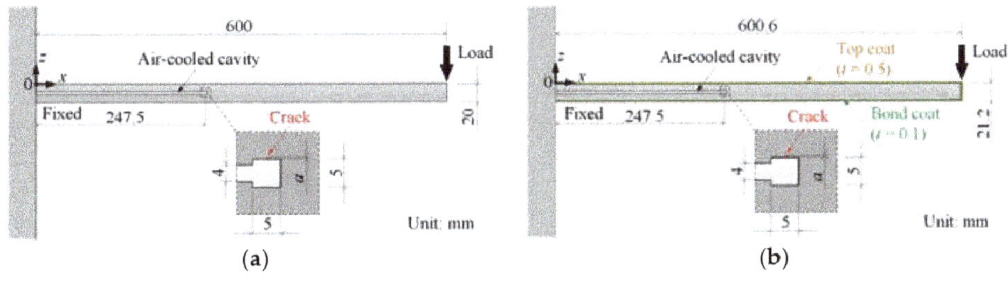

(a)    (b)

**Figure 1.** Shape and dimensions of the two-dimensional models: (a) Model 2D and Model 2D-Anisotropy; (b) Model 2D-Thermal barrier coating (TBC).

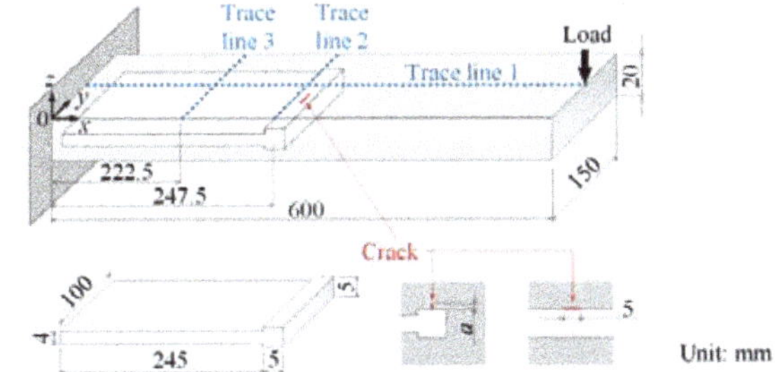

**Figure 2.** Shape and dimensions of the three-dimensional model.

Figure 3 shows the elements of Model 2D and Model 3D. Table 2 shows the number of nodes and number of elements in each model. In order to determine the element size, FE analysis was performed with a simple shape before analyzing the actual model, and the analysis values of the stress distribution near the crack were compared with theoretical values. To conduct the FE analysis calculation with high accuracy and efficiency, two types of elements were used for the models. A square eight-node element was used to model the element near the crack tip, and square four-node elements were used for other areas. The weight of the beam was not considered in the analysis because we focused on the difference of the deformation before and after the loading. There is no contact element in the FE analysis. In the FE analysis of all models, there is no contact on the crack surfaces because the crack opens due to the load. In the crack detection method, a load is applied to the actual turbine blade to open the presumed crack. By loading in such direction, the deformation can become larger and the crack detection can become easier.

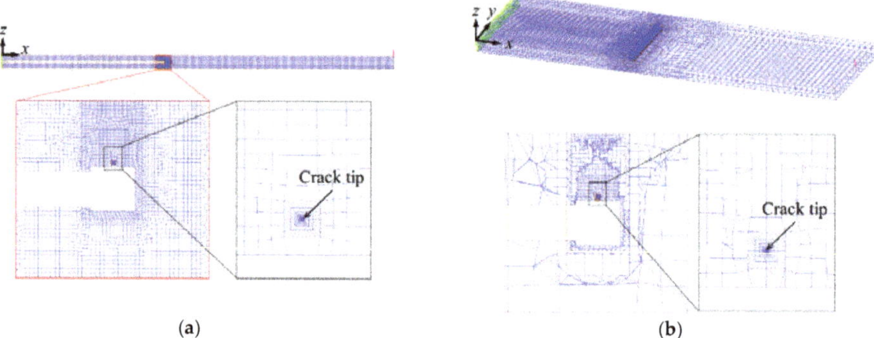


**Figure 3.** Mesh of the finite element (FE) models: (**a**) Model 2D, $a$ = 0.5 mm; (**b**) Model 3D, $a$ = 0.5 mm.

**Table 2.** Total number of nodes and elements of models.

| Model | Crack Size, $a$ (mm) | Total Number of Nodes | Total Number of Elements |
|---|---|---|---|
| Model 2D | 0.5 | 8044 | 6479 |
| | 1.0 | 7600 | 6403 |
| | 1.5 | 8043 | 6479 |
| Model 2D-Anisotropy | 0.5 | 8044 | 6479 |
| | 1.0 | 8057 | 6491 |
| | 1.5 | 7683 | 6359 |
| Model 2D-TBC | 0.5 | 20,181 | 14,559 |
| | 1.0 | 20,195 | 14,571 |
| | 1.5 | 20,181 | 14,559 |
| Model 3D | 0.5 | 318,023 | 189,422 |
| | 1.0 | 158,950 | 151,767 |
| | 1.5 | 318,118 | 189,891 |

## 3. Results and Discussion

### 3.1. Two-Dimensional Models

#### 3.1.1. Model 2D

Figure 4 shows the $z$-direction displacement of the surface of Model 2D. Macroscopically, no difference is observed between the four types of test specimens. Therefore, in order to reveal the influence of the existence and size of the crack on displacement in the $z$-direction, the difference in the $z$-direction displacement of the models with and without a crack is shown, as illustrated by Figure 5. In addition, it is found that the difference increases as the crack grows. Furthermore, it is found that there is a sharp change in the inclination near the position where $x$ = 247.5 mm, where the crack is introduced. These height distribution differences are in the order of several microns to several tens of microns, which are sufficiently large with respect to the vertical DHM resolution of 0.2 nm. Therefore, it may be possible to detect cracks by ascertaining this change by DHM.

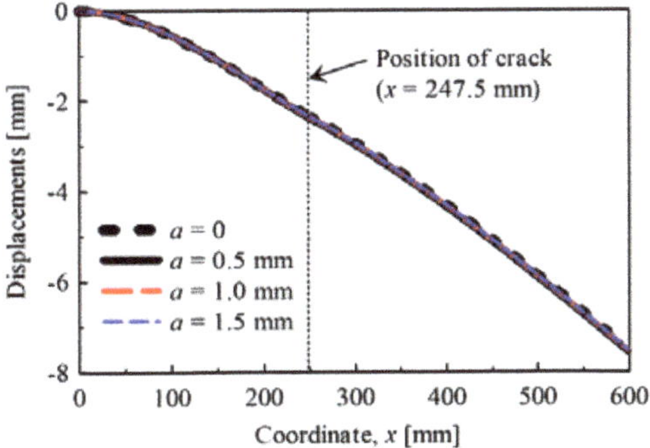

**Figure 4.** The $z$-displacement of the upper surface of Model 2D. The $z$-displacement implies the change in the $z$-direction due to loading.

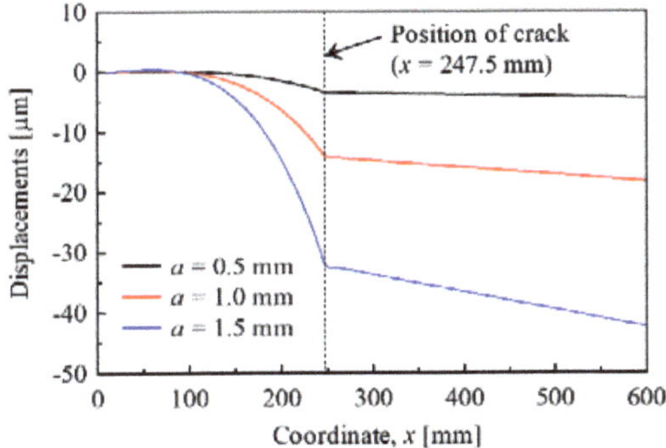

**Figure 5.** The difference in the $z$-displacement between the models with and without a crack.

The effects of internal cracks on the difference in the change of the top surface height are discussed based on the results of stress and strain analysis. First, regarding the stress around the crack, it is required to determine the dominant stress among the normal stress ($\sigma_x$), which causes Mode I type deformation, and the shear stress ($\tau_{xz}$), which leads to Mode II type deformation. Figure 6 shows the distribution of each stress above the cooling passage for each crack length. As shown in Figure 6a, when there is no crack, it can be observed that the proportion comprising the part with large tensile stress in the $x$-direction is higher than the stress in the $z$-direction and shear stress. As shown in Figure 6b–d, due to the introduction of the crack, the region where the tensile stress in the $x$-direction is 150 MPa or more is found to extend over a broad range compared to the tensile stress and shear stress in the $z$-direction in the vicinity of the crack tip. The ratio of $\sigma_x/\tau_{xz}$ at the same location near the crack tip in the same model, which roughly corresponds to the ratio of $K_I/K_{II}$, is approximately 2.3–2.4. $K_I$ and $K_{II}$ are the stress intensity factors for Model I and Mode II, respectively. From this, Mode I deformation is considered to have a greater effect than Mode II deformation. Additionally, as the crack length increases, the region in which each stress value is large expands. Subsequently, to clarify the influence of stress on the top surface height, Figure 7 shows the strain distribution in the $z$-direction above the cooling passage at each crack length. It is found that the strain distribution changes significantly around the crack in all the models. Additionally, in the range $x = 240$–245 mm and 250–255 mm, the compressive strain in the $z$-direction is found to increase as the crack length increases. This is considered to be the cause of the increase in the difference in change with the increase in the crack length. However, tensile strain in the $z$-direction is confirmed at the upper part of the crack tip at $x = 247.5$ mm in Figure 7. This is considered to be the cause of the inclination changing sharply in Figure 5. Additionally, as the crack length increases, the tensile strain at the tip of the crack spreads over a wider region. As described above, the stress distribution and strain distribution of the material under a low load is considered to change depending on the existence and size of the crack in the material. The existence and size of the crack change the cross-section. The cross-section change produces the macroscopic change in strain distribution. Moreover, the crack opens due to loading and the crack opening produces the microscopic change in strain distribution near the crack tip. The changes in macroscopic and microscopic strain distributions lead to the $z$-displacement change, which is expressed as the difference in the top surface height of the materials with and without a crack.

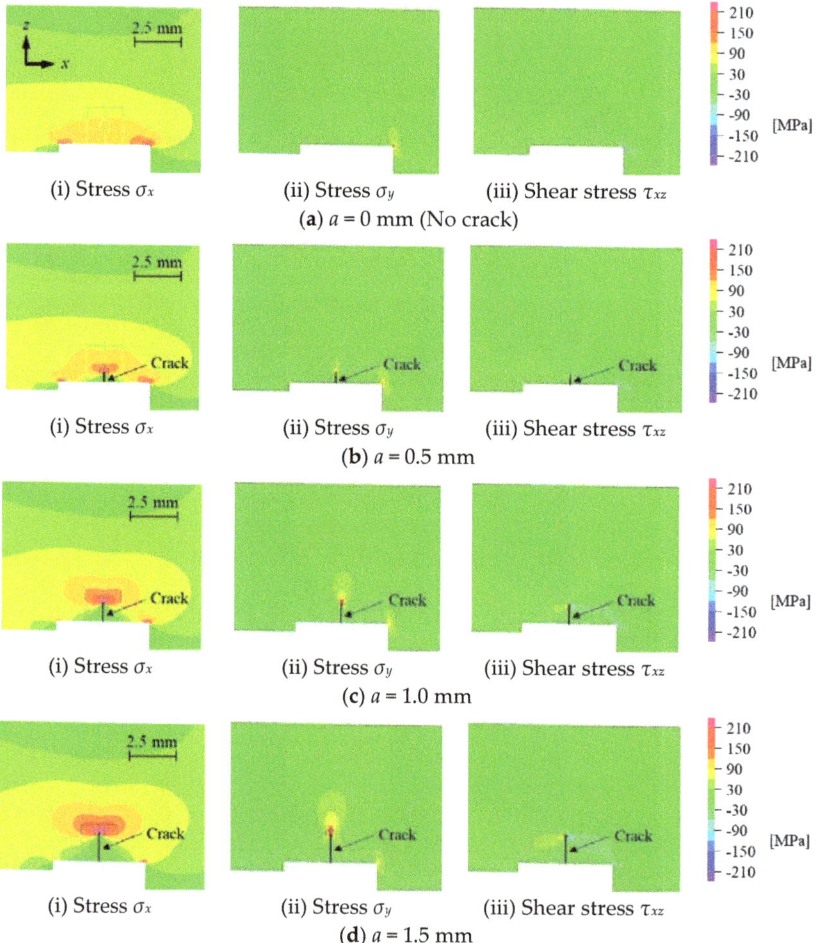

**Figure 6.** The stress $\sigma_x$, $\sigma_y$, and shear stress $\tau_{xz}$ distribution above the cooling passage of Model 2D, $a = 0$, 0.5, 1.0, 1.5 mm. (**a**) $a = 0$ mm (No crack); (**b**) $a = 0.5$ mm; (**c**) $a = 1.0$ mm; (**d**) $a = 1.5$ mm.

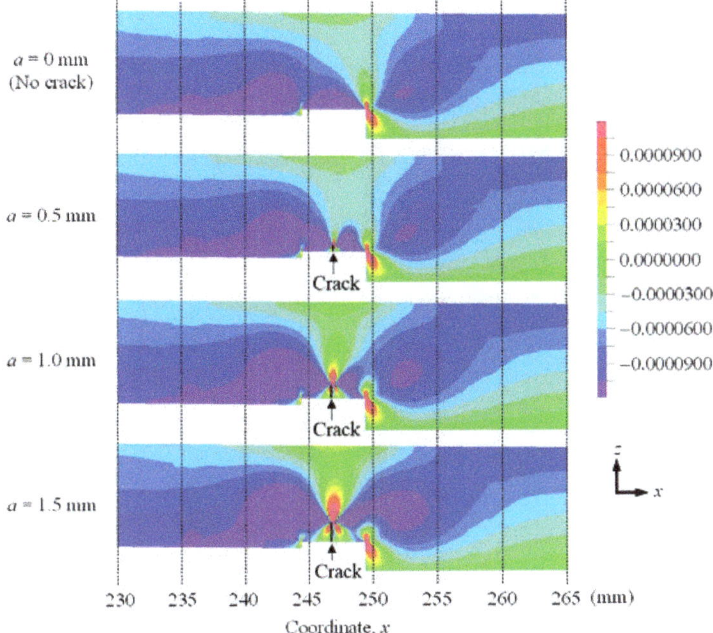



**Figure 7.** The strain distribution ($\varepsilon_z$) of Model 2D, $a$ = 0, 0.5, 1.0, 1.5 mm.

### 3.1.2. Model 2D-Anisotropy

Figure 8 shows the difference in the $z$-direction displacement of Model 2D-Anisotropy with and without a crack. The observed behavior is the same as that of Model 2D. Therefore, it is considered that cracks can be detected by using the difference in the top surface height of the materials with and without a crack, even in anisotropic materials. In addition, the displacement of Model 2D-Anisotropy is found to be slightly larger than that of Model 2D. This is because, as shown in Figure 9, the compressive strain in the $z$-direction is larger than that of Model 2D, shown in Figure 7. This may be because the Young's modulus of Model 2D-Anisotropy is less than that of Model 2D.

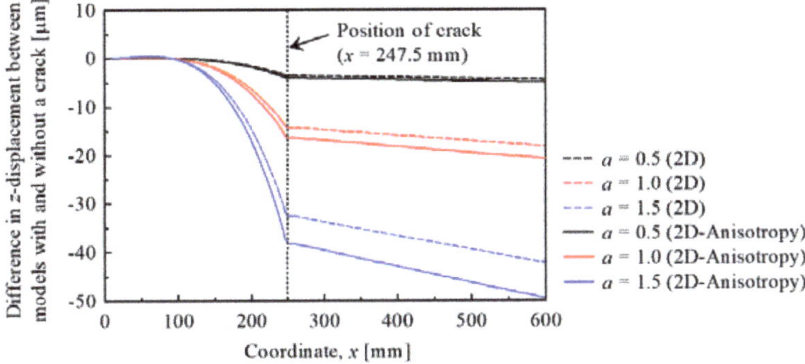



**Figure 8.** The difference in the $z$-displacement between the models with and without a crack of Model 2D-Anisotropy.

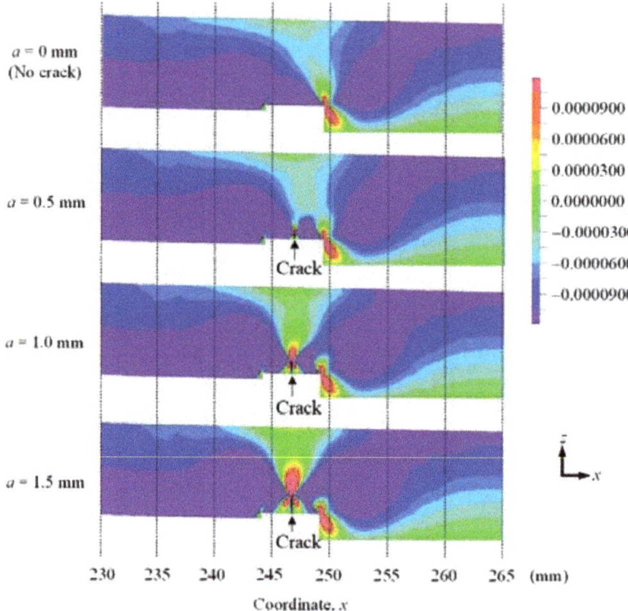



**Figure 9.** The strain distribution ($\varepsilon_z$) of Model 2D-Anisotropy, $a = 0$, 0.5, 1.0, 1.5 mm.

### 3.1.3. Model 2D-TBC

Figure 10 shows the difference in the z-direction displacement of Model-TBC with and without a crack. The observed behavior is the same as that of Model 2D and Model 2D-Anisotropy. Therefore, it is considered that cracks can be detected by using the difference in the top surface height of the materials with and without a crack, even in materials with a heat coating. In addition, the displacement of Model 2D-TBC is found to be slightly less than that of Model 2D. Figure 11 shows the z-direction strain distribution of Model 2D-TBC. It is found that the part where the compressive strain in the z-direction is large is slightly smaller compared to Figure 7. This may be because the applied load was the same as the other models, and the applied stress was reduced by the thickness of the coating.

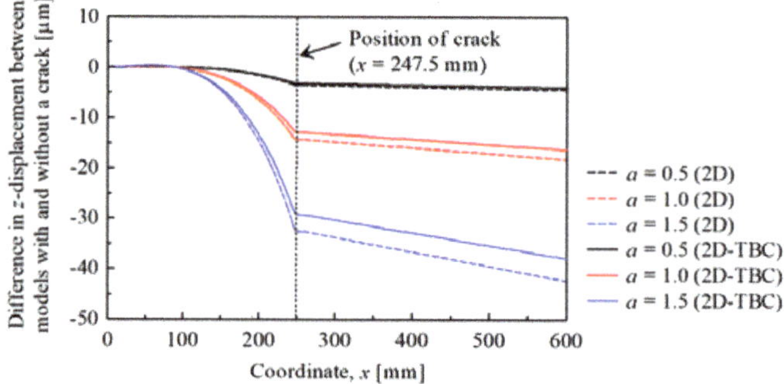



**Figure 10.** The difference in the z-displacement between the models with and without a crack of Model 2D-TBC.

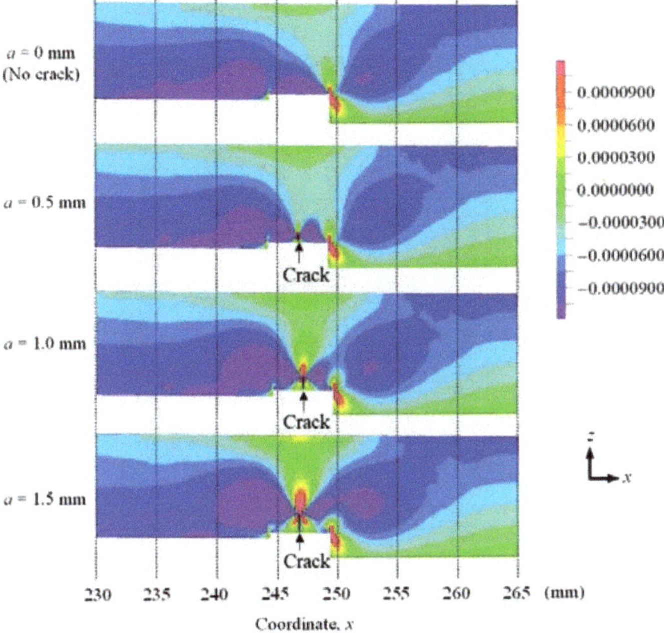



**Figure 11.** The strain distribution ($\varepsilon_z$) of Model 2D-TBC, $a = 0, 0.5, 1.0, 1.5$ mm.

### 3.2. Three-Dimensional Model

Figure 12 shows the $z$-direction displacement for trace line 1 of Model 3D. Macroscopically, no difference is observed between the four types of test specimens, the same as for Model 2D. Therefore, in order to reveal the influence of the existence and size of the crack on the $z$-direction displacement, the difference in the $z$-direction displacement of a model with and without a crack is shown in Figure 13. It is found that the difference increases as the crack grows. The difference in the $z$-direction displacement differs from that of the 2D model. In the 3D model, the height rises once at around $x = 200$ mm, and then decreases. The minimum height is at $x = 247.5$ mm, which is the location of the crack. Subsequently, it decreases gradually. These height distribution differences are in the order of several nanometers to several tens of nanometers, which are large with respect to the vertical DHM resolution of 0.2 nm. Therefore, it may be possible to detect cracks by ascertaining this change through DHM. The effects of internal cracks on the difference in the change of the top surface height are discussed based on the results of stress and strain analysis. First, similar to the 2D Model, regarding the stress around the crack, it is required to determine the dominant stress among the normal stress ($\sigma_x$), which causes Mode I type deformation, and the shear stress ($\tau_{xz}$), which leads to Mode II type deformation. Figure 14 shows the distribution of each stress above the cooling passage for the model with no crack and the models with cracks. The top surface of the model is trace line 1. As shown in Figure 14a, when there is no crack, it can be observed that the proportion comprising the part with a large tensile stress in the $x$-direction is higher than the stress in the $y$- and $z$-directions and the shear stress. As shown in Figure 14b, due to the introduction of the crack, the region where the tensile stress in the $x$-direction is 40 MPa or more, is found to extend more widely compared to the tensile stress in the $y$- and $z$-directions and shear stress in the vicinity of the crack tip. From this, the Mode I deformation is considered to have a greater effect than the Mode II deformation. Next, to clarify the influence of stress on the top surface height, Figure 15 shows the strain distribution in the $z$-direction above the cooling passage for the model with no crack and the model with a crack. The top surface of the model is trace line 1. The strain distribution in the range $x = 230$–$265$ mm is shown. It is found that the strain distribution changes significantly around the crack in the range $x$ = 245–250 mm. Furthermore, at the upper part of the crack tip at $x$ = 247.5 mm in Figure 15, tensile strain in the $z$-direction is confirmed. In addition, in the range $x$ = 240–245 mm and 250–255 mm, the part with a large compressive strain, which is lower than −0.0001050, in the $z$-direction of the model with a crack is found to be slightly broader than that in the model without a crack. This is considered to be the reason for the descent in the range $x$ = 240–245 mm and 250–255 mm and the inclination changing sharply in Figure 13.

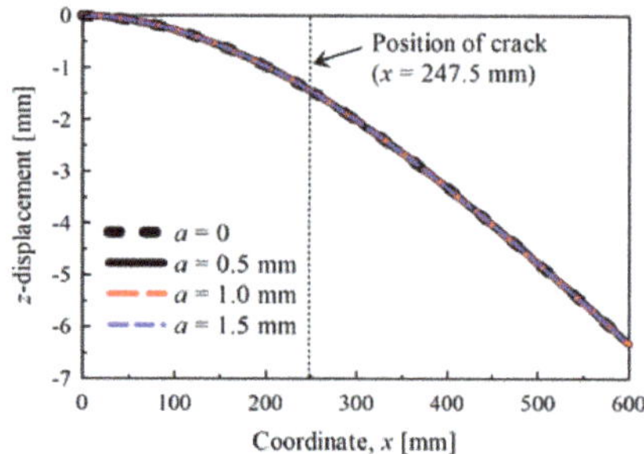


**Figure 12.** The $z$-displacement of the surface of Model 3D along trace line 1. The $z$-displacement implies the change in $z$-direction due to loading.

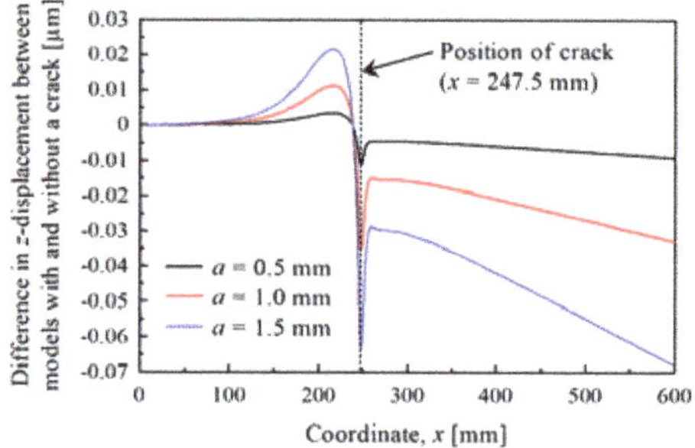


**Figure 13.** The difference in the $z$-displacement between the models with and without a crack of Model 3D along trace line 1.

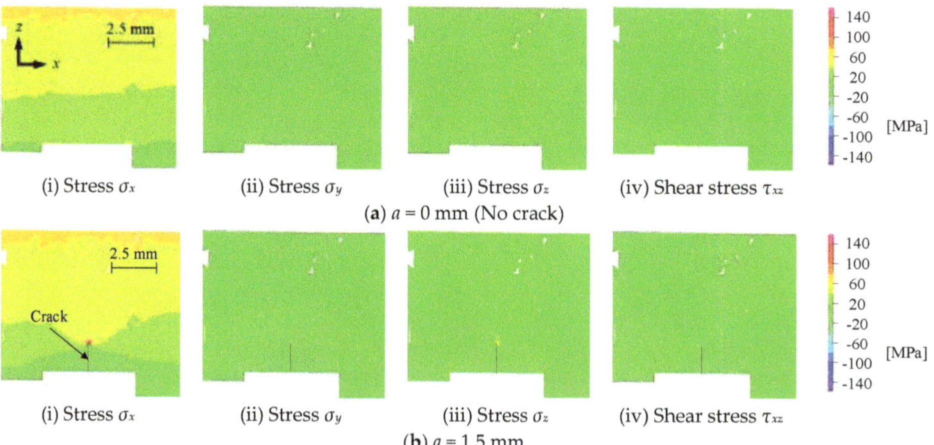


**Figure 14.** The stress $\sigma_x$, $\sigma_y$, $\sigma_z$, and shear stress $\tau_{xz}$ distribution above the cooling passage of Model 3D, $a$ = 0, 1.5 mm. (**a**) $a$ = 0 mm (No crack); (**b**) $a$ = 1.5 mm.

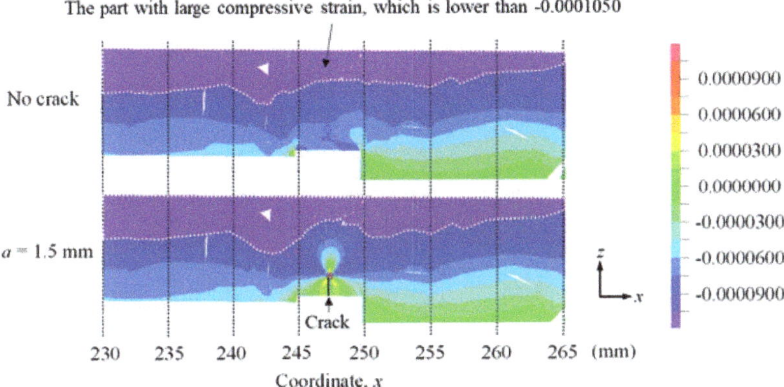



**Figure 15.** The strain distribution ($\varepsilon_z$) of Model 3D, $a = 0$, 1.5 mm.

In addition, the results of trace lines 2 and 3 are shown with the purpose of investigating whether crack detection was possible in the case where a crack generated in the turbine blade was traced in parallel, and in the case where a portion slightly away from the crack was traced in parallel with the crack. Figure 16 shows the $z$-direction displacement for trace line 2 of Model 3D. The largest displacement in the $z$-direction in the negative direction exists at $x = 75$ mm where the crack exists. There is no striking difference in the macro displacement depending on the presence or absence of the crack. Accordingly, Figure 17 shows the difference in displacement in the $z$-direction between the models with and without a crack. It can be observed that the displacement in the minus direction is large due to the presence of the crack, especially when $x$ is approximately 75 mm in the model with a crack. As the change is in the order of tens of nm, it is considered that the change can be detected by DHM. Figure 18 shows the $z$-direction displacement for trace line 3 of Model 3D. Similar to the results of trace line 2 shown in Figure 16, the displacement in the $z$-direction is the largest in the negative direction at $x = 75$ mm where the crack exists. There is no remarkable difference in the macro displacement according to the presence or absence of a crack. Accordingly, Figure 19 shows the difference in displacement in the $z$-direction between the models with and without a crack. Due to the crack, the displacement in the negative direction is large at the ends, but the displacement in the positive direction is large at the center. This corresponds to the raised part shown in Figure 13. As the change is in the order of tens of nm, it is considered that the change can be detected by DHM.

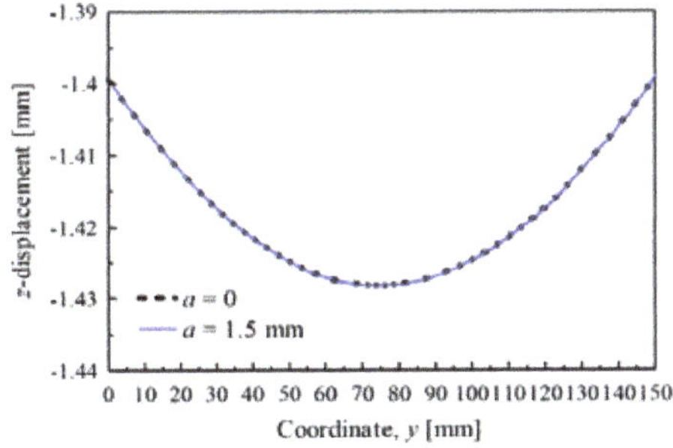



**Figure 16.** The $z$-displacement of the surface of Model 3D along trace line 2. The $z$-displacement implies the change in $z$-direction due to loading.

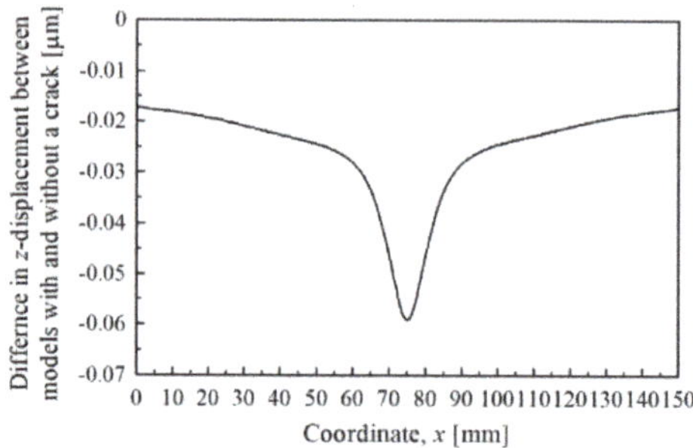

**Figure 17.** The difference in the *z*-displacement between the models with and without a crack of Model 3D along trace line 2.

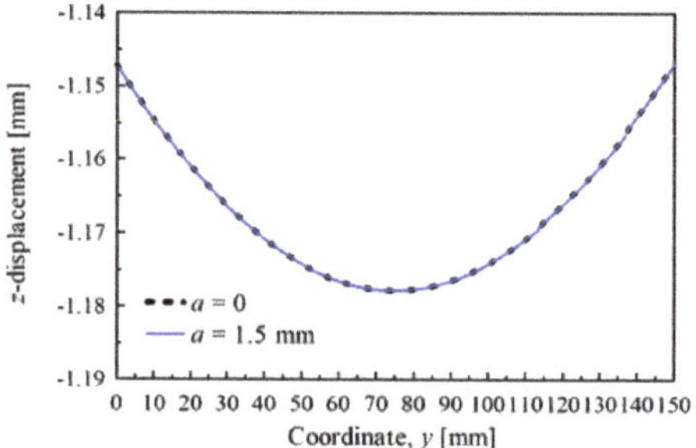

**Figure 18.** The *z*-displacement of the surface of Model 3D along trace line 3. The *z*-displacement implies the change in *z*-direction due to loading.

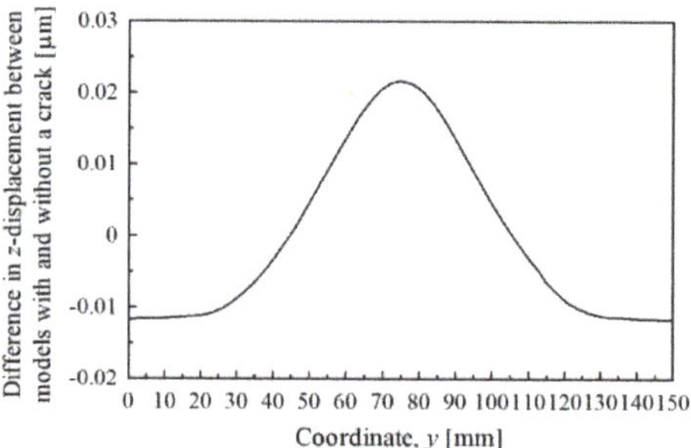

**Figure 19.** The difference in the *z*-displacement between the models with and without a crack of Model 3D along trace line 3.

As described above, the results of trace lines 1, 2, and 3 indicate that in all cases, i.e., when the trace is perpendicular, parallel to the crack, and slightly away from and parallel to the crack, the difference in displacement in the *z*-direction varies from nanometers to tens of nanometers according to the presence or absence of a crack of several millimeters. This suggests that DHM can be used to detect cracks in turbine blades.

## 4. Conclusions

In this study, we conducted FE analysis using 2D and 3D models that simulated turbine blades for the cases where an internal crack detection method using a digital holographic microscope and digital height correlation method were applied to turbine blades. The results acquired are shown below.

(1) Analysis of the 2D model clarified that the change in the surface properties under a small load differs according to the presence or absence of a crack, and elucidated the strain distribution that caused the difference in the change. In addition, analyses of the 2D models, taking into account the anisotropy of the material and the heat-resistant coating, were conducted. The difference in the change in the surface properties and strain distribution, according to the presence or absence of cracks in these models, was elucidated.

(2) Analysis of the 3D model clarified that the change in the surface properties under a small load differs according to the presence or absence of a crack, and elucidated the strain distribution that caused the difference in the change. In addition, it was found that the longer the crack, the greater the difference in the change in the top surface height distribution compared to the case without a crack. For a crack of approximately several millimeters, the difference in the change in the top surface height distribution is approximately several nanometers to several tens of nanometers, even if the measurement point is at a distance from the crack. As the value is large with respect to the DHM vertical resolution of 0.2 nm, it is clarified that the change can be detected by DHM.

**Author Contributions:** Conceptualization, N.T.; methodology, H.K., T.U. and N.T.; validation, H.K.; formal analysis, H.K. and J.S.; investigation, H.K. and J.S.; resources, N.T.; data curation, H.K. and J.S.; writing—original draft preparation, H.K. and J.S.; writing—review and editing, N.T. and T.U.; visualization, H.K. and J.S.; supervision, N.T.; project administration, N.T.; funding acquisition, N.T. All authors have read and agreed to the published version of the manuscript.

**Funding:** This research was funded by JSPS KAKENHI Grant Number JP18H01337.

**Conflicts of Interest:** The authors declare no conflict of interest.

## References


1. Gupta, S.; Chaube, A.; Verma, P. Review on heat transfer augmentation techniques: Application in gas turbine blade internal cooling. *J. Eng. Sci. Technol. Rev.* **2012**, *5*, 57–62. [CrossRef]
2. Effendy, M.; Yao, Y.; Sugati, D.; Tjahjono, T. Numerical study of pin-fin cooling on gas turbine blades. *AIP Conf. Proc.* **2019**, *2114*, 060022. [CrossRef]
3. Bose, S.; DeMasi-Marcin, J. Thermal barrier coating experience in gas turbine engines at Pratt & Whitney. *J. Therm. Spray Technol.* **1997**, *6*, 99–104.
4. Arai, M.; Suidzu, T. Porous ceramic coating for transpiration cooling of gas turbine blade. *J. Therm. Spray Technol.* **2013**, *22*, 690–697. [CrossRef]
5. Japanese Industrial Standards. *JIS Z 2300:2009*; Japanese Industrial Standards: Tokyo, Japan, 2009.
6. Yoshioka, Y.; Yamashita, A. Degradation/damage measurement and life assessment of high-temperature components III: Heavy-duty and aero gas turbines. *J. Soc. Mat. Sci. Jpn.* **2009**, *58*, 649–656. (In Japanese) [CrossRef]
7. Tada, N.; Hamada, S.; Teramae, T.; Yoshino, S.; Suzuki, T. A method of crack detection in the turbine blade using digital holographic microscopy (DHM). In Proceedings of the ASME 2011 Pressure Vessels and Piping Conference 2011, PVP2011-57299, Baltimore, MD, USA, 17–21 July 2011; American Society of Mechanical Engineers (ASME): New York, NY, USA, 2011; pp. 211–216.
8. Tada, N.; Uchida, M.; Uenoyama, Y. Non-destructive crack detection by nanometric change in surface profile using digital holographic microscope. In Proceedings of the ASME 2012 Pressure Vessels and Piping Conference 2012, PVP2012-78425, Toronto, ON, Canada, 15–19 July 2012; American Society of Mechanical Engineers (ASME): New York, NY, USA, 2012; pp. 251–257.
9. Tada, N.; Uchida, M.; Matsukawa, Y. Non-destructive detection of crack in HDPE plate by nanometric change in surface profile. In Proceedings of the ASME 2013 Pressure Vessels and Piping Conference 2013, PVP2013-97732, V005T10A012, Paris, France, 14–18 July 2013; American Society of Mechanical Engineers (ASME): New York, NY, USA, 2013.
10. Seth, B.B. Superalloys: The utility gas turbine perspective. In Proceedings of the 9th International Symposium on Superalloys, Champion, PA, USA, 17–21 September 2000; The Minerals, Metals & Materials Society: Warrendale, PA, USA, 2000.

11. Caron, P.; Khan, T. Evolution of Ni-based superalloys for single crystal gas turbine blade application. *Aerosp. Sci. Technol.* **1999**, *3*, 513–523. [CrossRef]
12. Xu, Q.; Yang, C.; Zhang, H.; Yan, X.; Tang, N.; Liu, B. Multiscale modeling and simulation of directional solidification process of Ni-based superalloy turbine blade casting. *Metals* **2018**, *8*, 632. [CrossRef]
13. Arai, M.; Shimizu, Y.; Suidzu, T. On damage process of ceramic thermal barrier coatings subjected to high-temperature tensile loading. *Trans. JSME* **2015**, *81*, 825. (In Japanese)

*Article*

# Experimental and Numerical Investigation of Striker Shape Influence on the Destruction Image in Multilayered Composite after Low Velocity Impact

Sebastian Sławski [1,*], Małgorzata Szymiczek [1], Jarosław Kaczmarczyk [1], Jarosław Domin [2] and Sławomir Duda [1]

1    Department of Theoretical and Applied Mechanics, Silesian University of Technology, Konarskiego 18a, 44-100 Gliwice, Poland; malgorzata.szymiczek@polsl.pl (M.S.); jaroslaw.kaczmarczyk@polsl.pl (J.K.); slawomir.duda@polsl.pl (S.D.)
2    Department of Mechatronics, Silesian University of Technology, Akademicka 10a, 44-100 Gliwice, Poland; jaroslaw.domin@polsl.pl
*    Correspondence: sebastian.slawski@polsl.pl; Tel.: +48-32-237-12-87

Received: 4 December 2019; Accepted: 27 December 2019; Published: 31 December 2019

**Abstract:** The paper presents results obtained by experimental and numerical research focusing on the influence of the strikers' geometry at the images of the destruction created in hybrid composite panels after applying impact load. In the research, the authors used four strikers with different geometry. The geometries were designed to keep the same weight for each of them. The composite panels used in the experiment were reinforced with aramid and carbon fabrics. An epoxy resin was used as a matrix. The experiments were carried with an impact kinetic energy of 23.5 J. The performed microscopy tests allowed for determination of destruction mechanisms of the panels depending on the geometry of the striker. The numerical calculations were performed using the finite element method. Each reinforcement layer of the composite was modeled as a different part. The bonded connection between the reinforcement layers was modeled using bilateral constraints. That approach enabled engineers to observe the delamination process during the impact. The results obtained from experimental and numerical investigations were compared. The authors present the impact of the striker geometry on damage formed in a composite panel. Formed damage was discussed. On the basis of the results from numerical research, energy absorption of the composite during impact depending on the striker geometry was discussed. It was noted that the size of the delamination area depends on the striker geometry. It was also noted that the diameter of the delamination area is related to the amount of damage in the reinforcing layers.

**Keywords:** hybrid composite; damage; aramid fiber; carbon fiber; finite element method; delamination

## 1. Introduction

Energy absorbing panels are installed wherever there is a risk of damage to important machine components, human life, or health risk. Nowadays, energy-absorbing covers are used in many fields, such as energy, automotive, rail, aviation, mining, and maritime industries [1–3]. The most demanding branch of industry in which we deal with energy-absorbing shields is the military industry [4]. In the case of the military application, energy absorbing panels depending on this purpose must meet the relevant requirements set out in the standards [5,6]. Depending on the application, the designed covers must meet a number of different requirements. The important criterion is of course the degree of energy absorption. However, an increasingly important aspect of the newly designed solutions is to minimize their weight [7,8]. Therefore, the demand for solutions using composite materials is growing [9]. Polymer composites in addition to high strength are also characterized by a low weight and high value of specific strength [4,10]. Depending on the application, composite shields can be subjected to the impacts of stones, hail, birds, bullets, or explosive debris. Therefore, resistance to impact loads perpendicular to the plane of laminate reinforcement layers is particularly important. Composite energy absorbing panels in the military application are used in various types of solutions. They are used both in personal protective equipment in the form of bulletproof vests or helmets [11–13], as well as in special purpose vehicles [9,12,14,15]. In the case of combining a polymer composite material with a ceramic material [7,11], the ceramic layer is responsible for defragmenting the striker and changing its trajectory, as a result of which a large part of the impact energy is absorbed. The layer made of fiber composite is designed to capture the remains of the striker, the defragmented ceramic layer, and the striker's inhibition. In the case where a multilayer panel with fiber reinforcement is used individually, it must stop the striker and absorb all of the impact energy. Impact loads cause exceeding the strength of the polymer matrix in the form of shear and bending. A delamination process is initiated between successive layers of reinforcing fabrics [16]. This is also a disadvantage of fibrous composites, consisting of the loss of connection between adjacent layers of reinforcing material, which leads to decreasing of the composite strength. During the continuous operation of composite products, this is particularly important because, in combination with fatigue strength [17], it leads to a significant weakening of the material. Delamination in the multilayered composites could be located using, for example, the wave propagation method [18,19]. The material susceptibility for delamination depends on many factors. Some of them could be mentioned, for example, the material used as an reinforcement, its properties, or bond quality between the fiber and the matrix [20]. Material delamination occurs because the energy threshold initiating this process is much lower than the energy threshold causing destruction of the composite reinforcement fibers [21]. In other words, delamination occurs because the strength of the matrix material is much less than the strength of the reinforcing material. Therefore, the largest part of the impact energy is absorbed as a result of fiber destruction [16,22].

The process of destroying the energy absorbing panel made of laminate can be divided into two phases. In the first phase, a high-speed striker hits the panel, causing fiber shear and matrix cracking. Primary yarns carry the highest loads. Secondary yarns are much less stressed. In the second phase, as the successive penetration of reinforcement layers progresses, the smaller and smaller energy of the striker is distributed over an ever larger surface, and finally it becomes insufficient to break the fibers in subsequent reinforcing layers [13,16]. The remains of the impact energy initiates the matrix delamination process. The reinforcement fibers in deeper layers are stretched under the influence of the impact energy concentration [12,16,23]. The microscopic image of damage in the composite with the hybrid reinforcement that was used in the conducted research is shown in Figure 1. The presented sample was hit by the hemispherical striker. The jagged fragments of the fibers at the side opposite to the impact side are the result of the cutting process. In Figure 1, it can be seen that the aramid fibers are deformed, broken, and sheared as a result of the striker impact. Around the impact area, accumulation of the damaged reinforcing layers could also be observed. The carbon fibers located between the aramid fibers owing to the high Young's modulus were deformed elastically and rebounded the striker.

The total energy at impact is the sum of the remaining kinetic energy of the striker and the energy absorbed by the composite panel. The total energy during impact could be divided at the energy of the moving striker, energy absorbed by shear plugging, energy absorbed by deformation of secondary yarns, energy absorbed by tensile failure of primary yarns, energy absorbed by delamination, energy absorbed by matrix cracking, and energy absorbed by friction [22].

Considering energy absorption, the most commonly used reinforced material is aramid fibers [16]. These fibers can absorb a large amount of the impact energy before their breaking. Furthermore, aramid fibers are more elastic than, for example, carbon fibers [24]. They are characterized by high rigidity and orientation, and they are connected to each other by strong, dense hydrogen bonds [25]. They have a good strength to density ratio (specific strength). Strength expressed in this way for Kevlar® fibers is greater; that is, five times greater than steel [16]. The tensile curves of aramid fibers, similar to glass and graphite fibers, are approximately linear to break. Aramid fibers have a density

43% lower than that of glass fibers. Because of that, aramid fibers are particularly attractive for the production of many composites [25]. Energy-absorbing panels built using aramid fibers provide excellent protection against pistol bullets, revolver bullets, or fragmenting debris. Aramid fibers are one of the most important materials used for the production of energy absorbing panels [16]. However, in the application of energy absorbing panels, stiffness is also very important and should be enough high to prevent too much deformation after impact. To minimize that deformation employment of carbon fibers between aramid fibers, reinforcement layers could be a solution. However, carbon fibers have some disadvantages. The first of them is the brittle damage [26] after impact, which can be dangerous for nearby people. It is worth mentioning that the brittle damage in some cases can be desirable, especially during modelling of the separation process [27–30]. In this case, the process should be designed in such a way that the equivalent stress should exceed the allowable stress that appears in the direct cutting zone. Otherwise, the material being removed would not be separated. The proper combination of different reinforcing materials allows the composite to take advantage of each of the applied fibers [16,31,32]. Undoubted advantages of the carbon fiber reinforced polymer composites are their low density, good static and fatigue strength, high modulus of elasticity, resistance to abrasion, and corrosion resistance [25]. Because they consist almost exclusively of graphite, they are non-melting and chemically resistant. The heat resistance of carbon fibers is unique and outperforms any known materials in this respect, except graphite. The high values of Young's modulus mean that this fiber is often used in hybrid composites to increase the stiffness of the structure [1,20].

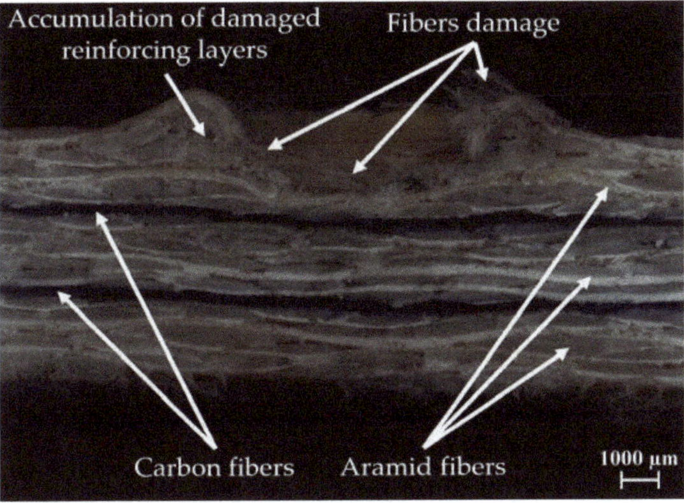


**Figure 1.** Microscopic image of damage formed in a multilayered composite with hybrid reinforcement after impact of the hemispherical striker.

Naik et al. [33] showed that the proper configuration of reinforcing layers in a glass–carbon epoxy composite could increase post-impact compressive force. It was also shown that the damage area in a hybrid composites depends on the reinforcing material configuration. Studies presented in [34] also show that the hybridization in glass–carbon epoxy composites provide greatly enhanced damage tolerance of these structures. The experimental results show that the hybrid composites can absorb more energy in the impact event compared with non-hybrid composites [35]. Aramid fibers are often used in the case of the personal protective equipment, such as helmets. In order to increase the amount of information about the examined object, experimental and numerical research is carried out [36–38]. In another paper [39], the impact resistance of composite panels reinforced with aramid fibers and the matrix made from various thermoplastic materials were compared. The authors of [40] compared the puncture resistance of the composites reinforced by 2D and 3D aramid fabrics. A further paper [41] presents the analysis of puncture resistance of aramid laminates on styrene–butadiene–styrene and epoxy resin matrix. The authors of [42] present the influence of the introduction of nanoclay into the resin on the increase of the maximum impact load of aramid fiber composite. Owing to the popularity of aramid fibers, research on composites with hybrid reinforcement (where aramid fibers were one of the used reinforcing materials) was also conducted. Research focused on a hybrid composite with reinforcement made of aramid and basalt fibers [43] shows that the use of both types of reinforcing materials in the proper configuration can increase the composite energy absorption during impact. The comparison of the hybrid composites with reinforced made of aramid and carbon fibers, based on the DI parameter (defined as the ratio between the damage propagation energy and the damage initiation energy), presented in the work of [35], showed that the highest value of this parameter was achieved for the composite in which reinforcing layers made of aramid fibers were used alternately with carbon fiber reinforcement layers. It was also noticed that the adjacent aramid fibers can also play a role in bridging the broken carbon fibers, which could improve the toughness of hybrid composites [35]. Studies on hybrid reinforcement made of aramid and carbon fibers in a sandwich structure [44] showed that the use of hybrid reinforcement increases the energy absorption of the composite during low speed impact. The authors of [44] obtained the highest values of absorbed energy in the case of a combination of three reinforcing layers of carbon fibers for one reinforcing layer of aramid fibers, and vice versa. However, the reduction of compressive strength after the impact was much smaller in the case where three reinforcing layers of aramid fibers for one reinforcing layer of carbon fibers was used. The damage area formed in the composite and its impact energy absorption during low velocity impact depend on the striker geometry [45]. Experimental and numerical research [45,46] showed that the influence of the striker geometry was changed with the impact velocity and the thickness of the composite.

The authors of that paper decided to assesses the influence of the striker geometry on the damage formed in the epoxy composite with hybrid reinforcement (made from aramid and carbon fabrics) after low velocity impact. Differences in the formed damage depending on the striker geometry were described. The authors of the present work decided to perform experimental and numerical research. The numerical calculations were performed in order to increase the amount of information about damage caused in the composite panel after impact of the strikers with different geometry; in particular, information about delamination between reinforcing layers. In Section 2, the authors present materials used as a reinforcement and as a matrix. The producing method of the composite and the adopted research methodology are also presented. The results of experimental research, which was carried out using four strikers with different geometries, are presented and discussed on the basis of microscopic images. In Section 3, the authors describe the process of preparation of the physical model. A methodology of modeling of the multilayered composite is presented. The damages caused in the composite plate as well as the phenomenon of delamination were presented and discussed. The authors present the values of the rebounded strikers' kinetic energy and the relationship between the amount of damage in the reinforcing layers and the occurring diameter of the delamination area. The experimental and numerical results were compared. The differences in obtained results were discussed. Composite panel wear was considered in the local response.

## 2. Experimental Research

### 2.1. Methodology

During the experiment, hybrid composite panels reinforced with aramid and carbon fabrics were used. Aramid twill weave fabric used in the experiment with weight of 300 g/m$^2$ was made from Twaron 2200 fibers. The second material used as a reinforcement was carbon twill weave fabric with a weight of 200 g/m$^2$. It was made from Pyrofil TR30 S fibers. Mechanical properties of used fibers [47,48] are shown in Table 1.

**Table 1.** Mechanical properties of used reinforcing fibers.

| Parameter | Unit | Twaron 2200 | Pyrofil TR30 S |
|---|---|---|---|
| Elongation at break [%] | % | 2.9 | 1.8 |
| Tensile strength [MPa] | MPa | 2930 | 4120 |
| Tensile modulus | GPa | 102 | 235 |

Hybrid panels consisted of 14 layers of reinforcing fabrics. They were arranged in the following combination: four layers of aramid fabric, one layer of carbon fabric, four layers of aramid fabric, one layer of carbon fabric, and four layers of aramid fabric. Epoxy resin LG285 was used as a matrix. The dedicated by manufacturer HG285 curing agent was mixed with resin with a 100:40 weight ratio. The epoxy matrix's properties declared by the manufacturer [49] are shown in Table 2.

**Table 2.** Properties of epoxy resin LG285 with HG285 curing agent.

| Parameter | Unit | Value |
|---|---|---|
| Flexural modulus | MPa | 2700–3300 |
| Tensile strength | MPa | 75–85 |
| Compressive strength | MPa | 130–150 |
| Elongation at break | % | 5–6.5 |
| Hardness in Shore D scale | - | 85 |

The composite panels were manufactured by the hand-laminating method with vacuum support (−68 kPa of vacuum pressure). On the basis of the literature [6,16], four striker geometries were developed. Geometries of those strikers are shown in Figure A1. The adopted geometries allow the strikers to maintain a constant mass and to simulate the impact of various splinters. The required repeatability of the strikers' impact kinetic energy can be achieved because of this. This repeatability is important because the researchers' main goal is to determine the destruction images depending on the geometry of the strikers. They could be compared only when the impact energy was the same. Steel strikers with a mass of 49 g are shown in Figure 2.

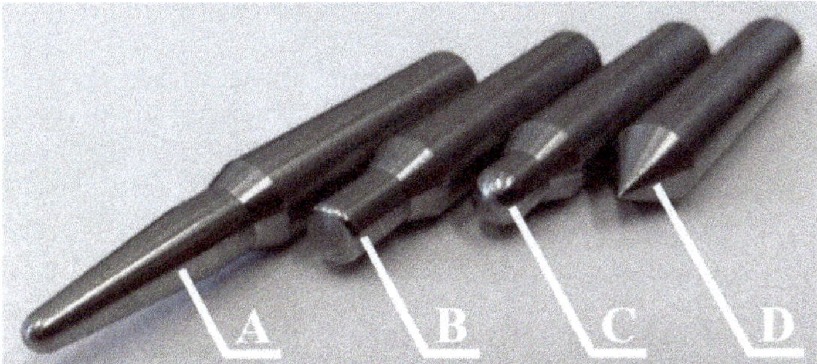


**Figure 2.** Strikers used in experiment: A—ogival striker, B—blunt striker, C—hemispherical striker, D—conical striker.

The tests were realized using a hybrid electromagnetic launcher with a pneumatic support [26,50]. Initial air pressure value in an air tank and air valves were set up using a control panel. The test stand is shown in Figure 3.

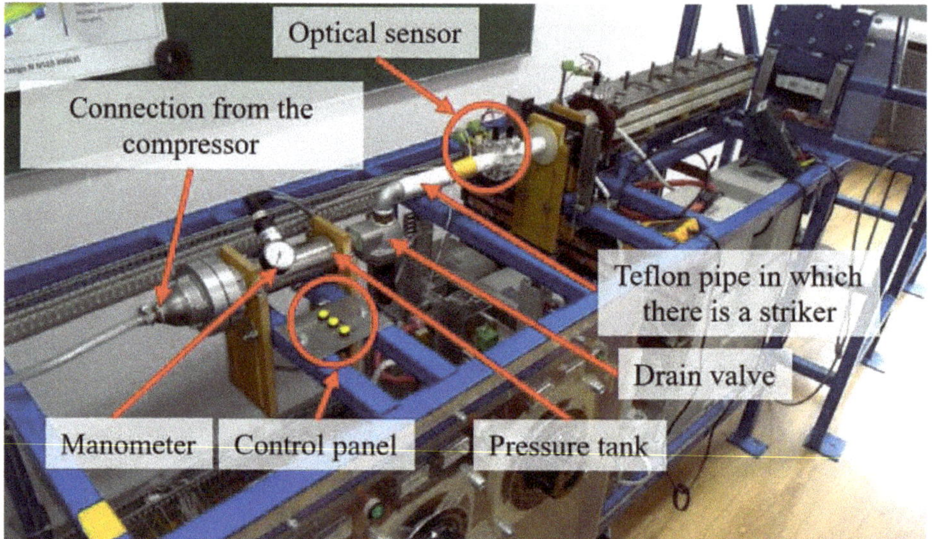


**Figure 3.** Test stand used in experiment.

The striker velocity was measured using an optical gate connected to the oscilloscope. When the striker breaks the optical beam, the voltage spike could be observed on the oscilloscope. The next voltage spike is observed when the striker ends, breaking the optical beam. The experiments were carried out with 23.5 J of impact kinetic energy. This energy refers to sub ballistic velocities of the impact—such as collisions with fast moving elements [16]. The samples were glued to the 10 mm polyethylene base plate and hit three times by each striker. The polyethylene plate was exchanged with each composite sample. There is no deformation and destruction in the polyethylene plates after impact.

*2.2. Results*

The impact striker geometry assessment on the composite panels' destruction was made on the basis of microscopic tests. The tests allowed the authors of this article to measure the diameter of the panel damages after impact. Because of the plastic character of aramid fiber destruction, the damages in the panel reflected the striker geometry, and the depth of the penetration could be calculated based on the cavity diameter and the geometry of the used striker. The results are shown in Table 3.

**Table 3.** Average experimental results, depending on striker geometry.

| Striker Geometry | Average Striker Velocity [m/s] | Average Kinetic Energy [J] | Average Cavity Diameter [mm] | Average Depth of Penetration [mm] |
|---|---|---|---|---|
| Conical | 30.99 | 23.5 | 5.9 | 5.1 |
| Hemispherical | 30.98 | 23.5 | 6.2 | 1.5 |
| Blunt | 30.92 | 23.4 | 7.0 | 0.1 |
| Ogival | 31.05 | 23.6 | 4.7 | 4.5 |

The microscope images of the damage in composite panels presented in Figure 4 show the significant influence of the striker geometry. During the same impact, the same values of impact kinetic energy and destruction level in composite panels were limited from the matrix cracking in the blunt striker case to the almost complete penetration in the conical striker case. For the blunt striker, fibers absorbed almost all impact energy—there was no fiber damage. The cracked matrix is only one visible sign after impact in this case. That fact proved the high resistance of fiber reinforced composites on the impact of blunt elements. Cross sections of the samples after impact of the strikers are presented in Figure A2.

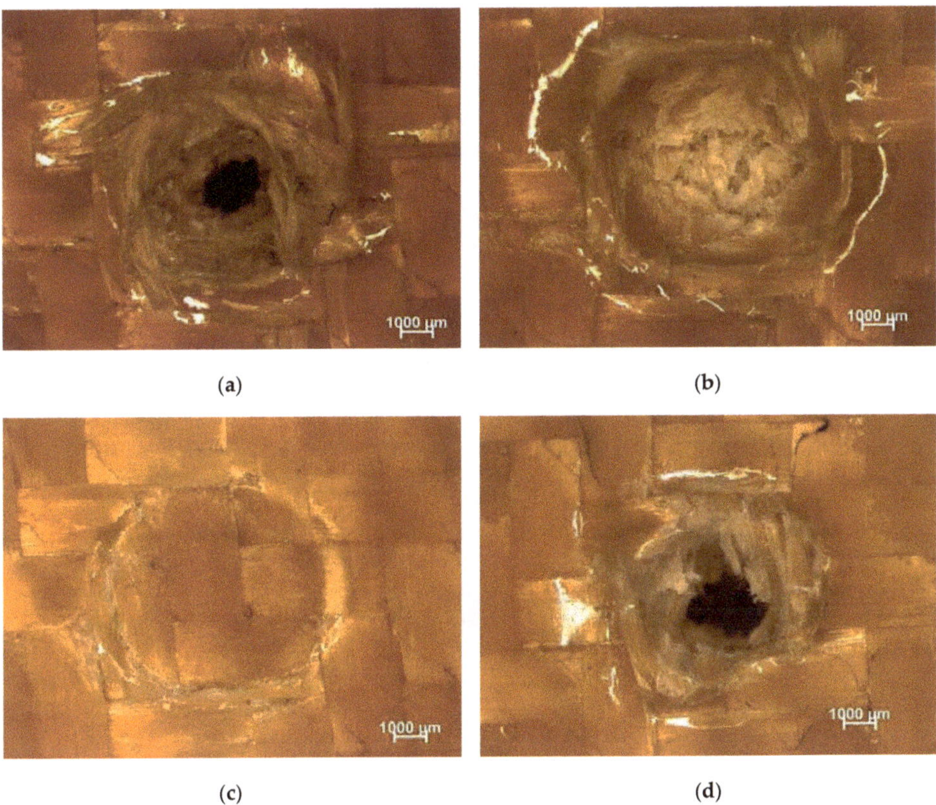

**Figure 4.** Selected samples after impact of the striker: (**a**) conical, (**b**) hemispherical, (**c**) blunt, (**d**) ogival.

The destruction caused by a hemispherical striker was characterized by a high amount of compressed fibers. Around a newly created cavity, the characteristic bulge was formed. Fibers that were initially in the first layer of the impact area (in the center of panel) were then ripped out from the matrix. The broken and compressed fibers in this formed cavity were also visible. However, their amount was low, which proved that only fibers from the first layers were damaged, and fibers in the next layers were in good shape. There were no carbon fibers in the cavity, which suggests that the panel penetration is limited to the first four layers of the reinforced material. This was also confirmed by the calculated depth of penetration. It was circa 1.5 mm, which corresponds to the thickness of approximately of three layers of used aramid fabric. The cross section of the sample (Figure A2) also confirms that the penetration of the composite stops before the reinforcing layer made from carbon fibers was damaged.

The destruction caused by an ogival striker suggests much more sensitivity of the hybrid composite to the impact of such elements. The penetration depth was much higher compared with the penetration depth in the case of a hemispherical striker. The analysis of the microscopic images shows a lot of broken fibers in the created cavity. The broken fibers were visible as sticking out parts of the reinforcing fabric directed towards to the cavity. These fibers were also compressed at the boundary of the cavity. Like in the previous case, the first layer damage around the impact zone was also visible. The penetration depth in this case was 4.5 mm, which corresponds to the depth of the second reinforcing layer made from carbon fibers. At the bottom of the created cavity, there were no carbon fibers, which suggests that the second carbon reinforced layer was not penetrated. The cross section of the formed damage presented in Figure A2 also confirms that the second reinforcing layer made from the carbon fibers was not damaged.

The conical striker caused the largest destruction in the tested material. In this case, both the cavity diameter and its penetration depth were the largest. The destruction mechanism was the same as in the case of the ogival striker. Inside the formed cavity, it was possible to observe the parts of broken fibers. In the case of this striker, fibers' deformation was also visible at the first layer around the impact area. This fact suggests that the conical geometry of the striker caused pushing fibers sideways of the creating cavity. The considered striker did not have enough energy to break fibers, so it stuck between them and pushed them sideways. The weave and weight of an applied reinforced fabric could have a critical meaning in strength aspect. As mentioned, fibers in the first layers were deformed, which confirm this fact. This also explained why damage caused by a conical and an ogival striker was largest compared with that of other strikers used.

## 3. Numerical Research

### 3.1. Preparation of Numerical Model

The numerical research was carried out using the finite element method and commercially available LS-PrePost/LS-Dyna (LSTC, Livermore, CA, USA) software. The multilayered composite could be modeled in a few scales. The modeling scale is related to the homogenization of the composite strength properties. In the case of modeling the composite as a one part, homogenization of the mechanical properties should be performed for the whole composite. Modeling in this scale makes it impossible to observe some processes occurring inside the composite material, for example, delamination, which, as mentioned in the introduction, is an energy absorbing process. A different approach is to model the composite with respect to the division on the reinforcing layers. In this approach, the strength properties should be homogenized in the level of the reinforced layer (used reinforcing fabric and the matrix). This modelling scale allows observing the delamination between the reinforcing layers. There is also the possibility to observe the movement between the adjacent reinforcing layers and the consideration of the friction between them. Woven composites could also be modeled with respect to the reinforcing fiber geometry. The mechanical properties of the fiber and the matrix are defined individually. Observation of the delamination process is also possible. Moreover, modelling in that scale allows to observe the deformation of each fiber and the friction between the fibers. If the modeling scale is more accurate, the physical model takes into account more energy absorbing phenomena. Therefore, the choice of the modeling scale and level of homogenization of strength properties has an impact on the obtained result. The selection of the modeling scale is related to use of the simplifications. Each simplification will be associated with the omission of some energy absorbing phenomenon. However, it should be remembered that, if the physical model is more advanced, the computational cost will be greater. Modeling of the composite with respect to the division on the reinforcing layers (homogenization of the strength properties in the reinforcing layer level) gives satisfactory results, which are in accordance with experimental research [51]. It was decided to model the composite material in a macro scale. It means that each reinforced layer was modeled as an individual part. This approach is widely used [51–54] in the case of impact analysis. A flat surface with dimensions of $50 \times 100$ mm (height × width) was created, and then it was discretized into the finite elements. The surface was divided into Belytschko–Tsay type shell elements with "hourglass" control based on stiffness. A total of 5000 Belytschko–Tsay shell elements were created in this way. It was assumed that the thickness of each layer, regardless of the reinforced material, was 0.5 mm, which allowed the physical model to achieve the overall thickness of the composite used in the experiment (7 mm). On the basis of this assumption, 13 subsequent surfaces were created, spaced 0.5 mm from the previous layer along the axis perpendicular to the surface of the layers. The composite model created in this way consists of 70,000 shell elements. As a composite material model, the MAT58 *MAT_LAMINATED_COMPOSITE_FABRIC [55] was used. This model requires the definition of homogeneous strength properties of the used materials. Taking into account the destruction of the physical model, the finite elements are controlled by the ERODS parameter. The ERODS parameter is calculated based on the deformation in defined fiber directions in the reinforced plane and on the shear deformation. In this material, model stress increases nonlinearly until the maximum strength is reached ($X_T$). When the maximum strength is reached, the stress is reduced by the SLIMx factor and held until material reaches the strain specified by the ERODS parameter. When strain reaches the value defined as an ERODS, the finite element is deleted [56]. The typical stress–strain curve for the selected MAT58 composite material model is shown in Figure 5.

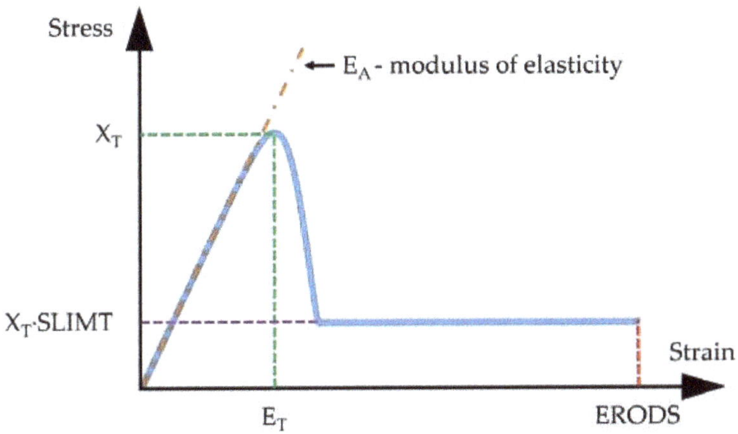


**Figure 5.** Typical stress—strain curve for the selected MAT58 composite material model.

On the basis of the research [56], ERODS = 0.4 was adopted. Because the values of SLIMx coefficients do not have their physical interpretation [57], their value was adopted in accordance with the recommendations [55]. The impact of individual factor values on the convergence with an experimental solution is presented, among others, in the work of [56]. The used material properties of epoxy composites with carbon and aramid reinforcement (in two directions) are shown in Table 4.

**Table 4.** Mechanical properties of epoxy composites reinforced by carbon and aramid fibers [58].

| Material Properties | Epoxy Resin/Carbon Fiber | Epoxy Resin/Aramid Fiber |
|---|---|---|
| Young's modulus $E_1 = E_2$ [MPa] | 70,000 | 30,000 |
| Shear modulus $G_{12}$ [MPa] | 5000 | 5000 |
| Tensile strength $X_T = Y_T$ [MPa] | 600 | 480 |
| Compressive strength $X_C = Y_C$ [MPa] | 570 | 190 |
| Shear strength S [MPa] | 90 | 50 |
| Tensile strain $\varepsilon_{XT} = \varepsilon_{YT}$ [%] | 0.85 | 1.6 |
| Compressive strain $\varepsilon_{XC} = \varepsilon_{YC}$ [%] | 0.8 | 0.6 |
| Shear strain $\varepsilon_S$ [%] | 1.8 | 1 |
| Density [g/cm$^3$] | 1.6 | 1.4 |
| Poisson's Ratio | 0.1 | 0.2 |

The delamination is an important energy absorption mechanism in the case of low velocity impact [51]. In case of the modeling approach used in research, the connection between successive layers of reinforcing material can be modeled in several ways [52,59–61]. Modeling of the bonding connection between adjacent reinforcing layers could be realized using cohesive elements [54,62]. This approach involves inserting additional cohesive elements between elements of adjacent reinforcing layers. In the initial phase, these elements may have zero thickness. The behavior of the cohesive elements can be described, for example, using a bilinear curve. These curves describe the dependence between the cohesive element stresses from its deformation [54,59,60,62]. If the cohesive element is not damaged, its stiffness is constant. However, if an element is damaged, its stiffness decreases as deformation increases [62]. The use of cohesive elements requires high computational costs [52]. A broader description of this type of connection has been described, among others, in the literature [54,59,60,62]. The second approach of modeling the bonding connection between the adjacent reinforcing layers is the use of the bilateral constraints (tiebreak contact type). This contact allows for the simplification of crack propagation based on the cohesive element [51]. After reaching normal and shear stresses, damage is a linear function of the distance between points that were initially in contact. After reaching the defined critical crack opening, the bonding connection is broken and further contact behaves like unilateral constraints. This approach does not require the use of additional elements and, shown in the work of [51], this bonding connection modelling method give results in accordance with experimental research. In the conducted research, it was decided to use the bilateral constraint contact with strength criterion (*AUTOMATIC_SURFACE_TO_SURFACE_TIEBREAK [63]). This contact type behaves like a bonded connection before the strength criterion is exceeded. After exceeding the strength criterion, this contact behaves like a unilateral constraint contact without a bonded connection. There are several possibilities to define the bonded connection using the selected contact type. In the case of shell elements, the definition OPTION = 8 is most commonly used [59]. This option requires to define the critical normal and shear stresses in a bonding connection. The critical normal stress assumed in the physical model equals $S_n$ = 75 MPa [49] and the critical shear stress equals $S_s$ = 44 MPa [59], respectively. The friction coefficient between reinforced layers was defined as 0.18 [52,64,65].

Additionally, one more part created in the presented model was a polyethylene base plate, to which the tested samples were glued. The plate was discretized using eight node solid elements with one integration point. The edge length of each element was 1 mm. The base plate model consisted of 50,000 solid elements. It was decided to choose the linear elastic material model (*MAT_ELASTIC [55]). The applied mechanical properties of polyethylene [66] are shown in Table 5. The contact between the last layer of the composite and the polyethylene base plate was modeled using the unilateral constraint contact (*AUTOMATIC_SURFACA_TO_SURFACE). The value of 0.29 was adopted as a friction coefficient [66].

**Table 5.** Mechanical properties of polyethylene.

| Young's Modulus [MPa] | Poisson's Ratio | Density [g/cm$^3$] |
|---|---|---|
| 701 | 0.4 | 0.946 |

During the experiment, four different types of strikers were used (Figure 2). Their geometries were made in CAD software (Autodesk Inventor Professional 2019), and then they were imported and discretized using eight nodal solid elements. These elements were given mechanical properties corresponding to the steel and treated as non-deformable using the *MAT_RIGID [55] material model. The strikers were placed opposite to the first layer of the laminate. An initial velocity of the striker $V_0$ = 31 m/s was set. The boundary condition assigned to the model blocked all degrees of freedom (translational and rotational) for all nodes of a polyethylene base plate at the side opposite to the side of contact with the composite. The contact between the non-deformable striker and the composite layers was defined as *AUTOMATIC_SURFACE_TO_SURFACE (unilateral constraint). The values of static and dynamic friction coefficients between the striker and the reinforcing layers were set as 0.18 [52,64,65]. The schematic diagram of the created model is shown in Figure 6a. The prepared model is shown in Figure 6b.

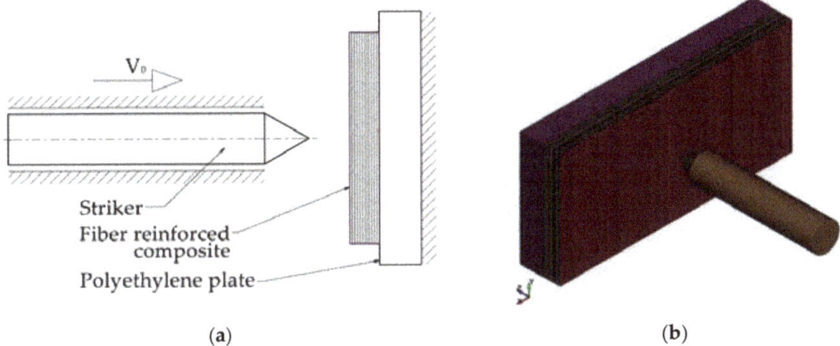



**Figure 6.** Numerical research: (**a**) schematic diagram, (**b**) physical model.

## 3.2. Numerical Results

Damage caused by the strikers with various geometries at different instances of the simulation (up to reach the maximum depth of the penetration) is shown in Figures A3–A6. Each reinforcing layer is marked by a different color. The cross sections of damaged areas owing to the impact of various strikers are shown in Figure 7. The direction of the impact is marked using a white arrow.

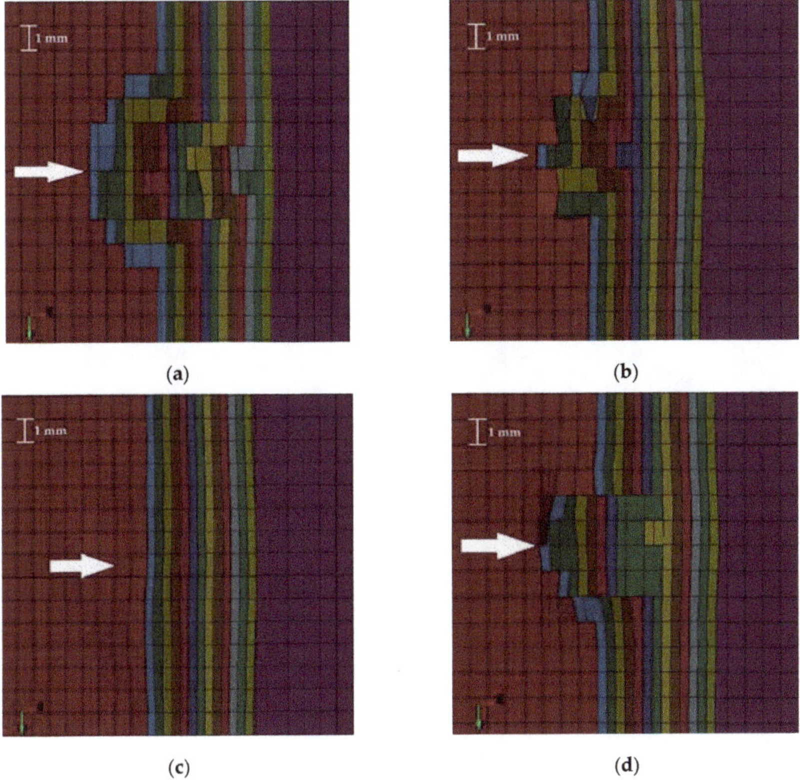



**Figure 7.** Cross sections of damaged areas caused by (**a**) conical striker, (**b**) hemispherical striker, (**c**) blunt striker, and (**d**) ogival striker.

The numerical results have shown that the largest damages (Figure 7a) occurred when a conical striker was used. The smallest damage occurred in the case of a blunt striker (Figure 7c), which corresponds to the experimental results. The area of damage for each striker geometry is different. The damage caused by impact of the striker with a conical end (Figure 7a) is characterized by the formation of a narrowing taper. In each subsequent damaged layer, the number of degraded elements decreased to form a characteristic cone. These damages are the largest among the analyzed cases. The sharp end of the conical striker penetrating successive layers of material caused rapid damage to the individual elements in subsequent reinforcing layers, which contributed to the weakening of the entire structure.

The damage caused by the hemispherical striker (Figure 7b) was less extensive compared with the damage caused by the conical striker. The resulting damage does not form a characteristic cone. The largest damages were observed in the first four layers of reinforcement. The elements that were not deleted are arranged in a chaotic manner. It can be observed that the elements in the second layer of reinforcement in the impact area are not completely destroyed, which can be interpreted as the fiber compression inside the formed crater (as mentioned in the chapter about experimental research). The other two damaged layers (5 and 6) were minimally damaged—individual elements were deleted. The blunt striker did not cause any visible damage in the reinforcement material (Figure 7c). The ogival striker caused the most damage (Figure 7d) in the first two layers of the material. Penetration of subsequent layers was characterized by the removal of the same number of finite elements until the striker energy was too low to damage the next layer. The last of the damaged layers degraded in the form of individual elements.

The colored map of Tresca stress presented in Figure 8 showed that the primary yarns (at the direction defined as a fiber direction—X and Y axis) carried the largest load. The secondary yarns were much less loaded. This conclusion was confirmed in the work of [16]. In addition to the analysis of destruction, which was performed based on the Figure 7c, the value of Tresca stresses shown in Figure 8 suggests that the matrix could be damaged. As shown in Table 2, the tensile strength of the matrix is 75–85 MPa. Tresca stress at the edge of impacting striker (on the first layer of reinforcement) was much higher (92.5–110 MPa). This analysis suggests that the matrix could be damaged, as in the case of the experimental research (matrix cracking). The average diameter of this area was 7.5 mm.

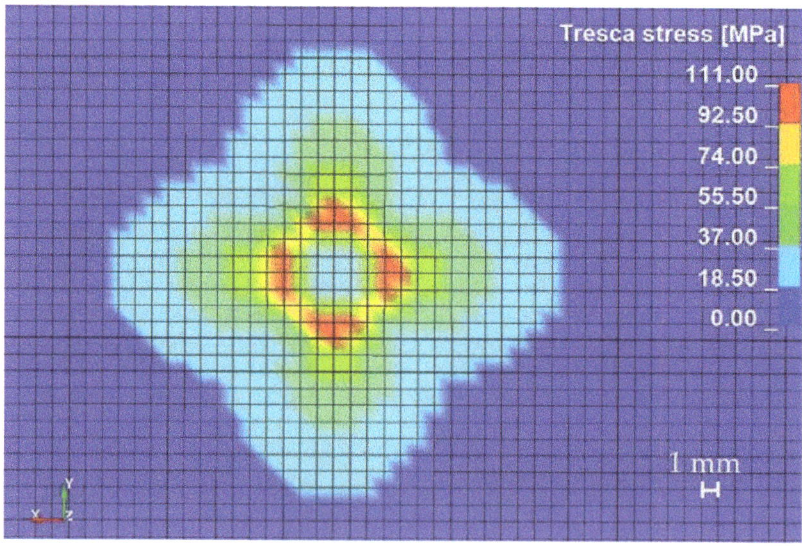


**Figure 8.** Colored map of Tresca equivalent stress on the first layer of reinforcement in case of blunt striker impact.

Table 6 shows the diameters of the cavities and depth of the penetration of the damages created by the impact of strikers with different geometries. The measurement of cavity diameters was done between the nodes of the elements at the first layer, which were not deleted after impact. The biggest damage in the first layer of reinforcing material was observed for the case in which the conical striker hits the composite. The second largest diameter of the damaged area was observed for the case in which a hemispherical striker was used. The diameter of the damaged area in this case was slightly larger than in the case of the impact of the ogival striker. The depth of penetration was based on the number of damaged reinforced layers—0.5 mm of damage for each damaged layer.

**Table 6.** Numerical results, depending on striker geometry.

| Striker Geometry | Cavity Diameter [mm] | Depth of Penetration [mm] |
| --- | --- | --- |
| Conical | 8 | 6 |
| Hemispherical | 6 | 2 |
| Blunt | 7.5 | 0.07 |
| Ogival | 5.5 | 4 |

The largest depth of penetration was observed for the case in which the composite panel was hit by a conical striker. As a result of its impact, 12 layers of reinforcing material were damaged. The impact with an ogival striker resulted in damage to eight layers of reinforcing material. In the case of a hemispherical striker, six layers of reinforcing material were damaged. However, these damages were different. The damage created did not resemble a cone geometry. As it was mentioned, the individual elements were deleted from 5 and 6 damaged layer (damage area was smaller than used striker). Due to that, this two layers were not considered in depth of penetration calculation. For the impact on the composite material with a blunt striker, no damage was observed in the reinforcing material, but small deformation was observed.

The graph representing kinetic energy of the striker depending on its geometry is shown in Figure 9. The fastest braking of the striker occurred with a blunted one. This striker was rebounded from the composite panel, obtaining the kinetic energy of about 16 J. This testified that a small amount of energy was absorbed by the composite panel. The hemispherical striker was rebounded from the panel, obtaining a value of the kinetic energy of about 8.5 J. The braking of the striker before it was rebounded proceeded with the same intensity as in the case of a blunt striker, despite the damage caused in the reinforcing layers. A smaller value of the rebounded kinetic energy of the striker is associated with a greater amount of energy absorbed by the composite. Conical and ogival strikers had the lowest kinetic energy after rebounded (approximately 2 J). Despite the differences in the strikers' braking intensity, they were stopped at the same time and had similar kinetic energy after being rebounded. The curve depicting the kinetic energy value of the ogival striker has a distinctive area of braking corresponding to the level of material penetration. The first change in the angle of inclination of this curve at the time t ≈ 0.1 ms is associated with the first damage to the material. Another angle change in the time t ≈ 0.2 ms is associated with the end of damaging of the reinforcement layers. In the next phase, material was deformed without damaging and striker was inhibited and rebounded. Braking of the conical striker was the gentlest, which was probably because of the fact that its pointed end easily damaged the elements in the subsequent layers of the composite. It is interesting that the conical and ogival strikers, despite large differences in the damage caused, were characterized by the same value of the kinetic energy after rebound.

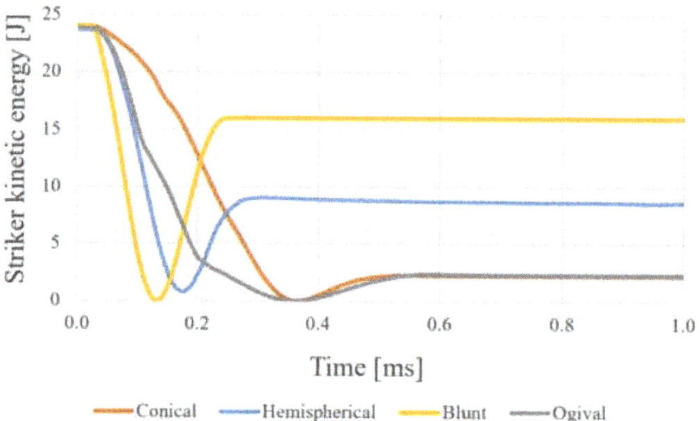



**Figure 9.** Kinetic energy of strikers during impact depending on their geometry.

### 3.3. Comparison of Experimental and Numerical Research

The results of the conducted experimental and numerical research were compared based on the diameters and depth of penetration of cavities formed after impact of the strikers. Figure 10 shows the comparison of the diameters of the cavities formed after the impact.

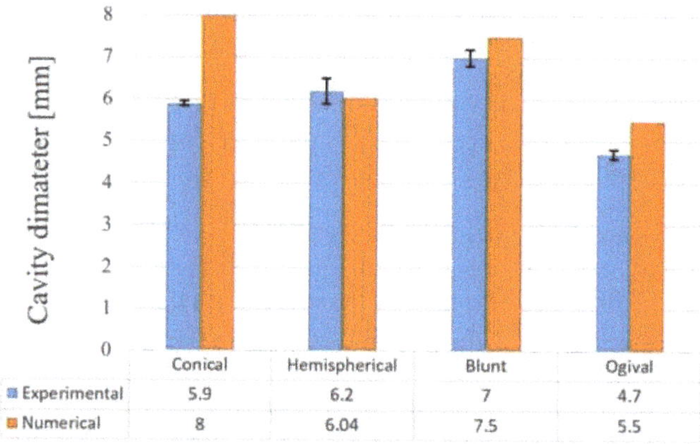



| | Conical | Hemispherical | Blunt | Ogival |
|---|---|---|---|---|
| Experimental | 5.9 | 6.2 | 7 | 4.7 |
| Numerical | 8 | 6.04 | 7.5 | 5.5 |

**Figure 10.** Comparison of cavity diameter formed after impact.

The damage caused by the blunt striker in the experiment (Figure 4c) is limited to the matrix cracking. The diameter of the resulted crack corresponds to the diameter of the edge of the impacting striker. The resulting damage did not damage the reinforcing layer in any way. Performing numerical research in the selected scale makes it impossible to capture phenomena like matrix cracking, but proper analysis of the stress map could bring valuable conclusions. As discussed in the numerical research results, the obtained Tresca stress value suggests that the matrix was damaged, as in the case of the experimental research. The measured diameter of damaged area was almost the same as in the case of the experimental research. Another discrepancy between the experimental and numerical research is in the case of hitting the composite by the conical striker. During the experimental research, the pointed tip of the cone stuck between the individual fibers, pushing them sideways. Fibers pushed sideways constituted resistance for the striker, and tightened on it, thus reducing damage caused in subsequent layers. In the case of numerical research, individual fibers were not considered. Each reinforcing layer was modeled as a single part with directionally assigned properties. The indicated directions corresponded to the directions of the fibers in the used reinforcing fabric, however, the modeling of the reinforcement in this way makes it impossible to imitate the mechanism of deformation of individual fibers. Finite elements hit directly by the tip of the cone were quickly removed. When an element was removed, it stopped absorbing the energy, so material did not resist as much energy as in the case of the experiment. This fact could have an effect for the larger amount of destruction in the case of the numerical research. The same remark applies to the discrepancy between the results obtained for the ogival striker.

Figure 11 shows the comparison of the depths of penetration obtained as a result of the provided experimental and numerical research. For conical and hemispherical strikers, the depths of penetration obtained as a result of numerical research were larger than in the experimental research. In the case of the experimental research, depths were determined indirectly based on the diameter of the formed cavity and the striker geometry. This method could be subject to a large error, resulting from the lack of information about the damaged fibers under the visible layer of damaged reinforcement. Additionally, in the case of the numerical research, elements removed during calculations do not constitute any resistance for the strikers' movement at a later stage. In real conditions, the damaged fibers were still in the forming cavity, clamping on the striker, or accumulating in front of its forehead, thus affecting the formation of further damage. Figure 12 shows the percentage variation between the results obtained from the experimental and numerical research.

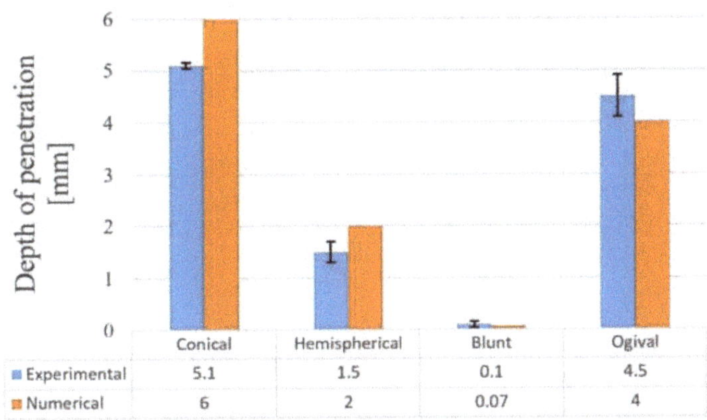

**Figure 11.** Comparison of depth of penetration after impact.

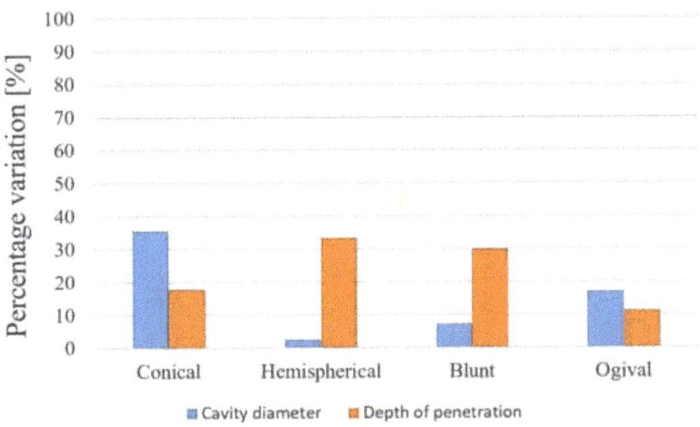

**Figure 12.** Percentage variation between the results obtained from experimental and numerical research.

### 3.4. Delamination

As mentioned in the Introduction, in the case of multilayer composite materials, the energy is absorbed not only as a result of the destruction of reinforcing fibers, but also as a result of the destruction of connections between successive reinforcing layers (delamination). The inclusion of this process in numerical simulations has an effect on the obtained results. Figures 13–15 show maps of the delamination areas between selected layers as an effect of the striker's impact. The areas marked in red (for which the parameter value was 1) were the areas where the bonded connection between layers was broken (delamination occurs).

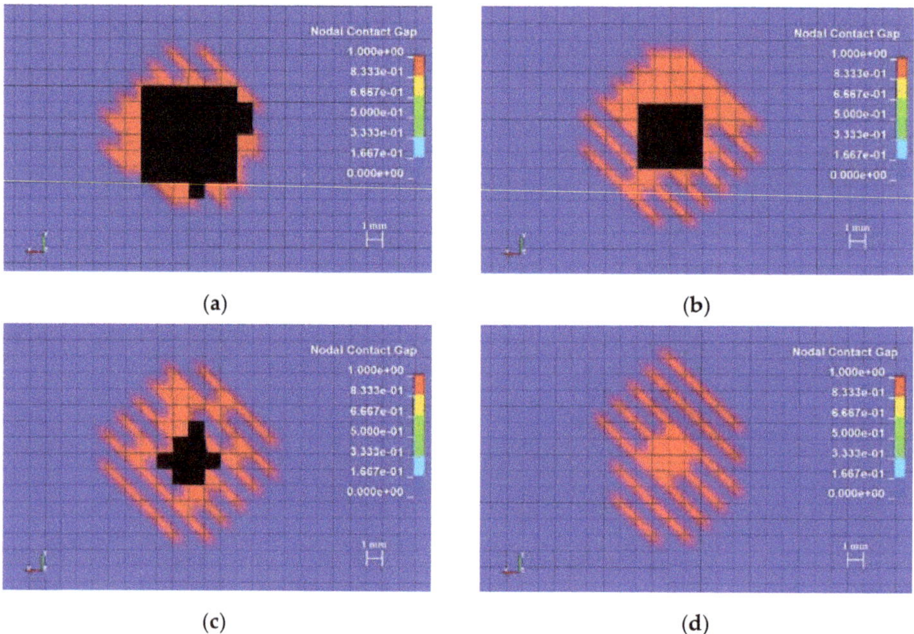

(a)          (b)

(c)          (d)

**Figure 13.** Colored maps of delamination areas after impact of the conical striker between layers: (**a**) 1 and 2, (**b**) 5 and 6, (**c**) 7 and 8, (**d**) 13 and 14.

The areas of material delamination resulting from impact by a conical striker indicate an increase of the diameter of delamination area in layers with a smaller amount of destruction in the reinforcing material. In the area of the first reinforcing layers, where the damage in the material is the biggest, the delamination covered a narrow area around the damage (Figure 13a). The delamination area increases with successive layers of reinforcing material, which is accompanied by a reduction in the number of deleted elements in reinforcing layers. In the case of the layers that were not damaged, as a result of the striker impact, the diameter of the area where delamination occurred was slightly increased.

In the case of the composite hit by an ogival striker, the delamination area occurring between the first and second reinforcement layer (Figure 14a) was smaller than in the case of the conical striker. The delamination area increased with each next reinforcing layer, until reaching a diameter of 14 mm between layer 13 and 14 of the reinforcing material. Concentration of the delamination was observed in places located in the axis of impact of the striker. As in the case of the conical striker, if more damage in reinforced layer was observed, then the less delamination occurs between this and next reinforcing layer.

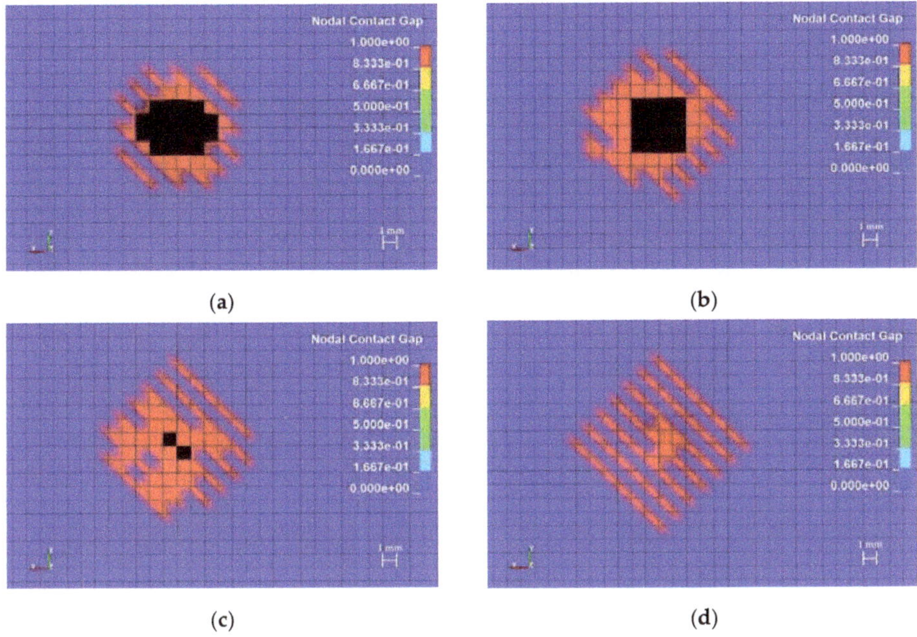

**Figure 14.** Colored maps of delamination areas after impact of the ogival striker between layers: (**a**) 1 and 2, (**b**) 5 and 6, (**c**) 7 and 8, (**d**) 13 and 14.

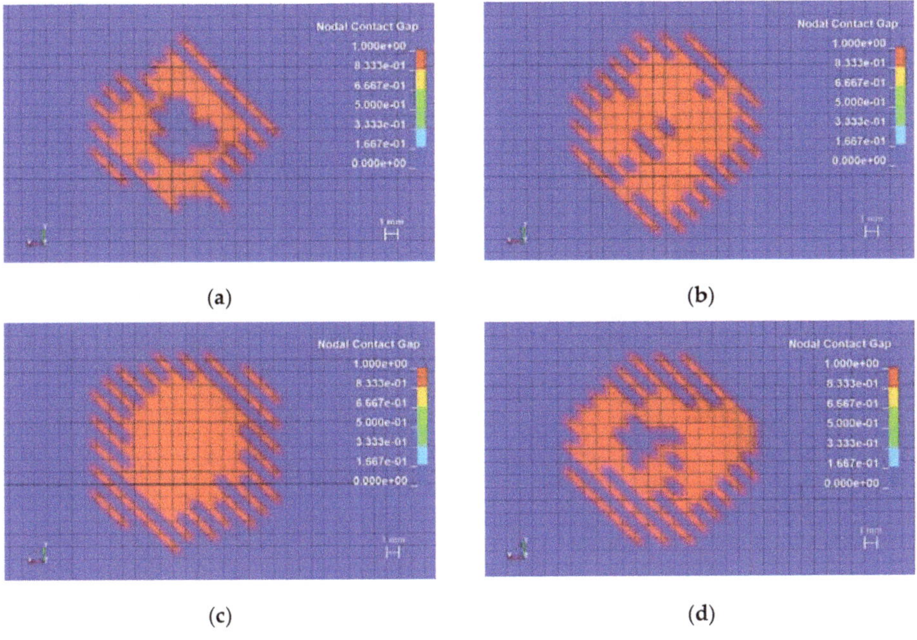

**Figure 15.** Colored maps of delamination areas after impact of the blunt striker between layers: (**a**) 1 and 2, (**b**) 2 and 3, (**c**) 9 and 10, (**d**) 13 and 14.

The colored maps presented in Figure 15 indicate that, in the composite material that was hit by a blunted striker, the internal structure was damaged—delamination between the reinforcement layers. The diameter of the delamination area is almost the same over the entire thickness of the composite.

The detection of areas in which the material structure is damaged, like a delamination, is particularly important because this damage changes the behavior of the entire composite material and affects its mechanical properties. In the case of the practical application of multilayered composite materials, the lack of detection of such material damage can lead to dangerous situations in which there may be a serious failure, loss of load capacity, or reduction of the strength of elements working under high pressure, which can directly lead to the health and life threat of people working near such elements. The delamination is particularly difficult to detect because they are usually not visible and their detection requires specialized tests. An interesting solution for monitoring such damage is the triboelectric sensor. These sensors do not require external power supply and they are mounted directly on the composite material. The results presented in the work of [67] showed it can be observed that the voltage outputs of the sensor are proportional to the extension of the damage in the composite. The triboelectric sensor can be used to predict the damage state of the composite plates and the size of the delamination caused by impacts of the strikers [67].

Figure 16 showed the diameters of the measured delamination areas between first and second reinforced layers, as well as the thirteenth and fourteenth. It is characteristic for all analyzed cases that the delamination areas increase with each subsequent reinforced layer. It is also worth noting the relationship between diameter of delamination area and the damage in the reinforced layer. If the damaged area was bigger, than the delamination area was smaller. The impact with the conical striker caused the largest damage in the reinforcing layers, but the delamination area was the smallest among the analyzed cases. The impact with the blunt striker caused the smallest damage in the reinforcing layers, but the delamination area was the biggest among the analyzed cases.

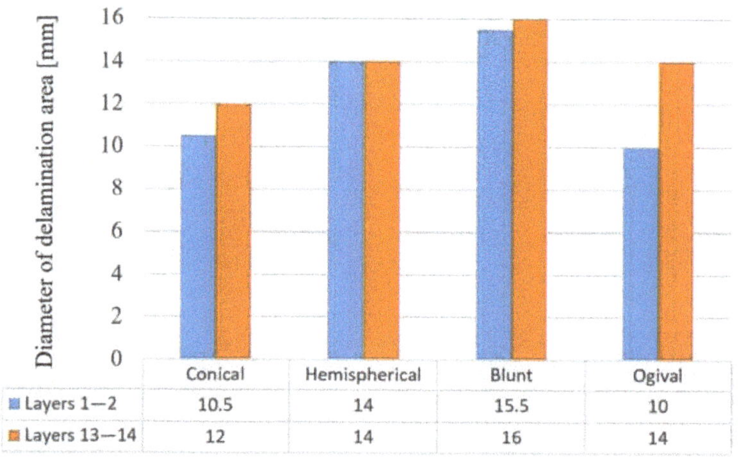

|  | Conical | Hemispherical | Blunt | Ogival |
|---|---|---|---|---|
| Layers 1—2 | 10.5 | 14 | 15.5 | 10 |
| Layers 13—14 | 12 | 14 | 16 | 14 |

**Figure 16.** Diameters of delamination areas between reinforcement layers 1 and 2 and 13 and 14.

## 4. Discussion

The blunt striker did not cause any visible damage in the reinforcing layers, however, the area of delamination was largest between analyzed cases. As mentioned, the delamination process absorbs part of the impact energy. The main part of the energy difference between the initial kinetic energy of blunt striker and the kinetic energy after its rebound is connected with the energy absorbing delamination process. Including the phenomenon of delamination in numerical studies allows the scientist to estimating the extent of the structure damage. It should be remembered that not every damage that was caused is immediately visible, as was the case of a panel impacted by the blunt striker. The reinforcing material was not damaged, however, the composite structure was damaged, which affects its strength. The same fact applies to cases in which other strikers were used. As shown by numerical research, the real areas of damage were much larger than that estimated on the diameter of the cavities formed after impact in experimental research. Experimental research presented in another paper [68] confirms that the delamination area grows with each next reinforcing layer from the impact point. The shape of the delamination areas was also similar to the shape that could be observed in experimental research [22]. It is worth to noting that delamination areas can be reduced by adding, for example, nanofibers. Research presented in another paper [62] showed that the incorporation of polycaprolactone nanofibers reduces the delaminated area by about 27% in glass fiber reinforced polymer composites. Research presented in another paper [62] also confirms that the epoxy glass composites reinforced with polycaprolactone nano fibers are less susceptible to impact damage than the same composites without polycaprolactone nano fibers.

The comparison of rebounded energy value in the case of conical and ogival strikers with damages caused by those strikers in composites suggests that the energy was absorbed in different ways, depending on the strikers' geometry. Rebounded energy was the same, but damaged was different. In the case of the conical striker, damage in reinforced layers was bigger. In the case of the ogival striker, the delamination areas was bigger, which suggests that the delamination absorbed a higher part of the impact energy than in the case of the conical striker, which causes the smallest damage in the reinforcement layers.

In the case of the composite made from glass fibers and polyester resin [69], it was noted that the largest areas of damage were obtained when the conical and hemispherical strikers were used. In the conducted research, the largest diameter of damage in the first reinforcing layer was also noted for these two strikers. In the case of the research presented in a further paper [69], the composite material absorbed different amounts of the impact energy depending on the striker geometry. The amount of absorbed energy in descending order was as follows: conical striker, ogival striker, hemispherical striker, and blunt striker. The same results were obtained for carbon epoxy composite (impact energy up to 6 J) in another paper [70]. In the conducted numerical research, the same order was observed (Figure 9). Carbon fiber reinforced composite in the case of the blunt striker impact (50 J of the impact energy) absorb almost the same amount of impact energy as in the case of the conical striker [45]. In the conducted research, the hybrid composite absorbs different amounts of impact energy in the case of those strikers. In the case of the conical striker, the composite absorbs a few times more impact energy than in the case of the blunt striker. In the case of the analyzed composites, the plastic behavior of aramid fibers causing the damage caused by the strikers with different geometries is much more diverse than in the case of, for example, the composite made from carbon fibers, the damage of which is brittle [69,70].

The conducted experimental and numerical research indicates that the process of energy absorption in composite panels is complex. In the case of the experimental research, energy was absorbed by all factors listed in the introduction. Owing to the adopted modeling scale, the impact energy in the case of numerical research could not be absorbed by friction between the fibers (inside the reinforcing layer). The energy was absorbed by the friction between the striker and the reinforcing layers and by the friction between the adjacent reinforcing layers. The numerical results present that the delamination absorbed some of the impact energy, which is best seen in the case of the blunt striker. The presented Tresca equivalent stress (Figure 8) indicates that the primary yarns were subjected to the highest stress; however, in the adopted modeling scale, it is difficult to clearly divide the energy absorbed by the reinforcing layer into the energy absorbed by the destruction of primary yarns, deformation of secondary yarns, and shear plugging. The difference between energy absorption in the case of experimental research and absorption in the numerical research (which results from the adopted modeling scale) obviously has an impact on the result. The lack of consideration of some energy absorbing phenomena increases the damage in the composite panel.

It was stated that carrying out numerical research simultaneously with experimental research can significantly affect the quality of the obtained conclusions. Appropriate model preparation and consideration of physical phenomena occurring in the analyzed material can provide a lot of important information. The calibration of the numerical model and its verification based on the performed experimental research is also important. If the model is considered as a satisfactory, it can be used to observe phenomena that are hard or even impossible to observe during the experiment. In the conducted research, such a phenomenon was delamination areas between reinforcing layers. In the case of multilayered composites, it is very important to use the right modeling scale, because observing some phenomena on the model prepared in the wrong scale is not possible. During the interpretation of the numerical research results, the simplifications used should be taken into account, particularly simplifications resulting from the used modeling scale. A compromise between which phenomena can be observed and the time of calculations should be found. Preparing an increasingly detailed model, one should take into account that the calculation time may increase several times.

## 5. Conclusions

On the basis of research, it was found the strikers' geometry has a significant effect on the damaged microscopic images. The greatest damage was observed for the conical striker (the panel was almost pierced). During the impact, those strikers stuck between the fibers, and pushed them sideways. In the case of such a material penetration mechanism, stretching of the fibers absorbs a smaller part of the impact energy. The fibers in this case are pushed sideways and pressed on the margin of the formed cavity. In the case of a hemispherical striker, much lower damage was observed. The damage caused by this striker was characterized by a large amount of crushed fibers, compressed inside the formed cavity. It is worth noting that the depth of penetration in the case of this type of striker was much smaller than in the case of strikers with the cone-like geometry. The smallest damage was observed for a blunt striker (matrix cracking only). In the analyzed case, there were no damages in the reinforcing layers, but only, as numerical research showed, the integrity of the bonded connection between the reinforcing layers was broken (delamination). The numerical investigations presented that the largest diameters of the delamination areas occur on the side opposite to the impact side. The measurements of the diameters of delamination areas in subsequent layers indicate that these delamination areas form a cone with a base located on the side opposite to the impact side. It was also noticed that the larger delamination areas between the reinforcing layers were between layers where there was smaller damage in the reinforcement. As could be observed in the experimental research, the penetration level in the case of two strikers stops at the depth corresponding to the place when carbon reinforcement layers have been used, which could testify that the use of hybrid structures with carbon and aramid fibers could increase the stiffness of the panels and increase the puncture resistance of the composite.

**Author Contributions:** Conceptualization, S.S. and M.S.; Formal analysis, S.S.; Funding acquisition, S.D.; Investigation, S.S., M.S., J.K., J.D., and S.D.; Methodology, S.S. and M.S.; Software, S.S.; Supervision, S.D.; Validation, S.S., M.S., J.K., and S.D.; Visualization, S.S.; Writing—original draft, S.S. and J.K.; Writing—review & editing, S.S. and J.K. All authors have read and agreed to the published version of the manuscript.

**Funding:** This work was financially supported by statutory funds from the Faculty of Mechanical Engineering of Silesian University of Technology in 2019.

**Acknowledgments:** Calculations were carried out using the computer cluster Ziemowit (https://www.ziemowit. hpc.polsl.pl) funded by the Silesian BIO-FARMA project No. POIG.02.01.00-00-166/08 in the Computational Biology and Bioinformatics Laboratory of the Biotechnology Centre in the Silesian University of Technology.

**Conflicts of Interest:** The authors declare no conflicts of interest.

## Appendix A

Strikers used in experimental research were manufactured from S235JR steel rod by machining.

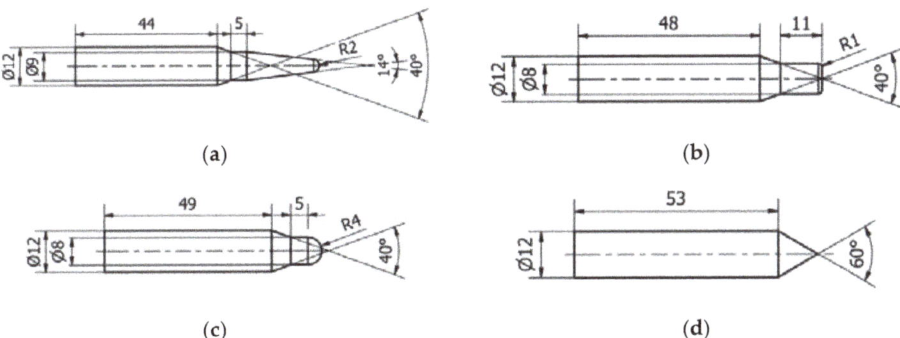

(a)           (b)

(c)           (d)

**Figure A1.** Geometries of developed strikers: (**a**) ogival, (**b**) blunt, (**c**) hemispherical, (**d**) conical.

## Appendix B

Microscopic images of damage in the composite panels with the hybrid reinforcement are shown in Figure A2. The jagged fragments of the fibers at the side opposite to the impact side are the result of the cutting process.

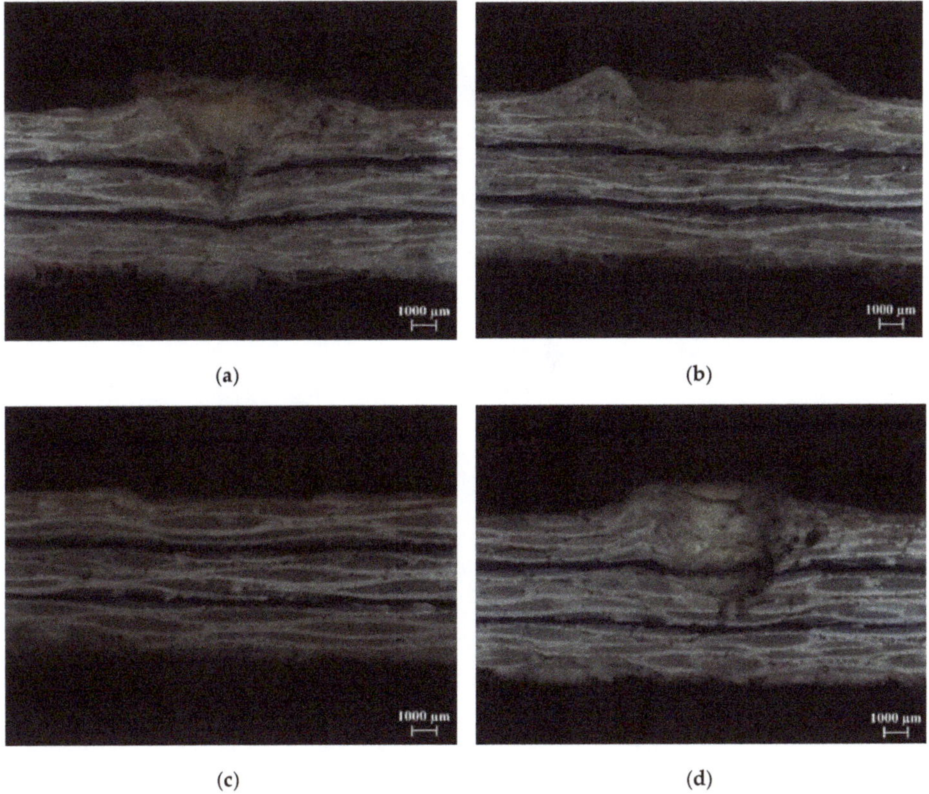

(a)           (b)

(c)           (d)

**Figure A2.** Cross sections of the damage formed in the composite panels after impact of the striker: (**a**) conical, (**b**) hemispherical, (**c**) blunt, (**d**) ogival.

## Appendix C

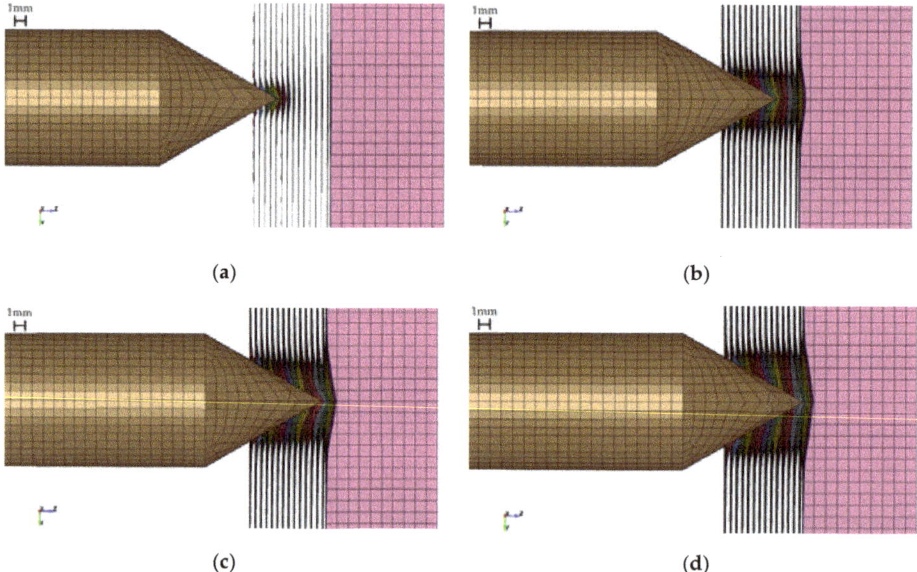

**Figure A3.** Damage caused by the conical striker at different instances of the simulation: (**a**) time = 0.1 ms, (**b**) time = 0.2 ms, (**c**) time = 0.3 ms, (**d**) time = 0.35 ms.

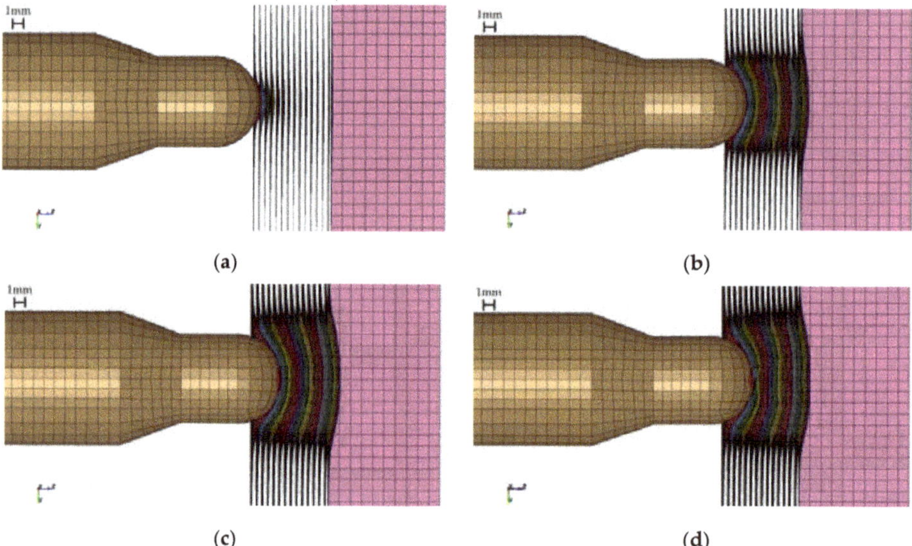

**Figure A4.** Damage caused by the hemispherical striker at different instances of the simulation: (**a**) time = 0.05 ms, (**b**) time = 0.1 ms, (**c**) time = 0.15 ms, (**d**) time = 0.2 ms.

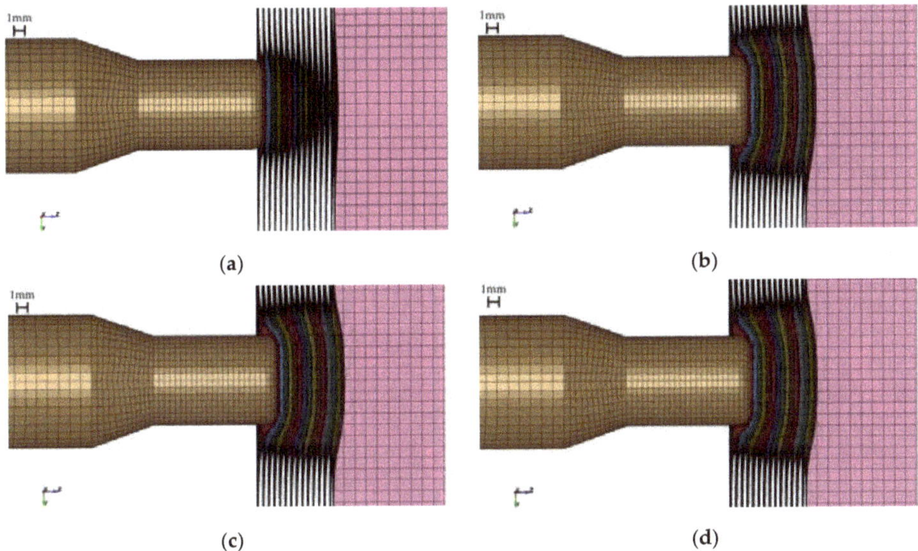

**Figure A5.** Damage caused by the blunt striker at different instances of the simulation: (**a**) time = 0.05 ms, (**b**) time = 0.1 ms, (**c**) time = 0.12 ms, (**d**) time = 0.15 ms.

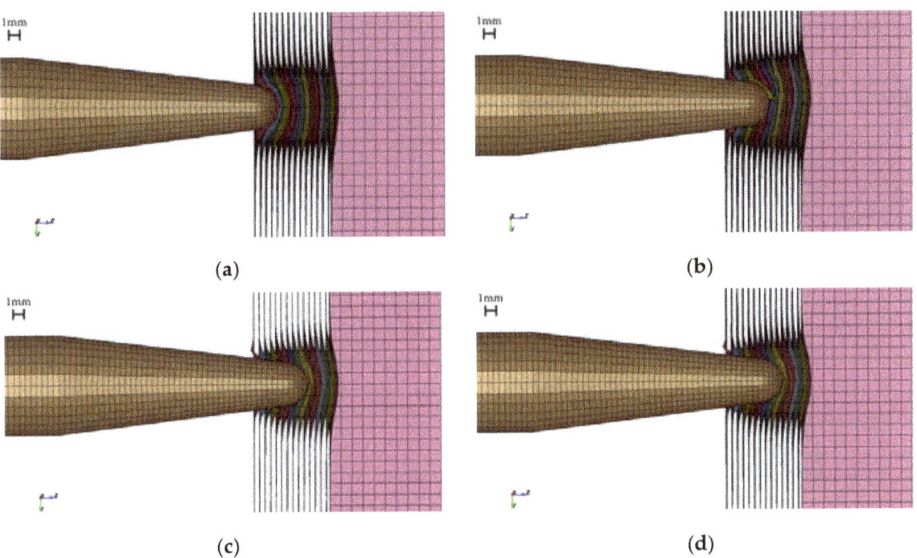

**Figure A6.** Damage caused by the ogival striker at different instances of the simulation: (**a**) time = 0.1 ms, (**b**) time = 0.2 ms, (**c**) time = 0.3 ms, (**d**) time = 0.35 ms.

## References

1.	Jureczko, M. Multidisciplinary Optimization of Wind Turbine Blades with Respect to Minimize Vibrations. In *Recent Advances in Composite Materials for Wind Turbine Blade*; Attaf, B., Ed.; The World Academic Publishing Co. Ltd.: Hong Kong, China, 2013; pp. 129–146.

2.  Pawlak, M. The Acceleration Severity Index in the impact of a vehicle against permanent road equipment support structures. *Mech. Res. Commun.* **2016**, *77*, 21–28. [CrossRef]
3.  Szymiczek, M. Selection of engineering materials on lightweight energy-intensive shields. *Poly. Process.* **2016**, *22*, 149–157.
4.  Chatys, R.; Kleinhofs, M.; Panich, A.; Kisiel, M. Modeling of mechanical properties of composite structures taking into account military needs. *AIP Conf. Proc.* **2019**, *2077*, 1–8. [CrossRef]
5.  *MIL-STD-662F Military Standard: V50 Ballistic Test for Armor*; Department of Defense: Washington, DC, USA, 1997.
6.  *STANAG 4569 Protection Levels for Logistic and Light Armoured Vehicle Occupants*; NATO/PFP Unclassified; NATO: Brussels, Belgium, 1998.
7.  Rojek, M.; Szymiczek, M.; Stabik, J.; Mężyk, A.; Jamroziak, K.; Krzystała, E.; Kurowski, J. Composite materials with the polymeric matrix applied to ballistic shields. *Arch. Mater. Sci. Eng.* **2013**, *63*, 26–35.
8.  Machoczek, T.; Mężyk, A.; Duda, S. Shaping the dynamic characteristics of military special vehicles. *Solid State Phenom.* **2016**, *248*, 192–203. [CrossRef]
9.  Krzystała, E.; Kciuk, S.; Mężyk, A. *Identification of Hazards for the Crew of Special Vehicles during an Explosion*; Wydawnictwo Naukowe Instytutu Technologii Eksploatacji: Gliwice, Poland, 2012.
10. Jureczko, M.; Mężyk, A. Multidisciplinary optimization of wind turbine blades. In Proceedings of the Fourth European & African Conference on Wind Engineering, EACWE 4, Prague, Czech Republic, 11–15 July 2005; Naprstek, J., Ed.; Institute of Theoretical and Applied Mechanics, Academy of Sciences of the Czech Republic: Prague, Czech Republic, 2005; pp. 40–46.
11. Cegła, M.; Habaj, W.; Stępniak, W.; Podgórzak, P. Hybrid ceramic-textile composite armour structures for a strengthened bulled-proof vest. *Fibres Text. East. Eur.* **2015**, *23*, 85–88.
12. Lane, R.A. High performance fibers for personnel and vehicle armor systems. Putting a stop to current and future threats. *Amptiac Q.* **2005**, *9*, 3–9.
13. Radziszewski, L. *Final Ballistics of Small-Caliber Ammunition When Shooting at Selected Targets*; Wydawnictwo Politechniki Świętokrzyskiej: Kielce, Poland, 2007.
14. Czapla, T.; Wrona, J. Technology development of military applications of unmanned ground vehicles. *Stud. Comput. Intell.* **2013**, *481*, 293–309. [CrossRef]
15. Nabaglo, T.; Kowal, J.; Jurkiewicz, A. Construction of a parametrized tracked vehicle model and its simulation in MSC.ADAMS program. *J. Low Freq. Noise Vib. Act. Control* **2013**, *32*, 167–173. [CrossRef]
16. Jamroziak, K. *Identification of Materials Properties in Terminal Ballistics*; Oficyna Wydawnicza Politechniki Wrocławskiej: Wroclaw, Poland, 2013.
17. Wersa, E.; Seweryn, A.; Szusta, J.; Rak, Z. Fatigue testing of transmission gear. *Maint. Reliab.* **2015**, *17*, 207–214. [CrossRef]
18. Ostachowicz, W.M.; Krawczuk, M.; Palacz, M. Detection of delamination in multilayer composite beams. *Key Eng. Mater.* **2003**, *245–246*, 483–490. [CrossRef]
19. Palacz, M. Spectral methods for modelling of wave propagation in structures in terms of damage detection—A review. *Appl. Sci.* **2018**, *8*, 1124. [CrossRef]
20. Boczkowska, A.; Kapuściński, J.; Lindemann, Z.; Witemberg-Perzyk, D.; Wojciechowski, S. *Composites*; Oficyna Wydawnicza Politechniki Warszawskiej: Warsaw, Poland, 2000.
21. Kang, T.J.; Kim, C. Impact energy absorption mechanism of largely deformable composites with different reinforcing structures. *Fibers Polym.* **2000**, *1*, 45–54. [CrossRef]
22. Naik, N.K.; Shrirao, P.; Reddy, B.C.K. Ballistic impact behaviour of woven fabric composites: Formulation. *Int. J. Impact Eng.* **2006**, *32*, 1521–1552. [CrossRef]
23. Hogg, P.J. Composites for ballistic applications. *Proc. Compos. Process.* **2003**, 1–11.
24. Sławski, S.; Szymiczek, M.; Domin, J. Influence of the reinforcement on the destruction image of the composites panels after applying impact load. *AIP Conf. Proc.* **2019**, *2077*, 1–10. [CrossRef]
25. Fejdyś, M.; Łandwijt, M. Technical fibres reinforcing the composite material. *Tech. Tex.* **2010**, *18*, 12–22.
26. Sławski, S.; Szymiczek, M.; Chmielnicki, B. Puncture resistance of epoxy-carbon composites subjected to ageing process in the thermal shock conditions. *Polym. Process.* **2018**, *24*, 52–60.
27. Kaczmarczyk, J.; Kozłowska, A.; Grajcar, A.; Sławski, S. Modelling and microstructural aspects of ultra-thin sheet metal bundle cutting. *Metals* **2019**, *9*, 162. [CrossRef]
28. Kaczmarczyk, J. Modelling of guillotine cutting of a cold-rolled steel sheet. *Materials* **2019**, *12*, 2954. [CrossRef]

29. Xiong, X.; Xiao, Q. Meso-scale simulation of concrete based on fracture and interaction behavior. *Appl. Sci.* **2019**, *9*, 2986. [CrossRef]

30. Cho, J.-R. A numerical evaluation of SIFs of 2-D functionally graded materials by enriched natural element method. *Appl. Sci.* **2019**, *9*, 3581. [CrossRef]

31. Wróbel, G.; Szymiczek, M.; Kaczmarczyk, J. Influence of the structure and number of reinforcement layers on the stress state in the shells of tanks and pressure pipes. *Mech. Compos. Mater.* **2017**, *53*, 165–178. [CrossRef]

32. Ochelski, S. *Experimental Methods of Construction Composites Mechanics*; Wydawnictwa Naukowo-Techniczne: Warsaw, Poland, 2004.

33. Naik, N.K.; Ramasimha, R.; Arya, H.; Prabhu, S.V.; ShamaRao, N. Impact response and damage tolerance characteristics of glass–carbon/epoxy hybrid composite plates. *Compos. Part B Eng.* **2001**, *32*, 565–574. [CrossRef]

34. Hosur, M.V.; Adbullah, M.; Jeelani, S. Studies on the low-velocity impact response of woven hybrid composites. *Compos. Struct.* **2005**, *67*, 253–262. [CrossRef]

35. Ying, S.; Mengyun, T.; Zhijun, R.; Baohui, S.; Li, C. An experimental investigation on the low-velocity impact response of carbon-aramid/epoxy hybrid composite laminates. *J. Reinf. Plast. Compos.* **2017**, *36*, 422–434. [CrossRef]

36. Rodríguez-Millán, M.; Ito, T.; Loya, J.A.; Olmedo, A.; Miguélez, M.H. Development of numerical model for ballistic resistance evaluation of combat helmet and experimental validation. *Mater. Des.* **2016**, *110*, 391–403. [CrossRef]

37. Tan, L.B.; Tse, K.M.; Lee, H.P.; Tan, V.B.C.; Lim, S.P. Performance of an advanced combat helmet with different interior cushioning systems in ballistic impact: Experiments and finite element simulations. *Int. J. Impact Eng.* **2012**, *50*, 99–112. [CrossRef]

38. Tham, C.Y.; Tan, V.B.C.; Lee, H.P. Ballistic impact of a KEVLAR® helmet: Experiment and simulations. *Int. J. Impact Eng.* **2008**, *35*, 304–318. [CrossRef]

39. Mayer, P.; Pyka, D.; Jamroziak, K.; Pach, J.; Bocian, M. Experimental and numerical studies on ballistic laminates on the polyethylene and polypropylene matrix. *J. Mech.* **2019**, *35*, 187–197. [CrossRef]

40. Bandaru, A.K.; Chavan, V.V.; Ahmad, S.; Alagirusamy, R.; Bhatnagar, N. Ballistic impact response of Kevlar® reinforced thermoplastic composite armors. *Int. J. Impact Eng.* **2016**, *89*, 1–13. [CrossRef]

41. Pach, J.; Mayer, P.; Jamroziak, K.; Polak, S.; Pyka, D. Experimental analysis of puncture resistance of aramid laminates on styrene-butadiene-styrene and epoxy resin matrix for ballistic applications. *Arch. Civ. Mech. Eng.* **2019**, *19*, 1327–1337. [CrossRef]

42. Reis, P.N.B.; Ferreira, J.A.M.; Zhang, Z.Y.; Benameur, T.; Richardson, M.O.W. Impact response of Kevlar composites with nanoclay enhanced epoxy matrix. *Compos. Part B Eng.* **2013**, *46*, 7–14. [CrossRef]

43. Sarasini, F.; Tirillò, J.; Valente, M.; Ferrante, L.; Cioffi, S.; Iannace, S.; Sorrentino, L. Hybrid composites based on aramid and basalt woven fabrics: Impact damage modes and residual flexural properties. *Mater. Des.* **2013**, *49*, 290–302. [CrossRef]

44. Gustin, J.; Joneson, A.; Mahinfalah, M.; Stone, J. Low velocity impact of combination Kevlar/carbon fiber sandwich composites. *Compos. Struct.* **2005**, *69*, 396–406. [CrossRef]

45. Ulven, C.; Vaidya, U.K.; Hosur, M.V. Effect of projectile shape during ballistic perforation of VARTM carbon/epoxy composite panels. *Compos. Struct.* **2003**, *61*, 143–150. [CrossRef]

46. Rodríguez Millán, M.; Moreno, C.E.; Marco, M.; Santiuste, C.; Miguélez, H. Numerical analysis of the ballistic behaviour of Kevlar® composite under impact of double-nosed stepped cylindrical projectiles. *J. Reinf. Plast. Compos.* **2016**, *35*, 124–137. [CrossRef]

47. Mitsubishi Chemical Corporation—Information Brochure. Available online: https://www.m-chemical.co.jp/en/products/departments/mcc/cfcm/product/CFtow_Jul2018en.pdf (accessed on 25 November 2019).

48. Twaron—Information Brochure. Available online: https://www.fibermaxcomposites.com/shop/datasheets/aramidFiberMDS.pdf (accessed on 25 November 2019).

49. GRM Systems sp. z o. o.—Product Datasheet: Epoxy Resin LG 285 + Hardeners HG 285, HG 286, HG 287. Date of Modification: 11 April 2012. Available online: http://www.grm-systems.cz/en/epoxy (accessed on 11 April 2012).

50. Kroczek, R.; Domin, J. Project of an pneumatic drive unit as initial decelerator of hybrid electromagnetic launcher. *Electr. Mach. Proc. Note* **2015**, *108*, 89–94.

51. Heimbs, S.; Heller, S.; Middendorf, P.; Hähnel, F.; Weiße, J. Low velocity impact on CFRP plates with compressive preload: Test and modelling. *Int. J. Impact Eng.* **2009**, *36*, 1182–1193. [CrossRef]

52. Bresciani, L.; Manes, A.; Ruggiero, A.; Iannitti, G.; Giglio, M. Experimental tests and numerical modelling of ballistic impacts against Kevlar 29 plain-woven fabrics with an epoxy matrix: Macro-homogeneous and meso-heterogeneous approaches. *Compos. Part B Eng.* **2016**, *88*, 114–130. [CrossRef]

53. Manes, A.; Bresciani, L.M.; Giglio, M. Ballistic performance of multi-layered composite plates impacted by different 7.62 mm calibre projectiles. *Proc. Eng.* **2014**, *88*, 208–215. [CrossRef]

54. Heimbs, S.; Bergmann, T.; Schueler, D.; Toso-Pentecôte, N. High velocity impact on preloaded composite plates. *Compos. Struct.* **2014**, *111*, 158–168. [CrossRef]

55. Hallquist, J.O. *LS-DYNA® Keyword User's Manual Volume II*; LS-DYNA R11; Livermore Software Technology Corporation: Livermore, CA, USA, 2018.

56. Jackson, K.E.; Littell, J.D.; Fasanella, E.L. Simulating the impact response of composite airframe components. In Proceedings of the 13th International LS-DYNA Users Conference, Hampton, VA, USA, 8–10 June 2014; pp. 1–26.

57. Cherniaev, A.; Montesano, J.; Butcher, C. Modeling the axial crush response of CFRP tubes using MAT054, MAT058 and MAT262 in LS-DYNA®. In Proceedings of the 15th International LS-DYNA®Users Conference, Detroit, MI, USA, 10–12 June 2018; pp. 1–17.

58. Performance Composites Limited. Available online: http://www.performance-composites.com/carbonfibre/mechanicalproperties_2.asp (accessed on 12 September 2019).

59. Dogan, F.; Hadavinia, H.; Donchev, T.; Bhonge, P. Delamination of impacted composite structures by cohesive zone interface elements and tiebreak contact. *Cent. Eur. J. Eng.* **2012**, *2*, 612–626. [CrossRef]

60. Elmarakbi, A.M.; Nu, H.; Fukunaga, H. Finite element simulation of delamination growth in composite materials using LS-DYNA. *Compos. Sci. Technol.* **2009**, *69*, 2283–2391. [CrossRef]

61. Muflahi, S.A.; Mohamed, G.; Hallett, S.R. Investigation of delamination modeling capabilities for thin composite structures in LS-DYNA®. In Proceedings of the 13th International LS-DYNA Users Conference, Detroit, MI, USA, 8–10 June 2014; pp. 1–14.

62. Garcia, C.; Trendafilova, I.; Zucchelli, A. The effect of polycaprolactone nanofibers on the dynamic and impact behavior of glass fibre reinforced polymer composites. *J. Compos. Sci.* **2018**, *2*, 43. [CrossRef]

63. Hallquist, J.O. *LS-DYNA® Keyword User's Manual Volume I*; LS-DYNA R11; Livermore Software Technology Corporation: Livermore, CA, USA, 2018.

64. Das, S.; Jagan, S.; Shaw, A.; Pal, A. Determination of inter-yarn friction and its effect on ballistic response of para-aramid woven fabric under low velocity impact. *Compos. Struct.* **2015**, *120*, 129–140. [CrossRef]

65. Rao, M.; Nilakantan, G.; Keefe, M.; Powers, B.; Bogetti, T. Global/Local Modeling of Ballistic Impact onto Woven Fabrics. *J. Compos. Mater.* **2009**, *43*, 445–467. [CrossRef]

66. INEOS Olefins & Polymers USA—Information Brochure. Available online: https://www.ineos.com/globalassets/ineos-group/businesses/ineos-olefins-and-polymers-usa/products/technical-information--patents/ineos-typical-engineering-properties-of-hdpe.pdf (accessed on 12 September 2019).

67. Garcia, C.; Trendafilova, I. Triboelectric sensor as a dual system for impact monitoring and prediction of the damage in composite structures. *NANO Energy* **2019**, *60*, 527–535. [CrossRef]

68. Will, M.A.; Franz, T.; Nurick, G.N. The effect of laminate stacking sequence of CFRP filament wound tubes subjected to projectile impact. *Compos. Struct.* **2002**, *58*, 259–270. [CrossRef]

69. Mitrevski, T.; Marshall, I.H.; Thomson, R.S.; Jones, R. Low-velocity impacts on preloaded GFRP specimens with various impactor shapes. *Compos. Struct.* **2006**, *76*, 209–217. [CrossRef]

70. Mitrevski, T.; Marshall, I.H.; Thomson, R.S.; Jones, R.; Whittingham, B. The effect of impactor shape on the impact response of composite laminates. *Compos. Struct.* **2005**, *67*, 139–148. [CrossRef]

*Article*

# Finite Element Modeling and Stress Analysis of a Six-Splitting Mid-Phase Jumper

**Pengfei Ma [1], Yongjie Li [2], Jiceng Han [3], Cheng He [2] and Wenkai Xiao [1],***

[1]    Power & Mech Engn, Wuhan University, Wuhan 430072, China; mpf01234@163.com
[2]    Electric Power Research Institute, State Grid Xinjiang Electric Power Company Limited, Xinjiang 830000, China; 13909972277@139.com (Y.L.); xjdkyhc@163.com (C.H.)
[3]    Electric Power Research Institute, State Grid Fujian Electric Power Company Limited, Fujian 350000, China; hanjiceng1989@163.com
*    Correspondence: xiaowenkai@whu.edu.cn

Received: 29 November 2019; Accepted: 13 January 2020; Published: 16 January 2020

**Abstract:** In this study, a finite element, fully three-dimensional solid modeling method was used to study the mechanical response of a steel-cored aluminum strand (ACSR) with a mid-phase jumper under wind load. A whole model (simplifying an ACSR into a solid cylinder) and a local model (modeling according to the actual structure of an ACSR) of the mid-phase jumper were established. First, the movement of the mid-phase jumper of the tension tower under wind load was studied based on the whole finite element model, and the equivalent Young's modulus of the whole model was adjusted based on the local model. The results of the whole model were then imported into the local model and the stress distribution of each strand of the ACSR was analyzed in detail to provide guidance for the treatment measures. Therefore, the whole model and the local model complemented each other, which could reduce the number of model operations and ensure the accuracy of the results. Through the follow-up test to verify the results of the finite element simulation and the comparison of the simulation and fatigue test results, the causes of the broken strand of the ACSR were discussed. Although this modeling method was applied to the stress and deformation analysis of a mid-phase jumper in this study, it can be used to study the bending deformation of rope structures with a complex geometry and the main bending deformation. In addition, the effect of the friction coefficient on the bending of the mid-phase jumper was studied.

**Keywords:** finite element modeling; aluminum conductor steel-reinforced cable; bend deformation; stress; friction coefficient; wind loads; fatigue fracture

---

## 1. Introduction

Fatigue of the wires in an overhead conductor occurs within the couplings that restrict the conductor vibration from incurring a vortex-induced oscillation [1–3]. At present, most of the overhead conductors are made of steel-cored aluminum strands (ACSRs), and ACSR is widely used in power delivery systems around the world [4,5]. An aeolian vibration is one of the causes of a fretting fatigue failure of a conductor. Such failures invariably occur in suspension clamps, dampers, and spacers [6,7].

The northwest part of China has several areas with a strong wind, which significantly impacts the operation of transmission and transformation lines. According to statistics, from July 2015 to September 2016, 750 kV ultra-high voltage (UHV) transmission lines undertaking the task of power transmission from the west to the east in a strong wind areas suffered three strand breaks of the mid-phase jumper in succession, resulting in significant economic losses. Therefore, it is very important to study the stress and deformation distribution of the mid-phase jumper in terms of the movement under strong winds to provide guidance for the treatment measures.

Rizzo et al. [8] described a method based on outlier analysis and the wavelet transform for structural damage detection based on guided ultrasonic waves. Furthermore, this method was applied to the detection of notch-like defects in a seven-wire strand by using built-in magnetostrictive devices for ultrasound transduction. Castellano et al. [9] presented an innovative non-destructive experimental technique based on ultrasonic waves for a quantitative determination of the damage level in structural components. Neslušan et al. [10] presented a systematic study using a magnetic non-destructive evaluation of corrosion attack in rope wires via Barkhausen noise emission. Because of the complexity of the scene, it is difficult to detect the damage of a six-splitting mid-phase jumper directly. Through the establishment of the finite element model six-splitting of a mid-phase jumper, the damage under wind load can be calculated. This paper presents an effective finite element modeling method for a large steel-cored aluminum strand when a main bending deformation occurs.

Because of the interaction between strands, multi-layer strands have a variable bending stiffness. With the bending of the strands, they begin to gradually slide with respect to each other, resulting in a decrease in the bending stiffness. However, the structural model of the conductor studied here is larger and has a spatial bending form. Therefore, it is extremely difficult to establish a finite element model of a mid-phase jumper with equal proportions.

Some scholars have proposed different modeling methods to study the mechanical response of steel strands during a bending deformation. Raoof and Hobbs [11] idealized a strand as a series of concentric orthotropic cylinders, each of which is associated with a specific layer and its corresponding mechanical properties. The bending process of the wire was also considered. In addition, Lanteigne [12] studied the mechanical behavior of cables under an arbitrary combination of tension, torsion, and bending. The force model of the cables is universal.

Leclair and Costello [13] applied the love crank balance equation to each conductor and deduced a mechanical conductor model that can predict the stress of single twisted strands under axial, bending, and torsion loads. Hong et al. [14] established a mechanical model of the conductor bending behavior under tension, although the model was significantly simplified. For example, the contact between adjacent cables of the same layer is neglected in the established model, and the contact line is used instead of the contact point of the superimposed layer. In addition, the cross-section of the twisted strand after bending is still planar.

Judge et al. [15] used LS-DYNA to establish a full three-dimensional, elastic–plastic finite element model of multi-layer helical cables under a quasi-static axial load. The model uses three-dimensional solid elements to discretize each strand and simulates the contact types between all strands. The model also uses the law of non-linear hardening to explain the plastic deformation. Zhang et al. [16] succeeded in analyzing the flexural stiffness of the strands using a solid three-dimensional finite element model, although their research was limited to single-pitch, single-layer cables. To summarize, the method of full three-dimensional solid modeling considers the contact between wires, among other factors, which results in a large number of computations, and thus the length of a fully three-dimensional solid element finite element model is usually very short. To reduce the computational complexity of the model, some scholars have used a beam element instead of solid element to establish a finite element model of a strand. Zhou et al. [17] used a beam element to model a single-layer conductor and managed the contact interaction between the lines using a coupling equation between the corresponding nodes, and thus the slip between the lines was not considered. The model was applied to the analysis of strands under bending loads, which is still limited to single-layer strands with a small deflection. Beleznai et al. [18] also modeled a conductor. Each line contact was simulated using a spring element. The stiffness of the spring element was based on the Hertzian contact theory. Although the correctness of this method has been proved, its application was still limited to single or double layers with a small displacement. Lalonde et al. [19,20] proposed a finite element modeling method for multi-layer steel strands under multi-axial loads. This method utilizes a beam element and a three-dimensional line contact, which greatly reduce the computational complexity of the model under the premise of ensuring a high level of accuracy.

In this study, a finite element, fully three-dimensional solid modeling method was used to study the mechanical response of a steel-cored aluminum strand with a mid-phase jumper under a bending deformation. A whole model (simplifying an ACSR into a solid cylinder) and a local model (modeling according to the actual structure of an ACSR) of the mid-phase jumper were established. First, the movement of the mid-phase jumper of the tension tower under wind load was studied based on the whole finite element model, and the equivalent Young's modulus of the whole model was adjusted based on the local model. The results of the whole model were then imported into the local model, and the stress distribution of each strand of the ACSR was analyzed in detail under a bending deformation to provide guidance for the treatment measures. Therefore, the whole model and the local model complemented each other, which could reduce the number of model operations and ensure the accuracy of the results. The feasibility of this method was verified through a comparison with the results of subsequent experiments. Although this modeling method was applied to the stress and deformation analysis of a mid-phase jumper in this study, it can be used to study the bending deformation of rope structures with a complex geometry and the main bending deformation. The finite element modeling method of a mid-phase jumper presented in this paper can be implemented in any general finite element software. Abaqus (2018, Dassault Systemes Simulia Corp, Johnston, RI, USA) was also used in this study.

## 2. Finite Element Modeling Approach

### 2.1. Spatial Structure and Motion Form of the Mid-Phase Jumper

Figure 1a shows the location of a mid-phase jumper on a tension tower and Figure 1b shows the lateral view of a mid-phase jumper on a tension tower. Because of the tension at both ends of the tower, large-span wires are subjected to a significant amount of tension. However, for a mid-phase jumper, owing to its different spatial structure, the force differs from that of a large-span wire. As shown in Figure 1a, b, the lower end of the mid-phase jumper is fixed by a tension clamp and the upper end is suspended by an insulator string, which shows a spatial tortuous distribution. Compared with a large-span wire, the initial tension applied to the mid-phase jumper is very small, and thus the initial tension applied to the mid-phase jumper can be neglected. Figure 1c shows an image of the motion of the lower end of a mid-phase jumper near a fixed position under level-7 wind. It is seen that the mid-phase jumper oscillates along the A-B direction under a strong wind, and its range of motion is slightly larger than the diameter of the six-splitting spacer (650 mm). In addition, the oscillation frequency is 5–8 times per minute. As shown in Figure 1d, the broken position of the mid-phase jumper is at the outlet of the tension clamp. The common motion forms of an overhead conductor under wind load are breeze vibration, subspan vibration, galloping, etc. [21–23]. The frequency of aeolian vibration is generally 3–150 Hz, and the amplitude is 0.01–1$d$ ($d$ is the diameter of the conductor), which occurs when the wind speed is small. The frequency of galloping is 0.08–3 Hz, and the amplitude is much larger than that of aeolian vibration, which often occurs on the iced conductor with a high wind speed. Subspan oscillation refers to the vibration of a conductor between adjacent spacer bars under wind load. The amplitude of subspan oscillation is generally 5–10 mm and the frequency of oscillation is 1–2 Hz. To sum up, the movement form of the mid-phase jumper is different from the breeze vibration, the subspan vibration, and the galloping, which is a large-scale cyclic swing under a strong wind.

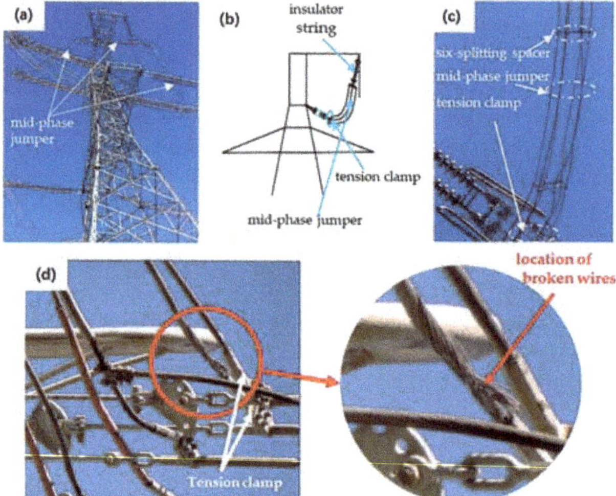


**Figure 1.** (**a**) Position of a mid-phase jumper on a tension tower, (**b**) lateral view of a mid-phase jumper on a tension tower, (**c**) movement of a mid-phase jumper near a tensile clamp, and (**d**) position of broken strands on the site of a mid-phase jumper.

## 2.2. Establishment of the Whole Model

Figure 2a shows the structure of the mid-phase jumper and the related data size (mm). It is seen that the mid-phase jumper includes a six-splitting, steel-cored aluminum strand and a six-splitting spacer. Figure 2b shows a side view of a mid-phase jumper on a tension tower. Figure 2c shows the solid model of the mid-phase jumper imported into Abaqus, which simplified the steel-core aluminum wire into a solid rod and was established in SolidWorks. (2016, Dassault Systemes Simulia Corp, Johnston, RI, USA) Because the lower end of the mid-phase jumper was rigidly connected through a tension clamp, all degrees of freedom of the lower end of the mid-phase jumper were fixed. As shown with the A arrows, the fixed ends of the lower of the mid-phase jumper were the outlet position of the tension clamp. In the actual structure of the mid-phase jumper, the top of the insulator string was fixed on the tension tower. According to the assembly of the insulator string and the mid-phase jumper, the maximum displacement of the middle section of the mid-phase jumper deviating from the east–west direction was limited to 30 mm, and other displacement degrees of freedom and all rotational degrees of freedom were limited, as shown with the B arrows. Because of the large deformation of the mid-phase jumper under a strong wind, geometric nonlinearity was opened during the analysis step of Abaqus. The quadratic tetrahedron stress element (C3D10) with a high accuracy was used in the whole model. The number of elements and nodes were approximately $3.5 \times 10^5$ and $6.6 \times 10^5$, respectively.

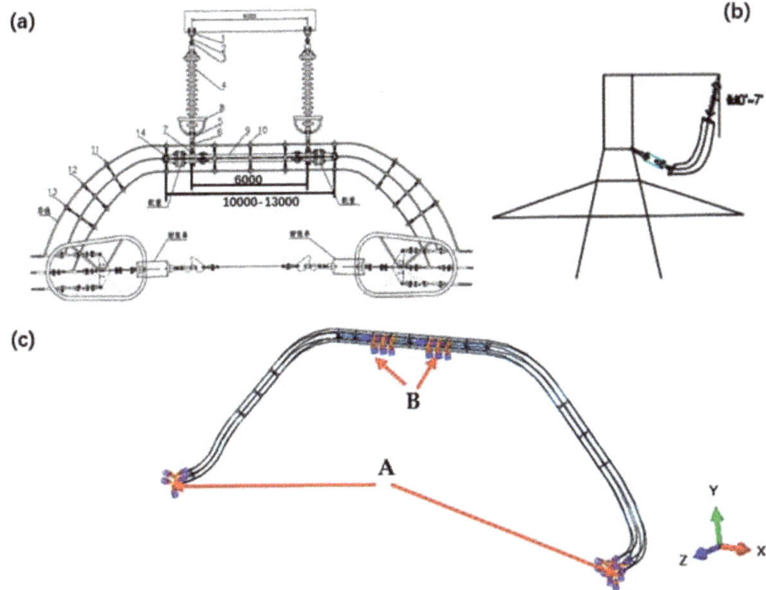

**Figure 2.** (**a**) Structural diagram of a mid-phase jumper, (**b**) side view of a mid-phase jumper on a tower, and (**c**) finite element model of a mid-phase jumper in Abaqus.

### 2.3. Establishment of Local Model

As shown in Figure 1d, the actual fracture position of the mid-phase jumper is the outlet position of the tension clamp. In order to accurately analyze the stress distribution of the actual fracture position, a local model was established. The local model established in this study consisted of four helical layers wound around a straight cylinder at the center. In addition, adjacent spiral layers were wound in the opposite directions to reduce the internal torque caused by the winding. The formula for calculating the radius of each helical layer is shown in Equation (1). The formula for calculating the helix angle is shown in Equation (2). In Equations (1) and (2), $R_i$, $d_i$, $d_{core}$, $m_i$, and $ai$ indicate the layer radius (mm), wire diameter (mm), diameter of the central steel wire (mm), pitch diameter ratio, and helical angles of the helical layers, respectively.

$$R_i = \frac{d_{core}}{2} + \frac{d_i}{2} + \sum_{k=1}^{k=i-1} d_k \tag{1}$$

$$a_i = \arctan(\frac{m_i}{\pi}) \tag{2}$$

Table 1 shows the mechanical properties of the materials and modeling parameters of the local finite element model. Figure 3a shows the sectional diagram of the local finite element model. As shown in Figure 3b, a section of all strands was fixed at one end of the local model. The fixed end was located at the outlet of the tension clamp. By setting the motion coupling constraints in Abaqus, the other end was coupled with the central point, all six degrees of freedom of the constrained area were selected to be coupled with the coupling points, and the constrained end face then became a rigid plane. There was no relative displacement between the nodes in the constrained area, only a rigid motion with the selected coupling point. In this case, the displacement boundary condition ($Y_a$) could be applied to the coupling point.

**Table 1.** Mechanical properties of materials and modeling parameters of the local finite element model.

| Line Strand of Position | Number of Strands | E (GPa) | Yield Strength (MPa) | Poisson Ratio | Winding Angle (°) | Pitch Ratio | Diameter (mm) |
| --- | --- | --- | --- | --- | --- | --- | --- |
| Outermost layer | 24 | 69 | 275 | 0.30 | 74.0 | 11 | 3.07 |
| Layer 3 | 18 | 69 | 275 | 0.30 | 75.3 | 12 | 3.07 |
| Layer 2 | 12 | 69 | 275 | 0.30 | 76.4 | 13 | 3.07 |
| Layer 1 | 6 | 207 | 68 | 0.28 | 81.5 | 21 | 3.07 |
| Central steel wire | 1 | 207 | 68 | 0.28 | 0 | — | 3.07 |

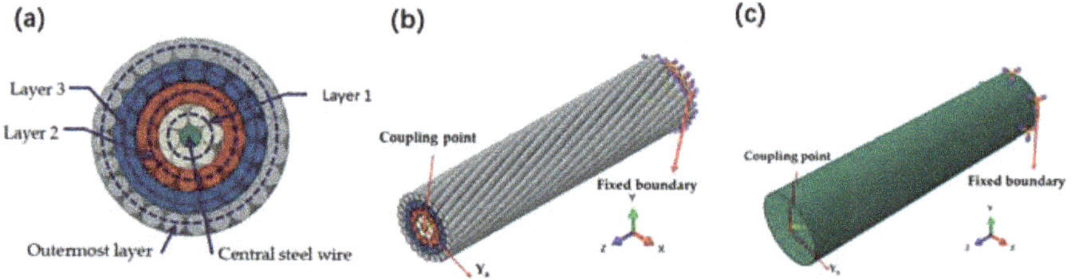


**Figure 3.** (a) Sectional diagram of the local finite element model, (b) 3-D local finite element model, and (c) the whole microsegment finite element model.

The hexahedral incompatible element (C3D8I) was used in the local model. The number of elements and nodes were approximately $1.3 \times 10^5$ and $1.8 \times 10^5$, respectively. The default contact behavior in the Abaqus/Standard was the "hard contact." The surface contact was set between each strand. Although numerous contact algorithms can be used to establish contact pairs, the contact is extremely complex owing to the large number of surface-to-surface contacts involved. Therefore, the contact algorithm was set as the augmented Lagrange algorithm, which improved the convergence efficiency of the model operation under the premise of ensuring the accuracy. The default value of the contact stiffness was adopted, and the expansion coefficient of the contact stiffness was 1.

Although Utting and Jones [24] explored the friction effect between steel wires, they were not specified regarding the friction coefficient. Judge et al. [15] and Lalonde et al. [19] applied a friction coefficient μ of 0.115 to all contact points. The present simulation used the same coefficient value. When the mid-phase jumper sways significantly under wind load, the main deformation of the mid-phase jumper is bending. Therefore, the bending stiffness of the whole model was made the same as that of the local model by adjusting the Young's modulus parameters, and the adjusted Young's modulus was then input into the whole model for the calculations. Finally, the motion behavior of the whole model under wind load was obtained. The adjustment process of the Young's modulus of the whole model is described in detail in Section 3.

*2.4. Application of Wind Load on the Whole Model*

A mid-phase jumper oscillates under wind load. According to Bernoulli's equation, the wind load on the conductor is as shown in Equation (3):

$$P = \frac{\gamma}{2g} v^2 \sin \theta. \tag{3}$$

In Equation (3), $P$, $\gamma$, $g$, $v$, and $\theta$ indicate the wind load (N), the air density (kg/m$^3$), acceleration of gravity (m/s$^2$), wind speed (m/s), and angle between the wind direction and conductor axis, respectively. For a large-span conductor, an aerodynamic coefficient and a nonuniformity coefficient were introduced. According to Equation (3), the main factors influencing the wind load on the conductor are the wind speed and angle $\theta$ between the wind direction and conductor. When $\theta$ is 90°, the formula for calculating the wind load is shown in Equation (4):

$$P_f = a_f \cdot K_a \cdot A_f \cdot \frac{v_f^2}{1600},\tag{4}$$

In Equation (4), $a_f$ indicates the asymmetrical coefficient of the wind pressure. When the wind speed is less than 20 m/s, $a_f$ is taken to be 1, and when the wind speed is between 20 and 30 m/s, it is taken to be 0.85. In addition, $K_a$ indicates the aerodynamic coefficient and is taken to be 1, $A_f$ represents the projected area of the conductor in the direction of the wind load, and $v_f$ indicates the speed (m/s). Finally, $P_f$ indicates the wind load (N). The amplitude change of the wind load applied on the mid-phase jumper is shown in Equation (5):

$$A_{mp} = \sin\left(\frac{\pi}{2}t\right)\tag{5}$$

In Equation (5), $A_{mp}$ indicates the amplitude of wind load applied, and $t$ indicates the time (s). The wind load is converted from the kinetic energy of the wind, and the kinetic energy consumption of the wind perpendicular to the conductor is the largest; therefore, the wind speed component perpendicular to the strand determines the actual wind load, and that parallel to the conductor has little effect. Table 2 shows the applied wind speed level to the whole model, the corresponding wind speed, and the magnitude of the applied wind load when $\theta$ was 90°. Westerly winds dominate the northwest part of China throughout the year. In this study, the mechanical response of the mid-phase jumper to a westerly wind was mainly considered. As shown in Figure 4, the wind load was applied to the whole model, and the arrow in the figure represents the wind load and wind direction.

**Table 2.** Magnitude of wind load applied to the surface of the conductor corresponding to different wind speeds.

| Wind Speed Level | Wind Speed (m/s) | Wind Load (MPa) |
|:---:|:---:|:---:|
| 1 | 1.0 | $2.39 \times 10^{-7}$ |
| 2 | 2.0 | $1.49 \times 10^{-6}$ |
| 3 | 3.5 | $4.83 \times 10^{-6}$ |
| 4 | 5.5 | $1.01 \times 10^{-5}$ |
| 5 | 8.0 | $2.15 \times 10^{-5}$ |
| 6 | 11.0 | $2.41 \times 10^{-5}$ |
| 7 | 14.0 | $3.90 \times 10^{-5}$ |

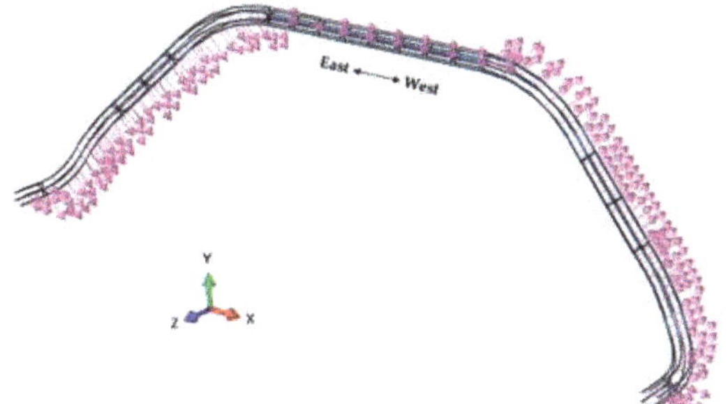


**Figure 4.** Wind load applied to the whole model.

*2.5. Model Solution*

The whole model used the Newton–Raphson algorithm to solve non-linear problems with a large displacement. The local model used the Newton–Raphson method to solve the contact and material nonlinearity problems. All simulations conducted in this study were implemented on a 2.3 GHz, 24-core CPU and a computer with 64 GB of memory.

## 3. Matching Adjustment of Whole Model and Local Model

On the one hand, in order to accurately simulate the deformation of a six-splitting mid-phase jumper under wind load, it is necessary to adjust the equivalent Young's modulus of the whole model to make it have the same bending resistance as the local model. On the other hand, in order to more accurately simulate the stress of each strand of a conductor at the outlet of the tension clamp, it is necessary to ensure the accuracy of the deformation of the whole model introduced into the local model, that is, to ensure that the whole model has the same bending resistance as the local model. Therefore, it is very important for the accuracy of the simulation results to adjust the Young's modulus of the whole model and obtain the equivalent Young's modulus of the whole model based on the local model.

The Young's modulus of the whole model is really the key of this paper, which was determined using the following method. In order to determine the Young's modulus of the whole model, as shown in Figure 3c, the whole microsegment model with a length of 120 mm was established, which was consistent with the cross section of the whole model (27.63 mm). By adjusting the Young's modulus of the whole microsegment model, the whole microsegment model and the local model had the same bending stiffness, and then the adjusted Young's modulus was input into the whole model such that the whole model had the same bending stiffness as the local model. A hexahedral incompatible element (C3D8I) was used in the whole microsegment model. The number of elements and nodes were approximately $2.4 \times 10^4$ and $2.6 \times 10^4$, respectively. One end of the whole microsegment model was completely fixed, while the other end was coupled with the central point, and the displacement boundary condition ($Y_a$) was applied to the coupling point. The boundary conditions of the microsegment whole model were in good agreement with those of the local model.

The $Y_a$ of the coupling point and the force (F) needed to achieve the $Y_a$ were obtained by changing $Y_a$. As shown in Figure 5a, for the local model, the relationship of $F - Y_a$ was approximately linear. By continuously adjusting the Young's modulus of the whole microsegment model, the relationship of $F - Y_a$ under different Young's moduli can be obtained, and the relationship of $F - Y_a$ was approximately linear. As shown in Figure 5a, when the Young's modulus of the whole microsegment model was 6.25 GPa, the $F - Y_a$ lines of the whole microsegment model and the local model almost coincided. Therefore, when the Young's modulus of the whole microsegment model was 6.25 GPa, the whole microsegment model and the local model had a similar bending stiffness. Then, 6.25 GPa was input into the whole model, which had the same bending stiffness as the local model.

As shown in Figure 5b, for the same $Y_a$, the difference in F between the whole microsegment model and the local model was defined as $\Delta F$. Based on an F corresponding to a $Y_a$ of the local model, the ratio of $\Delta F$ to F corresponding to $Y_a$ of this model was the error when applying this method. As shown in Figure 5b, $Y_a$ had a larger error of 0–0.25 mm and a smaller error of 0.25–0.6 mm. According to the simulation results in the fourth section, $Y_a$ corresponding to a strong wind was between 0.25–0.60; therefore, this study mainly considered the part of $Y_a$ between 0.25 and 0.60 mm. The absolute error of this part was less than 10%, which is within a reasonable range.

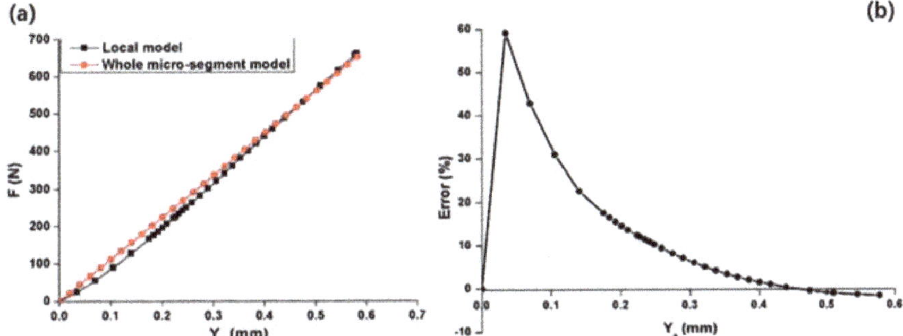

**Figure 5.** (**a**) F — $Y_a$ of the whole microsegment model and the local model, and (**b**) the error of the whole model.

## 4. Results and Discussion

Figure 6a shows a displacement nephogram of the whole model under a westerly, level-6 wind. Figure 6b shows the displacement nephogram of the whole model with a length of 1000 mm at the outlet of the tension clamp. The middle section of the mid-phase jumper swung back and forth along the east–west direction. The swing of the mid-phase jumper in this direction caused a greater bending moment at the lower area of the mid-phase jumper.

It was assumed that the deflection of the conductor with the maximum displacement at the lower area of the whole model under wind load was $Y_b$ at 89 mm away from the outlet of the tension clamp. When $Y_b$ corresponding to different wind loads was imported into the local model, the maximum equivalent stress ($\sigma_a$) corresponding to the outlet position of the tension clamp of the local model under different wind loads could be obtained. As shown in Figure 6c, $Y_b$ changed with the swing of the whole model. When the displacement of the middle part of the mid-phase jumper along the negative X direction reached the maximum, $Y_b$ also reached the maximum ($t = 3$ s). This maximum value of $Y_b$ (0.258 mm) was imported into the local model to obtain the stress distribution at the outlet of the tension clamp under the level-6 wind, as shown in Figure 6d. It is seen that the stress concentration appeared near the fixed end of the local model, that is, the stress concentration appeared near the outlet of the tension clamp and the maximum stress appeared here. This explains why the actual mid-phase jumper breakage occurred at the outlet of the tension clamp.

As shown in Figure 6e, with an increase in the wind load, $Y_b$ increased, accelerating rapidly after a level-4 wind, which was due to the rapid increase in the wind load applied to the whole model. As shown in Figure 6f, with an increase in the wind load, the value of $\sigma_a$ of the outermost, third, and second layers of the aluminum conductor increased, where the $\sigma_a$ of the third layer was the largest. The $\sigma_a$ of the outermost and third layers were higher than the yield strength of the conductor under the level-6 wind. The difference between $\sigma_a$ of the outermost layer and $\sigma_a$ of the third layer increased gradually with the wind load increases, and the difference was approximately 10 MPa under the level-6 wind load. As shown in Figure 6g, the steel layers had not reached their yield strength (275 MPa) when all three aluminum layers yielded under the level-7 wind load, although $\sigma_a$ on the first layer was the largest at this time. This was because the Young's modulus of the first layer was much larger than that of the aluminum because of its steel material, which resulted in a larger stress on the first layer of the steel core.

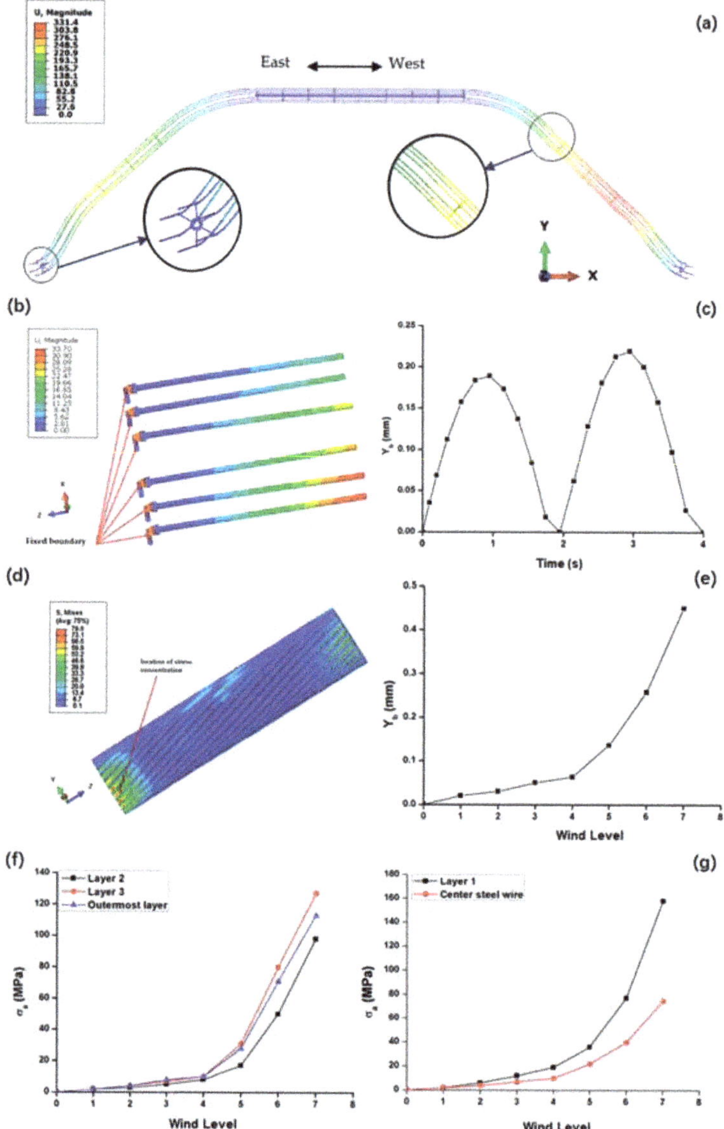


**Figure 6.** (**a**) Swing mode of the mid-phase jumper under the westerly wind, (**b**) the displacement nephogram of the whole model with a length of 1000 mm at the outlet of the tension clamp, (**c**) $Y_b$ value of the whole model under the level-6 wind, (**d**) stress nephogram of the local finite element model, (**e**) value of $Y_b$ corresponding to different wind loads, (**f**) $\sigma_a$ of three aluminum conductor layers corresponding to different wind loads, and (**g**) $\sigma_a$ of the corresponding steel layer under different wind loads.

## 5. Verification of Simulation Results

### 5.1. Verification of the Fatigue Test

Combined with the finite element simulation, the fatigue test was carried out on the fatigue test platform of an ACSR. Figure 7 shows the three-dimensional design of the fatigue test platform of an ACSR. The initial adjustment device was fixed to the tension clamp at the end of the ACSR, and the pretension and other parameters of the ACSR could be adjusted according to the working conditions on site. The axial adjustment device can exert axial tension on the ACSR. The device for transverse motion acted directly on the ACSR, causing it to move transversely, and the amplitude of the conductor swing can be adjusted. A steel-cored aluminum strand of LGJ-400/50 (Dazheng electric wire Corp, cangzhou, Hebei, China) was pressed in the tension clamp and assembled on the fatigue test platform of the ACSR. A displacement of 89 mm ($Y_b$) at the outlet of the tension clamp was measured using a dial indicator. The $Y_b$ simulated by the finite element method under the level-7 wind load was set to the maximum displacement ($Y_b$) of 89 mm at the outlet of the tension clamp during the fatigue test, and a cyclic fatigue test was then conducted. Three groups of experiments were conducted. Table 3 shows that the cyclic fatigue life of the conductor was between 150,000 and 300,000 times without or with very little tension.

**Table 3.** Results of fatigue tests.

| Specimen | $Y_b$ | Cycles (Number of Times) |
|---|---|---|
| Specimen A | 0.45 | 156,000 |
| Specimen B | 0.45 | 290,000 |
| Specimen C | 0.45 | 226,000 |

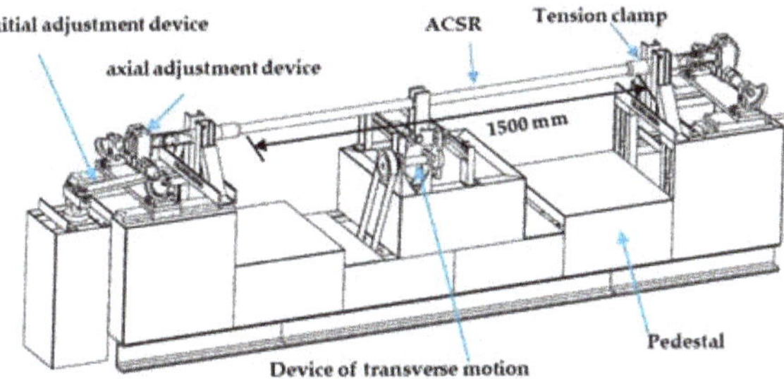


**Figure 7.** 3-D design drawing of the fatigue test platform of a steel-cored aluminum strand (ACSR).

As shown in Figure 8a, there was an obvious stress concentration at the outlet of the tension clamp for the outermost conductor and the third layer of conductor. As shown in Figure 8b,c, $\sigma_a$ of the third layer of the conductor (79.7 MPa) was greater than $\sigma_a$ of the outermost conductor (71.4 MPa), which was very interesting. Under the guidance of the finite element results, when one strand of the outermost conductor broke during the fatigue test, the outermost conductor was cut off completely, as shown by arrow A in Figure 9. Interestingly, at this time, the third layer of conductor had almost broken, as shown by arrow B in Figure 9. This indicates that at the outlet of the tension clamp, the crack of the third layer of conductor occurred earlier than the outmost conductor, which shows the accuracy of local model simulation results.

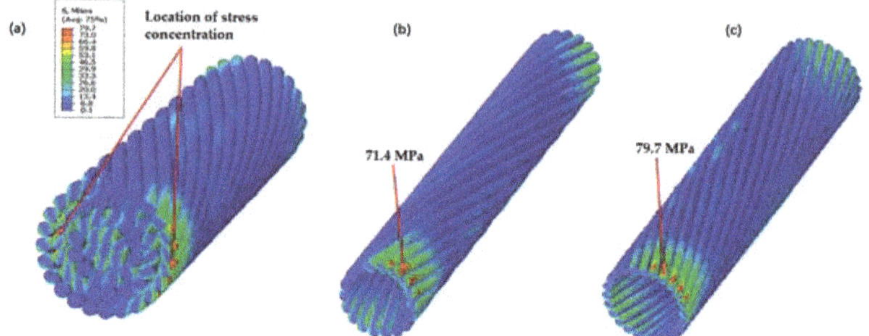

**Figure 8.** Stress distribution of local model at the outlet of the tension clamp: (**a**) the stress distribution of each layer, (**b**) the stress distribution of the third layer of the conductor, and (**c**) the stress distribution of the outmost conductor.

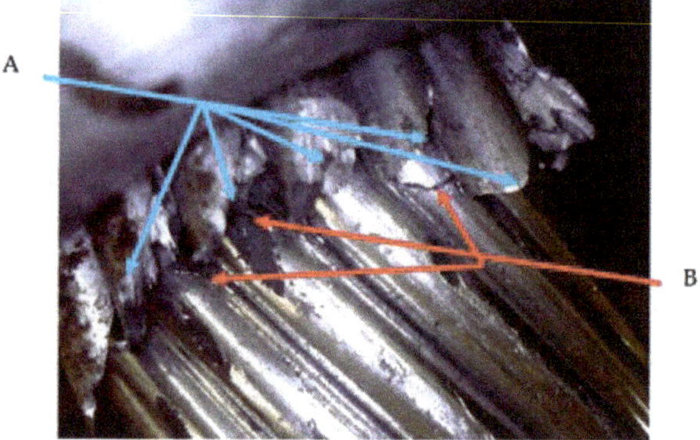

**Figure 9.** Fracture diagram of the third layer of conductor.

Figure 10a shows the macroscopic morphology of the fatigue fracture. As shown in Figure 10b, the fatigue band with the characteristics of bainite lines was observed. These fatigue bands were thick, short, deep, wide, and discontinuous, which obviously accorded with the characteristics of low cycle fatigue fracture bands. Therefore, under the level-7 wind load, the stress concentration of the third layer at the outlet of the tension clamp had yielded. To summarize, the results of the fatigue tests were in good agreement with the simulation results, which proved the accuracy of the local model simulation results.

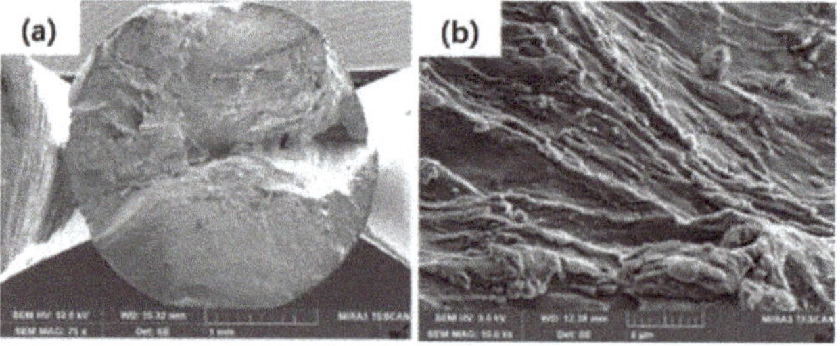

**Figure 10.** (**a**) The macroscopic morphology of the fatigue fracture, and (**b**) morphology of the fatigue fracture.

As shown in Figure 11a, after cutting the outermost conductor, it was found that the third layer of the conductor has obvious wear on the surface at the outlet of the tension clamp. As shown in Figure 11b, after the experiment, several wear defects could be seen on the surface of the third layer of conductor at the outlet of the tension clamp after cutting. The outer surface of the third layer led to wear defects due to the sliding of the conductor in the process of moving. These defects became the source of cracks in the third layer of conductor, and the cracks eventually continued to expand in the process of repeated cyclic movement of the conductor, resulting in the fracture of the third layer of conductor.

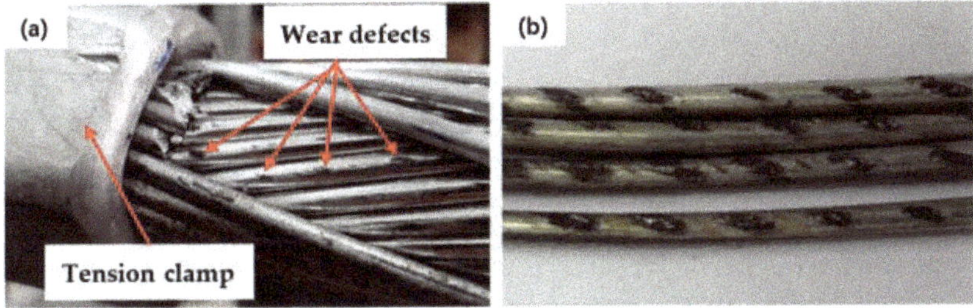


**Figure 11.** (**a**) Wear defects on the surface of the third layer of conductor, and (**b**) wear defects of the third layer of conductor at the outlet of the tension clamp after cutting.

### 5.2. Verification of the Strain Gauges Test

In order to verify the accuracy of the finite element model, strain gauges were pasted at the outlet of the clamp. When $Y_b$ changed, $\sigma_a$ was measured under different wind loads and compared with the simulated stress. Because the strain gauge needed to be pasted on a smooth surface, the surface of the single strand of aluminum wire to be pasted with the strain gauge was lightly polished with a file. The thickness of the polished part accounted for about one fifth of the diameter of the single strand. The model of static strain instrument used was YE2538A (Yutian Technology Company, Wuxi, China). The strain gauge used was a conventional small resistance strain gauge with the size of 0.5 mm × 0.5 mm and a sensitivity of 2.00 kV. Figure 12 shows the installation position of strain gauges.

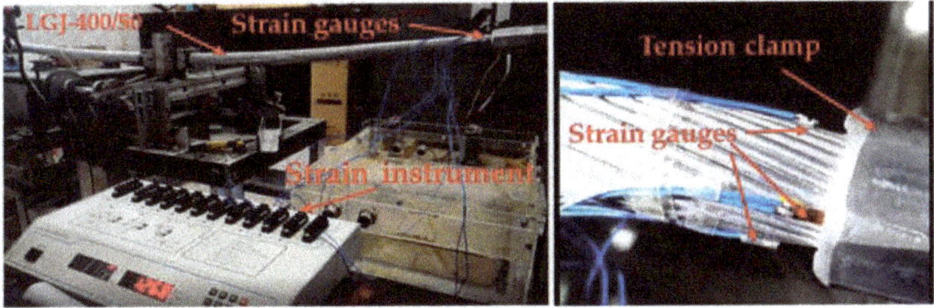


**Figure 12.** Installation positions of the strain gauges.

As shown in Figure 13, with the increase of the wind load, the $\sigma_a$ obtained using the simulation and $\sigma_a$ obtained using the test both increased, and the maximum error between $\sigma_a$ obtained using the simulation and $\sigma_a$ obtained using the test was about 20 MPa. On one hand, the error was the error of experimental measurement, on the other hand, it was the error of the finite element numerical calculation itself. In conclusion, although there was an error between $\sigma_a$ obtained using the simulation and $\sigma_a$ obtained using the test, it also proved that the accuracy of the finite element model results were within a reasonable range.

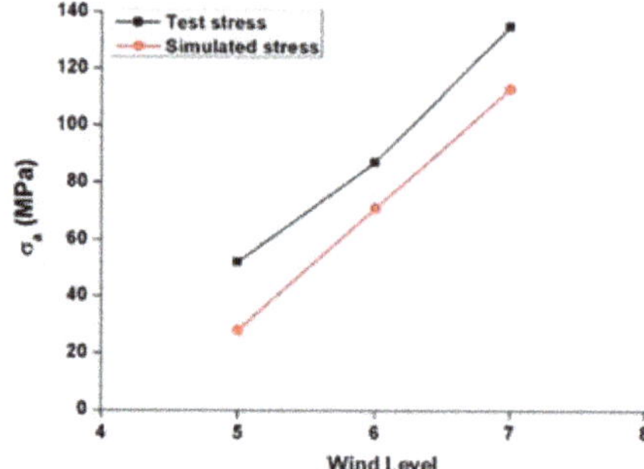

**Figure 13.** Comparison of the simulated stress and test stress under different wind loads.

## 6. Effect of the Friction Coefficient

Because the friction coefficient is affected by the metal type, chemical composition, surface state, deformation temperature, and contact pressure, the friction coefficient between aluminum strands is not easy to determine. This paper mainly studied the bending deformation of a conductor, and the contact between each strand of a conductor is extremely complex during the process of the bending deformation. Coulomb's law was introduced to study this problem. The tangential contact of each strand in the aluminum strand follows this law. The Coulomb friction model was used in Abaqus.

Papailiou et al. [25] proposed a conductor model that can withstand both the tension and bending loads, and considered the interlayer friction and slip during the process of wire bending. It was found that the bending analysis of the conductor was extremely sensitive to the selection of the friction coefficients between two adjacent layers of the conductors. The range of friction coefficients between cables was found to be between 0.55 and 0.9. Wharton et al. [26] measured the fretting fatigue strength of A1-4Mg-0.7Mn in contact with aluminum and copper. It was found that the friction coefficient of A1-4Mg-0.7Mn in contact with aluminum and copper was extremely low, but with an increase in the number of fatigue cycles, the friction coefficient increased to approximately 1.0 and remained unchanged at this value. Clearly, a conductor bending analysis is extremely sensitive to the choice of friction coefficient between two adjacent conductors. In the process of long-term service, the friction coefficient between strands of a conductor will change due to the wear between strands. In order to accurately evaluate the change of the friction coefficient in the service process of the mid-phase jumper, this section discusses the influence of the friction coefficient on the stress of the mid-phase jumper when other conditions remain unchanged. Thus, the friction coefficients between the steel-cored aluminum strands were set to 0.3, 0.5, and 0.7. At this time, the equivalent Young's modulus of the whole finite element model was 6.78, 6.68, and 7.67 GPa, respectively, and the other conditions remained unchanged. The influence of the friction coefficient is discussed next.

As shown in Figure 14, between level-1 and level-3 wind loads, the wind load was small, and thus $\sigma_a$ and $Y_b$ were extremely small; therefore, the friction coefficient had little effect on the corresponding $\sigma_a$ and $Y_b$. Between level-4 and level-7 wind loads, when the friction coefficient was 0.115 to 0.50, $\sigma_a$ and $Y_b$ first decreased and then increased with an increase of the friction coefficient under the same wind load; when the friction coefficient was 0.50 to 0.70, $\sigma_a$ and $Y_b$ decreased again with an increase of the friction coefficient under the same wind load. Therefore, the friction coefficient had a great effect on the corresponding $\sigma_a$ and $Y_b$ between level-4 and level-7 wind loads. The difference in $\sigma_a$ between the outermost layer and the third layer reached the maximum under a level-7 wind load when the friction coefficient took the above four parameters, and the maximum difference did not exceed 20 MPa. In conclusion, the friction coefficient influenced the bending of the mid-phase jumper, but the effect was not great.

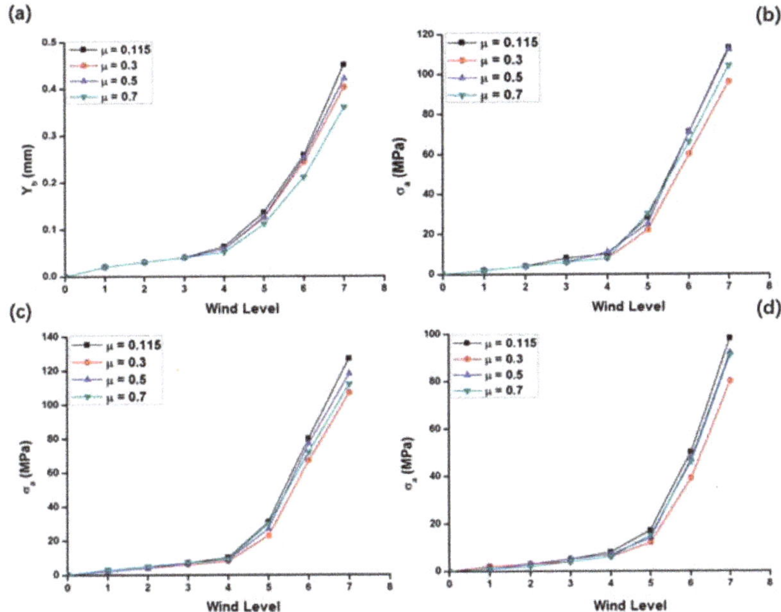

**Figure 14.** (**a**) Effect of the friction coefficient on $Y_b$, (**b**) $\sigma_a$ of the outermost layer, (**c**) $\sigma_a$ of the third layer, and (**d**) $\sigma_a$ of the second layer.

## 7. Conclusions

In this paper, a finite element, fully three-dimensional solid modeling method was used to study the mechanical response of a steel-cored aluminum strand (ACSR) with a mid-phase jumper under a wind load. The whole model (simplifying an ACSR into a solid cylinder) and a local model (modeling according to the actual structure of an ACSR) of the mid-phase jumper were established. First, the movement of the mid-phase jumper of the tension tower under a wind load was studied based on a whole finite element model, and the equivalent Young's modulus of the whole model was adjusted based on the local model. The results of the whole model were then imported into the local model, and the stress distribution of each strand of the ACSR was analyzed in detail to provide guidance for the treatment measures. Therefore, the whole model and the local model complemented each other, which could reduce the number of model operations and ensure the accuracy of the results. Although this method was applied to the stress and deformation analysis of a mid-phase jumper in this paper, it can be used to study the bending deformation of rope structures with a complex geometry and a main bending deformation.

The analysis showed that the swing of the mid-phase jumper in the east–west direction caused a greater bending moment at the lower area of the mid-phase jumper, which led to the stress concentration appearing near the outlet of the tension clamp. This explained why the actual mid-phase jumper breakage occurred at the outlet of the tension clamp. The maximum stresses of the outermost and third layers were higher than the yield strength under a level-6 wind. The difference in $\sigma_a$ between the outermost and third layers increased gradually when the wind load increased, and the difference was approximately 10 MPa under a level-6 wind load.

Interestingly, $\sigma_a$ of the third layer of the conductor (79.7 MPa) was greater than $\sigma_a$ of the outermost conductor (71.4 MPa), which was verified by the results of the fatigue tests. Based on the analysis of the surface wear of the third layer of the conductor, the reason for the fracture of the third layer of conductor was explained. The feasibility of the element modeling method was verified through a comparison with the results of a subsequent series of experiments.

In addition, the effects of the friction coefficient on the bending of a mid-phase jumper were studied. It was found that the friction coefficient influenced the bending of the mid-phase jumper, but the effect was not great.

**Author Contributions:** Methodology, W.X., P.M., and C.H.; validation, P.M. and W.X.; investigation, C.H.; data curation, P.M.; writing—original draft preparation, P.M.; writing—review and editing, W.X.; project administration, Y.L.; funding acquisition, J.H. All authors have read and agreed to the published version of the manuscript.

**Funding:** This work was supported by the State Grid of China Science and Technology Project, under the grant no. 52130417002U.

**Conflicts of Interest:** The authors declare no conflict of interest.

# References


1. *Transmission Line Reference Book: "The Orange Book"*; Electric Power Research Institute: Palo Alto, CA, USA, 1979.
2. Azevedo, C.R.F.; Henriques, A.; Filho, A.R.P.; Ferreira, J.L.A. Fretting fatigue in overhead conductors: Rig design and failure analysis of a grosbeak aluminium cable steel reinforced conductor. *Eng. Fail. Anal.* **2009**, *16*, 136–151. [CrossRef]
3. Zhou, Z.R.; Cardou, A.; Fiset, M.; Goudreau, S. Fretting fatigue in electrical transmission lines. *Wear* **1994**, *173*, 179–188. [CrossRef]
4. Harvard, D.; Bellamy, G.; Buchan, P.; Ewing, H.; Horrocks, D.; Krishnasamy, S.; Motlis, J.; Yoshiki-Gravelsins, K. Aged ACSR conductors. I. Testing procedures for conductors and line items. *IEEE Trans. Power Deliv.* **1992**, *7*, 581–587. [CrossRef]
5. Chen, G.; Wang, X.; Wang, J.; Liu, J.; Zhang, T.; Tang, W. Damage investigation of the aged aluminium cable steel reinforced (ACSR) conductors in a high-voltage transmission line. *Eng. Fail. Anal.* **2012**, *19*, 13–21. [CrossRef]
6. Zhou, Z.; Cardou, A.; Goudreau, S.; Fiset, M. Fundamental investigations of electrical conductor fretting fatigue. *Tribol. Int.* **1996**, *29*, 221–232. [CrossRef]
7. Ramey, G.E.; Townsen, J.S. Effects of clamps on fatigue of ACSR conductors. *Energy Div. ASCE* **1981**, *107*, 103–119.
8. Rizzo, P.; Sorrivi, E.; Di Scalea, F.L.; Viola, E. Wavelet-based outlier analysis for guided wave structural monitoring: Application to multi-wire strands. *J. Sound Vib.* **2007**, *307*, 52–68. [CrossRef]
9. Castellano, A.; Foti, P.; Fraddosio, A.; Galietti, U.; Marzano, S.; Piccioni, M.D. Characterization of Material Damage by Ultrasonic Immersion Test. *Procedia Eng.* **2015**, *109*, 395–402. [CrossRef]
10. Neslušan, M.; Bahleda, F.; Minárik, P.; Zgútová, K.; Jambor, M. Non-destructive monitoring of corrosion extent in steel rope wires via Barkhausen noise emission. *J. Magn. Magn. Mater.* **2019**, *484*, 179–187. [CrossRef]
11. Raoof, M.; Hobbs, R.E. The Bending of Spiral Strand and Armored Cables Close to Terminations. *J. Energy Resour. Technol.* **1984**, *106*, 349–355. [CrossRef]
12. Lanteigne, J. Theoretical Estimation of the Response of Helically Armored Cables to Tension, Torsion, and Bending. *J. Appl. Mech.* **1985**, *52*, 423–432. [CrossRef]
13. LeClair, R.A.; Costello, G.A. Axial, Bending and Torsional Loading of a Strand with Friction. *J. Offshore Mech. Arct. Eng.* **1988**, *110*, 38–42. [CrossRef]
14. Hong, K.-J.; Der Kiureghian, A.; Sackman, J.L. Bending Behavior of Helically Wrapped Cables. *J. Eng. Mech.* **2005**, *131*, 500–511. [CrossRef]
15. Judge, R.; Yang, Z.; Jones, S.; Beattie, G. Full 3D finite element modelling of spiral strand cables. *Constr. Build. Mater.* **2012**, *35*, 452–459. [CrossRef]
16. Zhang, D.; Ostoja-Starzewski, M. Finite Element Solutions to the Bending Stiffness of a Single-Layered Helically Wound Cable with Internal Friction. *J. Appl. Mech.* **2015**, *83*, 031003. [CrossRef]
17. Zhou, W.; Tian, H. A novel finite element model for single-layered wire strand. *J. Cent. South Univ. Technol.* **2013**, *20*, 1767–1771. [CrossRef]

18. Beleznai, R.; Páczelt, I. Design curve determination for two-layered wire rope strand using p-version finite element code. *Eng. Comput.* **2013**, *29*, 273–285. [CrossRef]
19. Lalonde, S.; Sébastien, R.; Guilbault, R.; Légeron, F. Modeling multilayered wire strands, a strategy based on 3D finite element beam-to-beam contacts—Part I: Model formulation and validation. *Int. J. Mech. Sci.* **2017**, *126*, 281–296. [CrossRef]
20. Lalonde, S.; Guilbault, R.; Langlois, S. Modeling multilayered wire strands, a strategy based on 3D finite element beam-to-beam contacts—Part II: Application to wind-induced vibration and fatigue analysis of overhead conductors. *Int. J. Mech. Sci.* **2017**, *126*, 297–307. [CrossRef]
21. Diana, G.; Belloli, M.; Giappino, S.; Manenti, A.; Mazzola, L.; Muggiasca, S.; Zuin, A. A Numerical Approach to Reproduce Subspan Oscillations and Comparison with Experimental Data. *IEEE Trans. Power Deliv.* **2014**, *29*, 1311–1317. [CrossRef]
22. Wang, J. Overhead Transmission Line Vibration and Galloping. In Proceedings of the 2008 International Conference on High Voltage Engineering and Application, Chongqing, China, 9–12 November 2008; Institute of Electrical and Electronics Engineers (IEEE): Piscataway, NJ, USA, 2008; pp. 120–123.
23. Ohkuma, T.; Kagami, J.; Nakauchi, H.; Kikuchi, T.; Takeda, K.; Marukawa, H. Numerical analysis of overhead transmission line galloping considering wind turbulence. *Electr. Eng. Jpn.* **2000**, *131*, 19–33. [CrossRef]
24. Utting, W.; Jones, N. The response of wire rope strands to axial tensile loads—Part I. Experimental results and theoretical predictions. *Int. J. Mech. Sci.* **1987**, *29*, 605–619. [CrossRef]
25. Papailiou, K. On the bending stiffness of transmission line conductors. *IEEE Trans. Power Deliv.* **1997**, *12*, 1576–1588. [CrossRef]
26. Wharton, M.; Waterhouse, R.; Hirakawa, K.; Nishioka, K. The effect of different contact materials on the fretting fatigue strength of an aluminium alloy. *Wear* **1973**, *26*, 253–260. [CrossRef]

 *applied sciences*

*Article*

# Finite Element Approaches to Model Electromechanical, Periodic Beams

**Wiktor Waszkowiak, Marek Krawczuk and Magdalena Palacz ***

Department of Mechatronics and High Voltage Engineering, Faculty of Electrical and Control Engineering, Gdańsk University of Technology, Narutowicza 11/12, 80-233 Gdańsk, Poland; wiktor.waszkowiak@pg.edu.pl (W.W.); marek.krawczuk@pg.edu.pl (M.K.)
* Correspondence: mpalacz@pg.edu.pl; Tel.: +48-58-347-25-08

Received: 6 February 2020; Accepted: 11 March 2020; Published: 14 March 2020

**Abstract:** Periodic structures have some interesting properties, of which the most evident is the presence of band gaps in their frequency spectra. Nowadays, modern technology allows to design dedicated structures of specific features. From the literature arises that it is possible to construct active periodic structures of desired dynamic properties. It can be considered that this may extend the scope of application of such structures. Therefore, numerical research on a beam element built of periodically arranged elementary cells, with active piezoelectric elements, has been performed. The control of parameters of this structure enables one for active damping of vibrations in a specific band in the beam spectrum. For this analysis the authors propose numerical models based on the finite element method (FEM) and the spectral finite element methods defined in the frequency domain (FDSFEM) and the time domain (TDSFEM).

**Keywords:** FEM; SFEM; active periodic structures; smart materials

## 1. Introduction

Periodic structures can be defined as structures consisting of a series of repeating segments with the same physical properties and sizes. Theoretical investigations of such structures have usually been carried out by the assumption of infinite dimensions [1–5]. However, certain features of periodic structures may also manifest even in the case of structures of finite dimensions if they include a sufficient number of segments.

One of specific features of periodic structures are band gaps in their frequency spectra. Band gaps define frequency ranges within which signals cannot propagate within these structures. The locations and the widths of these gaps in the frequency spectra are strongly dependent on the size of the unit cell and such material properties as modulus of elasticity [6–8]. These special features of periodic structures can be employed for very efficient vibration damping. On the other hand, active vibration damping methods include techniques that use piezoelectric materials. Structures with active piezoelectric elements enable one the conversion of mechanical vibrations to electrical vibrations and thus to control damping properties of the system. Therefore, only the balance between passive and active damping allows one to maximise the effectiveness of the damping process.

Additionally, while modeling periodic structures, the influence of features resulting from the application of a particular numerical model, on the results of calculations, should be taken into account carefully. Almost every computational model of a discretised structure (finite element method (FEM) or time domain spectral finite element method (TDSFEM) models), has certain characteristics of a periodic structure. Therefore, it is worth to analyse if certain features of periodic structures may be utilised in a directed manner in order to make practical use of the unusual behaviour of such structures. This approach may potentially allow to reduce, or enhance, periodic properties of the computational model.

In this paper, the authors propose to combine all the aspects mentioned above. They propose a special numerical model of the beam with active piezoelectric elements, by means of which the dynamic characteristics of the beam can be analysed and the width of band gaps can be controlled.

## 2. Numerical Model

The structure under consideration, presented in Figure 1, is a sequence of 50 unit cells, consisting of an aluminium beam with piezoelectric rectangular strips (from APC International, Ltd., Cat.No. 70-1000, item 721) attached on both sides. Each pair of piezoelements is connected to the RLC resonant circuit with a controlled inductance.

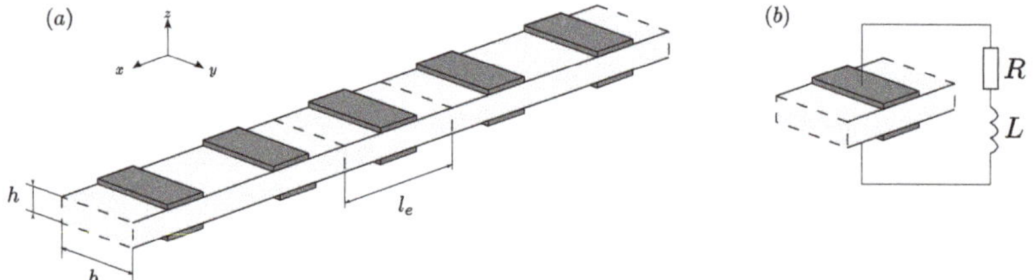


**Figure 1.** A concept of an electromechanical periodic structure (**a**), unit cell (**b**).

The material parameters taken into calculations were as follows: for aluminium E = 67.5 GPa, $\rho = 2700 \ \text{kg}/\text{m}^3$ and $v = 0.33$ and for piezoelectric material E = 63 GPa, $\rho = 7800 \ \text{kg}/\text{m}^3$, $v = 0.33$ and piezoelectric electro-mechanical coupling coefficient $k_{31} = 0.35$. The geometry of the analysed beam was as follows: the length $L = 1$ m, the width $b = 0.02$ m, the height $h = 0.01$ m, the single RLC element length was 0.01 m.

Numerical modelling of the piezoelectric material properties was based on the approach proposed in the literature [9,10]. The authors presented there a formulae for calculation of the effective Young's modulus of the piezoelectric element being an element of a resonant circuit. The piezoelectric material has frequency-dependent stiffness and damping, and the frequency itself depends on the parameters of the resonance circuit. Therefore, the effective Young's modulus of the piezoelectric material in the resonant circuit can be described by the equation:

$$E_p^{SU}(\omega) = E_p^D \left( 1 - \frac{k_{31}^2}{1 + i\omega C_p^\epsilon Z^{SU}(\omega)} \right),$$ (1)

where $E_p^{SU}$ is the effective Young's modulus of the piezoelectric material in a closed circuit mode, $E_p^D$ is the effective Young's modulus of the piezoelectric material in an open circuit mode, $k_{31}$ is the electro-mechanical coupling coefficient of thee piezoelectric material, $C_p^\epsilon$ is the capacitance of the piezoelectric element, and $Z^{SU}$ is the impedance of a resonant circuit. In the carried out numerical calculations, PZT impedance has been taken into account as an electrical circuit parameter. The impedance of the aluminium beam itself has been neglected because the aluminium beam is not a part of a controlled electrical circuit. In the presented paper, changes in the vibration characteristics of the aluminium beam with attached, actively controlled PZT elements have been analysed.

The displacement and deformation fields of the analysed beam structure have been assumed according to the Timoshenko theory. The mathematical formulae can be expressed by [11,12]:

$$\begin{cases} u(x) = z\phi(x) \\ w(x) = w_0(x), \end{cases}$$ (2)

$$
\begin{cases}
\epsilon_x = \dfrac{\partial u(x)}{\partial x} = z\dfrac{d\phi(x)}{dx} \\[2ex]
\gamma_{xz} = \dfrac{\partial w(x)}{\partial x} + \dfrac{\partial u(x)}{\partial z} = \dfrac{dw_0(x)}{dx} + \phi(x),
\end{cases}
\tag{3}
$$

where $u(x)$ and $w(x)$ are respectively the longitudinal and transverse components of the element displacements expressed in the global coordinate system, while the independent rotation $\phi(x)$ around the $y$ axis and the lateral displacement $w_0(x)$ are nodal displacements, defined in neutral element axis.

Following the standard FEM procedures, the inertia matrix $M$ and the stiffness matrix $K$ were evaluated:

$$
\mathbf{M} = \rho \iiint_V \mathbf{N}^t \mathbf{N} dV, \quad \mathbf{K} = \rho \iiint_V \mathbf{B}^t \mathbf{D} \mathbf{B} dV,
\tag{4}
$$

where $\rho$ is the density of the material, $\mathbf{D}$ is the matrix of elasticity coefficients, and $\mathbf{N}$ and $\mathbf{B}$ are the shape function and strain-displacement matrices, respectively.

The presented numerical simulations have been obtained by the use of the classical Finite Element Method (FEM), the Frequency Domain Spectral Element Method (FDSFEM, details have been widely presented by Doyle in [13], the interested Reader is encouraged to follow the source) or the Time Domain Spectral Finite element Method (TDSFEM) approach.

In the classical FEM approach, the unit cell has been divided into three finite elements while in the case of the spectral approach the unit cell has been represented by one finite element, as shown in Figure 2.

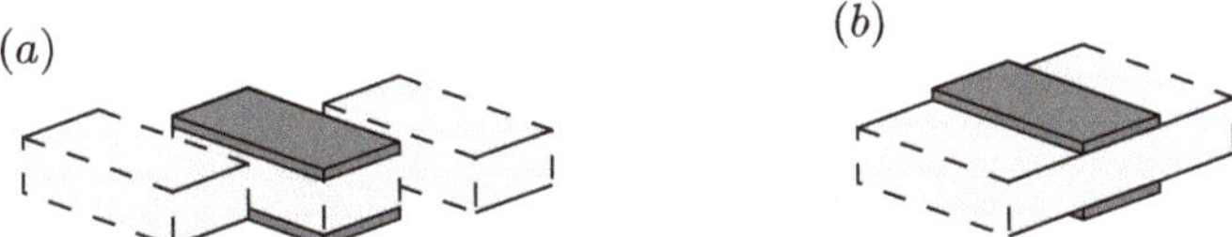


**Figure 2.** Modelling a unit cell of an electromechanical periodic structure: (**a**) by the finite element method (FEM), (**b**) by the spectral finite element method (SFEM).

The main difference of the TDSFEM in comparison to the FEM is that in the TDSFEM approach the element nodes are not equally distributed. Coordinates of the nodes are defined as roots of a certain orthogonal polynomial:

$$
T_p^c = (1 - \xi^2) U_{p-2}(\xi),
\tag{5}
$$

which in the analysed case has been $U_{p-2}(\xi)$—the second order Chebyshev polynomial. The element nodes in the element coordinate system may be calculated as follows:

$$
\xi_i = -\cos\frac{\pi(j-1)}{p} \quad j = 1, ..., p+1.
\tag{6}
$$

Such a definition of node distribution allows one to use higher order shape functions without the risk of causing the Runge effect. The node distribution used in the calculations performed for this paper has been shown in Figure 3.

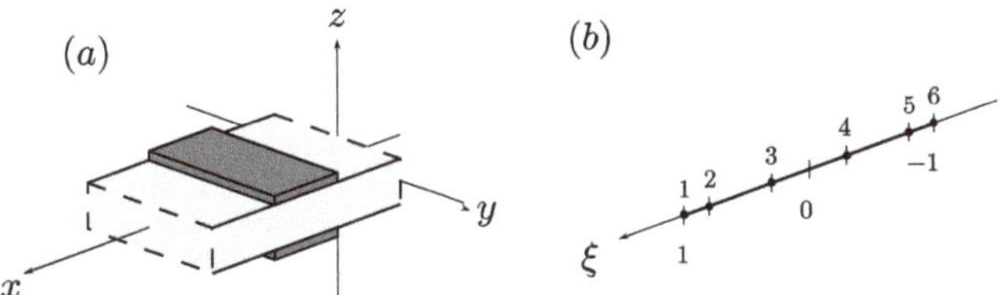


**Figure 3.** (**a**) A unit cell in the global coordinate system (**b**) A node distribution in element coordinate system.

The stiffness and inertia matrices corresponding to the piezoelectric element within the respective integration limits have been joined with the stiffness and inertia matrices of the aluminium element respectively, as shown in Figure 4. The procedure has been precisely described in [14] for the case of passive periodic structures being a beam and rod with a sequence of drilled holes.

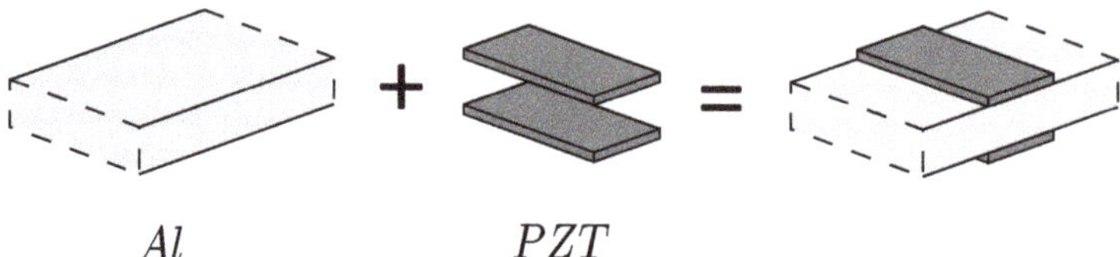


**Figure 4.** Construction of inertia and stiffness matrices in the case of the SFEM.

The aforementioned mathematical operations ensures that the stiffness matrix of the piezoelectric element is dependent on the frequency, therefore it is possible to actively control the mechanical responses of the analysed element by frequency variation.

## 3. Numerical Analysis

In order to examine whether the proposed numerical approach is appropriate, a series of numerical experiments were carried out to verify the impact of the resonance circuit parameters on the physical properties (the width and placement of band gaps) of the periodic beam. During calculations the periodic boundary conditions were assumed.

The graphs shown in Figure 5 represent frequency response functions in the ranges from 0 to 250 kHz for the periodic beam with a resonant circuit being: open, closed or tuned to the specific frequency. The left column of Figure 5 shows the results obtained by the use of the FEM, the right hand side column of this Figure—by the TDSFEM [12] respectively. It may be noticed that there appeared two natural band gaps in given frequency ranges for the passive structure. Tuning the PZT circuits to the resonant frequency introduced an artificial band gap in the range of that frequency. However tuning the circuits to the frequency in the range of the natural periodic beam band gap, with the inactive PZT element, significantly widened the band gap. It should be also mentioned that for a lower range of the frequency spectra both the FEM and the TDSFEM results were quite similar, but in a higher frequency range the FEM results were distorted. The reason for that has been widely discussed in [14], where several features of the numerical models have been addressed.

Figure 6 shows the influence of the PZT circuits resonance frequency on the width of beam band gaps. This example was calculated with the TDSFEM. In this case it has been demonstrated how changes in the resonant frequency of PZT circuits allows one to control the ranges of blocked frequencies in the case of forced vibrations. Red colour represents frequency ranges that will propagate freely in the structure, the other colours (blue, green and yellow) represent different levels of attenuation.

The graphs presented in Figure 7 illustrate the effect of changes in the electrical resistance on the active periodic structure frequency band gaps. Here the TDSFEM has been used. The figures shown on the left hand side present the results calculated for 1 $\Omega$ RLC circuit resistance, the right hand-side—5 $\Omega$ respectively. As it can be noticed higher values of the resistance in the RLC circuit increased the energy dissipation and, as a result, widens the band gap. This effect is more significant in case, when the resonant frequency of the RLC circuit is equal to the passive structure frequency band gap. It it may be noticed in the bottom right graph from Figure 7.

FEM                                TDSFEM

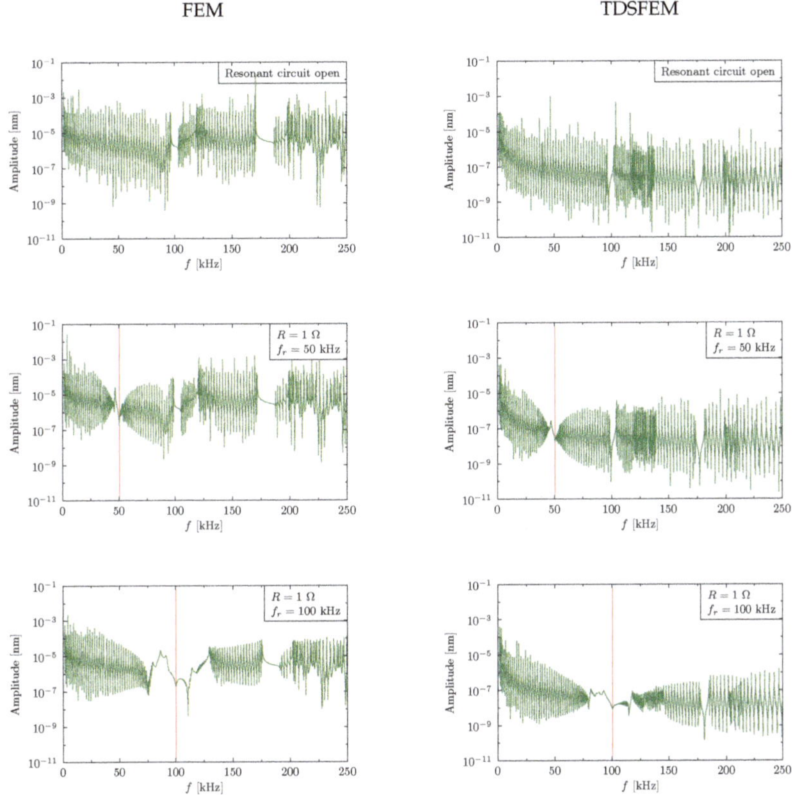

**Figure 5.** Frequency response functions of a passive and active periodic structure with the resonant circuit tuned to the frequency 50 kHz or 100 kHz (marked with a red line) and resistance 1 Ω.

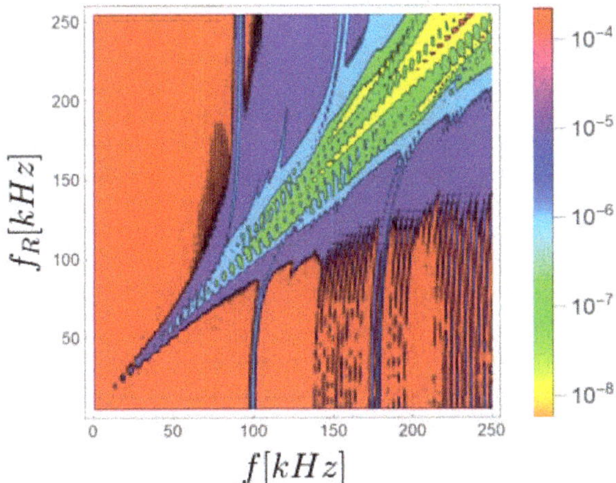

**Figure 6.** Dependence of the vibration amplitude on the resonant frequency of the RLC circuits.

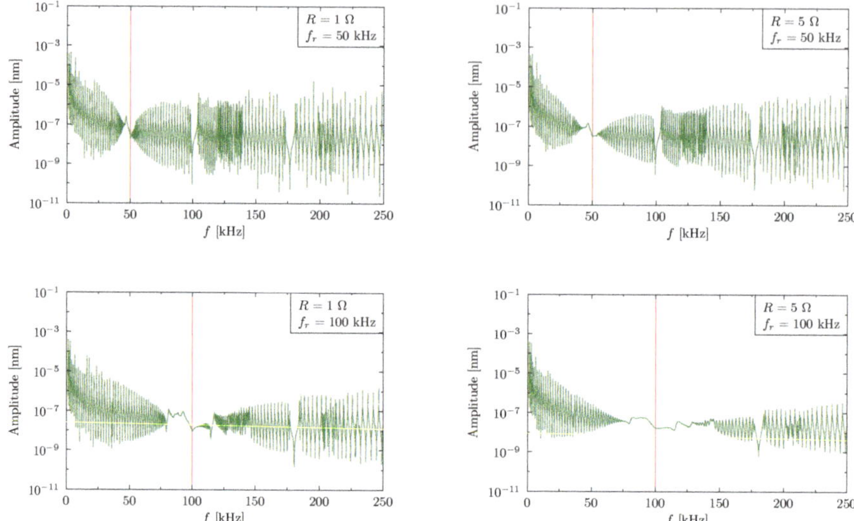

**Figure 7.** Frequency response functions of an active periodic structure with the resonant frequency of 50 kHz or 100 kHz (red line), and the resistance of 1 $\Omega$ (**left**) or 5 $\Omega$ (**right**).

*Wave Propagation*

This subsection presents the results of the analysis of changes in the propagating elastic waves in the modelled active periodical beam. For modelling the discussed structure the shape functions based on the FDSFEM [13] was used and the node distribution the same as in the TDSFEM [15] was adopted. Such a combination of the two totally different methods enabled a thorough analysis of the influence of frequency-dependent changes in the Young's modulus of piezoelectric material—Equation (1). The amplitude at both ends of the beam was determined. The aim of a such calculation programme was to analyse the changes of the propagating wave. The analysed periodic beam with active piezoelectric elements was excited to transverse vibrations at one end with the sinusoidal signal (eight pulses) modulated by the Hanning window, by the force of a $F = 1$ N amplitude. Two excitation frequencies $f = 50$ kHz and $f = 100$ kHz have been chosen; the first one from the frequency range of normal behaviour of the periodic beam, and the second within the passive band gap of the periodic structure.

Next Figure 8 shows the dispersion curves determined for the analysed periodical beam in the first Brillouin zone [16,17]. These curves have been determined by the use of the Bloch reduction method [18] by taking into account the relation from Equation (1). The graph in Figure 8a shows the first Brillouin zone of the passive system. One can notice the red lines meaning the vibrations of the propagating wave. The frequency ranges, marked with grey areas, at which there is no wave propagation are visible, i.e., there are no corresponding wave vectors. On the second chart in Figure 8b there is the first Brillouin zone of the system with each RLC circuit tuned to 100 kHz (frequency from the range of the band gap). RLC circuits create an area of anti-resonant vibrations independent of the wave vector that effectively widens the area of the grey fields of blocked wave propagation.

(a)                (b)

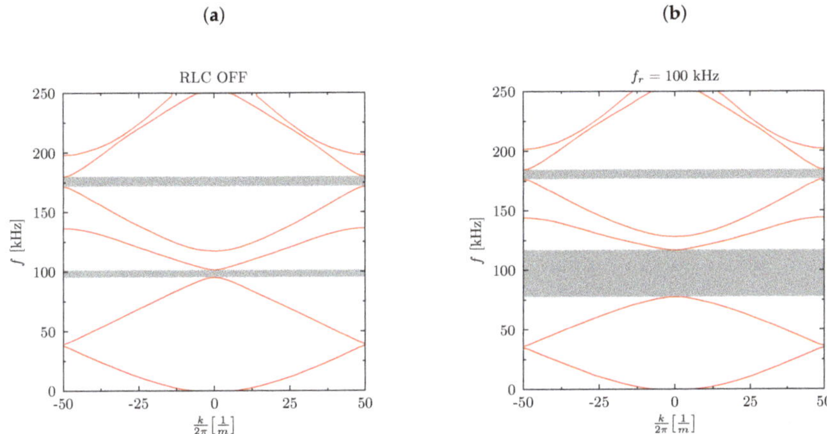

**Figure 8.** Dispersion curves for the analysed beam, (**a**) RLC circuits off, (**b**) RLC circuits tuned to 100 kHz.

Flexural wave propagation in an electromechanical periodic structure has been presented in Figure 9. The presented results represent six cases—the open RLC circuit (Figure 9a,b), the RLC tuned to the frequency of $f_R = 50$ kHz (Figure 9c,d) and the RLC tuned to the frequency of $f_R = 100$ kHz (Figure 9e,f) for both 50 kHz and 100 kHz excitation signals—left and right column, respectively. It may be concluded that the active RLC circuits with the resonant frequency equal to the excitation carrier frequency had the features of an active vibration damper.

To illustrate the damping character of an active periodic beam with the piezoelectric RLC circuits the following set of results was gathered (Figure 10). Here the vibration spectra before and after passing through the structure have been shown. The green colours show the spectra of vibration measured before passing through the structure ($P_1$), the blue after passing through the structure ($P_2$). The red line represents the tuned resonant frequency ($f_R$) of the RLC circuit. The left hand side column represents the data calculated for 50 kHz excitation signal, the right hand side column for 100 kHz respectively.

In the case of the passive system (Figure 10a,b), the wave of the carrier frequency of $f = 50$ kHz was free to propagate itself, there was no remarkable change in the amplitude magnitude (Figure 10a). In case of the wave of the carrier frequency of $f = 100$ kHz it may be noticed that some amount of the energy was blocked due to the presence of band gaps for this spectrum range as the band gap was a barrier for propagation of waves of these frequencies. However, the natural band gap was of relatively small width, therefore a certain amount of wave energy could propagate through the band gap.

The diagram below (Figure 10c,d) shows the changes in the excitation spectrum in the active periodic beam with the RLC circuits tuned to $f_R = 50$ kHz. As it can be noticed, the excitation wave (of the carrier frequency of $f = 50$ kHz) at that frequency was unable to propagate freely due to the dissipation of energy on the electrical resistance band gap that appears. Although the band gap was very narrow the amplitude of the wave decreased in a significant manner. On the other hand for this $f_R = 50$ kHz no significant changes were observed in the amplitude of the excitation signal of the carrier frequency of $f = 100$ kHz in comparison to the amplitude registered for the passive system.

Finally, tuning of the RLC circuit to the frequency of $f_R = 100$ kHz (Figure 10e,f) did not cause any changes in the wave propagation of the excitation wave of the frequency of $f = 50$ kHz (also comparing to the passive structure). However, this value of the $f_R = 100$ kHz caused a significant widening of the band gap. The amplitude of the wave after passing through the structure with band gap decreased tenfold. It was caused by the synergy of the natural periodic structure band gap with energy dissipation caused by the active RLC circuit and its electrical resistance.

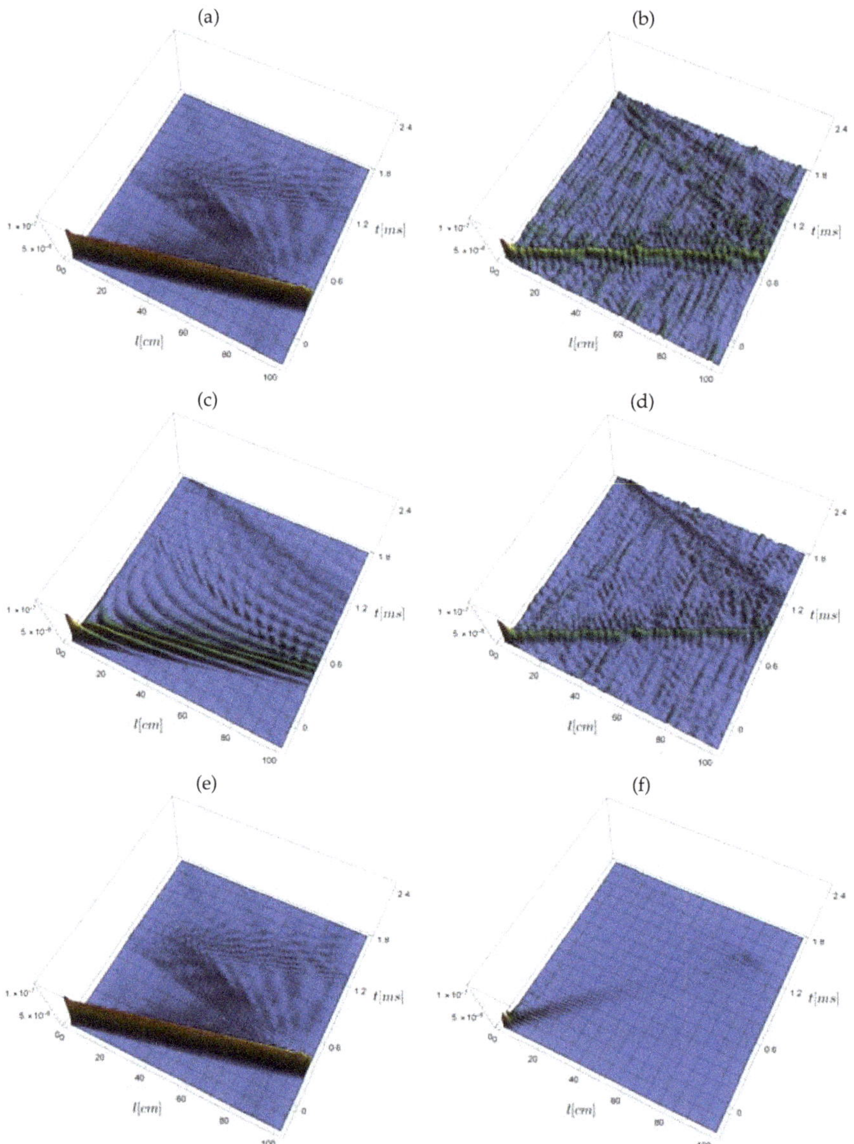

**Figure 9.** Patterns of flexural wave propagation in an electromechanical periodic structure.

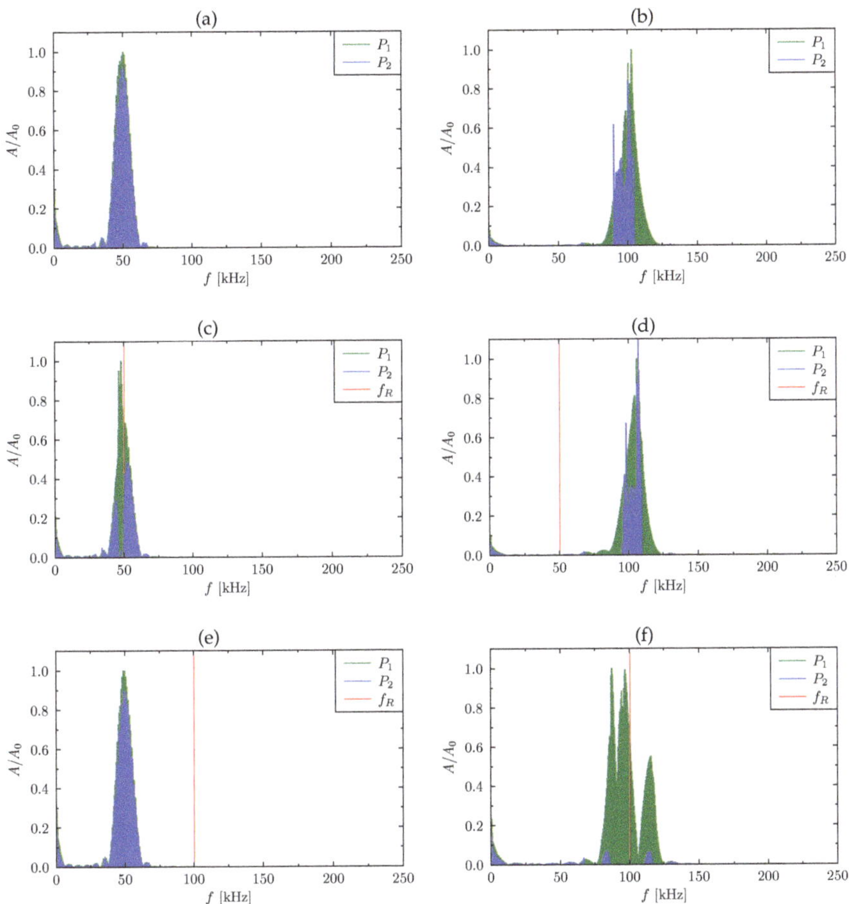

**Figure 10.** Spectra of flexural oscillations measured before passing through the structure (green) and after passing through the structure (blue). The excitation frequency equal to 50 kHz (**left**) and 100 kHz (**right**) marked with red line.

## 4. Conclusions

In this paper numerical investigations of a beam structural element built out of periodically arranged elementary cells with active piezoelectric elements has been performed. For this analysis the authors propose numerical models based on the use of the finite element method (FEM) and the spectral finite element methods defined in the frequency domain (FDSFEM) and the time domain (TDSFEM). The application of different modelling methods allow the authors to formulate conclusions that result from the calculations performed.

The FEM is the most common method and generally gives correct results. However, for high-frequency analysis problems it is necessary to use either a heavily dense grid or higher-order approximation polynomials. This leads to a correspondingly large sizes of the problems to be solved or a Runge effect. There is another reason why the FEM can be disadvantageous in the applications related to periodic structures—numerical models themselves show periodic characteristics [6,15]. Thus, it is easy to predict that the results obtained may have features typical to periodic structures resulting not only from the geometry of finite elements, but also from the features of the numerical models. The consequent misinterpretation of results may be simply dangerous.

Due to the above mentioned characteristics of the FEM, the study proposes to utilise the TDSFEM method. The unquestionable advantage of the TDSFEM method is its ability to employ higher order approximation polynomials, which results in higher calculation accuracy. Additionally, an un-uniform distribution of nodes in single finite elements enables one to obtain a diagonal form of the inertia matrix, which significantly reduces the time of numerical calculations. All presented amplitude-frequency characteristics of the analysed periodic beam structural element, taking into account the dependence of the PZT Young's modulus on the frequency, have been determined by this method.

While in the case of determination of the amplitude-frequency characteristics of the active periodic beam, the use of the TDSFEM allowed the authors to obtain results at a satisfactory level, the calculations for changes in the propagation of elastic wave required another modification of the modelling method. In order to precisely map changes in Young's modulus for the value of PZT material for each analysed frequency, it was necessary to use the FDSFEM method. Modification of the method involved the use of non-uniform mesh of nodes in the finite elements known from the TDSFEM and shape functions from the FDSFEM. In this way the changes in propagation of elastic waves in the active periodic beam structural element have been modelled.

After all numerical tests performed it may be concluded that periodic structures with active piezoelectric elements incorporated into the RLC resonance circuit can be successfully used to attenuate vibrations in a controlled manner. A resonant circuit with piezoelectric elements causes the appearance of an additional band gap in the spectrum of mechanical vibrations in the vicinity of the natural frequency of the RLC resonance system. Adjusting the resonance frequencies of the RLC systems to the frequency of the naturally present band gap in the spectrum of mechanical vibrations results in a significant widening of the band gap, which leads to effective vibration damping in this frequency range.

Although the proposed approach clearly demonstrates that there is a possibility of active control of band gaps, it should be added that the problem still requires a number of analyses and will definitely be the subject of further scientific considerations of the authors.

**Author Contributions:** Conceptualization, W.W., M.P. and M.K.; methodology, W.W., M.P. and M.K.; software, W.W.; validation, W.W., M.P. and M.K.; formal analysis, W.W., M.P. and M.K.; investigation, W.W., M.P. and M.K.; resources, W.W., M.P. and M.K.; data curation, W.W. and M.P.; writing—original draft preparation, M.P.; writing—review and editing, M.P.; visualization, W.W. and M.P.; supervision, M.K.; project administration, M.K. All authors have read and agreed to the published version of the manuscript.

**Funding:** This research received no external funding.

**Acknowledgments:** The authors would like to gratefully acknowledge the support of the Academic Computer Centre in Gdańsk, the provider of the software used for the research done by the author and described in this paper.

**Conflicts of Interest:** The authors declare no conflict of interest.

## References

1. Kushwaha, M.S.; Halevi, P.; Dobrzynski, L.; Djafari-Rouhani, B. Acoustic band structure of periodic elastic composites. *Phys. Rev. Lett.* **1993**, *71*, 2022–2025. [CrossRef] [PubMed]
2. Sigalas, M.M.; Economou, E.N. Elastic waves inplates with periodically placed inclusions. *J. Appl. Phys.* **1994**, *75*, 2845–2850. [CrossRef]
3. Sigalas, M.M.; Garcia, N. Theoretical study of three dimensional elastic band gaps with the finite-difference time domain method. *J. Appl. Phys.* **2000**, *87*, 3122–3125. [CrossRef]
4. Liu, Y.; Gao, L.T. Explicit dynamic finite element method for band-structure calculations of 2D phononic crystals. *Solid State Commun.* **2007**, *144*, 89–93. [CrossRef]
5. Narisetti, R.K.; Leamy, M.J.; Ruzzene, M. A perturbation approach for predicting wave propagation in one-dimensional nonlinear periodic structures. *J. Vib. Acoust.* **2010**, *132*, 031001. [CrossRef]
6. Żak, A.; Krawczuk, M.; Palacz, M. Periodic Properties of 1D FE Discrete Models in High Frequency Dynamics. *Math. Probl. Eng.* **2016**, *2016*, 9651430. [CrossRef]

7.  Wu, Z.J.; Wang, Y.Z.; Li, F.M. Analysis on band gap properties of periodic structures of bar system using the spectral element method. *Waves Random Complex Media* **2013**, *23*, 349–372. [CrossRef]

8.  Baz, A. Active control of periodic structures. *J. Vib. Acoust.* **2001**, *123*, 472–479. [CrossRef]

9.  Hagood, N.W.; von Flotow, A. Damping of structural vibrations with piezoelectric materials and passive electrical networks. *J. Sound Vib.* **1991**, *146*, 243–268. [CrossRef]

10. Airoldi, L.; Senesi, M.; Ruzzene, M. Piezoelectric Superlattices and Shunted Periodic Arrays as Tunable Periodic Structures and Metamaterials. In *Wave Propagation in Linear and Nonlinear Periodic Media*; Springer: Berlin/Heidelberg, Germany, 2012; pp. 33–108.

11. Zienkiewicz, O.C.; Taylor, R.L.; Nithiarasu, P.; Zhu, J. *The Finite Element Method*; McGraw-Hill: London, UK, 1977; Volume 3.

12. Ostachowicz, W.; Kudela, P.; Krawczuk, M.; Zak, A. *Guided Waves in Structures for SHM: The Time-Domain Spectral Element Method*; John Wiley & Sons: Hoboken, NJ, USA, 2011.

13. Doyle, J.F. Wave propagation in structures. In *Wave Propagation in Structures*; Springer: Berlin/Heidelberg, Germany, 1989; pp. 126–156.

14. Żak, A.; Krawczuk, M.; Palacz, M.; Doliński, Ł.; Waszkowiak, W. High frequency dynamics of an isotropic Timoshenko periodic beam by the use of the Time-domain Spectral Finite Element Method. *J. Sound Vib.* **2017**, *409*, 318–335. [CrossRef]

15. Żak, A.; Krawczuk, M.; Waszkowiak, W. Longitudinal, Torsional and Flexural Dynamics of 1-D Periodic Structures. In Proceedings of the 22nd International Congress on Sound and Vibration: Major Challenges in Acoustics, Noise and Vibration Research (22nd ICSV), Florence, Italy, 12–16 July 2015.

16. Brillouin, L. *Wave Propagation in Periodic Structures: Electric Filters and Crystal Lattices*; Courier Corporation: North Chelmsford, MA, USA, 2003.

17. Brillouin, L. Les électrons dans les métaux et le classement des ondes de de Broglie correspondantes. *Comptes Rendus Hebdomadaires des Séances de l'Académie des Sciences* **1930**, *191*, 292.

18. Farzbod, F.; Leamy, M.J. Analysis of Bloch's method in structures with energy dissipation. *J. Vib. Acoust. Trans. ASME* **2011**, *133*, 051010. [CrossRef]

*applied sciences*

*Article*

# Mechanical Integrity Analysis of a Printed Circuit Heat Exchanger with Channel Misalignment

Armanto P. Simanjuntak and Jae Young Lee *

Department of Mechanical and Control Engineering, Handong Global University, Pohang 37554, Korea;
armant.ost@gmail.com
* Correspondence: jylee7@handong.edu; Tel.: +82-54260-1392

Received: 2 March 2020; Accepted: 18 March 2020; Published: 22 March 2020

**Abstract:** Printed circuit heat exchangers (PCHEs), which are used for thermal heat storage and power generation, are often subject to severe pressure and temperature differences between primary and secondary channels, which causes mechanical integrity problems. PCHE operation may result in discontinuity, such as channel misalignment, due to non-uniform thermal fields in the diffusion bonding process. The present paper analyzes the mechanical integrity, including the utilization factors of stress and deformation under various channel misalignment conditions. The pressure difference of the target PCHE is 19.5 MPa due to the high pressure (19.7 MPa) of the steam channel in the Rankine cycle and the low pressure (0.5 MPa) of molten salt or liquid metal in the primary channel. Additionally, the temperature difference between channels is around 25 °C, however the average temperature is around 500 °C. The PCHE has a relatively large primary channel measuring approximately 3 x 3 mm, and a steam channel measuring 2 x 1.5 mm. The finite element method (FEM) is applied to determine the stress by changing the misalignment to below 30% of the primary channel width. It was found that the current PCHE is operable up to 700 °C in terms of the ASME code under these design conditions. Additionally, the change of utilization factor due to the misalignment increases, but is still under the ASME acceptance criteria of 700 °C; however, it violates the criteria at 725 °C, which is the allowable temperature condition. Therefore, the mechanical integrity of the PCHE with low-pressure molten salt or liquid metal and a high-pressure steam channel is acceptable in terms of utilization factor.

**Keywords:** PCHE; FEM; misalignment; stress; channel; utilization factor

---

## 1. Introduction

Heat exchangers are devices that facilitate the exchange of heat between two fluids at different temperatures while keeping them from mixing together [1,2]. Heat exchangers are mostly used in processing, power, petroleum, air conditioning, refrigeration, alternate fuels, and other industries. Heat transfer in a heat exchanger usually involves convection in each fluid and conduction through the wall separating the two fluids. In general industry, heat exchangers can be classified according to their construction, transfer process, degree of surface compactness, flow arrangement, pass arrangement, phase of the process fluids, and heat transfer mechanism. The shell–tube type is the most popular heat exchanger for industry. However, printed circuit heat exchangers (PCHE) have shown promise because of their advantages, such as having large area density and good pressure and temperature capabilities [3–5].

High-efficiency compact heat exchangers are being widely developed and becoming increasingly important for the nuclear industry. The main target of such development is improvement of the efficiency, economics, and safety. Depending on the intended scale and application, different types are being considered, using fluids such as helium, supercritical carbon dioxide (S CO2), mixed nitrogen and helium nitrogen, liquid metals, and molten salts. All of these reactors are operated at high temperature. For example, sodium fast-cooled reactors (SFRs) and supercritical carbon dioxide (S CO2) reactors are operated at maximum temperatures of 550 °C. The only difference is the working pressures, which are 0.1 and 20 MPa, respectively, for these reactor types [6]. For these reasons, safety is an important feature of PCHEs because some critical issues can occur. Additionally, the development of high-efficiency plate heat exchangers is also important for thermal energy storage in alternative energy. Alternative technologies are being developed in order to save fossil resources and reduce air pollution by capturing and using renewable sources of energy, such as solar, wind, and hydropower, or geothermal heat. Thermal energy storage (TES) is a technology that stocks thermal energy by heating or cooling a storage medium, so that the stored energy can be used at a later time for heating and cooling applications and power generation. One of the issues with this is the thermo-structural fatigue, as observed by Rakesh et al. [7]. The thermal stresses have been found to be more dominant than stresses due to pressure cycles. This is due to the higher temperature gradient. Hence, safety has a strong relationship with high mechanical integrity and long operating life.

Much effort has been given to studying the thermo-hydraulic characteristics of PCHE channels. The thermo-hydraulic performance of PCHEs for SCO2, which contains helium as the fluid, was studied by Mylavarapu et al. [8]. There was a difference in the critical Reynold numbers due to the difference in the shape of the cross-section between the circular and semi-circular ducts. The thermo-hydraulic performance of a newly developed PCHE, which had a longitudinal corrugation flow channel, was studied by Kim et al. [9]. The friction factors tended to be greater when the PCHE had a smaller hydraulic diameter and a larger angle of inclination. Sung et al. conducted a study on heat exchangers for thermoelectric power generation. The shape of the cross-section was a combined rectangular and circular shape, with a developed tangled channel. The obtained result showed the increase of the heat transfer at a relatively lower Reynold number [10]. Yoon et al. conducted an assessment study of straight, zigzag, s-shape, and airfoil PCHEs for intermediate heat exchanger (IHX) of high temperature gas reactors (HTGR) and sodium-cooled fast reactors (SFR). The study suggests a best option named the Kalimer-600 IHX, which is a straight PCHE [11]. For this reason, improvement of the cross-section shape may enhance the thermo-hydraulic performance. Therefore, the mechanical integrity of the channel shape effect needs to be studied. A preliminary study on the mechanical integrity of the proposed design is needed and is the main purpose of this paper. Some studies on the structural integrity of PCHEs have also been conducted. Lee et al. studied the structural integrity of an intermediate printed circuit heat exchanger for a SFR attached to supercritical carbondioxide. It was found that the mechanical stress concentration occurred at the PCHE channel tip [12]. Song et al. also conducted a study on the structural integrity evaluation of a lab-scale intermediate PCHE in a very-high-temperature reactor (VHTR). Under the test conditions, the maximum Tresca stress was far below the allowable stress limit [13]. Mizokami et al. conducted a structural design study of a plate–fin heat exchanger for HTGRs. In this study, we present a high-temperature structural design procedure for a plate–fin heat exchanger, which includes strength evaluation for primary stress and evaluation of creep–fatigue life [14]. Mochizuki and Takano conducted a study on heat transfer in heat exchangers in sodium fast-cooled reactor systems. The practical Nusselt numbers of a heat exchanger were derived in the MONJU fast breeder reactor which was a Japanese sodium-cooled fast reactor [15].

The present study examined the stress distribution in a newly designed printed circuit heat exchanger model. This model is different to the common designs, which mainly use a semicircle channel shape. This shape is a combination of rectangular and ellipsoidal shapes, as shown in Figure 1. This PCHE design is planned to be manufactured through chemical etching and diffusion bonding processes. First, the canals that become the channel will be manufactured by chemical etching. Then, all layers will be stacked through diffusion bonding. In this case, the successful incorporation of a heat exchanger in a high-temperature nuclear service requires the manufacturer to be able to reliably produce high-quality diffusion welds. Because inspecting each weld in the core stack is impractical, the quality depends on a good understanding of the essential welding variables and strict control during the welding process [16,17]. In the diffusion bonding process, misalignment may occur due to improper process control related to the temperature or pressure. Even though this misalignment can be controlled by strict variable controls during the bonding process, it is also essential to observe the effects of misalignment at the microscale, so that the limit corresponding to the safety factor can be set. In this paper, the misalignment phenomena that may occur in PCHE channels are modeled and simulated. A two-dimensional simulation using COMSOL Multiphysics is conducted to observe the effect of channel misalignment on the mechanical integrity of PCHE. Different misalignment conditions are modeled to compare the stress intensities.

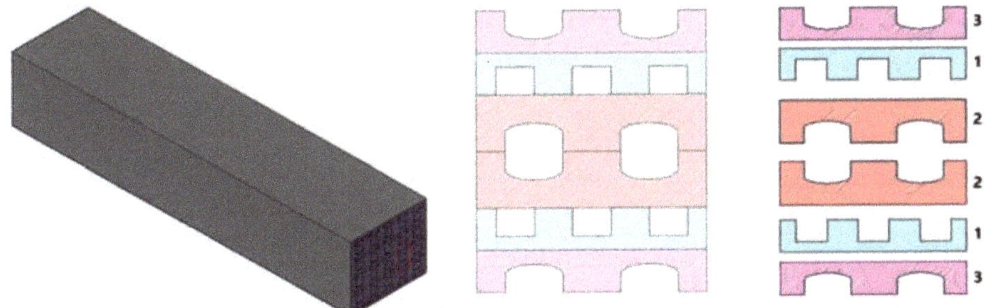

**Figure 1.** The schematic illustration of the printed circuit heat exchanger (PCHE).

In addition to the structural integrity study, several previous studies have examined this misalignment problem. Weld misalignment influences the structural integrity of a cylindrical pressure vessel; in the most dangerous analysis case, a crack depth of 7 mm and allowable centerline offset of 8 mm resulted in a pressure limit of 3.15 MPa, which is still 0.65 MPa greater than the hydro test pressure [18]. Finite element analysis was conducted on cylindrical pressure vessels with a misalignment in a circumferential joint by Brabin et al. [19]. The results showed good agreement with [20]. The structural integrity of a steam generator tube sheet with an incorrectly drilled hole was evaluated according to ASME Section III construction code. The presence of incorrectly drilled holes or locally thin ligaments was found to not affect the primary stress margin in the tube sheet or reduce its overall structural integrity [21]. The finite element method (FEM) was used to investigate local stress concentrations due to the misalignment of welded structures and determine the effect of geometric variables. The existence of a cope hole was found to influence the peak stress surrounding the transition zone [22]. From all the previous research mentioned above, there has not been a study on the misalignment effect in PCHEs under high-temperature conditions. For this reason, this study investigates the structural integrity of the proposed geometry and the effect of the misalignment condition along the diffusion bonding line on the structural integrity of the PCHE. The structural integrity is represented by the utilization factor, which is the ratio of stress intensity to the maximum allowable stress for design conditions. Besides the utilization factor, displacement is also a main observed variable, which further includes the displacement value and displacement behavior under the applied load.

ASME Section III is taken as the reference construction code. The obtained stress distributions are used to evaluate the compliance with design criteria by using ASME Boiler and Pressure Vessel Code (BPVC), Section III. This code contains specific rules for construction of nuclear facility components. ASME Section III, Division 5 [23] is taken as the reference construction code, specifically Subsection HB, which covers metallic pressure boundary components. Subsection HB consists of Subpart A (HBA) for low-temperature service components and Subpart B (HBB) for elevated-temperature service components. Table 1 shows the values of maximum temperature for various classes of permitted materials. Subpart HBB is applicable for materials that is used at temperatures above $T_{max}$. The maximum temperature of liquid sodium in SFR applications is 528 °C. Hence, subpart HBB is chosen as the construction code for this application. These rules are applicable to class A components, regardless of the type of contained fluid, such as water, steam, sodium, helium, or any other process fluids. The obtained stress distribution will be compared to the design loading parameters.

**Table 1.** Maximum temperature for various classes of permitted materials.

| No | Materials | $T_{max}$ (°C) |
|---|---|---|
| 1 | Carbon steel | 370 |
| 2 | Low alloy steel | 370 |
| 3 | Martensitic stainless steel | 370 |
| 4 | Austenitic stainless steel | 425 |
| 5 | Nickel–chromium–iron | 425 |
| 6 | Nickel–copper | 425 |

## 2. Numerical Simulation

### 2.1. Approach and PCHE Design

The nuclear industry uses PCHEs as reactor components. PCHEs are used in advanced nuclear reactors because they offer a high surface-area-to-volume ratio, high thermal effectiveness, and low overall pressure drop, as mentioned in [24].

A PCHE is a steam generator and an important component of a sodium-cooled fast reactor. In a steam generator system, a countercurrent flow is formed by the sodium and steam flow. Steam is produced at inlet and outlet temperatures of around 230 and 503 °C, respectively, with inlet and outlet pressures of 18 and 16.7 MPa, respectively. Sodium flows at inlet and outlet temperatures of around 528 and 322 °C, respectively, with a pressure of 0.5 MPa [25].

Here, a new design for the printed circuit steam generator is proposed. Figure 1 shows a schematic drawing of the geometry, which consists of water and sodium channels. The sodium channel is surrounded by the water channel. The water channel is rectangular with 0.1 mm fillets. The sodium channel is semi-rectangular with an ellipsoidal diameter ratio of 0.5 mm to 3 mm.

### 2.2. Boundary Conditions

A two-dimensional simulation using COMSOL Multiphysics was conducted with the proposed geometry. Table 2 lists the five different cases that were simulated to represent the possible misalignments due to a diffusion bonding discontinuity. The largest simulated misalignment was 1 mm, while the smallest was 0 mm, representing a perfect bond.

**Table 2.** Five Simulated misalignment cases. Misalignment is measured as the center-to-center distance along the x direction.

| Case # | Misalignment (m) | |
|---|---|---|
| | Right direction | Left direction |
| Case 1 | 0 mm | - |
| Case 2 | 0.25 mm | 0.25 mm |
| Case 3 | 0.5 mm | 0.5 mm |
| Case 4 | 0.75 mm | 0.75 mm |
| Case 5 | 0.85 mm | 0.85 mm |

Figure 2 shows the applied boundary conditions for the numerical simulation. The boundary pressure load was used as the boundary condition. The load-controlled stress limit was analyzed. For comparison to the construction code, only the primary stress produced by a mechanical load was considered. Boundary loads of 20 and 0.5 MPa were applied to the water and sodium channels, respectively.

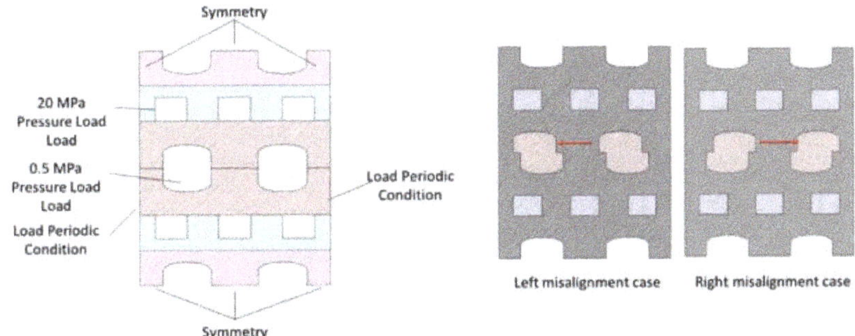


**Figure 2.** Applied boundary conditions for numerical simulation.

### 2.3. Mesh Independent Test

A mesh independent test was conducted to validate the geometry and mesh construction. Figure 3 shows the test and results. The number of elements was varied, and the surface average stress intensity was measured. The error was determined by subtracting the stress intensity from the highest stress intensity. The mesh sensitivity showed that the results converged for different numbers of elements.

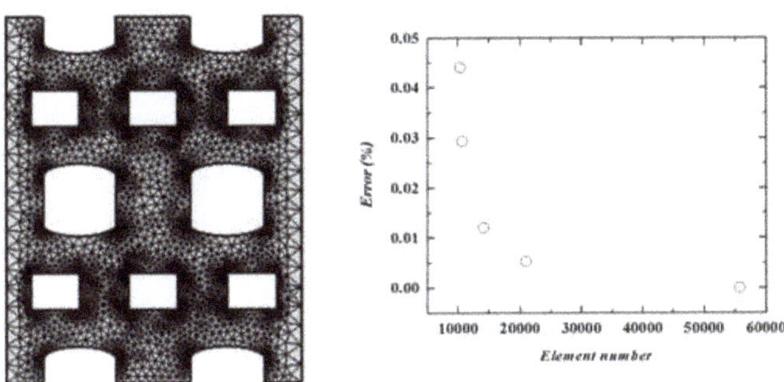


**Figure 3.** Mesh sensitivity test.

A mesh independent test was conducted to validate the geometry and mesh construction. Figure 3 shows the test and results. The number of elements was varied and the surface average stress intensity was measured. The error was determined by subtracting the stress intensity from the highest stress intensity. The mesh sensitivity showed that the results converged for different numbers of elements. Here, 10,705 mesh elements were used in this simulation, with an average element quality of 79.9%. The elements were triangular elements, with a minimum element size of 0.0045 mm.

## 3. Results and Discussion

### 3.1. Stress Distribution at the Surface

2D stress distribution resulted by FEA simulation in the misalignment case of 0.85 mm is shown in Figure 4. The simulation result indicates that the highest stress intensity occurs at the tip area of the water channel. In the ten cases, the highest stress was in the tip area of the water channel. This was due to the higher pressure load in the water channel area. The maximum equivalent surface von Mises stress is shown in Figure 5. The highest maximum stress intensity was observed under the misalignment condition of 0.85 mm, while the lowest intensity was observed under the misalignment condition of 0 mm. This result indicates that the maximum stress distribution increases with the misalignment. The maximum stress intensity in the case of 0.85 mm misalignment is 10% higher compared to the 0 mm misalignment condition.

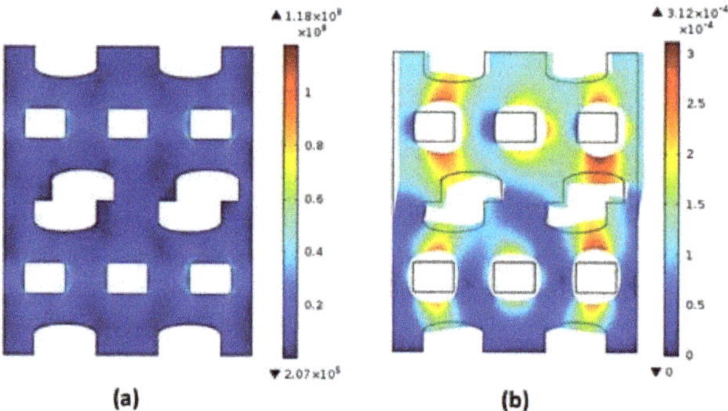

**Figure 4.** Huber–von Mises stress (**a**) and total displacement (**b**) distribution due to pressure load (at misalignment of 0.85 mm).

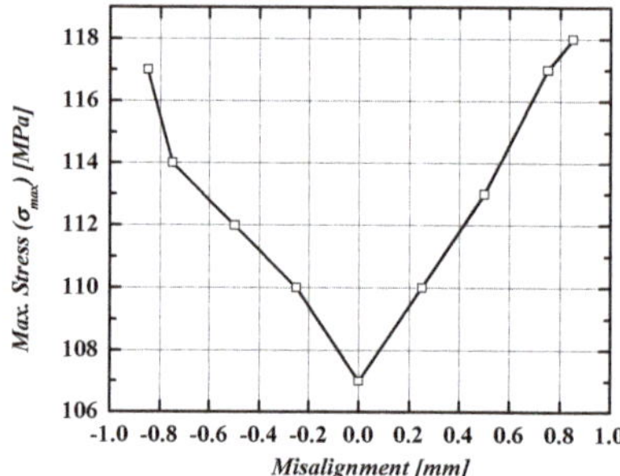

**Figure 5.** Maximum stress intensity for each misalignment condition.

### 3.2. Stress at Stress Classification Line (SCL)

The main standard for the design of nuclear equipment is American Society of Mechanical Engineers (ASME) Boiler and Pressure Vessel (BPVC) section III. ASME code section III has specific requirements on how to assess the result from the stress analysis to make the necessary verifications to avoid failure. In this work, PCHE is classified as "safety-related" components associated with metallic components used in the construction of high-temperature reactor systems. For this problem, ASME Section III, Division 5, Subsection HB, Subpart B, which covers elevated temperature service components, is considered as the construction code.

In this study, finite element analysis is utilized to analyze and obtain the stress intensities. In order to comply with design codes stated in ASME section III, the stress linearization method is needed. Stress linearization is a technique used to decompose actual stress distribution across thickness into membrane and bending stress. This technique is stated in ASME Section VIII, Division 2 [26]. The stress linearization method is done by following the statement given by ASME VIII below.

The membrane stress tensor is the tensor comprised of the average of each stress component along the stress classification line, or:

$$\sigma_{(ij,m)} = \frac{1}{t} \int_0^t \sigma_{ij}\, dx \tag{1}$$

The bending stress tensor is the tensor comprised of the linear varying portion of each stress component along

$$\sigma_{(ij,b)} = \frac{6}{t^2} \int_0^t \sigma_{ij} \left( \frac{t}{2} - x \right) dx \tag{2}$$

The equivalent stress can be calculated by

$$\sigma_e = \left[ \sigma_x^2 + \sigma_y^2 + \sigma_x\sigma_y + 3\tau_{xy}^2 \right]^{1/2} \tag{3}$$

To obtain the linearization result, stress classification lines (SCLs) are necessarily needed. Stress classification lines should represent the areas of interest of the proposed geometry, such as the sodium-to-sodium area, steam-to-steam area, and sodium-to-steam area. For this purpose, 10 SCLs have been taken throughout the geometry as shown in Figure 6 below.

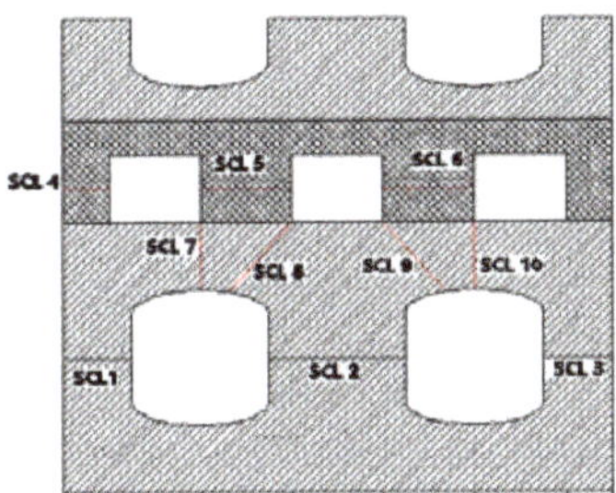


**Figure 6.** Stress classification lines (SCLs) for stress linearization method.

Numerical simulation using COMSOL Multiphysics 5.3 resulted in 10 linearized stress values as shown in Figure 6. Linearized stress values between sodium channels are represented by SCL 1, SCL 2, and SCL 3. The interest areas between steam channel stress values are represented by SCL 4, SCL 5, and SCL 6. SCL 7–10 represent areas between steam and sodium channels. The linearization results are shown in Figure 7.

The linearization results are compared with the ASME Section III stress value requirements for design load conditions. For design load conditions, a variable named $S_o$ is needed. $S_o$ is the maximum allowable value of the general primary membrane stress intensity for use as a reference for stress calculation under design loading. For the load-controlled stress limit under the design limit, two equations are used for compliance

$$P_m \leq S_o \tag{4}$$

$$P_L + P_b \leq 1.5 S_o \tag{5}$$

where Pm is the general primary membrane stress intensity resulted by linearization along the SCLs and PL + Pb is the combined primary membrane stress intensity and bending stress intensity. Therefore, values decrease with the increase of working temperature, as shown in Figure 8. The proposed system aims to be used in sodium fast-cooled reactor wherethe maximum working temperature of a steam generator heat exchanger is 550 °C. Therefore, for this purpose the maximum allowable stress (So) should be set at 88 MPa. Hence, the 1.5So will be 132 MPa. The maximum resulting value using the linearization method is 27.5 MPa for the primary membrane stress intensity and 32.8 MPa for the combined primary membrane and bending stress intensity. These values are obtained from linearization along SCL 6. Both conditions are found in the 0.85 mm misalignment condition. The condition without any misalignment resulted in the highest stress intensity values of 26.6 MPa and 31.5 MPa for primary membrane stress and combined primary and bending stress, respectively. Compared to the requirements of the design condition, all misalignment conditions for the sodium channel still comply with the acceptance criteria. This means the misalignment condition for the sodium channel does not have too much of an impact on the acceptance criteria. However, deeper analysis of the utilization factor is performed to observe the effects of this misalignment condition.

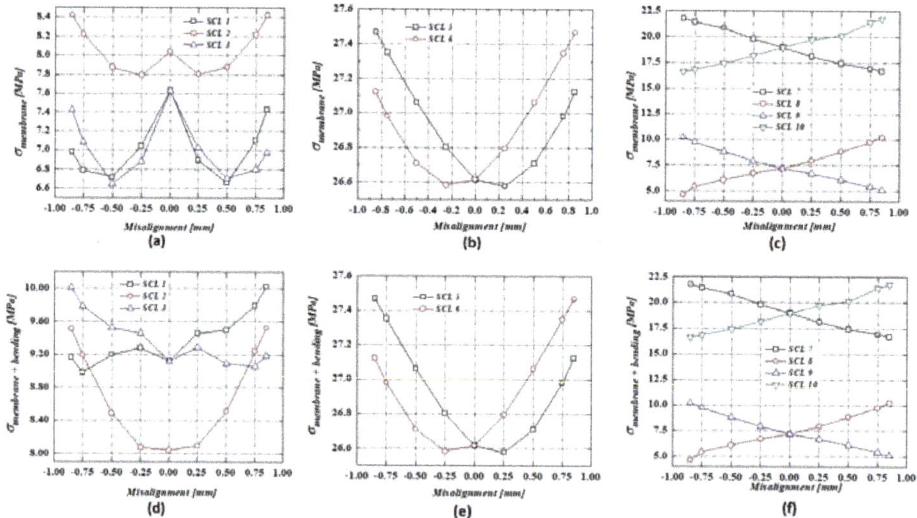

**Figure 7.** Primary membrane stress for various misalignment conditions: (**a**) sodium to sodium channel; (**b**) steam to steam channel; (**c**) between steam and sodium channels; And combined membrane and bending stress: (**d**) between sodium channel; (**e**) between steam channel; (**f**) between steam and sodium channels.

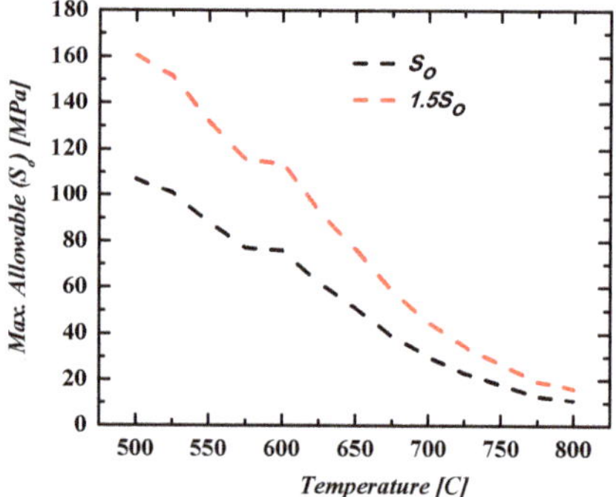

**Figure 8.** Stress classification line for stress linearization method.

### 3.3. The Utilization Factor Increases as the Misalignment Increases

The linearization method stress intensity values along the SCLs are still acceptable compared to the requirements of ASME III design conditions. However, deeper analysis should be done to observe the real effects of this misalignment condition. One of important factors in design and manufacturing is the utilization factor. Utilization factor is defined as the ratio of the worst condition to the acceptable condition. In this case, the ratio of the highest stress intensity resulting from the linearization method along SCLs and the $S_0$ value are considered. The utilization factor is the ratio of the stress intensity resulting from linearization method along SCLs to the $S_0$ as the minimum acceptable conditions.

$$U_f = \sigma/S_0 \tag{6}$$

The graphs shown in Figure 9 below indicate that the utilization factor increased with the increase of misalignment of sodium channel conditions. The maximum utilization factor is shown with 0.85 mm misalignment with both negative and positive shifting. The conditions still maintain the utilization factor at 30% of the allowable value, which is the safe condition for design load application.

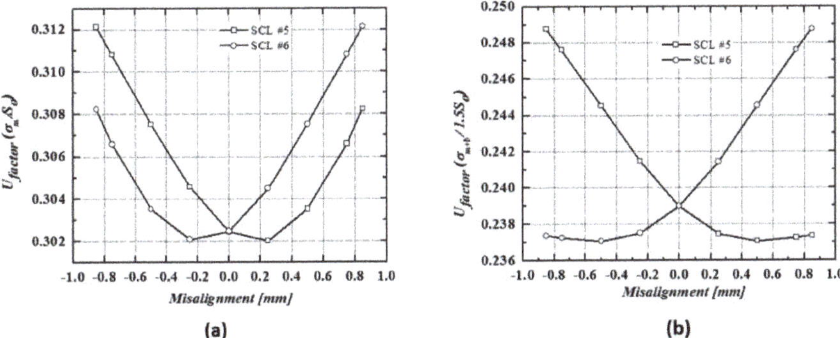

**Figure 9.** Increase of utilization factor as the increase of the misalignment condition: (**a**) compared to $S_0$; (**b**) compared to $1.5S_0$.

Since this proposed design is newly considered, the elevated temperature for other purposes should be considered. The utilization factor under elevated temperature (those exceeding 500 °C) is obtained, as shown in Figure 10. The figure shows that the condition of misalignment will be acceptable up to 725 °C. This means the application conditions above 725 °C cannot comply with the misalignment condition of 0.85 mm. This suggests that this proposed design can comply with the requirements of the ASME code for system applications under 725 °C.

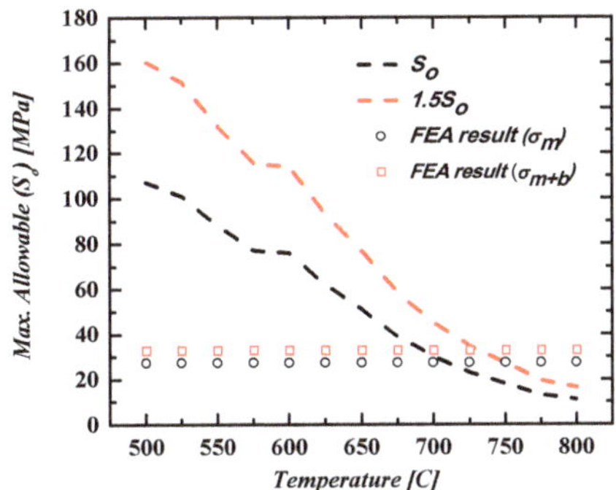

**Figure 10.** Maximum allowable misalignment (% of initial channel's width) under various temperature conditions.

### 3.4. Yielding Section under Design Loading

The yield strength or yield stress is a certain stress magnitude at which the material begins to deform plastically. Below the yield stress, the material deforms elastically and returns to its original shape when the applied stress is removed. Once the stress magnitude passes the yield stress, the material deforms permanently, which cannot be reversed. In this model, stainless steel 316 (SS31600) was set as the material. In this case, it is assumed that diffusion welds have the same material properties as the base metals. At a temperature of 550 °C, the yield stress ($S_y$) of SS31600 is 116 MPa as shown in Figure 11 [27]. This indicates that no section in the simulated cases yielded due to the applied pressure load under a working temperature of 550 °C. However, if a higher temperature condition is considered for application, yielding may occur on the surface. As shown in Figure 12, yielded section increases above 625 °C.

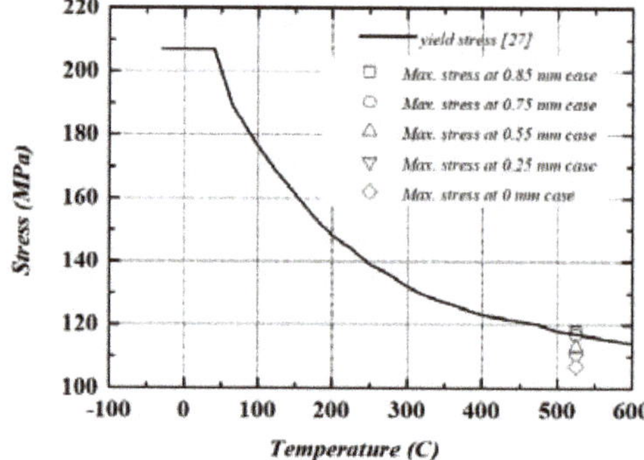

**Figure 11.** Maximum stress for each misalignment condition compared to yield stress of material SS316.

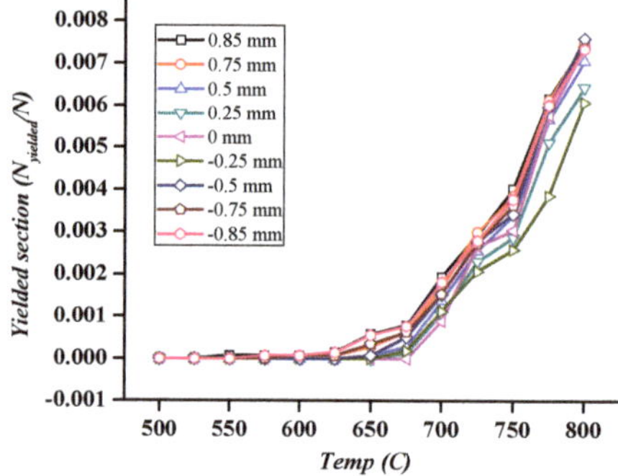

**Figure 12.** Yielded section under pressure loading for various temperature conditions.

## 4. Conclusions

In this study, two-dimensional FEM simulations were performed to observe the effects of primary channel misalignment in the PCHE under five different misalignment conditions.

1. The highest stress intensity was located at the tip edge of the water channel for the higher-pressure intensity compared to that in the sodium channel;
2. The stress intensity increases as the misalignment increases, causing the utilization factor to increase, which is less safe compared to the condition without misalignment;
3. The proposed geometry design can comply with the design limit condition up to 700° C in terms of the ASME code. This means the for the SFR application, which has an average temperature of 550° C, the geometry design can comply with the design limit conditions. However, the design can comply with the acceptance criteria up to 725° C under 30% misalignment condition;
4. The study shows that the design of PCHEs needs to be more precise regarding welding and assembly processes, due to decreasing utilization factor. More strict construction and assembly process monitoring, especially during stacking and welding processes, is needed to avoid this utilization factor degradation due to misalignment conditions.

**Author Contributions:** Conceptualization, A.P.S. and J.Y.L.; methodology, A.P.S.; software, A.P.S.; validation, A.P.S. and J.Y.L.; formal analysis, data curation, A.P.S.; writing—original draft preparation, A.P.S.; writing—review and editing, A.P.S. and J.Y.L.; visualization, A.P.S.; supervision, J.Y.L. All authors have read and agreed to the published version of the manuscript.

**Funding:** This work was supported by the National Research Foundation of Korea (NRF) grant, funded by the Korea government (MSIP) (No. 2017M2A8A4018624).

**Conflicts of Interest:** The authors declare no conflict of interest.

## Nomenclature

*Abbreviations*

| | |
|---|---|
| ASME | American Society of Mechanical Engineers |
| PCHE | Printed Circuit Heat Exchanger |
| FEM | Finite Element Method |
| SCO2 | Supercritical Carbon dioxide |
| SFR | Sodium Fast-cooled reactor |
| TES | Thermal Energy Storage |
| IHX | Intermediate Heat Exchanger |
| *HTGR* | High-Temperature Gas Reactor |
| *SCL* | Stress Classification Line |

*Symbol*

| | |
|---|---|
| $S_m$ | The Lowest Stress Intensity value at a given temperature |
| $K$ | Factor K |
| $S_o$ | Maximum allowable stress intensity |
| $S_{mt}$ | Maximum allowable stress intensity value time dependent |
| $P_m$ | Primary membrane stress intensity [MPa] |
| $P_L$ | Primary membrane stress local [MPa] |
| $P_b$ | Primary bending stress [MPa] |
| $U_f$ | Utilization factor as the ratio of stress intensity to the maximum allowable stress |
| $m$ | Misalignment case (mm) |
| $\sigma_{ij,m}$ | Membrane Stress |
| $\sigma_{ij,b}$ | Bending Stress |
| $\sigma_e$ | Equivalent Stress |
| $\sigma_x$ | Stress tensor in x direction |
| $\sigma_y$ | Stress tensor in y direction |
| $\sigma_z$ | Stress tensor in z direction |
| $S_y$ | Yield Stress |

## References

1. Çengel, Y.A. Heat Exchanger. In *Heat Transfer a Practical Approach*, 2nd ed.; McGraw Hill: New York, NY, USA, 2004; pp. 667–704.
2. Thulukkanam, K. Heat Exchangers Introduction, Classification, and Selection. In *Heat Exchanger Design Handbook*, 2nd ed.; CRC Press: New York, NY, USA, 2013; pp. 1–34.
3. Asadi, M.; Xie, G.; Sunden, B. A review of heat transfer and pressure drop characteristics of single and two phases microchannel. *Int. J Heat Mass Transf.* **2014**, *79*, 34–53. [CrossRef]
4. Chen, M.; Sun, X.; Christensen, R.N.; Shi, S.; Skavdahl, I.; Utgikar, V.; Sabharwall, P. Experimental and numerical study of a printed circuit heat exchanger. *Ann. Nucl. Energy* **2016**, *97*, 221–231. [CrossRef]
5. Chen, M.; Sun, X.; Christensen, R.N.; Shi, S.; Skavdahl, I.; Utgikar, V.; Sabharwall, P. Pressure drop and heat transfer characteristic of a high-temperature printed circuit heat exchanger. *Appl. Therm. Eng.* **2016**, *108*, 1409–1417. [CrossRef]
6. Li, X.; Le Pierres, R.; Dewson, S.J. Heat Exchanger for the next generation of Nuclear reactors. In Proceedings of the International Congress on Advances in Nuclear Power Plants, Reno, NV, USA, 4–8 June 2006; pp. 201–209.

7.  Patil, R.; Anand, S. Thermo- structural fatigue analysis of shell and tube type heat exchanger. *Int. J. Press. Vessel. Pip.* **2017**, *155*, 35–42. [CrossRef]
8.  Mylavarapu, S.K.; Sun, X.; Glosup, R.E.; Christensen, R.N.; Patterson, M.W. Thermal Hydraulic Performance testing of Printed Circuit heat exchangers in a high-temperature helium test facility. *Appl. Therm. Eng.* **2014**, *65*, 605–614. [CrossRef]
9.  Kim, J.H.; Baek, S.; Jeong, S.; Jung, J. Hydraulic Performance of a microchannel PCHE. *Appl. Therm. Eng.* **2010**, *30*, 2157–2162. [CrossRef]
10. Sung, J.; Lee, J.Y. Effect of tangled channels on the heat transfer in a printed circuit heat exchanger. *Int. J. Heat Mass Transf.* **2017**, *115*, 647–656. [CrossRef]
11. Yoon, S.H.; No, H.C.; Kang, G.B. Assessment of straight, zigzag, S-shape, and airfoil PCHEs for intermediate heat exchangers of HTGRs and SFRs. *Nucl. Eng. Des.* **2014**, *270*, 334–343. [CrossRef]
12. Lee, Y.; Lee, J.I. Structural assessment of intermediate printed circuit heat exchanger for sodium-cooled fast reactor with supercritical CO2 cycle. *Ann. Nucl. Energy* **2014**, *73*, 84–95. [CrossRef]
13. Song, K.N.; Hong, S.D. Structural Integrity Evaluation of a Lab-Scale PCHE Proto-type under the Test Conditions of HELP. *Sci. Tech. Nucl. Install.* **2013**, *2013*, 520145. [CrossRef]
14. Mizokami, Y.; Igari, T.; Kawashima, F.; Sakakibara, N.; Tanihira, M.; Yuhara, T.; Hiroe, T. Development of structural design procedure of plate-fin heat exchanger for HTGR. *Nucl. Eng. Des.* **2012**, *255*, 248–262. [CrossRef]
15. Mochizuki, H.; Takano, M. Heat transfer in heat exchangers of sodium cooled fast reactor systems. *Nucl. Eng. Des.* **2008**, *239*, 295–307. [CrossRef]
16. Nestell, J.; (MPR Associates, Inc., Washington DC, USA); Sham, T.L.; (Oak Ridge National Laboratory, Tennessee, USA). ASME Code Considerations for the Compact Heat Exchanger. Personal Communication. 2015.
17. Miwa, Y.; Noishiki, K.; Suzuki, T.; Takatsuki, K. Manufacturing technology of Diffusion-bonded Compact Heat Exchanger (DCHE). *Kobelco Tech. Rev.* **2013**, *32*, 51–56.
18. Kozak, D.; Konjatić, P.; Matejiček, F.; Damjanović, D. Weld Misalignment influence on the structural integrity of cylindrical pressure vessel. *Struct. Integr. Life* **2009**, *10*, 153–159.
19. Brabin, T.A.; Christoper, T.; Rao, B.N. Finite Element analysis of cylindrical pressure vessels having a misalignment in a circumferential joint. *Int. J. Press. Vessel. Pip.* **2010**, *87*, 197–201. [CrossRef]
20. Morgan, W.C.; Bizon, P.T. *Comparison of Experimental and Theoretical Stresses at a Mismatch in a Circumferential Joint in a Cylindrical Pressure Vessel*; Technical Note; Lewis Research Center: Cleveland, OH, USA, 1966.
21. Gomez, E. ASME Section III Stress Analysis of a Heat Exchanger Tube Sheet with a misdrilled Hole and Irregular or Thin Ligaments. In Proceedings of the ASME 2013 Pressure Vessel and Piping Conference, Paris, France, 14–18 July 2013; pp. 1–8.
22. Liu, X.; Song, W.; Yan, Z.; Qiang, W.; Pan, H. Misalignment effect on stress concentration of thickness mismatched plate structures. *Proc. Str. Integr.* **2016**, *2*, 2038–2045. [CrossRef]
23. ASME. *An International Code 2015 ASME Boiler & Pressure Vessel Code Section III. Rules for Construction of Nuclear Facility Components Division 5 High Temperature Reactors*; The American Society of Mechanical Engineer: New York, NY, USA, 2015.
24. Bartel, N.; Chen, M.; Utgikar, V.P.; Sun, X.; Kim, I.H.; Christensen, R.; Sabharwall, P. Comparative analysis of compact heat exchangers for application as the intermediate heat exchanger for advanced nuclear reactors. *Ann. Nucl. Energy* **2015**, *81*, 143–149. [CrossRef]
25. Yoo, J.; Chang, J.; Lim, J.Y.; Cheon, J.S.; Lee, T.H.; Kim, S.K.; Lee, K.L.; Joo, H.K. Overall System Description and Safety Characteristics of Prototype Gen IV Sodium Cooled Fast Reactor in Korea. *Nucl. Eng. Tech.* **2016**, *48*, 1059–1070. [CrossRef]
26. ASME. *An International Code 2015 ASME Boiler & Pressure Vessel Code Section VIII. Rules for Construction of Pressure Vessel Division 2*; The American Society of Mechanical Engineer: New York, NY, USA, 2015.
27. ASME. *An International Code 2015 ASME Boiler & Pressure Vessel Code Section II Part D. Materials*; The American Society of Mechanical Engineer: New York, NY, USA, 2015.

*Article*

# Numerical and Experimental Analysis of Torsion Springs Using NURBS Curves

**Young Shin Kim [1], Yu Jun Song [2] and Euy Sik Jeon [3,*]**

[1]  Industrial Technology Research Institute, Kongju National University, Cheonan-daero, Seobuk-gu, Cheonan-si 31080, Chungcheongnam-do, Korea; people918@kongju.ac.kr
[2]  International Testing and Evaluation Laboratory (ITEL), 53, Osongsaengmyeong 10-ro, Osong-eup, heungdeok-gu, Cheongju-si 28164, Chungcheongbuk-do, Korea; hpp44@naver.com
[3]  Department of Mechanical Engineering, Graduate School, Kongju National University, Cheonan-daero, Seobuk-gu, Cheonan-si 31080, Chungcheongnam-do, Korea
*  Correspondence: osjun@kongju.ac.kr; Tel.: +82-41-521-9284

Received: 28 March 2020; Accepted: 7 April 2020; Published: 10 April 2020

**Abstract:** Torsion springs, which transfer power through the twisting of their coil, provide advantages such as module simplification and efficient use of space. The design of a torsion spring has been formulated, but it is difficult to determine the local behaviors of torsion springs according to actual load conditions. This study proposes a torsion-spring design method through finite element analysis (FEA) using nonuniform-rational-basis-spline (NURBS) curves. Through experimentation, the angle and displacement values for the actual spring load were converted into useable data. Torsion-spring displacement values were obtained via experimentation and converted into coordinates that may be expressed using NURBS curves. The results of these experiments were then compared to those obtained via FEA, and the validity of this method was thereby verified.

**Keywords:** torsion springs; FEA; NURBS; applied load; local behaviors

---

## 1. Introduction

Torsion springs transfer power through the twisting of their coil, and provide advantages such as module simplification, efficient use of space, and reduction of overall product weight. As such, they have been widely used in various electronic products and industrial fields, including automobiles and machinery.

Research on these springs began in 1963 with Wahl's study on isotropic spiral compression and tension, and torsion springs [1,2]. Thus far, the spring design formula proposed by Wahl is used as a standard and has been extensively employed in industrial-machinery applications. This applied design formula can easily calculate the rotation angle of the spring for the applied load. However, as this formula simplifies the problem, it is difficult to determine the behavior or deformation of a torsion spring according to the change in load conditions with different angles and directions.

Case studies considering the structural analysis of such springs using finite-element analysis (FEA) have been proposed to solve this design problem [3–6]. This method allows the designer to visually confirm and design these springs considering the stress and displacement that locally occur in the spring.

As FEA expresses curves as a sum of subdivided linear elements, many elements are needed to increase the reliability of curve-structure analysis. As springs are curved structures, the number of elements used in these analyses must inevitably be increased to ensure accuracy, thereby lengthening analysis time. Due to this increase in spring-design time, as well as the aforementioned design problem, and considering the effective product applications of these springs, a more effective torsion-spring design method is required.

This study proposes the evaluation method of deformation behavior of torsion spring via FEA using nonuniform-rational-basis-spline (NURBS) curves [7,8]. The NURBS curve is a model that can accurately express a curved structure with a small amount of information and a simple calculation formula. These curves were applied to FEA, and through experimentation, the angle and displacement values for the actual spring load were converted into useable data. The torsion-spring displacement values measured through experiments were converted into coordinates that could be expressed using NURBS curves. Experiments were then conducted for comparison with the FEA results and to verify the validity of this method.

## 2. Torsion-Spring Design

### 2.1. NURBS Curve

A range of research has been conducted on curve expression according to polynomial theory. Among the numerous curve-expression methods, NURBS is the most accurate technique used to express complex organic shapes in two and three dimensions [9,10]. Due to its simple calculation method, NURBS is employed in a variety of industrial fields that require curved shapes. A NURBS curve can be expressed through a combination of parameters, including knot, control point, degree, and weight, in basis functions with a relatively simple calculation algorithm. In NURBS curves, the initial basis function can be expressed as in Equation (1).

$$N(u) = \sum_{i=0}^{n} N_{i,p}(u),\tag{1}$$

where $N_{i,p}(u)$ is the B-spline basis function. The equation applied to the algorithm differs with the number and value of the knot vectors; Equation (1) may also be expressed as either Equation (2) or Equation (3). To define B-spline basis functions, we need one more parameter: the degree of these basis functions, $p$. The i-th B-spline basis function of degree $p$, written as $N_{i,p}(u)$, is defined recursively as follows:

$$N_{i,0}(u) = \begin{cases} 1 \; if \; u_i \leq u \leq u_{i+1} \\ 0, \; Otherwise \end{cases}\tag{2}$$

$$N_{i,p}(u) = \frac{u - u_i}{u_{i+p} - u_i}N_{i,p-1}(u) + \frac{u_{i+p+1} - u}{u_{i+p+1} - u_{i+1}}N_{i+1,p-1}(u),\tag{3}$$

where $i$ is the degree plus 1 ($i = p + 1$), $u_i$ is the knot, and $u$ is the knot vector. The following basis function expresses the curve of a geometric shape as the degree ($p$) increases; however, as the function increases in complexity and decreases in accuracy, a suitable degree should be used according to the desired curve shape. The NURBS basis function can express the NURBS curve as a combination of the weight ($w_i$) and control point ($P_i$), as shown in Equation (4).

$$C(u) = \sum_{i=0}^{n} \frac{N_{i,p}w_i}{\sum_{j=0}^{n} N_{j,n}w_i}P_i = \frac{\sum_{j=0}^{n} N_{i,p}w_iP_i}{\sum_{j=0}^{k} N_{i,p}w_i},\tag{4}$$

where $P_i$ is the control point and $w_i$ is the weight of control point. This equation can then be simplified into Equation (5).

$$C(u) = \sum_{i=0}^{n} R_{i,p}(u)P_i\tag{5}$$

The NURBS curve expressed by Equation (5) can be freely varied as the weight ($w_i$) increases and the control point ($P_i$)shifts, demonstrating the relationship between the NURBS curve and these parameters.

## 2.2. Expression of Torsion Spring Through NURBS Curves

To express the shape of a torsion spring with a constant curvature using the NURBS curve, Equations (6) to (8) are used for the $x$, $y$, and $z$ planes on the global coordinate system.

$$x(u) = \sum_{i=0}^{n} R_{i,p}(u)P_x \tag{6}$$

$$y(u) = \sum_{i=0}^{n} R_{i,p}(u)P_y \tag{7}$$

$$z(u) = \sum_{i=0}^{n} R_{i,p}(u)P_z, \tag{8}$$

where $P_N$ are the coordinates of the interpolation point generated through the initial NURBS curve, and $P_{Ndisp}$ are the coordinates through a combination of the knot, control point, and weight (Equations (6) to (8)). The spring shape can be expressed on three planes in 3D space, as shown in Figure 1.

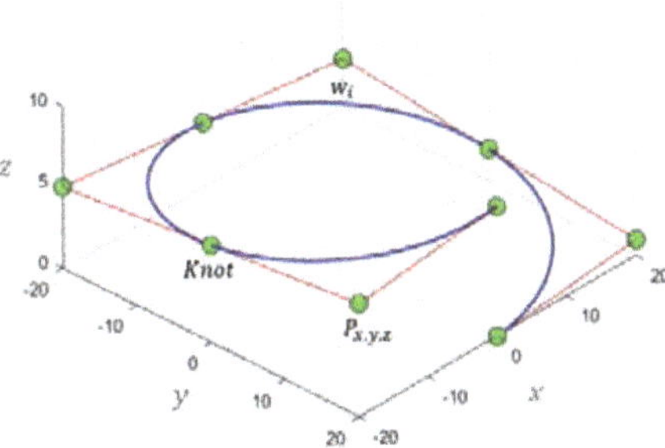


**Figure 1.** Representation of nonuniform-rational-basis-spline (NURBS) curves coupled in 3D space.

## 3. Torsion-Spring Displacement Analysis

### 3.1. Torsion-Spring Displacement Analysis

To verify the correlation between the interpolation point and control point, the coordinates of the interpolation point generated through the NURBS curve are applied to finite-element analysis to create the torsion-spring shape for analysis [11–13]. Figure 2 shows the finite-element model of the torsion spring that was used to derive the displacement of the control point as it shifted according to the external force [14,15]. Table 1 shows the parameters of the model used for analysis. Commercial software HyperWorks Optistruct (Altair, United States) was used for finite-element analysis, and the shifted displacement data were acquired for generated moment $M_z$. The displacement values shifted after the generation of the initial coordinates, and loads were analyzed for a total of 100 elements ($N_i$).

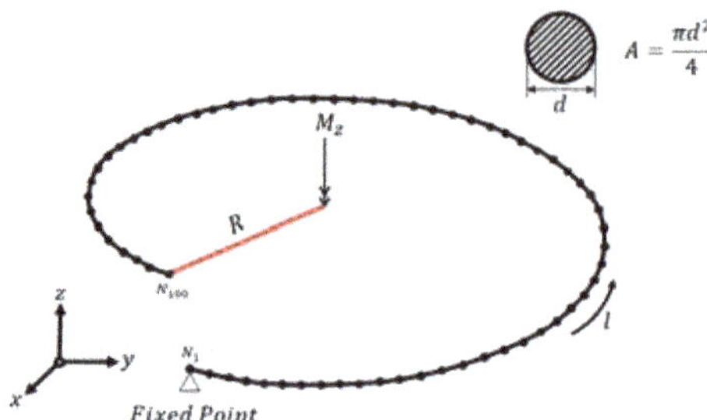


**Figure 2.** Finite-element model of torsion spring.

**Table 1.** Design parameters.

| Parameters | Unit | Dimension |
|---|---|---|
| Spring cross-section diameter ($d$) | mm | 3 |
| Spring radius ($R$) | mm | 20 |
| Spring length ($l$) | mm | 125.6 |
| Moment in the $z$-axis direction ($M_z$) | N m | 1000 |
| Elastic modulus | GPa | 206 |

The displacement of the element determined through finite-element analysis ($x_{disp}$, $y_{disp}$, $z_{disp}$) was added to the coordinates of the interpolation point generated through the NURBS curve to express the deformed shape of the spring. This is expressed using Equations (9) to (11).

$$P_{Ndisp}(x) = P_N(x) + x_{disp} \tag{9}$$

$$P_{Ndisp}(y) = P_N(y) + y_{disp} \tag{10}$$

$$P_{Ndisp}(z) = P_N(z) + z_{disp} \tag{11}$$

where $P_N$ are the coordinates of the interpolation point generated through the initial NURBS curve, and $P_{Ndisp}$ are the coordinates of the interpolation point of the NURBS curve after deformation.

The coordinates of $P_{Ndisp}(x, y, z)$, derived from Equations (9) to (11), were next applied to the inverse method to determine the displacement of the control point. This process is shown in Equation (12), which was derived with reference to Equations (4) and (5). Figure 3 shows the control point derived using Equation (12),

$$P_{(x,y,z)} = \frac{\sum_{i=1}^{n} R_{i,p}(u)}{P_{Ndisp}(x, y, z)}, \tag{12}$$

where $P_x$, $P_y$ and $P_z$ are the coordinates of the control point, and $R_{i,p}(u)$ is the NURBS curve-based function.

In Figure 3, $(x, y, z)_{disp}$ represents the $x$, $y$, and $z$ displacements of the point; $P_N(x, y, z)$ represents the coordinates of the point before interpolation; and $P_{Ndisp}(x, y, z)$ represents the coordinates of the point after interpolation. The inverse method can be applied to determine the displacement and load as $K_{NURBS}$.

$$a[D] = a[K]^{-1}[F], \tag{13}$$

where $[D]$ is the displacement matrix of the control point derived from the existing stiffness matrix, $a$ is the stiffness-correction constant, $[K]^{-1}$ is the inverse of the stiffness matrix, and $[F]$ is the external force matrix. Stiffness-correction constant a can be expressed as follows:

$$a = [a_i]^{T}. \tag{14}$$

Thus, the following is obtained by applying Equation (14) to Equation (13).

$$[a_i]^T [D] = [a_i]^T [K]^{-1} [F]$$ 

(15)

Equation (15) can then be expressed as in Equation (16).

$$[D_{\epsilon v}] = [K_{NRBS}]^{-1} [F]$$ 

(16)

where $[D_{\epsilon v}]$ is the control point displacement determined using the inverse method, and $[K_{NRBS}]^{-1}$ is the inverse of the stiffness matrix of the control point.

Using Equation (16), the $M_z$ load was applied to the NURBS curve-based spring structure that was composed of a total of nine control points, and analysis was conducted.

Using this method, the graphs in Figure 4 were derived for loads $F_x$, $F_y$, $F_z$, $M_x$, $M_y$, and $M_z$.

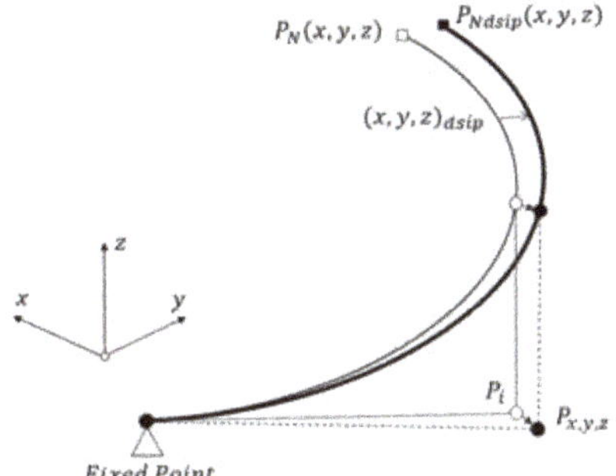


**Figure 3.** Derivation of control point using inverse method.

### 3.2. Experiment-Equipment Setup

Experiments were conducted to measure the displacement, load, and angle using a torsion-spring measuring device. Torsion-spring parameters were selected for a total of six types of specimens. Five experiments were conducted for each specimen, resulting in a total of 30 experiments. Table 2 shows the design parameters of the six spring specimens. The experiment equipment was configured to measure the torsion spring. Figure 5 shows the actual torsion spring and the device used to measure the spring displacement, load, and angle. The rotation angle was set to 0°, 20°, and 40°, considering the linear component of the load applied to the torsion spring. To determine torsion-spring displacement, certain sections were marked, and displacement and load values were measured at 20° and 40°. Images were taken of the front of the system to determine the x and y coordinates, and the z-axis was measured through images of the sides and Vernier calipers.

**Table 2.** Specimen types.

| Type | Spring Cross-Section Diameter, d (mm) | Spring Diameter, D (mm) | Spring Turns, N |
| --- | --- | --- | --- |
| 1 | 2.6 | 32.6 | 1.5 |
| 2 | 2.6 | 32.6 | 2.5 |
| 3 | 2.8 | 32.8 | 1.5 |
| 4 | 2.8 | 32.8 | 2.5 |
| 5 | 3.0 | 33.0 | 1.5 |
| 6 | 2.0 | 33.0 | 2.5 |

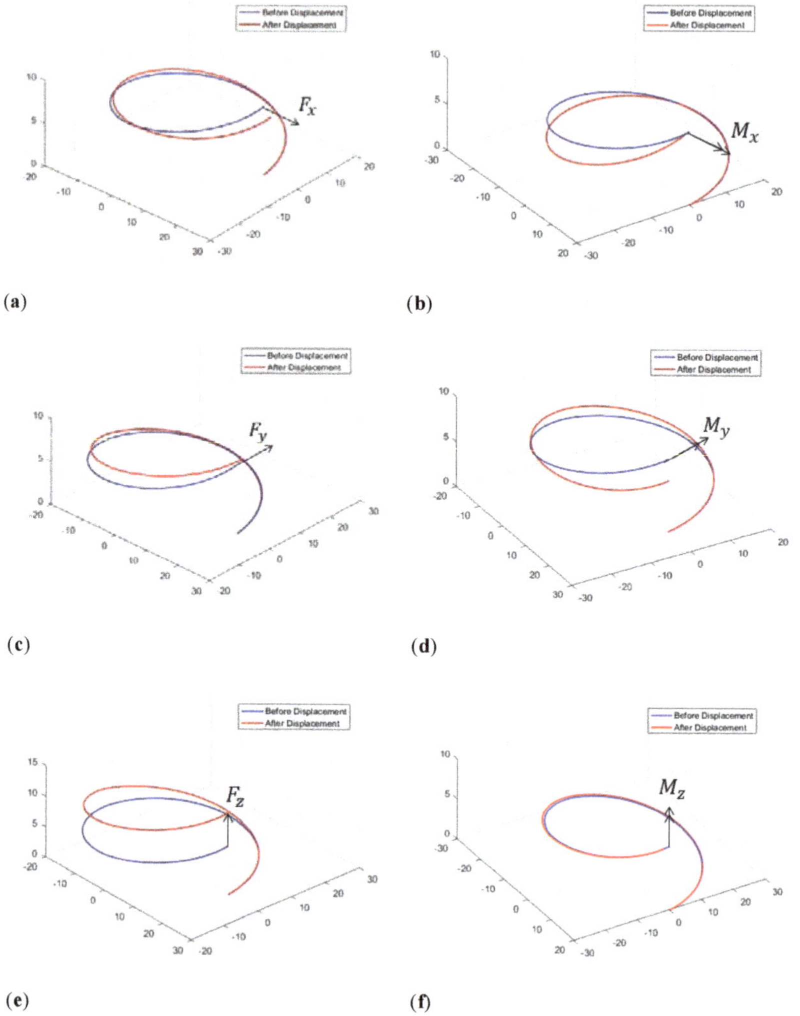

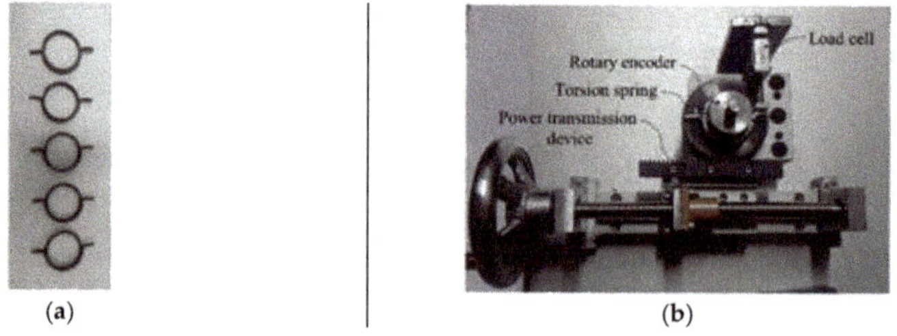

**Figure 4.** Direction-displacement graphs for load conditions (**a**) $F_x$, (**b**) $M_x$, (**c**) $F_y$, (**d**) $M_y$, (**e**) $F_z$, (**f**) and $M_z$.



**Figure 5.** Springs and experiment setup: (**a**) torsion springs; (**b**) experiment setup.

## 4. Results and Discussion

### 4.1. Experiment-Data Analysis

The force according to the rotation angle was measured using a load cell and rotary encoder. Figure 6 shows images of the x- and y-coordinate measurements obtained according to the angle of the torsion spring, and Figure 7 shows a load graph according to the rotation angle. This graph was drawn using the average data from the results of five replicates. Analytical results indicated that the torsion-spring load was generated nonlinearly according to the rotation angle. The regression equation was derived using the measured data and was expressed as a quadratic polynomial in the graph. R-square values of the regression equations were 98% or more. This regression equation had a high adjustment. In addition, it was found that the load increased as the cross-sectional diameter (d) of the torsion spring increased, and that the load decreased as the number of spring turns (N) increased.

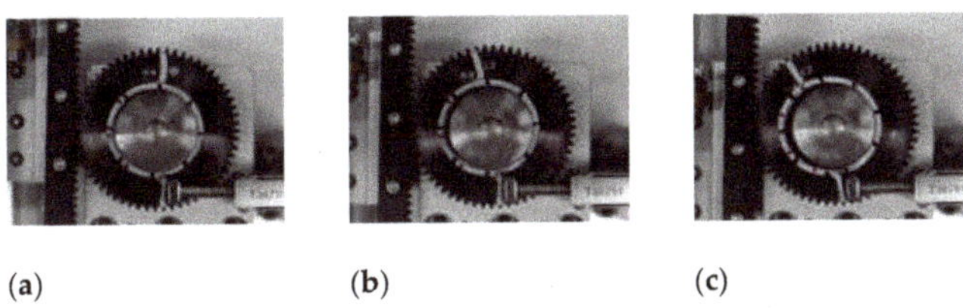

(a)      (b)      (c)

**Figure 6.** Measurement of x- and y-coordinates of torsion springs: (**a**) $0°$; (**b**) $20°$; (**c**) $40°$.

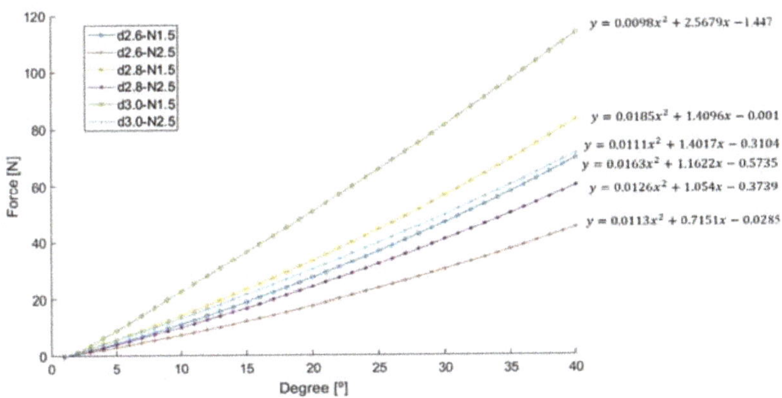


**Figure 7.** Force vs. degree graph of torsion springs.

### 4.2. Comparative Analysis of Analytical Results and Experiment Values

We performed comparative analysis of the deformed shape of the torsion spring, measured through the experiments, and the analytical results obtained using Equation (16). Figure 8 shows graphs of the analytically and experimentally obtained data. We then performed comparative analysis of the deformed shape of the torsion spring at 20° and 40°, measured through the experiments and Equation (16). Under the same load conditions, the displacement determined by the proposed equation exceeded that of the torsion spring measured through the experiments. However, by modifying the load parameters, similar behaviors and displacements were observed, thereby verifying the use of the proposed equation. The maximal error rate ranged from 5% to 6%, which could be attributed to the error that can occur during experiment measurements.

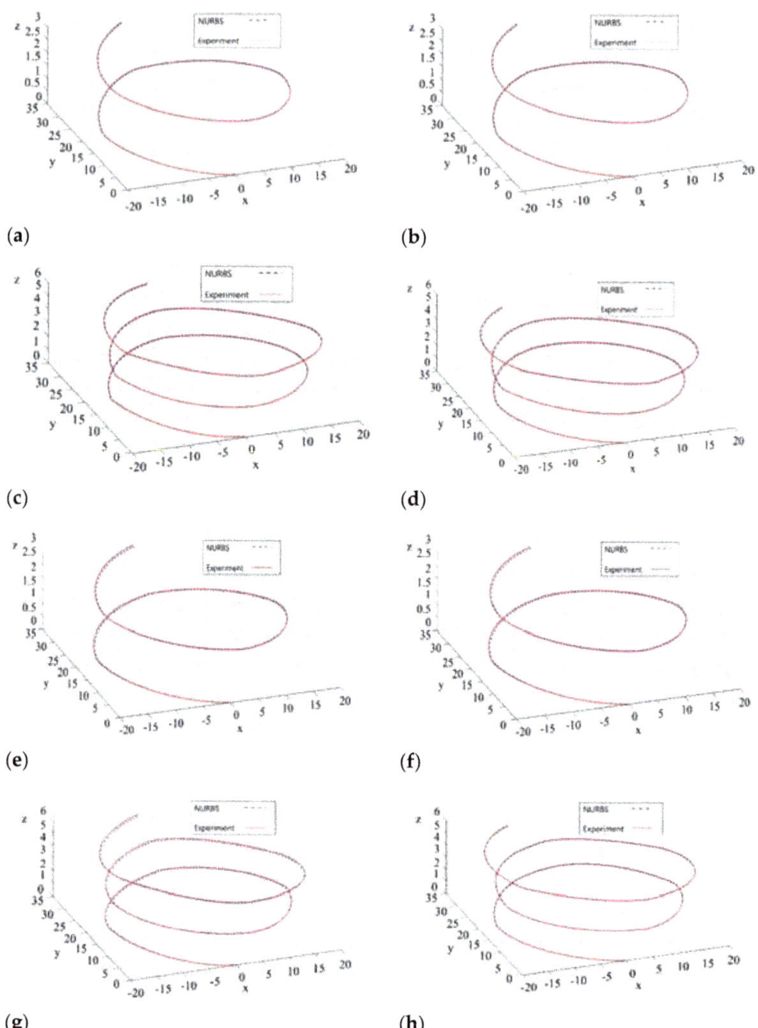



**Figure 8.** *Cont.*

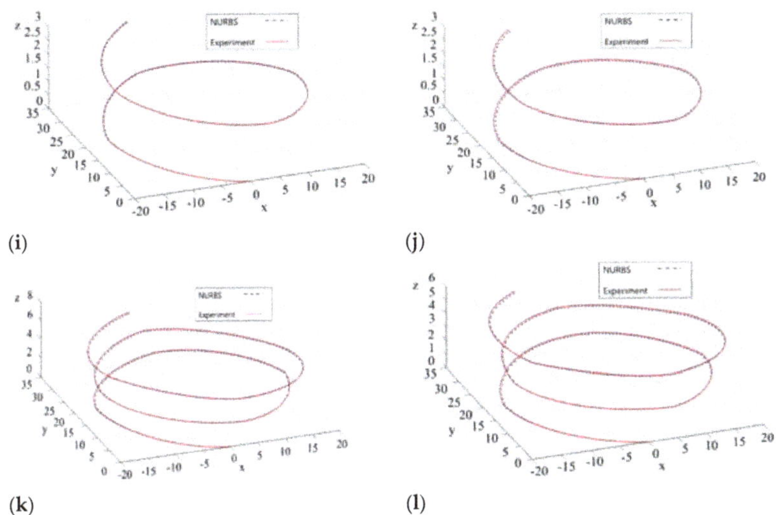

**Figure 8.** Comparative analysis of torsion springs: (**a**) d = 2.6 mm, D = 32.6 mm, n = 1.5 (20°); (**b**) d = 2.6 mm, D = 32.6 mm, n = 1.5 (40°); (**c**) d = 2.6 mm, D = 32.6 mm, n = 2.5 (20°); (**d**) d = 2.6 mm, D = 32.6 mm, n = 2.5 (40°); (**e**) d = 2.8 mm, D = 32.8 mm, n = 1.5 (20°); (**f**) d = 2.8 mm, D = 32.8 mm, n = 1.5 (40°); (**g**) d = 2.8 mm, D = 32.8 mm, n = 2.5 (20°); (**h**) d = 2.8 mm, D = 32.8 mm, n = 2.5 (40°); (**i**) d = 3.0 mm, D = 33.0 mm, n = 1.5 (20°); (**j**) d = 3.0 mm, D = 33.0 mm, n = 1.5 (40°); (**k**) d = 3.0 mm, D = 33.0 mm, n = 2.5 (20°); (**l**) d = 3.0 mm, D = 33.0 mm, n = 2.5 (40°).

## 5. Conclusions

This study applied nonuniform-rational-basis-spline (NURBS) curves for the design of torsion springs, analyzed the displacements of these springs using finite-element analysis, and verified the design of these springs through experimentation.

(1) A method was proposed for deriving the coordinates of a control point for shifted elements by applying the inverse method on the basis of data derived through finite-element analysis. In addition, the relationship between the movement of the control point and stiffness matrix was identified and formulated by varying the torsion-spring parameters.

(2) A method was proposed for deriving the torsion-spring shape by converting the torsion-spring displacement measured through experiments into coordinates that could be expressed using NURBS curves.

(3) Comparative analyses between the results of the proposed analytical method and the experiment measurements demonstrated that the proposed method is valid within a satisfactory range of error.

**Author Contributions:** Y.S.K., Y.J.S., and E.S.J. conceived and designed the experiments; Y.S.K., Y.J.S., and E.S.J. performed the experiments; Y.S.K., Y.J.S., and E.S.J. analyzed the data; Y.S.K., Y.J.S., and E.S.J. contributed reagents/materials/analysis tools; and Y.S.K., Y.J.S., and E.S.J. wrote the paper. All authors have read and agreed to the published version of the manuscript.

**Funding:** This research received no external funding.



## References

1.　Yuldirim, V. Exact determination of the global tip deflection of both close-coiled and open-coiled cylindrical helical compression spring having arbitrary doubly-symmetric cross-section. *Int. J. Mech. Sci.* **2016**, *115–116*, 280–298. [CrossRef]

2.    Wahl, A.M.; Bisshopp, K.E. Mechanical Springs (Second Edition). *J. Appl. Mech.* **1964**, *31*, 159–160. [CrossRef]

3.    Pollanen, I.; Martikka, H. Opimal re-design of helical spring using fuzzy design and FEM. *Adv. Eng. Softw.* **2010**, *41*, 410–414. [CrossRef]

4.    Chaudhury, A.N.; Datta, D. Analysis of prismatic springs of non-circular coil shape and non-prismatic springs of circular coil shape by analytical and finite element methods. *J. Comput. Des. Eng.* **2017**, *4*, 178–191. [CrossRef]

5.    Gzal, M.; Groper, M.; Gendelman, O. Analytical, experimental and finite element analysis of elliptical cross-section helical spring with small helix angle under static load. *Int. J. Mech. Sci.* **2017**, *130*, 476–486. [CrossRef]

6.    Chassie, G.; Becker, L.; Cleghorn, W. On the buckling of helical springs under combined compression and torsion. *Int. J. Mech. Sci.* **1997**, *39*, 697–704. [CrossRef]

7.    Gao, J.; Luo, Z.; Xiao, M.; Gao, L.; Li, P. A NURBS-based Multi-Material Interpolation (N-MMI) for isogeometric topology optimization of structures. *Appl. Math. Model.* **2020**, *81*, 818–843. [CrossRef]

8.    Alderson, T.; Samavati, F. Multiscale NURBS curves on the sphere and ellipsoid. *Comput. Graph.* **2019**, *82*, 243–249. [CrossRef]

9.    Timoshenko, S.P. *Mechanics of Materials*; Chaoman & Hall: London, UK, 1911.

10.   Piegl, L.; Tiller, W.; Piegl, L. *The NURBS Book*; Springer: Heidelberg, Germany, 1997.

11.   Ferreira, A.J.M. *MATLAB Codes for Finite Element Analysis: Solid Mechanics and its Applications*; Springer: Heidelberg, Germany, 2008; Volume 157.

12.   Shimoseki, M.; Hamano, T.; Inaizumi, T. *FEM for Springs*; Springer Science & Business Media: Berlin, Germany, 2013.

13.   Zhao, X.-Y.; Zhu, C. Injectivity of NURBS curves. *J. Comput. Appl. Math.* **2016**, *302*, 129–138. [CrossRef]

14.   Zhang, B.; Li, C.; Wang, T.; Wang, Z.; Ma, H. Design and experimental study of zero-compensation steering gear load simulator with double torsion springs. *Meas.* **2019**, *148*, 106930. [CrossRef]

15.   Kilic, M.; Yazicioglu, Y.; Kurtulus, D.F. Synthesis of a torsional spring mechanism with mechanically adjustable stiffness using wrapping cams. *Mech. Mach. Theory* **2012**, *57*, 27–39. [CrossRef]

*Article*

# Design Optimization for the Thin-Walled Joint Thread of a Coring Tool Used for Deep Boreholes

Yu Wang [1,2], Chengyuan Qian [1,3,*], Lingrong Kong [1,2,*], Qin Zhou [1,2] and Jinwu Gong [4]

[1] School of Engineering and Technology, China University of Geosciences, Beijing 100083, China; wangyu203@cugb.edu.cn (Y.W.); zhqtg@cugb.edu.cn (Q.Z.)
[2] Key Laboratory on Deep Geo-Drilling Technology of the Ministry of Natural Resources, China University of Geosciences, Beijing 100083, China
[3] Beijing Aidi Geological Investigation & Foundation Construction Company, Beijing 100041, China
[4] Shanxi Huanjie Petroleum Drilling Tools Co., Ltd., Jinzhong 045499, China; gongjinwu@huanjie.com.cn
[*] Correspondence: daiyuan917@126.com (C.Q.); lrkong@cugb.edu.cn (L.K.)

Received: 26 February 2020; Accepted: 10 April 2020; Published: 13 April 2020

**Abstract:** Threaded joints are key components of core drilling tools. Currently, core drilling tools generally adopt the thread structure designed by the API Spec 7-1 standard. However, fractures easily occur in this thread structure due to high stress concentrations, resulting in downhole accidents. In this paper, according to the needs of large-diameter core drilling, a core barrel joint was designed with an outer diameter of Φ135 mm and a trapezoidal thread profile. Subsequently, a three-dimensional simulation model of the joint was established. The influence of the external load, connection state and thread structure on the stress distribution in the joint was analyzed through simulations, from which the optimal thread structure was determined. Finally, a connection test was carried out on the threaded joint. The stress distribution in the joint thread was indirectly studied by analyzing gas leaks (i.e., the sealing effect) under axial tension. According to the test data and the simulation results, the final joint thread structure was optimized, which lays a good foundation for the design of a core barrel.

**Keywords:** drill pipe joint; design; sealing properties; experiment

---

## 1. Introduction

A threaded joint is one of the weakest and most critical components of a core drilling tool [1]. A threaded joint is a typical thin-walled structure that operates in a harsh working environment and is mainly subjected to tensile loads and torsional loads. During core drilling, downhole accidents often occur as a result of thread breakage. The most common reason for joint failure is fatigue, which is affected by the maximum stress [2]. Ensuring the joint strength and airtightness of the thread is the key to the success of the coring process; however, due to limited radial dimensions, it has always been difficult to achieve such goals in the study of core drilling tools [3–5].

At present, research on joint threads mainly focuses on oil drill strings [6–8], which have a large wall thickness and a large thread design margin; hence, these structures cannot provide a direct reference for the core drilling thread design [9–11]. Wireline core drill strings are thin-walled structures that can be used as a reference for the thread design of core drilling tools.

Many researchers have used model parameterization to study the mechanical characteristics of rotary shouldered connections (RSCs). Studies have shown that the 2D finite element method (FEM) is a powerful numerical method for solving this particular problem [12–14]. Feng Qingwen [15] studied the relationship between the thread taper and strength and the corresponding stress distributions through a range of lab tensile, torsional and optical-elastic stress tests. Su Jijun [16] optimized the pitch and length parameters of the thread for diamond wireline core drilling tools. Their results showed that an appropriate increase in the pitch and length of a thread could effectively improve the stress distribution inside a joint thread.

Owing to the complex structure, finite element simulation is the most effective method for studying threads. Tadeusz Smolnicki et al. [17] established the finite element modeling of fatigue loaded bolted flange joints and analyzed the mechanism of bolt fracture. Owing to the complex structure, finite element simulations are the most effective method for studying threads. Tadeusz Smolnicki et al. [18] established finite element modeling of fatigue-loaded bolted flange joints and analyzed the mechanism of bolt fracture. Based on the nonlinear finite element theory, Liang Jian [19] carried out a simulation analysis to assess problems involving a threaded joint in a drill pipe joint and the pull-out of a rod during deep-hole drilling. They used finite element software to analyze the pull-out ability of a thread with different wireline core drill strings under different parameters. Cui Chengmin [20] used ANSYS software to carry out a 3D simulation of a $\Phi$71 mm wireline core drill string and joint and determined that the thread root had the highest stress. Yin Feng [21] established a quasi-three-dimensional model for a $\Phi$89 mm wireline core string joint. Under the condition of neglecting the influence of the helix angle, ANSYS was used to analyze the nonlinear mechanical dynamics of a thread. Finally, the stress nephogram and the deformation nephogram at a joint thread were obtained; however, other loads and boundary conditions were ignored, making the results irregular. Gao Shenyou [22] used static mechanics to analyze the stress state of the negative angle thread of a $\Phi$74 mm wireline core drill string and calculated the torsional resistance of the thread. Moreover, theoretical calculations and static torsion tests were performed on negative angle threads with different taper and thread heights. Gao Jianlong [23] calculated the force in a thread from the influence of the thread structure and the drilling parameters on the thread of the joint. ANSYS/Workbench software was used to simulate the stress and fatigue of a joint thread, and the fatigue life of a drill pipe joint was predicted.

Current core drilling tools mostly use American Petroleum Institute (API) standards [24]. Although API RP 7G provides the ultimate working torque for API RSCs, the relationship between the ultimate working torque and axial tension is simplified to be linear, which is only true when the axial tension is not too large [25]. The stress distribution in API joints is not reasonable, which results in a short service life [26]. For a drilling pipe, taking the XT-M tool joint developed by GRANT Company as an example, the torsional strength of XT-M was higher than that of API joints, and the gas sealing performance was satisfactory, which was confirmed through an experiment in which XT-M joints sealed an external pressure load [27]. Threads designed in accordance with API standards are prone to downhole accidents owing to stress concentrations during coring operations, indicating that these threads cannot meet the needs of deep core drilling. At present, there is almost no research on the thread of a large-diameter thin-walled core barrel, which makes it impossible to meet the needs of the deep-hole coring process in many cases.

In this paper, according to the need for large-diameter deep-hole coring, a core barrel joint is designed with an outer diameter of $\Phi$135 mm, a trapezoidal thread profile, and flush internal and external faces. Subsequently, a three-dimensional finite element model of the threaded joint is established. The effects of the external load, connection state and thread structure on the joint connection are established via simulation, from which the thread structure is optimized. Finally, a threaded joint connection test is designed. The stress distribution in the threads is experimentally determined by testing the airtightness effect of the joints. Combined with the above simulation model, the final joint thread structure is optimized, which lays a good foundation for the design of the deep-hole core barrel.

## 2. Design of the Core Barrel

### 2.1. Structure Parameters of the Thread

The thread is the most important part of a joint. The thread connecting the drill is generally a conical thread structure, which has a self-locking effect and good sealing performance, so it is widely used in connecting liquid and gas pipelines. Other thread forms include trapezoidal threads, gear style threads, and symmetrical trapezoidal threads (Figure 1). In the picture, $P$ is the pitch, $H$ is the height, $\alpha$ is the thread angle, and $T$ is the taper. Trapezoidal threads are studied in this paper.

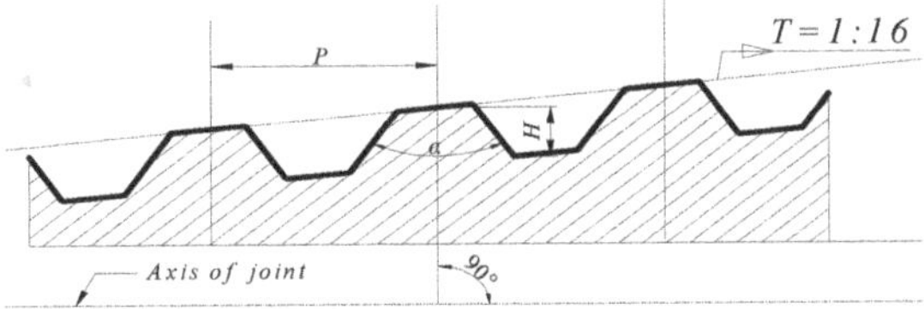


**Figure 1.** Schematic diagram of the thread profile.

The structural parameters of the conical thread include the taper $T$, pitch $p$, lead $S$, helix angle $\theta$, thread angle $\alpha$, and thread height $h$. The taper is the ratio of the difference between the diameters of the large end and small end to the length of the thread. A tapered thread can produce a tight fit between threads. Therefore, a tapered thread has a good sealing capacity and good pressure capacity. The pitch refers to the axial distance between two adjacent threads. The strength of the thread increases as the pitch increases, but the self-locking effect decreases as the pitch increases. The lead is the axial distance at which a point on the thread is rotated 360 degrees along the helix. In a single-helix thread, the lead is equal to the pitch. The thread of a drilling joint is generally a single-helix thread.

The helix angle is the angle between the tangent of the helix and the plane of the vertical helix axis. The helix angle determines the number of turns of the thread. The smaller the helix angle is, the greater the number of turns in the same distance and the better the self-locking effect of the thread. The thread angle is the angle between the two faces of the thread profile. The smaller the thread angle is, the larger the required upper buckle torque, and the easier it is for the thread to loosen. However, if the thread angle is too large, the thread strength will decrease, and the buckle torque will decrease. The thread height is the difference between the large diameter and the small diameter of the thread. When the thread height is too small, thread tripping can easily occur, and the stress in the thread is concentrated. When the thread height is too large, it is easy for the thread to buckle and break.

### 2.2. Design of the Joint

According to the need of large-diameter cores for deep holes, a preliminary design for a core barrel is shown in Figure 2. The core barrel consists of a pin joint, a box joint and a body, which are machined from a single cylinder. The wall thickness of the core barrel joint is only 7 mm. Through preliminary design, the joint threads are trapezoidal threads with a tooth profile angle of 30 degrees, a taper of 1:32, a tooth height of 2.24 mm, and a pitch of 8.4 mm. The threads cooperate to form a sealing surface by means of a shoulder face and an end face.

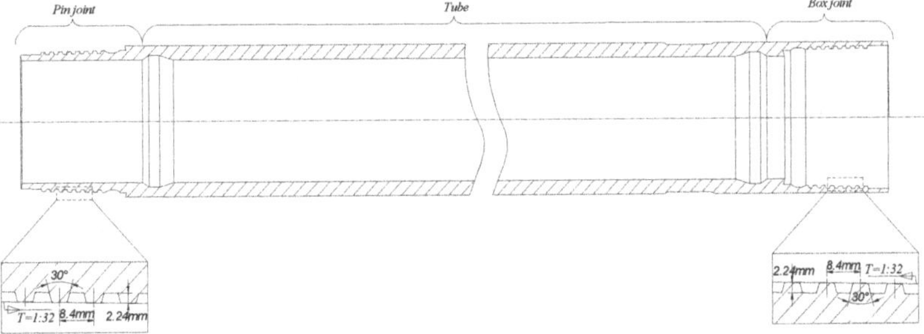


**Figure 2.** Structure of the coring tool.

## 3. Simulation and Optimization of the Joint

### 3.1. Finite Element Modelling

Previous finite element simulation studies on threads have the following limitations: (1) they are mostly 2D models that are unable to reflect the true three-dimensional force state and (2) transforming an asymmetrical model into a symmetrical model ignores the effect of the helix angle. Although this can simplify the finite element calculation process, the results do not truly reflect the stress variation in the thread. In this paper, a three-dimensional asymmetric model is used to study the conical thread of a core joint. This three-dimensional model can more accurately reflect the connection state of the thread and visually reveal the thread stress distribution.

A 1:1 three-dimensional model of the threaded joint is established, as shown in Figure 3. The joint material is AISI 4145H. The material parameters are as follows: the elastic modulus is $2.05 \times 10^{11}$ Pa, the Poisson's ratio is 0.29, the tensile strength is 965 MPa, and the yield strength is 820 MPa.

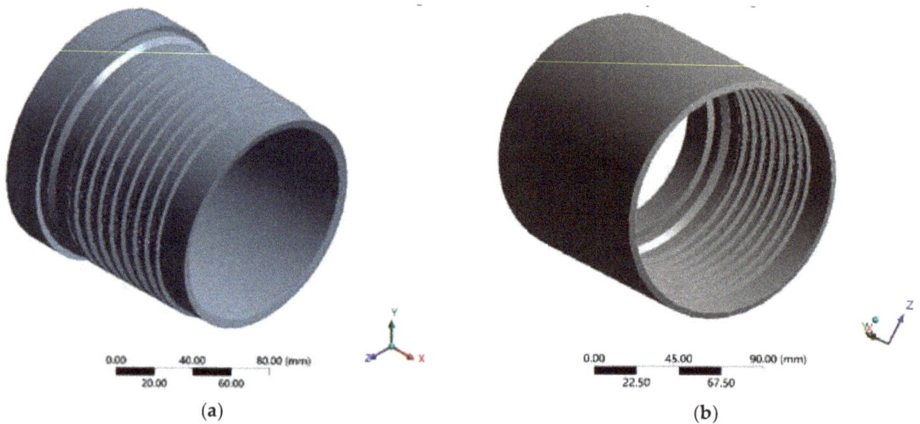

(a)          (b)

**Figure 3.** Three-dimensional models of the (**a**) pin joint and (**b**) box joint.

### 3.2. Element Selection and Meshing

To make the calculation results accurate and credible, it is necessary to analyze the independence of the mesh. The mesh independence was verified under a load of 500 kN. As shown in Figure 4, when the number of elements is greater than 56,713, the average stress in the thread no longer changes significantly as the number of elements increases. Therefore, it can be considered that the calculation results are independent of the mesh size when the number of elements is greater than 56,713.

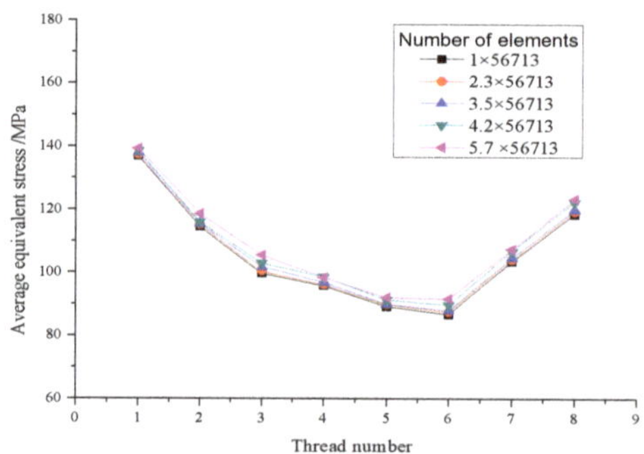


**Figure 4.** Mesh independence analysis results.

In view of the complexity of the thread structure and the contact behavior in the analysis, a mesh that is too dense will produce a considerable computational cost, so the mesh size cannot be too small. Moreover, mesh generation needs to have a certain degree of precision, and it is necessary to improve the correlation of the elements. The 3D model was meshed by tetrahedral elements and local mesh refinement was performed at the thread. The total number of model units is 56,713; the number of nodes is 101,147. The average element quality is 0.7. The grid model is shown in Figure 5.

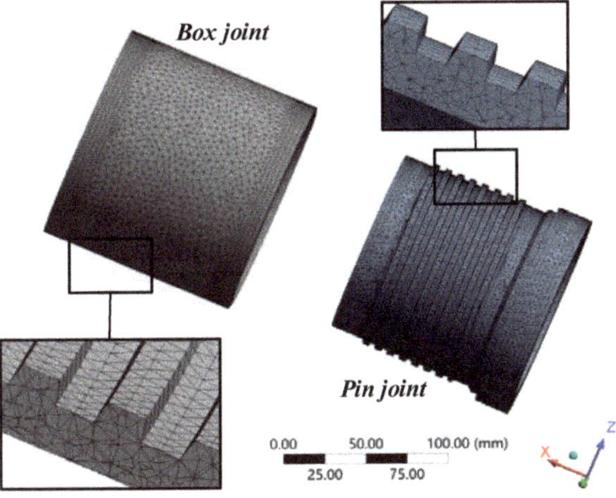


**Figure 5.** Meshed models of the pin joint and box joint.

To quantitatively analyze the stress variation along the axis of the thread, the thread is numbered as follows: from the end near the inner-shoulder face, a 360-degree rotation of the thread is defined as one turn, as shown in Figure 6.

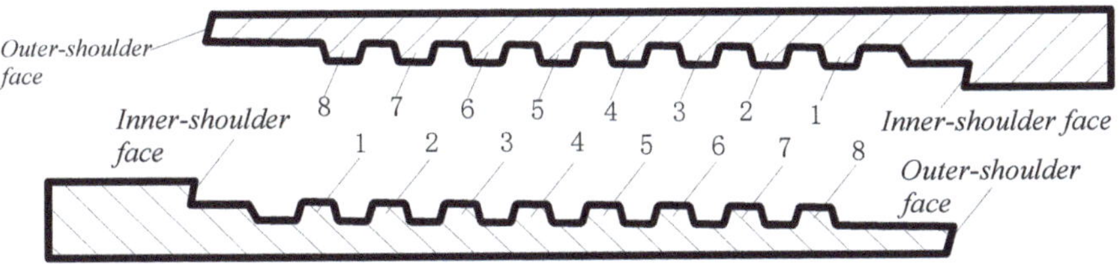


**Figure 6.** Schematic diagram of thread numbering.

### 3.3. Contact and Loading Conditions

ANSYS provides five contact modes: *bonded, frictionless, no separation, rough,* and *frictional.* The *bonded* contact model completely binds two contact surfaces, thereby preventing both separation and sliding. The *frictionless* model defines the friction coefficient between the contact surfaces as zero and allows normal separation. The *no separation* model allows frictionless sliding over a small area. The *rough* model is similar to the *frictionless* model, except that contact sliding between contact surfaces is not allowed. The *frictional* model defines that there will be shear forces based on the friction coefficient between the contact surfaces. The latter three models are nonlinear contact models. The *frictional* model was selected in the simulation, as shown in Figure 7. In the simulation, the coefficient of friction between the threads of the pin joint and box joint is set to 0.1, which is the actual coefficient of friction between steel and steel.

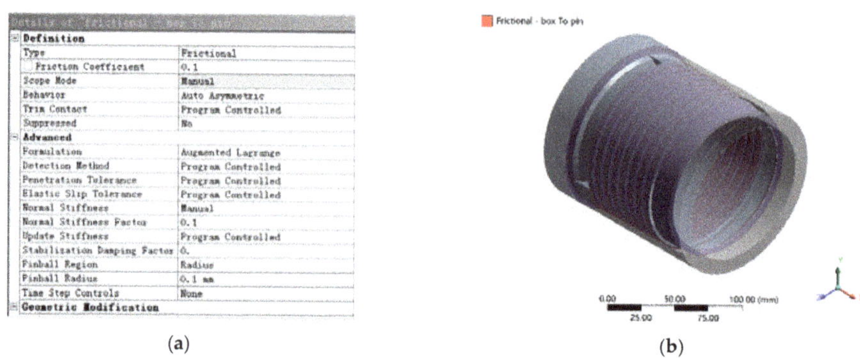

(a)                                                                    (b)

**Figure 7.** (**a**) Contact settings and (**b**) contact surfaces.

After the joints are connected, it is necessary to continue tightening at a certain angle to produce an interference fit, achieve sealing and prevent thread tripping. In the previous study, the interference fit between the threads was achieved by setting the contact offset. In this simulation, the interference fit between the threads was achieved by rotating the pin joint by a certain angle, which is consistent with the actual working conditions. Moreover, the circular cross section of the pin joint is set to the cylinder connection, that is, both the movement and rotation in the x and y directions are prohibited (wherein UX, UY, RX, and RY are constrained). After completing the above settings, the model can be loaded, as shown in Figure 8.

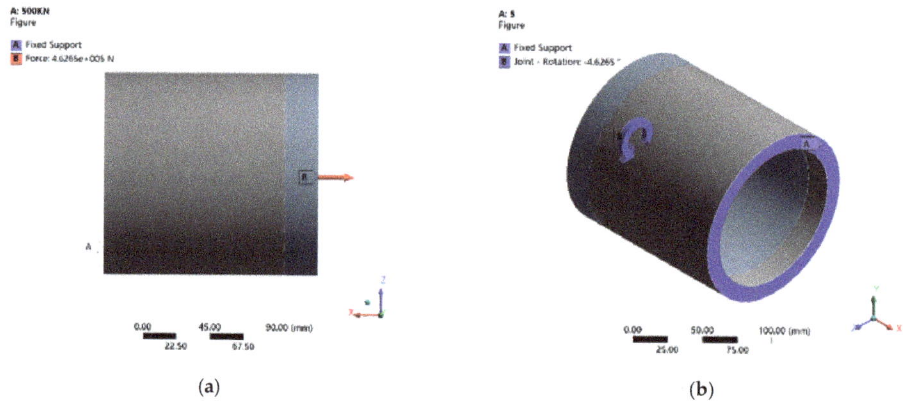

(a)                                                                    (b)

**Figure 8.** Loading conditions: (**a**) tensile load and (**b**) torsional load.

### 3.4. Effect of the Axial Tensile Load

Of all the loads, the axial load is an important factor affecting the strength of the joint. A. Tafreshi and W.D. Dover [6] compared stress concentration factors values for pin and box under bending, axial and torsional loading. The results showed that stress concentration factors values of axial loading were greater than bending, while for torsion they were very small. A. Baryshnikov et al. [27], through a series of full-scale fatigue tests, indicated that tension loads have a negative effect on the drill pipe and tool joint fatigue behavior.

In actual work, the force acting on the joint is very complicated, including the pressure exerted by the rig, the weight of the joint itself, the buoyancy of the medium and the friction of the core. The force in the joints of a core barrel is generally between 100 and 150 kN. In the case of pulling out in a stuck state, the load may exceed 300 kN. To cover the actual force range, the range of axial tension during the simulation is 100–500 kN. In the simulation analysis of the equivalent stress distribution under the axial load, the constraint condition of the box joint is a fixed constraint, and the constraint condition of the pin joint is a uniform tensile load. The pin joint is subjected to a uniformly distributed tensile load. Figure 9 shows the equivalent stress distribution in the thread when the tensile load is 500 kN.

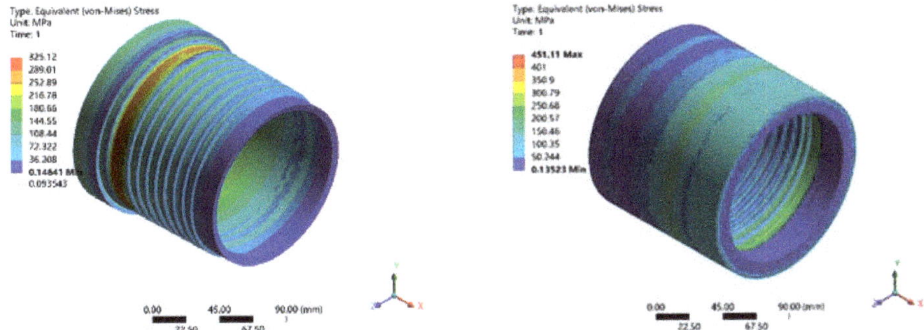

**Figure 9.** Stress distributions in the (**a**) pin joint and (**b**) box joint.

Figure 9 shows that the equivalent stress concentration occurs at the relief notch because the relief notch is thinner than the threaded section. The equivalent stress in the pin and box joint assemblies decreases from the outside to the inside. For a single joint, the equivalent stress is reduced from the side closer to the inner-shoulder face to the side of the outer-shoulder face.

In order to quantitatively study the stress distributions in the thread, the average equivalent stress and the maximum equivalent stress of each thread are compared. An equivalent stress path, as shown in Figure 10, is inserted into the pin joint under the tensile load. The stress path starts at point 1 along the helix to point 2. Each 360 degree rotation along the stress path is counted as one turn. Then, the sum of the stress values at the nodes is calculated for each turn. These nodes fall on the stress path when meshing. The sum of stress is then divided by the number of nodes to obtain the average equivalent stress on each thread, the results of which are is shown in Figure 11a. The maximum equivalent stress of each circle is shown in Figure 11b. The ordinate represents the average equivalent stress for each turn of the thread, and the number of threads is shown in Figure 6. Figure 11 shows that the distribution characteristics of the average equivalent stress in the thread under different axial tensile loads are as follows: the stress is high at the two ends and low in the middle, and the stress is highest near the inner-shoulder surface. A comparison of the curves shows that when the axial tensile load is low, the stress in each thread is relatively uniform. However, as the tensile load increases, the distribution of the thread stress becomes more uneven. The maximum equivalent stress of the thread has a similar distribution pattern as the equivalent stress, but the fluctuation is significantly larger than the average stress. Both the average stress distribution and the maximum stress distribution indicate that the surface near the first thread is a dangerous section.

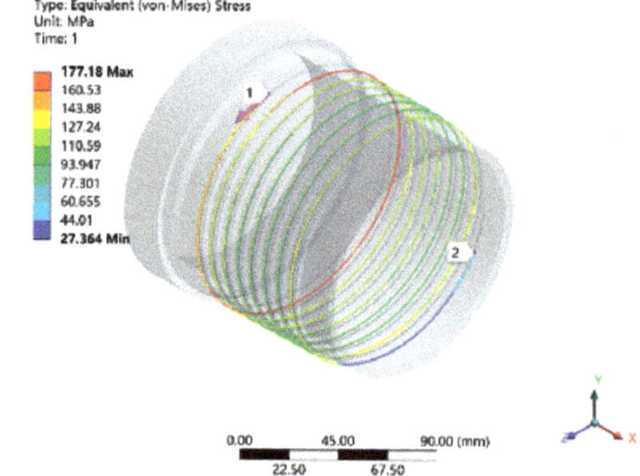

Figure 10. Stress path setting.

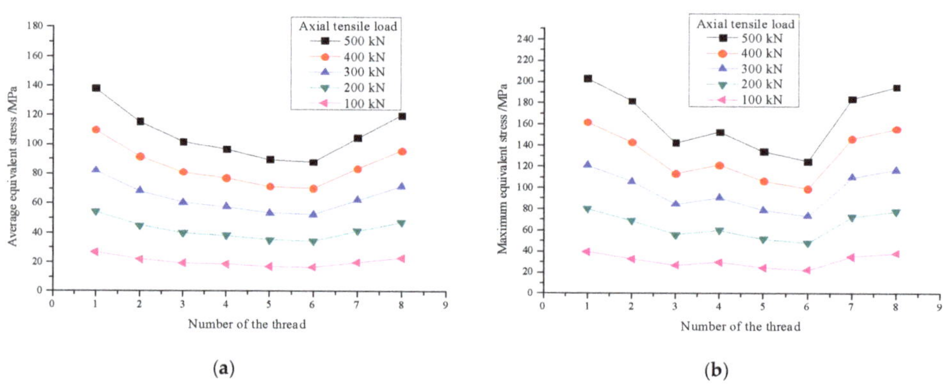

(**a**)                    (**b**)

**Figure 11.** Equivalent stress distribution in the thread: (**a**) average equivalent stress and (**b**) maximum equivalent stress.

When analyzing the stress distribution in the joint under the axial load, the friction between the thread surfaces can be ignored, and the contact surface is considered to have only normal stress. The results shown in Figure 12 were obtained under a tensile force of 500 kN. The compressive stress distribution in the contact surface is shown in Figure 12a. Similarly, the normal stress on the contact surface is also reduced from the inner-shoulder face side to the outer-shoulder side. A cross section of the assembly is shown in Figure 12b. This figure shows that the stress on the inner wall of the pin-joint end is higher than the stress in the thread. Similarly, the stress on the outer wall of the box-joint end is higher than the stress in the thread. Therefore, it can be inferred that when the thread is subjected to an excessive axial tensile load, the inner wall of the pin-joint end and the outer wall of the box-joint end are most likely to experience plastic deformation.

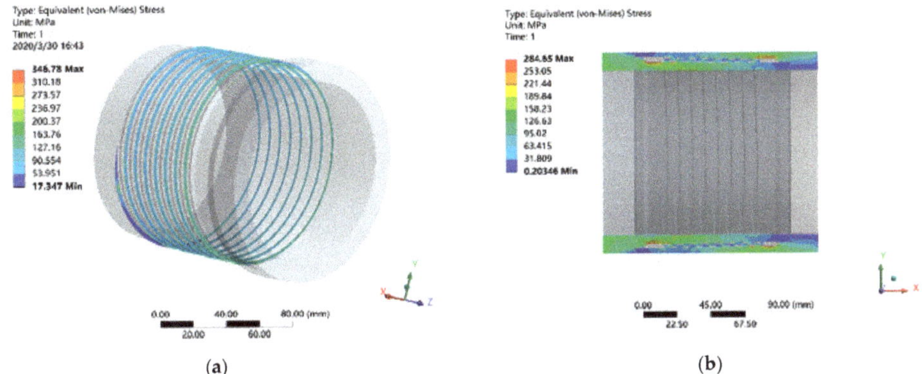

**Figure 12.** Stress distribution on the thread contact surface under a tensile load of 100 kN: (**a**) contact surface stress and (**b**) cross-sectional stress distribution.

Many scholars have investigated fractured joints at drilling sites and found that joint fractures caused by thread fracture under the combined action of static and alternating loads are the main causes of downhole accidents [28,29]. Further research found that thread fractures generally occurred at the root of the first turn of the pin joint or the root of the last turn of the box joint, indicating that those two parts are prone to stress concentrations [30–32]. These results are in good agreement with the above simulation results.

### 3.5. Effect of the Connection Status

The connection state of the drill pipe joint is an important factor affecting the thread stress distribution. Because the thread has a certain taper, it can be tightened to a greater extent, which will deform the threads and achieve an interference fit, which cannot be achieved by other threads. After the joints are connected, it is necessary to continue tightening at a certain angle to produce an interference fit, achieve sealing and prevent tripping. In this study, the magnitude of the interference of the joint tightened by one turn is 0.099 mm. After trial and error, it is finally determined that the maximum tightening angle is eight degrees, and the pretightening angle actually applied in the simulation is one to eight degrees.

Figure 13 shows the stress distribution in the joint when the pretightening angle is five degrees. Unlike the tensile load, the pretightening angle primarily affects the stress distribution in the joint shoulder. Figure 14 shows a line graph of the average stress in each thread for different pretightening angles. The results show that when the pretightening angle is small, the stress in the thread is relatively uniform. As the tightening angle increases, the thread stress becomes more concentrated at both ends, whereas the thread stress remains lower in the central portion. The results in Figures 11 and 14 are due to stress concentrations caused by the relief notch near the inner-shoulder face. There are relief notches near the inner-shoulder face of the pin and box joints. Therefore, the equivalent stress in the thread presents the following distribution characteristics: high stress at the two ends and low stress in the middle. This shows that the thread near the inner face is the weak point of the joint, which is consistent with previous research results.

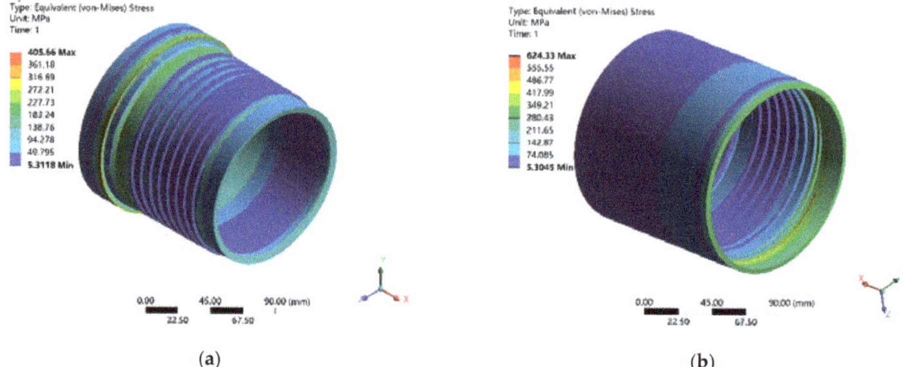

(a)                      (b)

**Figure 13.** Thread stress nephograms with a pretightening angle of five degrees: (**a**) pin joint and (**b**) box joint.

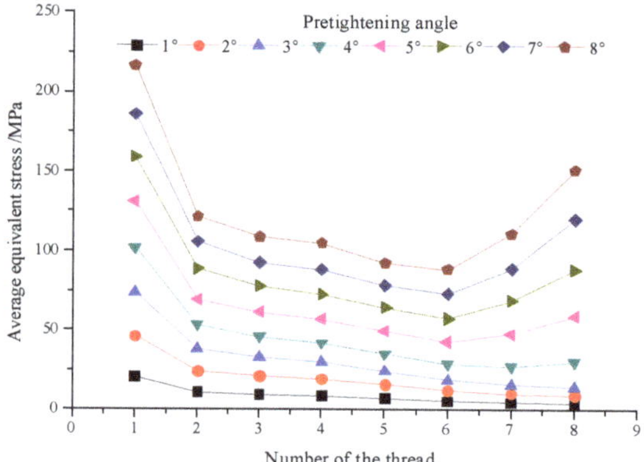

**Figure 14.** Relationship between the average equivalent thread stress and the pretightening angle.

Under different pretightening angles, the maximum and minimum axial displacements of the threaded surface, the outer-shoulder surface and the inner-shoulder surface of the box joint are obtained, the results of which are shown in Figure 15. Through an analysis, the maximum displacement of the inner shoulder and the outer shoulder is their interference. The interference of the thread surface is in the opposite direction of the axial displacement, so the absolute value of the minimum axial displacement is the interference.

Figure 15 shows that when the pretightening angle is less than 3.6 degrees, the interference of the outer-shoulder surface is higher than the interference of the thread surface. As the pretightening angle increases, the interference of the thread surface gradually exceeds the interference of the outer-shoulder surface. From these results, it can be inferred that the thread surface is more likely to deform than the shoulder surfaces at both ends. This difference in deformability produces a thread stress that exhibits the distribution characteristics described above.

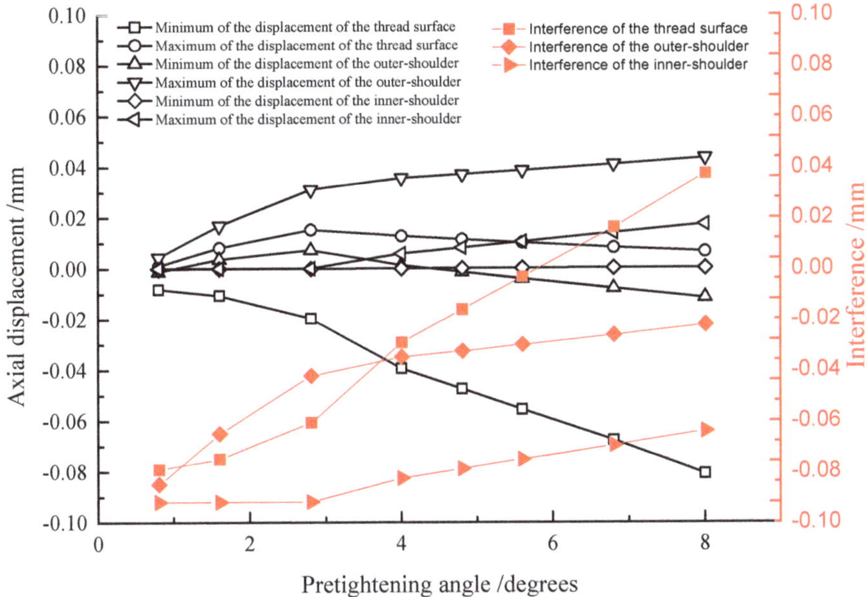



**Figure 15.** Relationship between the displacement and interference of each surface with respect to the pretightening angle.

## *3.6. Effect of the Structural Parameters*

### 3.6.1. Effect of the Structural Parameters

The structural parameters of the tapered thread include the taper, pitch, lead, helix angle, thread angle, and thread height. The strength of the thread is mainly affected by the taper, thread angle and thread height. The pitch and lead mainly affect the self-locking effect of the thread. Therefore, the effects of three parameters were analyzed: taper, thread height and thread angle. To further explore the degrees of influence of the three parameters on the equivalent stress, an orthogonal analysis was performed for each parameter with the equivalent stress concentration factor as the evaluation standard. The stress concentration factor is defined as

$$K = \frac{M_s}{A_s} \tag{1}$$

where $K$ is the stress concentration factor; $M_s$ is the maximum equivalent stress, MPa; and $A_s$ is the average equivalent stress, MPa.

The maximum equivalent stress and equivalent stress concentration factor under different structural parameters are shown in Table 1. The data in Table 1 were obtained by extracting data from the stress path.

The data in Table 1 show that the maximum stress in the thread is the smallest when the taper is 1:32, and the stress concentration factor is smaller. Therefore, the optimization scheme (i.e., the settings that produce the lowest stress concentration factor) uses a taper of 1:32, a thread height of 2.44 mm, and a thread angle of 15 degrees.

**Table 1.** Stress states under different structural parameters.

| Serial Number | Taper | Thread Height (mm) | Thread Angle (degrees) | Maximum Stress (MPa) | K |
|---|---|---|---|---|---|
| 1 | 1:4 | 2.24 | 30 | 454.74 | 3.68 |
| 2 | 1:6 | 2.24 | 30 | 447.41 | 3.52 |
| 3 | 1:16 | 2.24 | 30 | 449.13 | 3.46 |
| 4 | 1:32 | 2.24 | 30 | 445.98 | 3.42 |
| 5 | 1:32 | 2.04 | 30 | 523.64 | 3.69 |
| 6 | 1:32 | 2.24 | 30 | 445.98 | 3.42 |
| 7 | 1:32 | 2.44 | 30 | 396.21 | 3.39 |
| 8 | 1:32 | 2.24 | 15 | 419.55 | 2.89 |
| 9 | 1:32 | 2.24 | 30 | 445.98 | 3.42 |
| 10 | 1:32 | 2.24 | 45 | 485.59 | 3.59 |

### 3.6.2. Optimization Scheme of the Joint

The simulation results show that the structural parameters of the thread have a great influence on the stress distribution. Therefore, these three parameters—the taper, thread height, and thread angle—are optimized to produce a more uniform stress distribution.

- **Optimum scheme 1—Compound-tapered thread:** The simulations show that the stress during the first turn of the thread is the highest in any case. Therefore, reducing the stress in the first turn can make the overall stress distribution more even. The smaller the taper is, the smaller the stress, so the thread near the inner shoulder is designed to create a smaller taper. Cut the top surface of the thread near the inner shoulder to the same height so that the taper is zero and the rear thread taper is unchanged, as shown in Figure 16a. This turns the thread into a compound taper of 0 + 1:32. This turns the thread into a compound taper. The simulation results show that the thread stress concentration factor is 2.97, which is 15% lower than that before optimization.
- **Optimum scheme 2—Increased thread height:** The simulation results show that the stress concentration factor of the thread decreases as the thread height increases. By increasing the thread height to 2.44 mm while keeping the other parameters constant, as shown in Figure 16b, the thread stress concentration factor is 3.39, which is 1% lower than that before the optimization.
- **Optimum scheme 3—Reduced thread angle:** According to the simulation results, the stress concentration factor of the thread decreases as the thread angle decreases. By decreasing the thread angle from 30 degrees to 15 degrees while keeping the other parameters constant, as shown in Figure 16c, the thread stress concentration factor is 2.89, which is 18% lower than that before optimization.

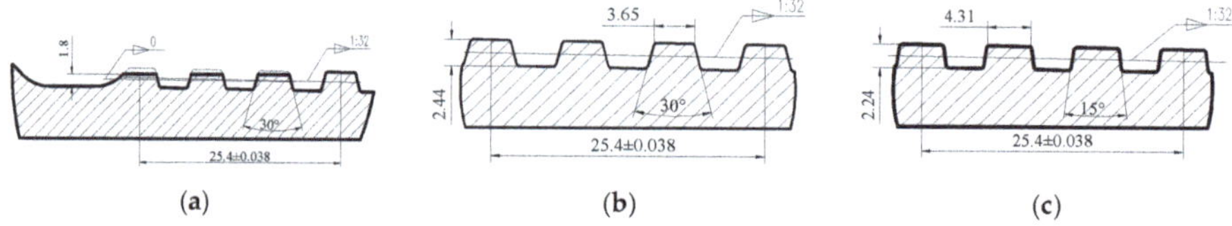


**Figure 16.** Three optimized thread structures: (**a**) optimized thread taper, (**b**) optimized thread height and (**c**) optimized thread angle.

## 4. Experiment and Discussion

### 4.1. Experimental Platform

Owing to the complicated structure of the thread, the stress distribution in the engaging surface of the thread cannot be measured by a strain gauge. However, the thread will deform after being loaded.

When the deformation reaches a certain level, the sealing of the thread will fail. If a pressurized gas is filled in the cavity of the joint, the gas will leak through the spiral passage of the thread. By measuring the pressure in the chamber, the small thread deformation can be converted into a change in the pressure reading. Accordingly, the sealing performance of the thread can be evaluated by the change in the pressure reading. Therefore, the relationship between the axial load and the seal of the thread can be obtained.

Figure 17 shows a schematic diagram of the experimental platform used to perform the tensile tests. The tests are performed on a WAW-2000 DL electrohydraulic servo-controlled universal testing machine, which can provide a maximum tensile force of 2000 kN. The air compressor (model: W-1/8) can provide a maximum air pressure of 1.5 MPa. To ensure the safety of the experiment, the pressure of the gas in the joint cavity during the tensile test is 0.6 MPa. Thus, the air compressor can meet the requirements of the experiment. The tensile load and crosshead displacement of the universal testing machine are collected and processed by the computer. The test piece subjected to the tensile test is shown in Figure 18. The test piece consists of two parts: a pin joint and a box joint. When the two parts are connected, a closed cavity is formed inside. An air injection port is provided in the wall of the box joint, through which a certain air pressure can be injected into the cavity. A clamping surface, which is designed to be gripped by the universal testing machine, is provided at both ends of the joint. In this study, four different threaded joints were tested: *optimization scheme 1*, *optimization scheme 2*, *optimization scheme 3* and the *original design*. For better sealing, grease is evenly applied to the threads before joining. The prepared test samples are shown in Figure 19. The main parameters of each group of threads are shown in Table 2.

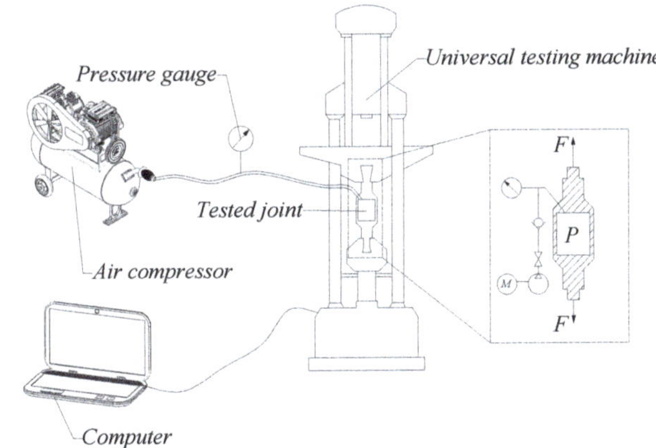


**Figure 17.** Tensile test platform.

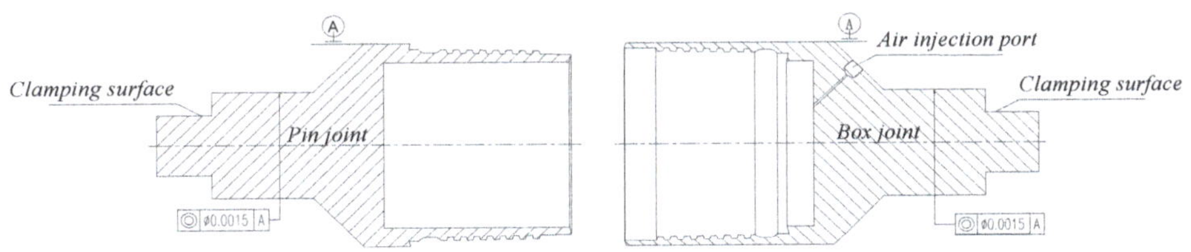


**Figure 18.** Structure of the joint sample.

**Figure 19.** Test samples.

**Table 2.** Structural parameters of the test samples.

| Group | Taper | Thread Height (mm) | Thread Angle (degrees) |
|---|---|---|---|
| Sample 1 | 1:32 | 2.24 | 30 |
| Sample 2 | 0 + 1:32 | 2.24 | 30 |
| Sample 3 | 1:32 | 2.24 | 15 |
| Sample 4 | 1:32 | 2.44 | 30 |

*4.2. Experimental Process*

Owing to the dry surface of the joint threads, a small amount of threading dope needs to be applied for lubrication before connection. In order to minimize the influence of the thread dope on the experiment, the thread dope was evenly applied and then wiped clean with a dry cloth.

It is important to check the sealing and leak points of the joint before performing the tensile test. The check method is to immerse the connected, gas-filled joint in a water tank and observe whether bubbles emerge. If a leak occurs, this sample will be discarded and replaced with a new sample and test the new joint again. The tensile test can be performed after the sample is successfully prepared. The experimental process is shown in Figure 20. The test procedure is as follows:

(1) Adjust the two grips of the universal testing machine to the appropriate distance and install the sample.

(2) Connect the air injection port on the joint to the outlet of the air compressor and fill the cavity with compressed air. After the pressure gauge reading stabilizes at 0.6 MPa, close the valve between the air pressure gauge and the air compressor.

(3) Allow the sample to stand for 10 min while continuously observing the pressure readings. If there is no significant change in the pressure gauge reading, indicating that no leak has occurred, and then proceed to the next step. However, if the pressure is significantly reduced, indicating a leak, replace the joint.

(4) Set the tensile force of the universal testing machine to 500 kN, the crosshead displacement rate to 5 mm/min, and the saturation loading time to 5 min.

(5) Start the universal testing machine. As the tensile force increases from 0 to 500 kN, the pressure gauge value is read every 5 s. During the loading phase, a 0.02 MPa change in the pressure gauge reading indicates the formation of the initial leakage. Therefore, the tensile force that corresponds to this stress reduction is recorded as the initial leakage tensile force. During the load saturation phase, the gauge values are read every 30 s.

(6) After reaching the saturated loading time, the universal testing machine is unloaded, after which the grips are opened and the sample is replaced. Repeat steps 1 through 5 for the next set of experiments.

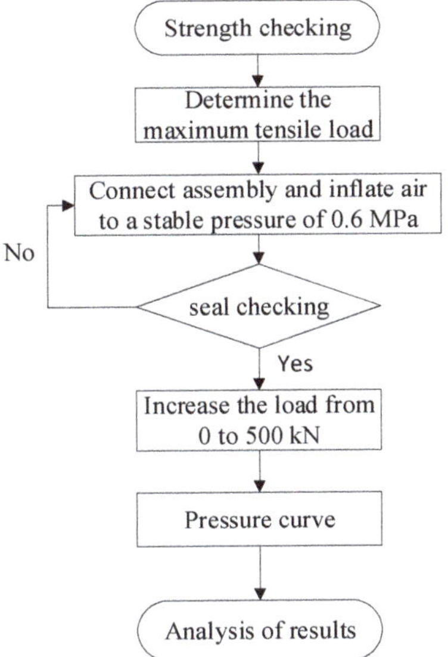

**Figure 20.** Tensile test flow.

## 4.3. Experimental Results and Discussion

The seal of the joint is achieved by an interference fit between the inner and outer shoulders. When the joint thread is subjected to a tensile load, if the mating surface is separated, gas leakage will occur. The greater the gas leakage, the greater the deformation of the thread.

Figure 21 shows the relationship between the change in the gas pressure over time during loading. It can be seen from the figure that in the initial stage, only the pressure of sample 2 did not decrease significantly, while the other samples showed a significant decrease. Comparing the entire loading phase, it can be seen that the curve of sample 2 is the most gradual, while the curves of sample 1 and sample 4 are steep. This shows that the thread with the composite taper structure has the best sealing performance, which indirectly reflects that the composite taper thread has a good ability to resist the axial tensile load.

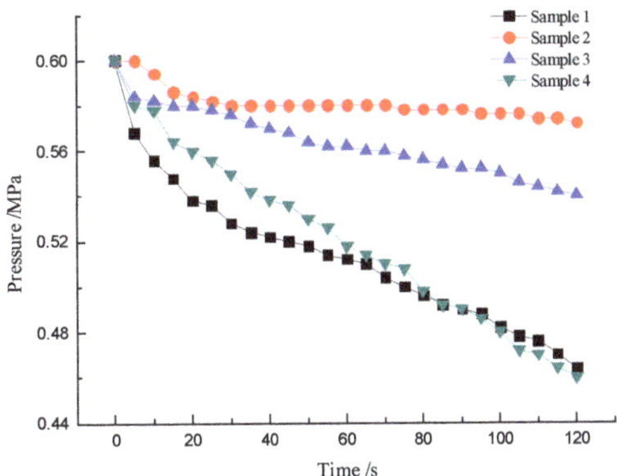


**Figure 21.** The relationship between the air pressure and time during axial stretching.

## 5. Conclusions

Combining the results of the simulation analysis with the experimental data, the following conclusions can be drawn from this study:

(1) Under a tensile load, stress concentrations occur at both ends of the thread, whereas the stress is low in the middle of the thread. As the tensile load increases, the stress at both ends increases faster than the stress in the middle, and as the number of turns increases, the stress concentration at both ends decreases.

(2) When the threaded joint is subjected to a tensile load, the stress concentration factor increases as the thread angle or taper increases, whereas the stress concentration factor decreases as the thread height increases; however, in the latter case, the overall decrease is small.

(3) Through simulations and experiments, four structures of the thread are optimized. The results show that the best performance is provided by compound taper thread, which has the following structural parameters: a thread height of 2.24 mm, a thread angle of 30 degrees, and a 0 + 1:32 compound taper.

**Author Contributions:** Conceptualization, Y.W. and C.Q.; methodology, Y.W.; software, C.Q.; validation, Q.Z.; formal analysis, L.K.; investigation, Y.W.; resources, Y.W.; data curation, C.Q.; writing—original draft preparation, C.Q.; writing—review and editing, L.K.; visualization, L.K.; supervision, Y.W.; project administration, J.G.; funding acquisition, Y.W. All authors have read and agreed to the published version of the manuscript.

**Funding:** This work was supported by the National Key R&D Program of China (No. 2018YFC0603404) and the National Natural Science Foundation of China (Nos. 41672366 and 41872180).

**Acknowledgments:** The authors would also like to thank former researchers for their excellent work. Their results were fundamental for the presented academic study.

**Conflicts of Interest:** The authors declare that there are no conflicts of interest regarding the publication of this paper.

## References

1. Santus, C.; Bertini, L.; Burchianti, A.; Inoue, T.; Sakurai, N. Fatigue resonant tests on drill collar rotary shouldered connections and critical thread root identification. *Eng. Fail. Anal.* **2018**, *89*, 138–149. [CrossRef]
2. Sajad, M.Z.; Sayed, A.H.T.; Hassan, S. Failure analysis of drill pipe: A review. *Eng. Fail. Anal.* **2016**, *59*, 605–623.
3. Shugen, X.; Chong, W.; Shengkun, W.; Lan, Z.; Xinchun, L.; Honghai, Z. Experimental study of mechanical properties and residual stresses of expandable tubulars with a thread joint. *Thin Walled Struct.* **2017**, *115*, 247–254.
4. Shuai, L.; Sujun, W. Effect of stress distribution on the tool joint failure of internal and external upset drill pipes. *Mater. Des.* **2013**, *52*, 308–314.
5. Yu, W.; Bairu, X.; Zhiqiao, W.; Chong, C. Model of a new joint thread for a drilling tool and its stress analysis used in a slim borehole. *Mech. Sci.* **2016**, *7*, 189–200.
6. Tafreshi, A.; Dover, W.D. Stress analysis of drill string threaded connections using finite element method. *Int. J. Fatigue* **1993**, *15*, 429–438. [CrossRef]
7. Baryshnikov, A.; Baragetti, S. Rotary shouldered thread connections: Working limits under combined static loading. *J. Mech. Des.* **2001**, *123*, 456–463.
8. Shahani, A.R.; Sharifi, S.M. Contact stress analysis and calculation of stress concentration factors at the tool joint of a drill pipe. *Mater. Des.* **2009**, *30*, 3615–3621. [CrossRef]
9. Liangliang, D.; Xiaohua, Z.; Desheng, Y. Study on mechanical behaviors of double shoulder drill pipe joint thread. *Petroleum* **2018**, *5*, 102–112.
10. Yosuke, O.; Masaaki, S.; Yoshinori, A.; Taizo, M.; Ryosuke, K.; Daisuke, T.; Masanobu, K. Fretting fatigue on thread root of premium threaded connections. *Tribol. Int.* **2017**, *108*, 111–120.
11. Sorg, A.; Utzinger, J.; Seufert, B.; Oechsner, M. Fatigue life estimation of screws under multiaxial loading using a local approach. *Int. J. Fatigue* **2017**, *104*, 43–51. [CrossRef]
12. Tafreshi, A. SIF evaluation and stress analysis of drillstring threaded joints. *Int. J. Press. Vessel. Pip.* **1999**, *76*, 91–103. [CrossRef]

13. Bahai, H. A parametric model for axial and bending stress concentration factors in API drillstring threaded connectors. *Int. J. Press. Vessel. Pip.* **2001**, *78*, 495–505. [CrossRef]
14. Baets, J.D.; Waele, P.D. Nonlinear contact analysis of different API line pipe coupling modifications. *J. Press. Vessel. Technol.* **2010**, *132*, 1–7.
15. Feng, Q. Study of relationship between wireline drill pipe thread taper and strength and its stresses. *Explor. Eng. Rock Soil Drill. Tunn.* **1995**, *6*, 24–26.
16. Su, J.; Yin, K.; Guo, T. Optimization of the joint-thread of diamond wire-line coring drill pipe. *J. Jilin Univ. Earth Sci. Educ.* **2005**, *35*, 677–680.
17. Smolnicki, T.; Rusiński, E.; Karliński, J. FEM modelling of fatigue loaded bolted flange joints. *J. Achiev. Mater. Manuf. Eng.* **2007**, *22*, 69–72.
18. Liang, J.; Guo, B.; Sun, J.; Zhang, Y. Finite element method of resistance to pull-off for deep hole wire-line drill rod. *Coal Geol. Explor.* **2013**, *41*, 90–93.
19. Chui, C.; Mei, D. Analysis on the stress on wire-line core drilling pipe and connection joint based on ANSYS. *Explor. Eng. Rock Soil Drill. Tunn.* **2014**, *41*, 61–63.
20. Yin, F.; Zhang, Y.; Xiong, J.; Xiong, L. Simulation analysis on structural mechanism of the thread for wire-line drill pipe. *Explor. Eng. Rock Soil Drill. Tunn.* **2014**, *41*, 66–69.
21. Gao, S.; Sun, J.; Cai, J.; Liu, D. Calculation analysis on negative angle thread torque of wire-line coring drill pipe and test research. *Explor. Eng. Rock Soil Drill. Tunn.* **2016**, *43*, 45–49.
22. Gao, J.; Ma, Y.; Wang, D.; Ji, S. Fatigue analysis on wire-line coring drill pipe joints in tonghua well-1 based on ansys workbench. *Explor. Eng. Rock Soil Drill. Tunn.* **2017**, *44*, 70–78.
23. API recommended practise API spec7-1. In *Recommended Practice for Drill Stem Design and Operating Limits*, 15th ed.; American Petroleum Institute: Washington, DC, USA, 1995.
24. Feng, C.; Di, Q.F.; Li, N.; Wang, C.S.; Wang, W.C.; Wang, M.J. Determination of operating load limits for rotary shouldered connections with three-dimensional finite element analysis. *J. Pet. Sci. Eng.* **2015**, *133*, 622–632.
25. Yong, Z.; Gao, L.X.; Yuan, P.B. Force analysis and tightening optimization of gas sealing drill pipe joints. *Eng. Fail. Anal.* **2015**, *58*, 173–183. [CrossRef]
26. Brock, J.N.; Jellison, M.J. Development of a gas-tight, pressure-rated, third-generation rotary shouldered connection for 20,000-psi internal and 10,000-psi external service. In Proceedings of the IADC/SPE Drilling Conference, Orlando, FL, USA, 4–6 March 2008; p. 112547.
27. Baryshnikov, A.; Calderoni, A.; Ligrone, A.; Ferrara, P. A new approach to the analysis of drillstring fatigue behavior. *SPE Drill. Complet.* **1997**, *12*, 77–84. [CrossRef]
28. Wang, Y.; Gao, L.X.; Yuan, P.B. Cause analysis of longitudinal cracking of internal thread joints of double shoulder drill pipes. *Mach. Tool Hydraul.* **2015**, *43*, 183–186.
29. Peng, C.Y.; Lou, Y.S.; Cao, Y.P.; Sun, W.H.; Liu, X.F. Research on fatigue failure of drill thread connections. *Pet Drill. Technol.* **2006**, *6*, 20–22.
30. Macdonald, K.A.; Deans, W.F. Stress analysis of drill string threaded connection using the finite element method. *Eng. Fail. Anal.* **1995**, *2*, 1–30. [CrossRef]
31. Moradi, S.; Ranjbar, K. Experimental and computational failure analysis of drill strings. *Eng. Fail. Anal.* **2009**, *16*, 923–933. [CrossRef]
32. Rahman, M.K.; Hossain, M.M.; Rahman, S.S. Survival assessment of die-marked drill pipes: Integrated static and fatigue analysis. *Eng. Fail. Anal.* **1999**, *6*, 277–299. [CrossRef]

 *applied sciences*

*Article*

# Finite Element Method and Cut Bar Method-Based Comparison Under 150°, 175° and 310 °C for an Aluminium Bar

José Eli Eduardo Gonzalez Duran [1,†], Oscar J. González-Rodríguez [2,†], Marco Antonio Zamora-Antuñano [3,†], Juvenal Rodríguez-Reséndiz [4,*,†], Néstor Méndez-Lozano [3], Domingo José Gómez Meléndez [4,‡] and Raul García García [5]

[1] Instituto Tecnológico Superior del Sur de Guanajuato, Guanajuato 38980, Mexico; je.gonzalez@itsur.edu.mx
[2] Centro Nacional de Metrología (CENAM), Querétaro 76246, Mexico; ogonzale@cenam.mx
[3] Departamento de Ingenieria, Universidad del Valle de Mexico, Querétaro 76230, Mexico; murck22@gmail.com (M.A.Z.-A.); nestor.mendez@uvmnet.edu (N.M.-L.)
[4] Facultad de Ingenieria, Universidad Autónoma de Querétaro, Querétaro 76010, Mexico; juvenal@uaq.edu.mx
[5] Departamento de Ingenieria, Universidad Tecnológica de San Juan del Río, San Juan del Río 76800, Querétaro, Mexico; rgarciag@utsjr.edu.mx
[*] Correspondence: juvenal@uaq.edu.mx; Tel.: +52-442-192-12-00
[†] These authors contributed equally to this work.
[‡] This author have passed away.

Received: 5 December 2019; Accepted: 27 December 2019; Published: 31 December 2019

**Abstract:** Analyses were developed using a finite element method of the experimental measurement system for thermal conductivity of solid materials, used by the Centro Nacional de Metrología (CENAM), which operates under a condition of permanent heat flow. The CENAM implemented a thermal conductivity measurement system for solid materials limited in its operating intervals to measurements of maximum 300 °C for solid conductive materials. However, the development of new materials should be characterised and studied to know their thermophysical properties and ensure their applications to any temperature conditions. These task demand improvements in the measurement system, which are proposed in the present work. Improvements are sought to achieve high-temperature measurements in metallic materials and conductive solids, and this system may also cover not only metallic materials. Simulations were performed to compare the distribution of temperatures developed in the measurement system as well as the radial heat leaks, which affect the measurement parameters for an aluminium bar, and uses copper bars as reference material. The simulations were made for measurements of an aluminium bar at a temperature of 150 °C, in the plane and 3D, another at 175 °C and one more known maximum temperature reached by a sample of the aluminium bar with a new heater acquired at 310 °C.

**Keywords:** cut bar method; thermal conductivity; finite element method; steady-state; heat lakes

---

## 1. Introduction

Thermal conductivity is a physical property of materials that measures heat conduction capacity. In other words, thermal conductivity is also the ability of a substance to transfer the kinetic energy of its molecules to adjacent ones or to substances with which it is in contact. In the International System of Units, the thermal conductivity is measured in W/(m K) equivalent to J/(m s K). There are several methods to measure the thermal conductivity of materials: the most conventional method for measuring thermal conductivity consists of two concentric metal spheres, of very small thickness to minimise the heat capacity of the system. It has not been used with measurements greater than

300 °C [1]. The parametric study consists in obtaining the temperature distribution in the most insulating composite bar system for different operating conditions. As operating conditions, it refers to the temperature difference at the ends of the system, characteristics of the reference material, aspects of the insulating material and features of the sample materials [1,2]. The determination of the thermophysical properties of materials is essential in all processes where energy exchanges occur, in particular, heat. For the design, operation and maintenance of systems and equipment where the temperature is present, it is essential to know the value of these properties in particular of thermal conductivity. This property has an important effect on solid thermal conductive materials such as aluminium, iron, copper, its alloys, and new materials that are used to build equipment and machinery parts, such as automotive vehicle engines [1]. Thermal conductivity is also an issue related to the second law of thermodynamics or the law of entropy, which governs most of the phenomena that occur in the universe, by which it is estimated that any process that involves work increases the entropy of the universe (increases the disorder and chaotic movement of atoms and the temperature of existing molecules and grains). Thermal energy always flows spontaneously from highest to lowest concentration, or from hot to cold. This implies that heat transfer by conduction occurs from one body to another at a lower temperature or between areas of the same material but with a different temperature. Heat transmission involves an internal energy exchange, which combines potential energy and kinetic energy of electrons, atoms, and molecules: the higher thermal conductivity, the better the heat conduction. The inverse property is the thermal resistivity, which indicates that, at lower thermal conductivity, more heat insulation (more resistivity). Concerning potential energy, we can say that it is the mechanical energy associated with the location of a body in a field of forces or the presence of an area of effects within the body itself. The potential energy is the result that the system of forces that affects a given body is conservative, then, the total work on a particle is zero. The kinetic energy of a body, meanwhile, is what it has thanks to its movement. It is the work needed to achieve its acceleration from rest to a given speed. When the body reaches this energy throughout the acceleration, it maintains it unless it alters its speed. To return to the resting state, it is necessary to perform a dangerous job with the same magnitude. By heating matter, the average kinetic energy of its molecules increases, and this increases its level of agitation. At the molecular level, heat conduction occurs because the molecules interact with each other by exchanging kinetic energy without making global movements of matter. It should be mentioned that at the macroscopic level, it is possible to model this phenomenon by means of Fourier's law.

The CENAM implemented a system to measure thermal conductivity of solid conductive materials, and the design criteria were developed for the construction of the measurement system, which operates under the condition of heat flow in a permanent state. The system uses a reference material, which limits the accuracy of the method. An analysis of the system is carried out considering that there is axial and radial heat flow. In addition, the solid bar of material that can be evaluated, it has a hollow bar of insulating material. The problem to be solved is a bar composed of a reference conductor material at the longitudinal ends, and a test material depicted in the centre, the entire bar an element by an insulating material, it is considered that it is axial and radial flow and the physical dimensions of the problem are shown in Figures 1a, 2 and 3. Using the apparatus developed in the CENAM, two concentric cylinders are used, housing the material to be tested between them. Inside the smaller diameter cylinder is placed the heating resistance, which is covered with another cylinder to standardise the surface temperature. The temperature measurement is carried out on the outer and inner cylinders, using thermocouples for this. The method is used to measure thermal conductivity in materials solid conductors. To meet this need for measurement, the CENAM developed a system for measuring thermal conductivity in thermally conductive solid materials employing a secondary method. This work presents a comparison of certain experimental results using the cut bar and the finite element method (FEM), to obtain information that serves in the development of the new cut bar system, to extend its operating range up to 600 °C, under optimal operating conditions [2].

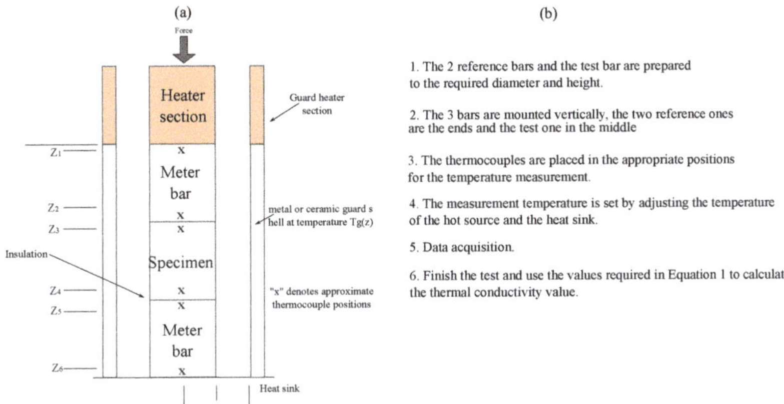



**Figure 1.** (**a**) Schematic of a Comparative-Guarded-Longitudinal Heat Flow System, indicating possible locations of temperature sensors (**b**) methodology for the experiment used in this work [1].

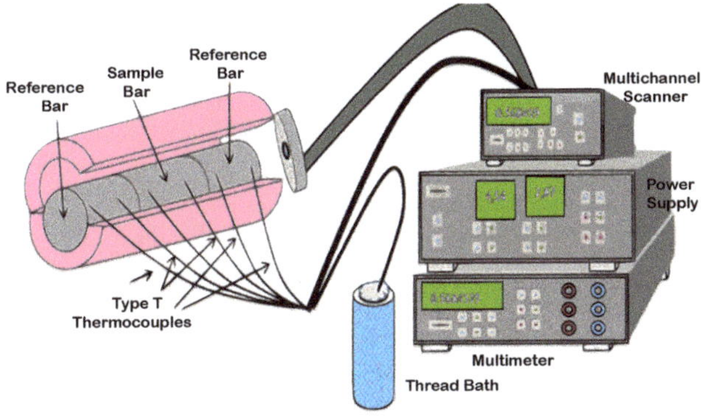



**Figure 2.** Cut bar method system [2].

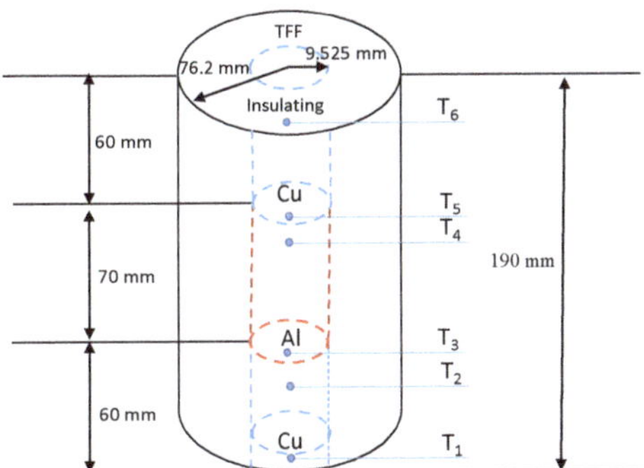



**Figure 3.** Diagram of the experimental model used.

## 2. Technique Background

The method determines the thermal conductivity of a sample using a reference material by a permanent state technique known as the concentric bar cut method [3]. The system consists of a bar with well-known properties, called reference bar, another bar with conductivity to be determined, called sample o test bar, and another reference bar. The composite metal bar is covered with an insulating material to prevent heat flow in the radial direction. At one end of the composite bar, a heat source is placed, and at the opposite end, there is a heat sink or cold source [4,5]. Then, employing temperature and length measurements, the conductivity of the sample material can be determined. Figure 1 shows a diagram of the composite bar system.

The arrangement diagram of the bars for the method used in this work is shown in Figure 1a. The reference bars are located at the ends and the test bar in the center of both. At one end of the bar array, a heat sink or cold source is located; at the other end, a heater that allows generating a temperature gradient, necessary for the determination of the thermal conductivity value. Marked with $x$ in Figure 1a is where the thermocouples indicated at a certain height in mm and designated according to the letter $z$ are located; the thermocouples type T were a fine wire of 0.6 mm diameter from OMEGA brand, calibrated by CENAM. $r_A$ indicates the radius of the bars used for the test and $r_B$ the radius of the insulation used in the test. A force is applied axially to improve the contact between the bars axially.

A brief description of the process performed to carry out the test is named in Figure 1b. Before starting the test the bars to be used need to meet the necessary diameter and height, as well as some flatness on the flat faces of the bars, after these perforations are made on the cylindrical face a few millimeters deep to house the thermocouples later. After finishing the bars, they are placed one above the other in the order, as appear in Figure 1a. Then the thermocouples are placed in each one of the sweepers made in the bars, then it is surrounded with the insulating material, and the guard is added; the axial force is applied to improve the contact. The next step is to adjust the operating temperature of the hot and cold source according to the measurement temperature at which the test is required to reach the operating temperature was used as a power supply, which supplies the necessary voltage to an electrical heater until it reaches temperature operation, which is registered by the thermocouple located in the hot source. For example, if a test temperature of 100 °C is required, the average temperature of the hot and cold source must be sought to be 100 °C. For example, the temperature of the hot source at 150 °C and the cold source at 50 °C, so the average temperature is 100 °C. In this way, several combinations can be generated. Once the operating temperatures have been adjusted, data acquisition begins, by a program in LabView developed by CENAM, where the signal of thermocouples are read by a multimeter and send to a PC to register its values. The values will be adequate when a the steady-state regime has been reached, it is known, because charts of temperature from thermocouples do not change, reach a constant temperature throw experiment in time. With the acquired data, Equation (1) is used to calculate the thermal conductivity value.

From the work in [2] it was found that the thermal conductivity of the sample is given by

$$\lambda_M = \frac{Z_4 - Z_3}{T_4 - T_3} \left[ \frac{\lambda_{R_1}}{2} \left( \frac{T_2 - T_1}{Z_2 - Z_1} \right) + \frac{\lambda_{R_2}}{2} \left( \frac{T_6 - T_5}{Z_6 - Z_5} \right) \right] \tag{1}$$

where $\lambda_M$ is the thermal conductivity of the sample. $\lambda_{R1}$, and $\lambda_{R2}$ are the thermal conductivity of reference materials 1 and 2. $T_i$ is the temperature in each of the $Z_i$ positions where the thermocouples are placed. Subscripts 1 and 2 refer to the first reference bar, 3 and 4 to the sample under measurement and 5 and 6 to the second reference bar.

If the distances between the thermocouples of each bar are equal and the reference material is the same for the two bars, so from Equation (1) which the thermal conductivity of reference materials leave the equation as a common term. Taking into account that the distances are also equal, $(Z_2 - Z_1)$ is the same that $(Z_6 - Z_5)$ then leave the parenthesis so with $(Z_4 - Z_3)$ obtain unity. Then, to simplify is rewritten $(T_4 - T_3)$ such as $\Delta T_2$, rewrite $(T_6 - T_5)$ such as $\Delta T_3$ and $(T_2 - T_1)$ such as $\Delta T_1$. Then, Equation (1) is reduced to

$$\lambda_M = \frac{\lambda_{R_2}}{2} \left( \frac{\Delta T_1 + \Delta T_3}{\Delta T_2} \right) \tag{2}$$

where $\lambda_{R_2}$ is the same the $\lambda_{R_1}$ because the reference material are equal. $\Delta T_1$ and $\Delta T_3$ are the difference among each reference bar and $\Delta T_2$ is the difference of temperature of the test bar. The cold source or heat sink is constituted by a 10 cm diameter copper plate that has a 10 mm diameter copper tube coil welded through which a fluid such as ethylene glycol flows from a bath of controlled temperature. One of the surfaces is in contact with one end of a reference bar and the other part in an insulated container [6]. The recirculation bath can maintain the temperature of the cold source between $-30\,^\circ\text{C}$ and $60\,^\circ\text{C}$.

The reference material bars are 99.999% high purity copper with a diameter of 19.1 mm and a length of 60 mm. The composite bar is surrounded by a 100 mm diameter polyvinyl chloride (PVC) pipe, the inside of which contains 50.8 mm thick fiberglass.

The measurement system has seven calibrated type T thermocouples. The electromotive force (EFM) of each thermocouple is measured with a digital high-accuracy multimeter of $8\frac{1}{2}$ digits model 3458A from Agilent Technologies and aided by an 8-channel scanner keithley 7001 both manufactured in USA. CENAM developed a computer program for the control, reading, and recording of data. The program to acquire data works with a graphic interface developed in LabView, where is registered tension measurement of each thermocouple and through of coefficient obtained from calibration and with the Newton–Raphson method is converted tension measurement to a temperature value. The EFM was measurement by scanner and multimeter connected to a PC. That value is introduced in a subprogram, where is converted to each temperature value from each thermocouple used in the experiment. With the distance between thermocouples of each bar and the thermal conductivity from reference bar is calculated the thermal conductivity of the bar under test. In the front panel of Labview developed by CENAM, are showed temperature of each thermocouple used, temperature of hot and cold source, constants from calibration of thermocouples used, graphs of $\Delta T_1$, $\Delta T_3$ and $\Delta T_2$ as well as the thermal conductivity value. Figure 2 shows a schematic of the measurement system [7].

## 3. Methodology

The development of the experiment was carried out using the following parameters. Two copper bars of 60 mm in length were used, and one aluminium bar of 70 mm length. The three bars have a diameter of 19.05 mm.

As an insulator to reduce radial heat leaks, glass fiber with thermal conductivity of 0.046 W/mK was used. The thermocouples were placed in such a way that there was a distance of 40 mm for the copper bars and 50 mm for the aluminium bar [8]. These were placed on the outside of the bars, on the surface, based on the work in [9], the standard test method for thermal conductivity of solids by means of the guarded-comparative-longitudinal heat flow technique. The configuration of the bars is depicted in Figure 3.

*Computational Model Setup*

It was made a model by aided computer design under dimensions illustrated in Figure 3, and due to its symmetry, a 2D model was done and another with azimuthal symmetry [10]. The temperature of the hot source (HST) was extracted from the data of experimental results; the same for the cold source temperature (CST) which were introduced as boundary conditions in the finite element model, the properties of the material, in the case copper, being a reference material, Equation (3) was used

$$\lambda_{Cu} = 416.3 - 0.05904T + 7.087 \times 10^7 / T^3 \tag{3}$$

Equation (3) was obtained from [9] because in the standard are published thermal conductivity of some materials considered such as meter bar reference materials for the cut bar method.

Mesh for the Models Used

ANSYS 18 software with the Mechanical APDL (Parametric Design Language) user interface and thermal module for the simulations of this work was chosen. For the flat model, it was used a PLANE 77 element of eight nodes, and it has one degree of freedom, temperature, at each node, and applies to a 2-D, steady-state or transient thermal analysis and a SOLID 90 element of 20 nodes with a single degree of freedom, temperature, at each node for the 3D model. The mesh for 3D and 2D models are shown in Figure 4a,b, respectively [11,12].

(a)      (b)

**Figure 4.** Final mesh (**a**) for the 3D model and (**b**) the 2D model used in this work.

The mesh generated for the 3D model is depicted in Figure 4a, where azimuthal symmetry was implemented. For the 2D model, the mesh that was generated is shown in Figure 4b, where symmetry was used on the $y$ axis. For loads, were used temperature values from the experimental results and heat flux 0 on the boundaries where the insulating material is presented. To reduce resources and computational time.

## 4. Results

The results for different temperatures of the aluminium sample selected in the experiment are described below, which were at 50 °C and one at 175 °C and the behaviour at an HST of 600 °C. It is because the heater that was acquired reaches a maximum operating temperature value of 600 °C. Only the points where the thermocouples are located were compared, both in copper and aluminium bars. The tables include material properties and initial conditions applied to the model of the finite element method. Also, this work shows a comparison of the results obtained in the simulation using ANSYS and the data acquired experimentally [13,14].

### 4.1. 2d and 3d Analysis at a Temperature of 150 °C

In Table 1 appears the boundary conditions applied for the 150 °C test. Also, in that table the material properties were introduced in the simulation the values, like 214.4 W/mK for aluminium,

386 W/mK for copper and 0.044 W/mK for fibreglass. The values for temperature loads used in the model in ANSYS were 279.5 °C for the hot source and 20 °C for the cold source.

**Table 1.** Boundary conditions applied for $T_{BAR} = 150\ °C$.

| Aluminum (Al) | → | 214.4 W/mK | $T_{FC} \rightarrow 279.5\ °C$ |
|---|---|---|---|
| Copper (Cu) | → | 386 W/mK | $T_{FF} \rightarrow 20\ °C$ |
| Fiberglass | → | 0.044 W/mK | |

Figure 5 shows the results obtained for the 150 °C test in the 2D model. In the distribution of temperature, it can be seen where the hot source, in red colour, and the cold source, in blue colour, are, in Figure 5a, the path begins with the origin and 289.6 °C on the coordinate axis, which corresponds to the point located in the hot source. The graphic represents the vertical line, where the model presents symmetry [15]. The graph in Figure 5b points out three slope changes, indicating the two types of materials since their thermal conductivities are different. Then the first slope from left to right is equal as the third slope since they are the same material and therefore have the same thermal conductivity value.

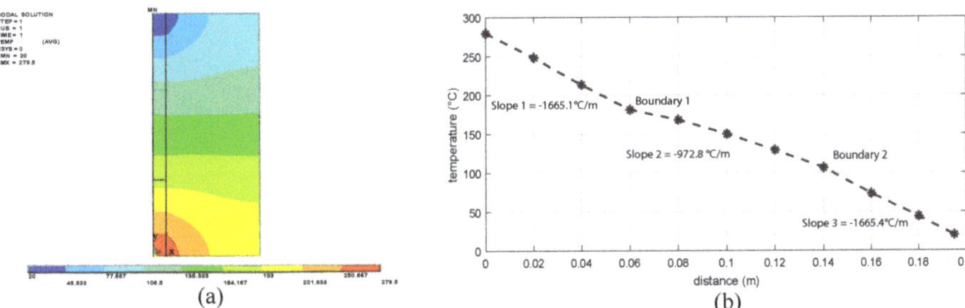

**Figure 5.** Results for 2D analysis at a test temperature of 150 °C. (**a**) temperature distribution in °C. (**b**) graphic of temperatures of the symmetry line.

The results of using a 3D model with azimuthal symmetry are illustrated in Figure 6; the differences are notorious compared with the temperature distribution image concerning the 2D model. In the graph, the changes are smoother; however, changes in the slopes are more defined. Also, we can see the point where contact exists in each metallic bar [16].

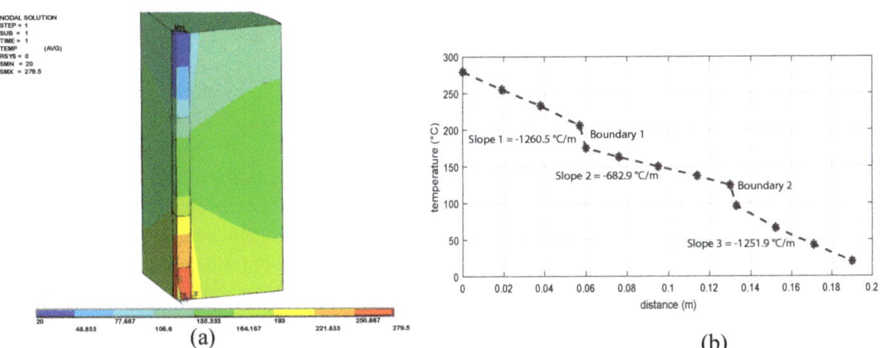

**Figure 6.** Results for 3D analysis at a test temperature of 150 °C. (**a**) Temperature distribution in °C. (**b**) Graphic of temperatures of the symmetry line.

In Figure 6b it can be observing a better definition of slope change between each bar. Even with that inflection point in the graphs, it can be obtained the temperature reached the junction of each bar. The transition between the boundary of each material is due to in analysis 2D, the interface is a line, but in 3D analysis, there are two surfaces, so in this case, ANSYS take in account radial heat transfer through surfaces.

However, Figure 7 shows the graph where the results are compared between 2D, 3D models, and the points that represent the thermocouples where experimentally are located in the metallic bars. The deviations are more significant near the borders, where the cold and hot sources are located. From the 2D and 3D simulations performed, the temperature values were extracted at the points where the thermocouples are experimentally located. The differences between these values were calculated to obtain the maximum and the minimum deviation between the results obtained from the simulation and the experiment. Therefore, the most substantial variance was around 20 °C, and the lowest was 1.6 °C.

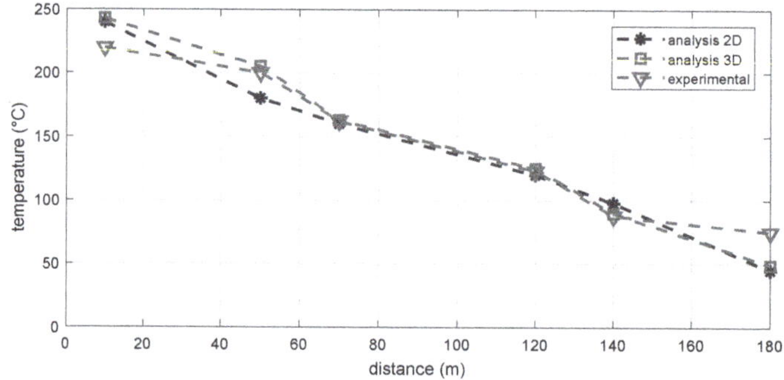


**Figure 7.** Comparison of results for 2D, 3D and experimental analysis at a test temperature of 150 °C in thermocouple positions.

### 4.2. 2d and 3d Analysis at a Temperature of 175 °C

The boundary conditions applied for the 175 °C test are shown in Table 2.

**Table 2.** Boundary conditions applied for $T_{BAR}$ = 175 °C.

| | | | |
|---|---|---|---|
| Aluminum (Al) | → | 214.4 W/mK | $T_{FC}$ → 339 °C |
| Copper (Cu) | → | 386 W/mK | $T_{FF}$ → 20 °C |
| Fiberglass | → | 0.044 W/mK | |

Figure 8 indicates the results obtained for the 175 °C test in the 2D model. In the temperature distribution, it is possible to observe where the hot source, in red, and the cold source, in blue, whose gradients are very similar to the test temperature at 150 °C. Figure 8b shows a graph where begins with the origin and 339 °C on the ordinates axis, which corresponds to the point where the hot source is, which represents a maximum temperature reached by the hot source. The graph describes the vertical line where the model presents symmetry. The graph represents the nodes that are on the vertical line of symmetry of the model. As in the previous case, it points out three slopes because there are two section changes, in this case, the slopes are of higher value because the operating temperature for the heater was higher than in the previous case.

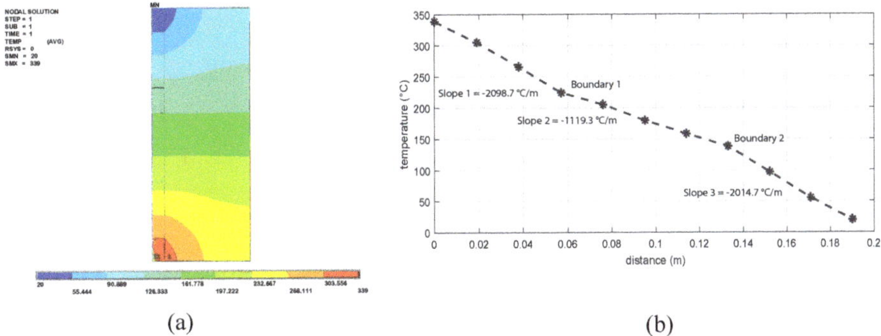

(a)                                                             (b)

**Figure 8.** Results for 2D analysis at a test temperature of 175 °C. (**a**) temperature distribution in °C. (**b**) graphic of temperatures of the symmetry line.

Figure 9 indicates the results of using a 3D model with azimuthal symmetry. The differences are notoriously comparing the temperature distribution image concerning the 2D model. In the graph, the changes are smoother, showing the variation of the section between the copper reference bar and the aluminium test. The changes are due to the temperature gradient and the different values of the thermal conductivity of each bar [17,18]. In this case and the previous one, the deviations concerning the experimental results are more significant near the cold source, and, in both comparisons for the case of the hot source, the finite element method predicts and for the cold source sub predicts the actual values according to those obtained in the experiment, which means that heat leaks are present.

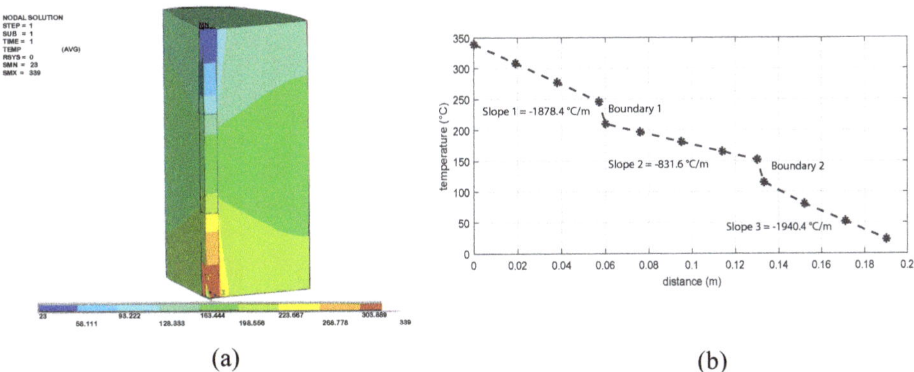

(a)                                                             (b)

**Figure 9.** Results for 3D analysis at a test temperature of 175 °C. (**a**) Temperature distribution in °C. (**b**) Graphic of temperatures of the symmetry line.

However, Figure 10 shows a result comparison between 2D, 3D models, and the experiment. Where the deviations are more significant near the borders, where the cold and hot source are located. Because in that zones exists the most more significant gradients with the surroundings, because temperature laboratory is 22 °C. The most significant deviation for this case was around 37 °C and less than 7 °C.

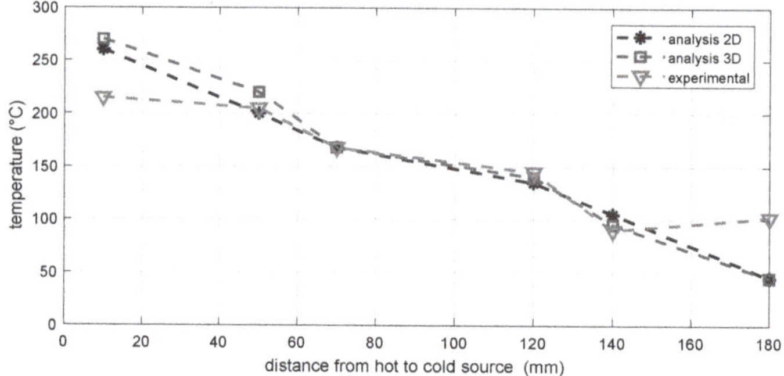

**Figure 10.** Comparison of results for 2D, 3D and experimental analysis at a test temperature of 175 °C in thermocouple positions.

*4.3. 2D and 3D Analysis at a Temperature of 310 °C*

Because a new heater was purchased, which operates at a maximum temperature of 600 °C, it is essential to know the temperature that the aluminium bar reaches, and by consequently evaluate if the conditions of the equipment are adequate to this new working temperature [19]. In Table 3 appears the boundary conditions applied for the 310 °C test, where it is observed that the maximum temperature reached by the new heater at 600 °C.

**Table 3.** Boundary conditions applied for $T_{BAR}$ = 310 °C.

| Aluminum (Al) | → | 200 W/mK | $T_{FC} \rightarrow$ 600 °C |
|---|---|---|---|
| Copper (Cu) | → | 365.74 W/mK | $T_{FF} \rightarrow$ 20 °C |
| Fiberglass | → | 0.044 W/mK | |

Figure 11 indicates the results obtained for the 310 °C test in the 2D model. In the distribution of temperature, it is possible to observe the hot source, in red, and the cold source, in blue, whose gradients are equal to the test temperature at 150 °C. However, the values of temperature in each zone are higher than the last case. In Figure 11b, the graph begins with the origin and 600 °C on the ordinate axis, which corresponds to the point located in the hot source as we can see slopes are greater than the last case because the temperature is higher. The graph represents the vertical line where the model presents symmetry.

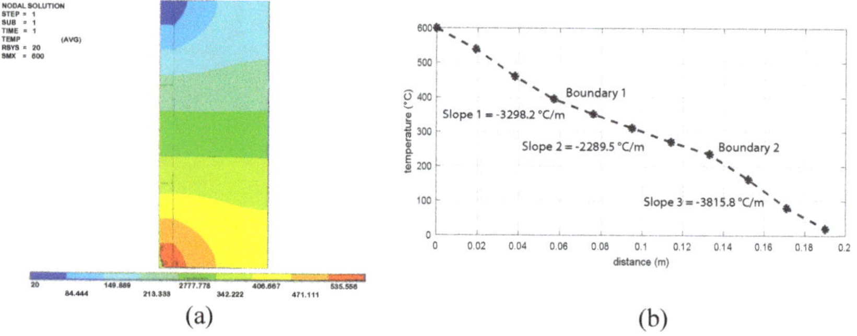


**Figure 11.** Results for 2D analysis at a test temperature of 310 °C. (**a**) Temperature distribution in °C. (**b**) Graphic of temperatures of the symmetry line.

Figure 12 indicates the results from the 3D model with azimuthal symmetry. The differences are notorious by comparing the temperature distribution image concerning the 2D model. The graph points out the temperature reached by the sample aluminium bar is 310 °C. Ideally, the temperature of the sample bar reached with the new heater.

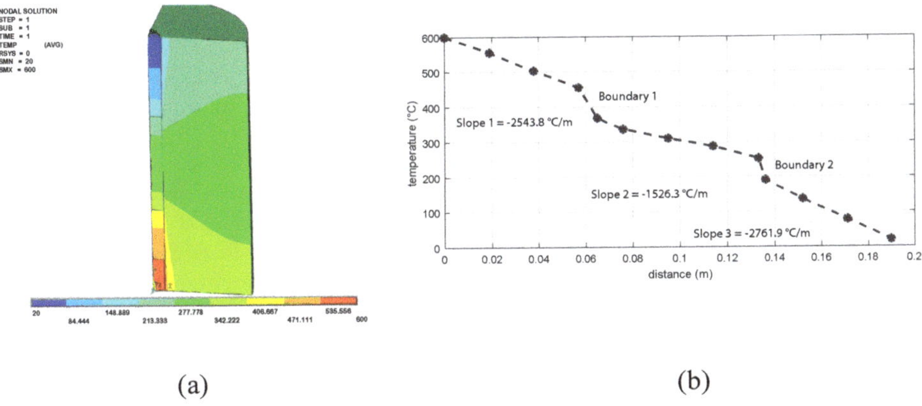


(a)        (b)

**Figure 12.** Results for 3D analysis at a test temperature of 310 °C. (**a**) Temperature distribution in °C. (**b**) Graphic of temperatures of the symmetry line.

However, the Figure 13 shows the comparison between 2D and 3D models. It was found the maximum difference is 40 °C, and the minimum is 0.6 °C in the two analyses. In this case, there is no experimental evaluation, because with the information obtained in this work, it is possible to evaluate if it is necessary to make modifications to the existing bar system to implement the new heater, due to the temperature reached in the system that includes the bars and the insulator [20,21].

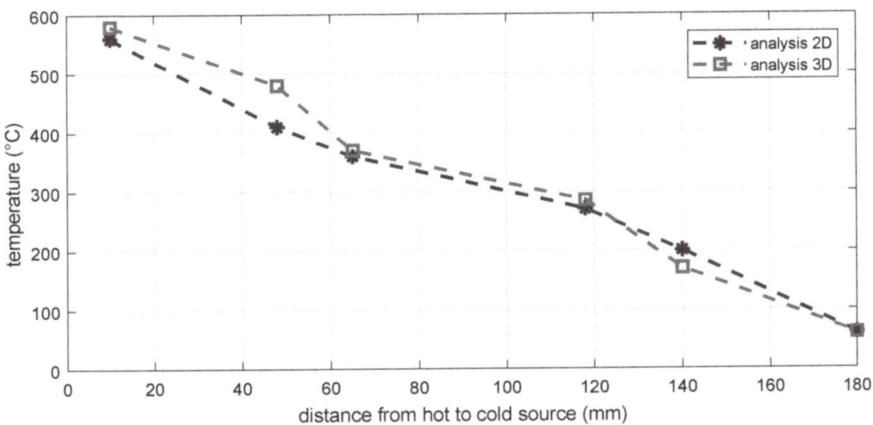


**Figure 13.** Comparison of results for 2D and 3D at a test temperature of 310 °C in thermocouple positions.

## 5. Discussion and Conclusions

### 5.1. Discussion

The vast majority of technological advances achieved in modern society have been supported by the discovery and development of engineering materials and manufacturing processes used to obtain them. An adequate selection of materials and methods guarantee the designers of mechanical parts their correct functioning, i.e., the performance of the designed components [22]. By means of FEM, it was intended to develop a model that could verify an adequate integration of all the necessary inputs in the heat transfer analysis. An analysis was carried out where the whole model consisted of different materials, and after applying the calculated thermal limit conditions, values were found that produces a close approximation to the experimental results. Some observed discrepancies can be attributed to inaccuracy in thermocouple locations [23–25]. Most methods are based on the availability of a wide range of materials, which must be analysed and refined, either with the help of recommendations, i.e., traditional methods, material maps with graphic method or information found in bibliographic sources or in software by virtual databases, type of material, which should result in the most appropriate for the intended purpose.

In this work, the levels of correspondence of the experimental results concerning those obtained by numerical simulation are outstanding. According to work in [26–28], the uncertainty that was reached in the measurement of thermal conductivity is less than 5%, of which more than 90% of the contribution to the uncertainty corresponds to the reference material; therefore, the temperature measurement does not contribute significantly with the final value of the uncertainty of the thermal conductivity of the material under test. Therefore, the results obtained are acceptable due to their little impact on the total value of the uncertainty of thermal conductivity [29].

The results obtained with the designed equipment have been validated using a comparative analysis with the values obtained according to ASTM E1225-99 "Standard Test Method for thermal conductivity of solids by the guarded comparative longitudinal heat flow" [9]. These tests were carried out at three test temperatures:

- 1–310 °C, to achieve this, the temperature of the hot source was set at 600 °C and the temperature of the cold source at 20 °C.
- 2–175 °C, to reach this value, the temperature of the hot source was set at 339 °C and the temperature of the cold source at 20 °C.
- 2–150 °C, to achieve this, the temperature of the hot source was set at 279.5 °C and the temperature of the cold source at 20 °C.

Regarding the work in [30,31], the sources of the uncertainty values are compared with the graphs obtained in this work. It is shown that the temperature difference near the heat source and the cold source are those that present a more significant deviation concerning the experimental results. A correlation can be inferred for the contribution of uncertainty. According to [32,33], the simulations by FEM performed, where the gradients are more significant, and strictly the heat leaks in the numerical model are not being considered, which could be taken as a reference to calculate heat losses and add a correction in the final uncertainty value [34]. On the other hand, the prediction of the temperature values that the CENAM cut-off bar equipment reaches when the heater operates maximum temperature makes it possible for an adequate selection of material for the fibreglass surface, because the area with higher temperature can reach up to 386°C according to the error obtained in this work [35].

*5.2. Conclusions*

Thermal conductivity is important in several applications to different temperatures, i.e., aerospace industry; nuclear industry; nuclear control rods; radioactive waste containment; the phase change material; or items such as bearings, piston parts, pumps, compressor plate valves, cable insulation and medical implants used in different applications. Therefore, it is essential to measure this thermal property with the most accurate as possible.

The percentage error obtained by ANSYS was 13.5% for the robust model, averaging the 4 volumes (2 copper bars, 1 aluminium bar, and the fibreglass insulator).

The union of elements that interact between the interfaces of the materials is essential and considerably affects the results. In spite of this, it is possible to know with a 13.5% error the temperature distribution inside the system of cut bars [35].

The values near the borders are very far from the experiment; however, the values near the sample bar are too close to those obtained experimentally.

The temperature deviation obtained through simulation and experimental work of the cold source is affected by the contact between it and the copper bar. On the other hand, the same is valid for hot source contact. In the simulation, losses due to bad contact or heat leakage to the environment are not considered. Other methods proposed in the literature to characterise materials have allowed us to verify experimental results [36,37].

The bars cut in CENAM must be designed to prevent radial heat leakage because, according to the simulation results, there is heat leakage in this direction. With the results obtained, a guard can be proposed that balances the gradients generated in the system. Because three distinct sections are noted for the latter case, it would be 213.33 °C, 277.77 °C, and 342.22 °C. As a proposal for improving the design of the CENAM cut bar system.

According to the analysis of the results and the simulations obtained, the following design criteria are proposed. Improve the thermal contact between the hot source and the cold source with the reference bars, which can be achieved by a system that compresses the three bars. Implement a guard with a control system in the hot source and in the cold source, which, although it does not eliminate the radial temperature gradient, reduces it to a minimum. From simulation realised, it is possible to obtain the location of the temperature sensor for the control guard system.

## 6. Future Work

ASTM standard E1225-99 establishes that the measuring equipment by the cut-car method can operate at 1000 °C with a fairly acceptable uncertainty of less than 2%. Then the equipment used by CENAM needs major adjustments and, most likely, a redesign because other critical heat transfer phenomena such as radiation have to be considered. Therefore, with the support of the finite element method, it is intended to analyse the behaviour of a new system, but at an operating temperature of 1000 °C to develop a measuring device that operates at that temperature and can perform measurements of thermal conductivity at temperatures of 500 °C.

There is another problem that affects the accuracy of the results obtained for thermal conductivity value by the method presented in this work, and they are the radial heat leaks. With the use of the finite element method, we will try to minimise these heat leaks to increase accuracy. Another future work is to try to calculate the heat losses by comparing the experimental method and the simulation to obtain an estimate of the heat losses in the experimental system and to find the cause of them. It will serve to make corrections when calculating the uncertainty and implement improvements to the system to reduce these heat leaks.

**Author Contributions:** Conceptualization, J.E.E.G.D. and O.J.G.-R.; Methodology, M.A.Z.-A. and J.R.-R.; Writing–original draft preparation, N.M.-L., R.G.G. and J.R.-R.; Writing–review and editing, M.A.Z.-A., J.E.E.G.D., J.R.-R. and O.J.G.-R.; Supervision, J.R.-R. and J.E.E.G.D.; Data curation, D.J.G.M. All authors have read and agreed to the published version of the manuscript.

**Funding:** This research was partial funded by CONACYT and PRODEP.

**Acknowledgments:** The authors appreciate the support of Centro Nacional de Metrología (CENAM). The authors appreciate Master César Javier Ortiz Echeverria for his support during the revision of the project.

**Conflicts of Interest:** The authors declare no conflicts of interest.

## Nomenclature

| | |
|---|---|
| $\lambda_{C_u}$ | Thermal conductivity of the copper material |
| $\lambda_M$ | Thermal conductivity of the sample |
| $\lambda_{R_1}$ | Thermal conductivity of the reference material 1 |
| $\lambda_{R_2}$ | Thermal conductivity of the reference material 2 |
| $\lambda_z$ | Thermal conductivity of any material |
| $\Delta_{T_i}$ | Temperature gradient $\Delta T$ through an area A (the area through which heat flows) |
| $\Delta_{T_1}$ | Temperature difference among reference material 1 |
| $\Delta_{T_3}$ | Temperature difference among reference material 2 |
| $\Delta_{T_2}$ | Temperature difference among sample bar |
| °C | Celsius degrees |
| Al | Aluminium |
| ASTM | American Society for Testing and Materials |
| CENAM | Centro Nacional de Metrologia |
| CST | Cold Source Temperature |
| Cu | Copper |
| emf | electromotive force |
| FEM | Finite element method |
| HST | Temperature of hot source |
| m | meter |
| mm | millimetre |
| PVC | Polyvinyl chloride |
| $r_A$ | Bar radius |
| $r_B$ | Guard radius |
| TFF | Heat sink Temperature |
| $T_i$ | Temperature in each of the $z_i$ positions where thermocouples are placed |
| $x$ | Denotes approximate thermocouple positions |
| $y$ | Denotes axis $y$ in cartesian coordinate system |
| $z_i$ | Reference distance for thermocouple location $T_i$ in the system |

## References


1. Melchers, R.E.; Beck, A.T. *Structural Reliability Analysis and Prediction*; John Wiley & Sons: Hoboken, NJ, USA, 2018.
2. Guo, Y.; Liu, C. Mechanical properties of hardened AISI 52100 steel in hard machining processes. *J. Manuf. Sci. Eng.* **2002**, *124*, 1–9. [CrossRef]
3. Kobayashi, S.; Thomsen, E. Some observations on the shearing process in metal cutting. *J. Eng. Ind.* **1959**, *81*, 251–262. [CrossRef]
4. DebRoy, T.; Wei, H.; Zuback, J.; Mukherjee, T.; Elmer, J.; Milewski, J.; Beese, A.M.; Wilson-Heid, A.; De, A.; Zhang, W. Additive manufacturing of metallic components–process, structure and properties. *Prog. Mater. Sci.* **2018**, *92*, 112–224. [CrossRef]
5. Górka, J. Assessment of steel subjected to the thermomechanical control process with respect to weldability. *Metals* **2018**, *8*, 169. [CrossRef]
6. Wan, L.; Huang, Y. Microstructure and mechanical properties of al/steel friction stir lap weld. *Metals* **2017**, *7*, 542. [CrossRef]
7. Huang, S.; Chen, R.; Xia, K. Quantification of dynamic tensile parameters of rocks using a modified Kolsky tension bar apparatus. *J. Rock Mech. Geotech. Eng.* **2010**, *2*, 162–168. [CrossRef]
8. Karavaeva, M.; Abramova, M.; Enikeev, N.; Raab, G.; Valiev, R. Superior strength of austenitic steel produced by combined processing, including equal-channel angular pressing and rolling. *Metals* **2016**, *6*, 310. [CrossRef]
9. ASTM. *Standard Test Mmethod for Thermal Conductivity of Solids by Means of the Guarded-Comparative-Longitudinal Heat Flow Technique*; ASTM International: West Conshohocken, PA, USA, 2009.

10. Hsia, S.Y.; Chou, Y.T.; Lu, G.F. analysis of sheet metal tapping screw fabrication using a finite element method. *Appl. Sci.* **2016**, *6*. [CrossRef]

11. Liu, Q.; Li, B.; Schlangen, E.; Sun, Y.; Wu, S. Research on the mechanical, thermal, induction heating and healing properties of steel slag/steel fibers composite asphalt mixture. *Appl. Sci.* **2017**, *7*, 1088. [CrossRef]

12. Grajcar, A.; Skrzypczyk, P.; Kozłowska, A. Effects of temperature and time of isothermal holding on retained austenite stability in medium-Mn steels. *Appl. Sci.* **2018**, *8*, 2156. [CrossRef]

13. Lin, Y.J.; Hwang, S.J. Temperature prediction of rolling tires by computer simulation. *Math. Comput. Simul.* **2004**, *67*, 235–249. [CrossRef]

14. Svetlík, J.; Brestovič, T.; Král', J.; Buša, J.; Kováč, J.; Štofa, M. Numerical Calculation of Oil Dispersion through the Air Flow Applied to the Inner Surface of Slim Tubes. *Appl. Sci.* **2019**, *9*, 2399. [CrossRef]

15. GuoMing, H.; Jian, Z.; JianQang, L. Dynamic simulation of the temperature field of stainless steel laser welding. *Mater. Des.* **2007**, *28*, 240–245. [CrossRef]

16. Yang, X.; Guo, X.; Ouyang, H.; Li, D. A Kriging Model Based Finite Element Model Updating Method for Damage Detection. *Appl. Sci.* **2017**, *7*. [CrossRef]

17. Zain-Ul-Abdein, M.; Ijaz, H.; Saleem, W.; Raza, K.; Mahfouz, A.S.B.; Mabrouki, T. Finite element analysis of interfacial debonding in copper/diamond composites for thermal management applications. *Materials* **2017**, *10*, 739. [CrossRef]

18. Courbon, C.; Mabrouki, T.; Rech, J.; Mazuyer, D.; D'Eramo, E. On the existence of a thermal contact resistance at the tool-chip interface in dry cutting of AISI 1045: Formation mechanisms and influence on the cutting process. *Appl. Therm. Eng.* **2013**, *50*, 1311–1325. [CrossRef]

19. De Santis, S.; Ceroni, F.; de Felice, G.; Fagone, M.; Ghiassi, B.; Kwiecień, A.; Lignola, G.P.; Morganti, M.; Santandrea, M.; Valluzzi, M.R.; et al. Round robin test on tensile and bond behaviour of steel reinforced grout systems. *Compos. Part B Eng.* **2017**, *127*, 100–120. [CrossRef]

20. Xing, Y.; Qin, M.; Guo, J. A Time Finite Element Method Based on the Differential Quadrature Rule and Hamilton's Variational Principle. *Appl. Sci.* **2017**, *7*. [CrossRef]

21. Yang, W.r.; He, X.j.; Zhang, K.; Yang, Y.; Dai, L. Combined effects of curing temperatures and alkaline concrete on tensile properties of GFRP bars. *Int. J. Polym. Sci.* **2017**, *2017*, 4262703. [CrossRef]

22. Mariner, J.T.; Sayir, H. High Thermal Conductivity Composite and Method. US Patent 5,863,467, 26 January 1999.

23. Ameen, J.G.; Mortimer, W.P., Jr.; Yokimcus, V.P. Thermally Conductive Interface. US Patent 5,545,473, 13 August 1996.

24. Andalib, Z.; Kafi, M.A.; Kheyroddin, A.; Bazzaz, M.; Momenzadeh, S. Numerical evaluation of ductility and energy absorption of steel rings constructed from plates. *Eng. Struct.* **2018**, *169*, 94–106. [CrossRef]

25. Kakaç, S.; Yener, Y.; Naveira-Cotta, C.P. *Heat Conduction*; CRC Press: Boca Raton, FL, USA, 2018. [CrossRef]

26. Sun, L.; Park, S.S.; Sheberla, D.; Dincă, M. Measuring and reporting electrical conductivity in metal–organic frameworks: Cd2 (TTFTB) as a case study. *J. Am. Chem. Soc.* **2016**, *138*, 14772–14782. [CrossRef] [PubMed]

27. Peng, L.; Xu, Z.; Liu, Z.; Guo, Y.; Li, P.; Gao, C. Ultrahigh thermal conductive yet superflexible graphene films. *Adv. Mater.* **2017**, *29*, 1700589. [CrossRef] [PubMed]

28. Ashrafi, H.; Bazli, M.; Vatani Oskouei, A.; Bazli, L. Effect of sequential exposure to UV radiation and water vapor condensation and extreme temperatures on the mechanical properties of GFRP bars. *J. Compos. Constr.* **2017**, *22*, 04017047. [CrossRef]

29. Purbolaksono, J.; Ahmad, J.; Khinani, A.; Ali, A.; Rashid, A. Failure case studies of SA213-T22 steel tubes of boiler through computer simulations. *J. Loss Prev. Process Ind.* **2010**, *23*, 98–105. [CrossRef]

30. Yu, M.; Saunders, T.; Su, T.; Gucci, F.; Reece, M. Effect of heat treatment on the properties of wood-derived biocarbon structures. *Materials* **2018**, *11*, 1588. [CrossRef]

31. Hosseini, V.; Karlsson, L.; Wessman, S.; Fuertes, N. Effect of sigma phase morphology on the degradation of properties in a super duplex stainless steel. *Materials* **2018**, *11*, 933. [CrossRef]

32. Carollo, L.F.D.S.; Lima e Silva, A.L.F.D.; Lima e Silva, S.M.M.D. A Different Approach to Estimate Temperature-Dependent Thermal Properties of Metallic Materials. *Materials* **2019**, *12*, 2579. [CrossRef]

33. Charles Murgau, C.; Lundbäck, A.; Åkerfeldt, P.; Pederson, R. Temperature and Microstructure Evolution in Gas Tungsten Arc Welding Wire Feed Additive Manufacturing of Ti-6Al-4V. *Materials* **2019**, *12*. [CrossRef]

34. Branco, R.; Antunes, F.; Costa, J. A review on 3D-FE adaptive remeshing techniques for crack growth modelling. *Eng. Fract. Mech.* **2015**, *141*, 170–195. [CrossRef]

35. Bugelnig, K.; Germann, H.; Steffens, T.; Sket, F.; Adrien, J.; Maire, E.; Boller, E.; Requena, G. Revealing the Effect of Local Connectivity of Rigid Phases during Deformation at High Temperature of Cast AlSi12Cu4Ni(2,3)Mg Alloys. *Materials* **2018**, *11*, 1300. [CrossRef]
36. Berto, F.; Lazzarin, P.; Marangon, C. Brittle fracture of U-notched graphite plates under mixed mode loading. *Mater. Des.* **2012**, *41*, 421–432. [CrossRef]
37. Bernal, S.A.; de Gutiérrez, R.M.; Pedraza, A.L.; Provis, J.L.; Rodriguez, E.D.; Delvasto, S. Effect of binder content on the performance of alkali-activated slag concretes. *Cem. Concr. Res.* **2011**, *41*, 1–8. [CrossRef]

*Article*

# 3D-Based Transition *hpq/hp*-Adaptive Finite Elements for Analysis of Piezoelectrics


**Grzegorz Zboiński [1,2,*] and Magdalena Zielińska [1]**

[1] Faculty of Technical Sciences, The University of Warmia and Mazury, Oczapowskiego 11, 10-719 Olsztyn, Poland; m.nosarzewska@uwm.edu.pl

[2] Institute of Fluid Flow Machinery, Polish Academy of Sciences, Fiszera 14, 80-231 Gdańsk, Poland

* Correspondence: zboi@imp.gda.pl; Tel.: +48-58-522-5219



**Abstract:** This paper concerns the algorithm of transition piezoelectric elements for adaptive analysis of electro-mechanical systems. In addition, effectivity of the proposed elements in such an analysis is presented. The elements under consideration are assigned for joining basic elements which correspond to the mechanical models of either the first or higher order, while the electric model is of arbitrary order. In this work, three variants of the transition models are applied. The first one assures continuity of displacements between the basic models and continuity of electric potential between these models, as well. The second transition piezoelectric model guarantees additional continuity of the stress field between the basic models. The third transition model additionally enables continuous change of the strain state between the basic models. Based on the mentioned models, three types of the corresponding transition finite elements are introduced. The applied finite element approximations are *hpq/hp*-adaptive ones, which allows element-wise changes of the element size parameter *h*, and the element longitudinal and transverse orders of approximation, respectively, *p* and *q*, depending on the error level. Numerical effectiveness of the models and their approximations is investigated in the contexts of: ability to remove high stress gradients between the basic and transition models, and convergence of the numerical solutions for the model problems of piezoelectrics with and without the proposed transition elements.

**Keywords:** electro-mechanical systems; piezoelectrics; hierarchical models; first-order models; transition models; *hpq/hp*-approximations; adaptivity; stress gradients; convergence



**Citation:** Zboiński, G.; Zielińska, M. 3D-Based Transition *hpq/hp*-Adaptive Finite Elements for Analysis of Piezoelectrics. *Appl. Sci.* **2021**, *11*, 4062. https://doi.org/10.3390/app11094062

Academic Editor: Marek Krawczuk

Received: 9 April 2021
Accepted: 26 April 2021
Published: 29 April 2021




## 1. Introduction

More and more common application of piezoelectrics in contemporary technology requires more and more efficient and accurate methods of their modeling and analysis. In this work, we develop the new transition elements which possess unique set of features enabling joining mechanical shell models of the first order and higher orders. Such models are incompatible due to the assumptions of the plane stresses and no elongation of the normals to the shell mid-surface, both present in the first-order model. In order to join such models, the transition models are necessary. This refers also to the piezoelectricity case, when the mechanical field modeling needs simultaneous application of the first- and higher-order mechanical models.

Adaptive capabilities of *hpq/hp*-type are the second important feature of the new elements, where *h* is the element size parameter, while *p* and *q* are the element longitudinal and transverse orders of approximation. As our implementation of these capabilities is based on the standard *hpq*-approximation rules, presented, respectively, in Reference [1,2] for 2D and 3D, we will not elucidate this aspect in this paper.

### 1.1. State of the Art

To the best of the authors' knowledge, the transition piezoelectric elements have not been proposed yet, both in the classical (non-adaptive) and adaptive versions. Because of

295

this, our literature survey can only be focused on two topics closest to the paper theme. The first topic concerns mathematical models of piezoelectrics and their finite element approximations. The second topic is the elastic transition models and the corresponding finite element approximations, as well.

### 1.1.1. Piezoelectric Models, Approximations and Finite Elements

#### Low-Order Models and Elements for Monolithic Piezoelectrics

Let us start this brief survey with the first-order and second-order models of piezoelectricity, where the model order concerns transverse displacement and electric potential fields. In Reference [3], the authors present classical (non-adaptive) finite element approximation and its features for the two-dimensional conventional (mid-surface unknowns) Reissner-Mindlin piezoelectric plate model including membrane effects. The finite element approximation is of mixed type with shear stresses treated as independent unknowns. The Kirhchoff piezoelectric plate model can be obtained as a special case. The latter model was also analyzed in Reference [4]. The simple flat shell element based on the conventional (mid-surface) displacements, rotations, and electric potential for plane stress and neglect of the in-plane electric field and electric displacements is presented in Reference [5]. The element is based on the assumed natural strain formulation in order to overcome numerical locking phenomena. Subsequently, in Reference [6], the Reissner-Mindlin mechanical model is combined with the first-order and second-order electric potential field in the transverse directions. Two formulations are applied, the displacement-like and hybrid with the electric displacements treated as independent unknowns. In Reference [7], the second-order transverse mechanical field is applied in the frame of three-dimensional elasticity for thin-walled piezoelectric plates. The electric potential field is also of the second-order in the transverse direction. The balanced reduced integration is applied as a numerical technique which serves removing the locking phenomena. The analytical conventional model of the Reissner-Mindlin plates with non-constant electric field distribution is presented in Reference [8].

#### Hierarchical Models and Elements for Monolithic Piezoelectrics

The hierarchical two-dimensional conventional (mid-surface unknowns) models of piezoelectric plates were introduced and mathematically substantiated in Reference [9]. The mixed formulation is applied there with stresses and electric displacements completing displacements and electric potential. Carrera and co-workers [10] proposed hierarchic plate models up to the fourth-order. The formulation is displacement-based in the case of the higher-order models, and mixed in the case of the Reissner-Mindlin model. The models employ three-dimensional degrees of freedom (dofs) for the top and bottom surfaces, and conventional (mid-surface) higher-order dofs. In addition, a thermal field can be included into this formulation. In Reference [11], the 3D-based (three-dimensional unknowns only) piezoelectric hierarchy of solid, first-order (Reissner-Mindlin) and hierarchical shell, and solid-to-shell and shell-to-shell transition piezoelectric models are formulated. Their *hpq*-adaptive finite element approximations are presented in Reference [12]. The formulation is displacement-like with displacements and electric potential as unknowns of the problem. In the case of the electric field, this approach allows for the three-dimensional and 3D-based symmetric-thickness models of dielectricity. The approach is based on the 3D-based hierarchical models [13] and approximations [14] for elasticity, and the corresponding models [15] and approximations [16] for dielectricity.

#### Models for Piezoelectric Composites

In the case of laminated piezoelectric structures, it is worth mentioning the following works. In Reference [17], the authors presented Reissner-Mindlin-type composite plate model with piezoelectric layers. The direct piezoelectric and pyroelectric effects are taken into account. In turn, the work of Reference [18] proposed the plain strain shell model with higher-order shear and normal deformation effects for analysis of cylindrical laminated piezoelectric shells with actuating and sensing piezoelectric layers. The work of Reference [6], mentioned in the context of the monolithic piezoelectrics, should be mentioned here as applied to tree-layered shell structures with two outer piezoelectric layers and an electrically neutral core. A similar sandwich plate structures are modeled as two Reissner-Mindlin piezoelectric layers and one neutral layer of the same character [19], with the mid-surface interdependent displacement dofs. The electric fields are of mixed character and are defined by the outer layers' mid-surface electric potentials and independent electric field mid-surface values of the outer layers. In order to remove locking, the reduced integration method is applied. The generalized analytical Reissner-Mindlin model of laminated piezoelectrics was presented in Reference [21]. It allows analyses of the problems where the electric potential is not prescribed through the thickness. The method assumes small thickness and thus allows decomposition of the three-dimensional Reissner-Mindlin description into two-dimensional in-plane and one-dimensional through-the-thickness problems. Subsequently, the already cited work of Reference [10] concerning hierarchical models of single layered piezoelectrics can also be applied to the numerical analysis of hierarchically modeled multilayered structures, with piezoelectric layers included. The recent example [22] of analysis of multilayered shells concerns the first-order model of the functionally graded layers. In turn, the piece-wise second-order finite element model for multilayered structures with delamination and debonding was proposed in Reference [23].

### 1.1.2. Transition Piezoelectric, Dielectric and Elastic Models And Elements
#### Transition Elements for Three Physical Classes of Problems

It can be concluded from the previous subsection that the piezoelectric transition models and elements that serve joining the three-dimensional (or hierarchical shell) piezoelectric models with the first-order piezoelectric model are rare. The only works, which we have cited, concern the idea [15] and variational [11] and finite element [12] formulations of the same classical transition piezoelectric model and element. They guarantee continuity of the displacement and transverse deformation fields between the mentioned models. They also guarantee continuity of the electric potential between the piezoelectric models.

In Reference [16], it has been demonstrated that, in the case of dielectric field of electric potential used in electrostatics, no transition elements between the three-dimensional model and symmetric-thickness hierarchical models of the first and higher orders are necessary if the consistent finite element approximations of non-adaptive or adaptive character are applied. This is because the same definition of electric field (intensity) is applied within both dielectric models.

In the case of the elastic mechanical fields, two groups of the transition models can be distinguished. The first one guarantees continuity of the inconsistent generalized displacement fields of the neighboring models. In such inconsistent models, the kinematic unknowns are different or of different order, e.g., in the transverse direction. In the second group, additionally the definitions of the derivatives of the kinematic unknowns, e.g., strains and/or stresses, are different in the neighboring models and the corresponding finite elements. In this context, one may deal with the neighborhood of the three-dimensional elasticity and shell or plate theories, the three-dimensional theory, and beam or truss theories, as well as the shell/plate theories and the one-dimensional theories of beams or trusses. Due to the paper scope, only the first case will be elucidated here. A general review, including all three cases, can be found in the thesis [24].

#### Non-Adaptive Transition Elements for Elasticity

We start with the transition elements of the classical (non-adaptive) finite element methods where the neighboring elements are joined through the sides of the same size and finite element interpolations on these sides are consistent. In addition, location of three-dimensional, shell/plate, and transition domains is fixed and strictly determined by the structure geometry. The first examples of the transition elements of this type can be attributed to Surana [25], who took into account the consideration of Ahmad [26] on thick shell elements and his own research [27] on general and axi-symmetric shell elements. The work, in Reference [25], of Surana presents solid-to-shell elements which enable joining the thick-shell element with the linear, quadratic or cubic solid elements. The elements possess the bigger shell part equipped with the mid-surface dofs, and the smaller solid part limited to one side of the element and equipped with the tree-dimensional dofs. Plane stress is assumed in the entire transition element. Applications of the elements of this type to axi-symmetric [28] and three-dimensional [29] problems are presented. Extension onto thermo-mechanical analysis was done in Reference [30]. Note that the independent research of Bathe and Bolourchi [31], based on the mentioned work of Ahmad, led to the analogous transition elements. Gmür and Schorderet [32] suggested application of the three-dimensional or plane stress states in the Surana's element depending on the geometrical shape of the domain the element is located in. Liao and others [33] extended application of the Surana's elements onto laminated composite structures. Both, the shell and solid parts of the transition element are modeled as such composites. In the work of Reference [34], in order to avoid the numerical locking, the method based on strain interpolation (with the points where the strain values are known) was applied instead of the reduced integration.

In the work of Dávila [35], the opposite idea is applied, i.e., the elements consist of the larger solid part and smaller shell part. The latter part comprises one or two lateral sides. The top and bottom three-dimensional dofs and mid-surface dofs are applied within the solid and shell parts, respectively. Three-dimensional stress state is assumed throughout single-layered shells or each layer of composite shells. Yet another idea was suggested in Reference [36], by Gong, who proposed generalization of the solid-to-shell transition elements for the elements joining two-dimensional and three-dimensional elements. In this approach, the shell or two-dimensional parts consist of one or two lateral sides of the transition element. The longitudinal order of approximation can be higher than three. In the entire element, either three-dimensional or plane stress state can be applied.

Adaptive Transition Elements for Elasticity

In the case of the adaptive methods, finite elements of different sizes, orders of approximation (non-conforming approximations), and mechanical models are present in finite element meshes. In the case of different element sizes resulting from *h*-adaptivity based on remeshing, continuity of the field of unknowns may be based on introduction of the distorted elements on the boundaries between the mesh parts of various density. In the case of different sizes resulting from *h*-adaptivity based on local element refinement, continuity of the fields of unknowns requires introduction of the distorted elements [37] or is based on transition elements for monolithic [38] or multilayered [39] structures. Such transition elements deliver piece-wise linear [40] or approximate [41] continuity along the entire common edge or face, or at the common nodes only, e.g., within triangles [42], quadrilaterals [43] and hexahedrons [44]. Note that application of the smart idea of the constrained approximation, proposed by Demkowicz for 2D [1] and 3D [2], removes the necessity of introduction of the transition elements of this type as the continuity is automatically guaranteed by the constraints on the boundary between the elements of different sizes.

As far as the different approximation orders of the elements are concerned, transition elements may be introduced with the consistent approximation on the boundary between specific 2D [20] or general 3D [45] elements, with the continuity guaranteed again on the entire common edge or face, or at the common nodes only. A smart alternative is the idea of 2D [1] and 3D [2] hierarchical shape functions, defined independently in the element vertices, on its edges and sides, and in the interior of the element, which allows for the same approximation order on the common edge or face of the neighboring elements and different orders in their interiors.

In the case of different models of the neighboring elements, the continuity of the field of unknowns may require transition elements which conform the model of the larger number of nodal dofs with some of these dofs constrained to zero on the boundary with the model of the smaller number of nodal dofs. The 3D-2D [46], 2D-2D [47], and 2D-1D [48] versions of this approach can be found. The alternative is the transition elements equipped with the degrees of freedom of both models (see comments in Reference [49]). Note that such transition elements are not necessary if the uni-dimensional, e.g., three-dimensional, approach is applied within each of the neighboring models. The idea lies in using the same, e.g., three-dimensional, dofs for each model. This usually requires internal constraints in the model simplified (e.g., 3D-based shell models) with respect to the three-dimensional one. The linear [50] or higher [51] order of approximation can be applied in the transverse direction on the common face of the three-dimensional and 3D-based shell elements. Even though this technique guarantees continuity of the field of unknowns, high gradients of the derivatives on the boundary between the models appear. If one wants to get rid of such gradients, the transition elements removing stress [52] and, additionally, strain, Reference [24] gradients are necessary between the basic models being joined.

### 1.1.3. Conclusions

In this paper, we will use piezoelectric transition elements combining the ideas present in the following works concerning the uncoupled problem of elasticity. Firstly, we will apply the 3D-based approach [13], where the solid [53], hierarchical shell [14], first-order shell [54], and transition [51] models are equipped with the three-dimensional dofs only. This approach will also be extended onto the dielectricity models [16]. We will also apply the idea of Dávila [35], where one deals with larger 3D part and smaller thin-walled part of the transition element. The stress state within the element will be either three-dimensional, as in Reference [35], or transition (from three-dimensional to plane one) [24]. The $hp$-adaptive capabilities of the elements will be based solely on the ideas of Demkowicz and others [2]. Our 3D-based hierarchical finite element models will resemble the ideas presented, respectively, in Reference [55,56] for the cases of the three-dimensional and conventional (mid-surface) dofs.

### 1.2. The Scope and Novelty of The Paper

This paper concerns the algorithm of the solid-to-shell and shell-to-shell transition piezoelectric elements for adaptive analysis of electro-mechanical systems. Such systems can be composed of piezoelectric transducers and elastic structural elements, both of arbitrary shapes. In addition, effectivity of application of the proposed elements in such an analysis is of our interest in the paper. The elements are assigned for joining basic elements, namely piezoelectric shell elements of the first-order mechanical field and arbitrary-order electric field and piezoelectric elements of the hierarchical shell model (second- or higher-order) within the mechanical field and arbitrary-order within the electric field. Alternatively, the latter model can be replaced with the piezoelectric description based on hierarchical models of three-dimensional elasticity and three-dimensional dielectricity.

In this work, three variants of the transition models are applied. The first one, called classical, assures continuity of displacements between the basic models and continuity of electric potential between these models, as well. The second, modified transition piezoelectric model guarantees additional continuity of the stress field between the basic models. The third, enhanced model additionally enables continuous change of the strain state between the basic models. All three models are 3D-based ones, namely only three-dimensional degrees of freedom are applied within the mechanical and electric fields. Based on the mentioned models, three types of the corresponding transition finite elements are introduced. The applied approximations are $hpq/hp$-adaptive ones, with $h$ being the element size parameter, $p$ standing for the longitudinal order of approximation, and $q$ denoting the transverse order of approximation.

Due to basic character of the presented research, numerical effectiveness of these models and their finite element approximations is limited to two crucial aspects, namely ability to remove high stress gradients between the basic and transition models, and convergence of the numerical solutions for the model problems of piezoelectrics. These solutions are obtained with and without the proposed transition elements. Error estimation and adaptivity control matters are out of scope of this work and will be presented in a separate paper.

Novelty of the paper results from the above unique features of the proposed transition elements and the orignal results concerning their effectiveness.

## 2. Model Problems of Stationary Piezoelectricity

First, we present the strong (or local) formulation of the model problem of stationary piezoelectricity. Due to further application of the matrix notation, convenient for finite elements, this formulation is expressed in the matrix form.

The mechanical equilibrium equations and geometrical (kinematic) relations read

$$\left. \begin{array}{l} \mathrm{div}\,\underline{\sigma} + f = 0 \\ \underline{\varepsilon} = \frac{1}{2}[u \otimes \mathrm{grad} + (u \otimes \mathrm{grad})^T] \end{array} \right\}, \quad x \in V, \tag{1}$$

where the matrix representations $\underline{\sigma}$ and $\underline{\varepsilon}$ of the symmetric, stress, and strain tensors, $\tilde{\sigma}$ and $\tilde{\varepsilon}$, of the second rank are employed. Both relations are valid for each point $x$ of the volume domain $V$ of the piezoelectric. The vector $f$ represents given mass loading vector, while $u$ is the unknown vector of (mechanical) displacements.

Within the electric field, the electric equilibrium equation, namely the Gauss law with the assumed zero volume charges, holds together with the electric field vector $E$ definition, i.e.,

$$\left. \begin{array}{l} \mathrm{div}\,d = 0 \\ E = -\mathrm{grad}\,\phi \end{array} \right\}, \quad x \in V. \tag{2}$$

Above, the vector of electric displacements $d$, resulting from electrical charging of the piezoelectric, and the unknown scalar electric potential $\phi$ appear.

In order to express the stress tensor by the strain tensor, and the electric displacement vector by the electric potential one, the constitutive relations have to be added to the systems (1) and (2). The general form of these relations is consistent with the work of Reference [57]. The matrix form of them corresponds to the isotropic elasticity, orthotropic dielectricity, and orthotropic piezoelectricity. Because of this, the general, constitutive tensors of the elasticity, dielectricity, and piezoelectricity, $\tilde{D}, \tilde{\gamma}, \tilde{C}$, of the fourth, second, and third ranks, respectively, can be represented by the matrices $D, \gamma$, and $C$, of the sizes: $6 \times 6$, $3 \times 3$, and $6 \times 3$, respectively, i.e.,

$$\left. \begin{array}{l} \sigma = D\varepsilon - CE \\ d = C^T\varepsilon + \gamma E \end{array} \right\}, \quad x \in V, \tag{3}$$

where the coupling of the mechanical and electric fields results from the presence of the matrix $C$, while $\sigma$ and $\varepsilon$ stand for the six-component vectorial representations of the stress and strain tensors $\tilde{\sigma}$ and $\tilde{\varepsilon}$.

The above set of the partial differential, governing equations of piezoelectricity (1)–(3) has to be completed with the boundary conditions. The Neumann and Dirichlet boundary conditions, concerning, respectively, unknown functions derivatives and unknown functions themselves, become stress and displacement boundary conditions in the mechanical case, namely:

$$\underline{\sigma}n = p, \quad x \in P \tag{4}$$

$$u = w, \quad x \in W, \tag{5}$$

where $p$ and $w$ are the given stress and displacement vectors, respectively, while $P$ and $W$ denote the parts of the surface $S \equiv \partial V$ of the piezoelectric body $V$, with the given values of the mentioned vectors. The vector $n$ represents the unit vector, outward and normal to the boundary.

In the case of the electric field, the Neumann boundary conditions express equality of the internal charges (defined with the unknown electric displacements $d$) and given external charges of the surface density $c$, while the Dirichlet boundary conditions describe the given values $\omega$ of the unknown field of electric potential $\phi$, i.e.,

$$d^T n = -c, \quad x \in Q, \tag{6}$$

$$\phi = \omega, \quad x \in F. \tag{7}$$

Above, $Q$ and $F$ represent the parts of the surface $S$ of the piezoelectric domain $V$, where the electric charge and electric potential are known, respectively.

It should be mentioned that the general constitutive relations (3) allow any piezoelectric model based on coupling of the linear elastic and linear dielectric models. In our proposition, we apply the 3D-based hierarchical piezoelectric models proposed in Reference [15,58] and elucidated in Reference [11,12]. These models employ three-dimensional degrees of freedom within the electric and mechanical fields. The 3D-based elastic models may represent: three-dimensional elasticity, hierarchical and first-order shell models, as well as the solid-to-shell and shell-to-shell transition models. The 3D-based dielectric models include: three-dimensional dielectricity model and the hierarchical symmetric-thickness models. These two groups of models were introduced and thoroughly described in Reference [13,14] and Reference [16], respectively. Note that the applied 3D-based approach allows analysis of complex piezoelectric structures within the above Formulations (1)–(3). In such structures, more than one piezoelectric model meet, including transition ones.

Thanks to the reciprocity theorem [59] of linear piezoelectricity, the above strong Formulations (1)–(7) can be converted into the corresponding variational (or weak) formulation. For the model problems of linear stationary piezoelectricity considered in this paper, the weak formulation reads

$$\begin{cases} B(v, u) - C(v, \phi) = L(v) \\ -C(\psi, u) - b(\psi, \phi) = -l(\psi) \end{cases} \tag{8}$$

for all admissible displacements $v \in V$ and all admissible potentials $\psi \in \Psi$, with the space definitions: $V = \{ v \in (H^1(V))^3 : v = 0 \text{ on } W \}$ and $\Psi = \{ \psi \in H^1(V) : \psi = 0 \text{ on } F \}$. The solution functions belong to the spaces: $u \in w + V$ and $\phi \in \omega + \Psi$, where the given values of $w$ and $\omega$, consistent with the Dirichlet boundary conditions (5) and (7), are called lifts (compare Reference [1]) within the displacements field and potential field, respectively.

The bilinear and linear forms of the first line of the relation (8) denote the virtual strain energy of the elastic field characterized with the matrix $D$ of elasticities and the virtual work of the external body $f$ and surface $p$ loadings. The following definitions of these forms hold:

$$B(v, u) = \int_V \varepsilon^T(v) D\, \varepsilon(u)\, dV$$
$$L(v) = \int_V v^T f\, dV + \int_P v^T p\, dS \tag{9}$$

In the second Equation (8), the bilinear form represents the virtual energy of the electric field characterized with the dielectricity constant matrix $\delta$ under constant strain, while the linear form is equal to the virtual work of the surface charges of the density $c$, namely

$$b(\psi, \phi) = \int_V E^T(\psi)\, \delta E(\phi)\, dV$$
$$l(\psi) = -\int_Q \psi\, c\, dS \tag{10}$$

The mixed bilinear forms $C$ stand for the virtual energies corresponding to the electromechanical (piezoelectric) coupling. Their definitions read:

$$\begin{aligned} C(v,\phi) &= \int_V \varepsilon^T(v)\, CE(\phi)\, dV \\ C(\psi,u) &= \int_V E^T(\psi)\, C^T \varepsilon(u)\, dV \end{aligned} \tag{11}$$

where, as before, $C$ is the piezoelectric constant matrix under constant strain.

The tensorial version of the above Formulations (8)–(11) can be found in Reference [11,57]. The existence and uniqueness theorem for the above linear piezoelectricity problem can be found in Reference [60]. The proof takes advantage of the generalized Lax-Milgram lemma [61,62].

## 3. Transition Models For Piezoelectricity

As said in the previous sections, the transition elements under consideration serve joining the first-order shell mechanical field of a piezoelectric with the higher-order hierarchical shell models or the three-dimensional elasticity model of this field within the piezoelectric. In the electric field corresponding to the mechanical fields of both types, the same hierarchical symmetric-thickness or three-dimensional dielectric model is applied. In order to introduce the original transition models, the basic piezoelectric models that are to be joined together will be presented first.

### 3.1. Basic Models of Linear Piezoelectricity

The starting point for two basic models of piezoelectricity, based on the first-order and higher-order mechanical fields, is the model corresponding to the three-dimensional piezoelectricity. The matrix form of the constitutive relations of linear three-dimensional piezoelectricity, expressed by the experimentally obtainable constitutive constants, is presented in Reference [57]:

$$\begin{cases} \varepsilon &= D^{-1}\sigma - cE \\ d &= -c^T\sigma + \delta E. \end{cases} \tag{12}$$

Above, $D^{-1}$, $\delta$, and $c$ denote the isotropic inverse matrix of elasticities (or compliance matrix), the orthotropic matrix of dielectric constants under constant stress, and the coupling orthotropic matrix of piezoelectric constants under constant stress, respectively. The six-component column vectors $\sigma$, $\varepsilon$ represent the symmetric stress and strain tensors, $\sigma_{ij}$ and $\varepsilon_{ij}$, $i,j = 1,2,3$, while $d$ and $E$ are the column three-component electric displacement and electric field vectors, with components $d_j$, $E_j$, $j = 1,2,3$. The mentioned tensors and vectors are defined locally, namely in the directions consistent with the axes of piezoelectric orthotropy. In the presentation below, the third local direction is the polarization direction of the piezoelectric. In the case of the thick- or thin-walled structures, this direction coincides with the transverse one.

The standard definition of the isotropic elasticity matrix reads:

$$D = \begin{bmatrix} D_{11} & D_{12} & D_{13} & 0 & 0 & 0 \\ D_{21} & D_{22} & D_{23} & 0 & 0 & 0 \\ D_{31} & D_{23} & D_{33} & 0 & 0 & 0 \\ 0 & 0 & 0 & D_{44} & 0 & 0 \\ 0 & 0 & 0 & 0 & D_{55} & 0 \\ 0 & 0 & 0 & 0 & 0 & D_{66} \end{bmatrix} = D \begin{bmatrix} 1-v & v & v & 0 & 0 & 0 \\ v & 1-v & v & 0 & 0 & 0 \\ v & v & 1-v & 0 & 0 & 0 \\ 0 & 0 & 0 & (1-2v)/2 & 0 & 0 \\ 0 & 0 & 0 & 0 & (1-2v)/2 & 0 \\ 0 & 0 & 0 & 0 & 0 & (1-2v)/2 \end{bmatrix}, \tag{13}$$

where $D = E/[(1+v)(1-2v)]$. As it can be seen above, the non-zero terms of the matrix are expressed by Young's modulus $E$ and Poisson's ratio $v$.

The dielectricity constant matrix under constant stress reads:

$$\delta = \begin{bmatrix} \delta_{11} & 0 & 0 \\ 0 & \delta_{22} & 0 \\ 0 & 0 & \delta_{33} \end{bmatrix}, \tag{14}$$

where $\delta_{ii}$, $i = 1, 2, 3$ stand for the orthotropic permittivity constants.

In the case of the piezoelectricity constant matrix under constant stress, one has

$$c = \begin{bmatrix} 0 & 0 & c_{13} \\ 0 & 0 & c_{23} \\ 0 & 0 & c_{33} \\ 0 & 0 & 0 \\ 0 & c_{52} & 0 \\ c_{61} & 0 & 0 \end{bmatrix}, \tag{15}$$

where $c_{13}$, $c_{23}$, $c_{33}$, $c_{52}$, and $c_{61}$ are non-zero piezoelectric constants within this matrix.

### 3.1.1. Hierarchical or Three-Dimensional Models

For both the cases, when the transverse mechanical and electric fields are polynomially constrained through the thickness of the thin-walled piezoelectric member or the transverse direction is treated in the same way as the longitudinal directions of the three-dimensional piezoelectric member, the same constitutive relations hold.

In order to present these relations, let us convert the Equation (12) to the form suitable for the variational and finite element formulation of the general piezoelectric problem. This needs inversion of the first Equation (12) and its substitution into the second one. The resultant form reads:

$$\begin{cases} \sigma &= D\varepsilon - CE \\ d &= C^T \varepsilon + \gamma E, \end{cases} \tag{16}$$

where the following definitions of the orthotropic matrix $C$ of piezoelectric constants under constant strain and the isotropic matrix $\gamma$ of dielectric constants under constant strain were introduced:

$$\begin{aligned} \gamma &= \delta - c^T D c \\ C &= -Dc. \end{aligned} \tag{17}$$

The first of the above definitions gives the following non-zero terms of the matrix $\gamma$:

$$\gamma = \begin{bmatrix} \gamma_{11} & 0 & 0 \\ 0 & \gamma_{22} & 0 \\ 0 & 0 & \gamma_{33} \end{bmatrix} = \begin{bmatrix} \delta_{11} - c_{61} D_{66} c_{61} & 0 & 0 \\ 0 & \delta_{22} - c_{52} D_{55} c_{52} & 0 \\ 0 & 0 & \begin{matrix} \delta_{33} - c_{13}(D_{11}c_{13} + D_{12}c_{23} + D_{13}c_{33}) \\ -c_{23}(D_{21}c_{13} + D_{22}c_{23} + D_{23}c_{33}) \\ -c_{33}(D_{31}c_{13} + D_{32}c_{23} + D_{33}c_{33}) \end{matrix} \end{bmatrix}. \tag{18}$$

The second definition leads to the following structure of the matrix $C$:

$$C = \begin{bmatrix} 0 & 0 & C_{13} \\ 0 & 0 & C_{23} \\ 0 & 0 & C_{33} \\ 0 & 0 & 0 \\ 0 & C_{52} & 0 \\ C_{61} & 0 & 0 \end{bmatrix} = - \begin{bmatrix} 0 & 0 & D_{11}c_{13} + D_{12}c_{23} + D_{13}c_{33} \\ 0 & 0 & D_{21}c_{13} + D_{22}c_{23} + D_{23}c_{33} \\ 0 & 0 & D_{31}c_{13} + D_{32}c_{23} + D_{33}c_{33} \\ 0 & 0 & 0 \\ 0 & D_{55}c_{52} & 0 \\ D_{66}c_{61} & 0 & 0 \end{bmatrix}. \tag{19}$$

### 3.1.2. First-Order Piezoelectric Model

Stress State

In the case when the mechanical field of the symmetric-thickness thick- or thin-walled member is linearly constrained through the thickness, the plane stress assumption of the first-order shell model have to be introduced into the three-dimensional piezoelectric constitutive relations. This means that the third row equation of the first constitutive relation (16), namely:

$$\sigma_{33} = \frac{E}{(1+v)(1-2v)}\left[v\varepsilon_{11} + v\varepsilon_{22} + (1-v)\varepsilon_{33}\right] - C_{33}E_3, \tag{20}$$

has to be combined with the assumption $\sigma_{33} = 0$, so as to give

$$\varepsilon_{33} = -\frac{v}{1-v}(\varepsilon_{11} + \varepsilon_{22}) + \frac{(1+v)(1-2v)C_{33}}{E(1-v)}E_3. \tag{21}$$

Now, the condition (20) can be rewritten in the following alternative identity-equation form

$$\sigma_{33} = \frac{E}{(1+v)(1-2v)}\left\{v\varepsilon_{11} + v\varepsilon_{22} + (1-v)\left[\frac{-v}{1-v}(\varepsilon_{11}+\varepsilon_{22}) + \frac{(1+v)(1-2v)C_{33}}{E(1-v)}E_3\right]\right\} - C_{33}E_3 = 0. \tag{22}$$

Substitution of the relation (21) into the first relation (16), arithmetic manipulations on the terms containing $E_3$ (with use of isotropic definitions of the non-zero terms of $D_{ij}$, $i,j = 1,2,\ldots,6$ from (13)) lead to:

$$
\begin{bmatrix} \sigma_{11} \\ \sigma_{22} \\ \sigma_{33} \\ \sigma_{12} \\ \sigma_{23} \\ \sigma_{31} \end{bmatrix}
=
\begin{bmatrix}
D_{11} & D_{12} & D_{13} & 0 & 0 & 0 \\
D_{21} & D_{22} & D_{23} & 0 & 0 & 0 \\
D_{31} & D_{23} & D_{33} & 0 & 0 & 0 \\
0 & 0 & 0 & D_{44} & 0 & 0 \\
0 & 0 & 0 & 0 & D_{55} & 0 \\
0 & 0 & 0 & 0 & 0 & D_{66}
\end{bmatrix}
\begin{bmatrix} \varepsilon_{11} \\ \varepsilon_{22} \\ \frac{-v}{1-v}(\varepsilon_{11}+\varepsilon_{22}) \\ \varepsilon_{12} \\ \varepsilon_{23} \\ \varepsilon_{31} \end{bmatrix}
-
\begin{bmatrix}
0 & 0 & C_{13}-\frac{v}{1-v}C_{33} \\
0 & 0 & C_{23}-\frac{v}{1-v}C_{33} \\
0 & 0 & 0 \\
0 & 0 & 0 \\
0 & C_{52} & 0 \\
C_{61} & 0 & 0
\end{bmatrix}
\begin{bmatrix} E_1 \\ E_2 \\ E_3 \end{bmatrix}. \tag{23}
$$

The alternative form of the above equation can be obtained after arithmetic manipulations on the terms containing $\varepsilon_{11}$ and $\varepsilon_{22}$. This leads to the change of the constitutive constants of the matrix $D$, with zero contributions of its third row and third column. This alternative form requires either retaining $\varepsilon_{33}$ as the third term of the strain vector or removal of: the third row and the third column in $D$, the third row in $\varepsilon$, and the third row in $C$.

Taking the plane stress assumption into account in the second constitutive relation (16) needs: introduction of (21) into this relation, arithmetic manipulations on the terms containing $E_3$, and summation of the contributions from the terms containing $\varepsilon_{11}$ and $\varepsilon_{22}$. The latter operation leads to the change of the piezoelectric constants in $C^T$, including zero contributions of the third column of this matrix. All these lead to:

$$
\begin{bmatrix} d_1 \\ d_2 \\ d_3 \end{bmatrix}
=
\begin{bmatrix}
0 & 0 & 0 & 0 & 0 & C_{61} \\
0 & 0 & 0 & 0 & C_{52} & 0 \\
C_{13}-\frac{v}{1-v}C_{33} & C_{23}-\frac{v}{1-v}C_{33} & 0 & 0 & 0 & 0
\end{bmatrix}
\begin{bmatrix} \varepsilon_{11} \\ \varepsilon_{22} \\ \varepsilon_{33} \\ \varepsilon_{12} \\ \varepsilon_{23} \\ \varepsilon_{31} \end{bmatrix}
+
\begin{bmatrix}
\gamma_{11} & 0 & 0 \\
0 & \gamma_{22} & 0 \\
0 & 0 & \gamma_{33}+\frac{C_{33}^2}{D(1-v)}
\end{bmatrix}
\begin{bmatrix} E_1 \\ E_2 \\ E_3 \end{bmatrix}. \tag{24}
$$

Note that the third column of $C^T$ and the third row of $\varepsilon$ can be removed from the above relation. The alternative form of (24) does not need any changes in $C^T$ from (16) but requires replacement of $\varepsilon_{33}$ with $\frac{-v}{1-v}(\varepsilon_{11}+\varepsilon_{22})$, as in the Equation (23).

The chosen and written down forms (23) and (24) of the constitutive relations are suitable for further considerations in the next sections of the paper.

Kinematics

The second group of assumptions required by the first-order piezoelectric symmetric-thickness model is of kinematic nature and requires deformation of the normals to the mid-surface onto normals, not necessarily perpendicular to the mid-surface, and no elongation of these normals during deformation. The first assumption reads:

$$u'_j = \frac{1}{2}(u'^b_j + u'^t_j) + \frac{x'_3}{t}(u'^t_j - u'^b_j), \qquad x \in V, \tag{25}$$

where $V$ represents the symmetric-thickness domain or such a part of the complex domain. Above, the local displacements $u'$ can be transformed into the global ones $u$ typical for the applied 3D-based approach, with use of the transformation matrix $\theta^T$ such that $u'_j = \theta_{ij}u_i$. It is obvious that $i = 1, 2, 3$ represent the global directions, while $j = 1, 2, 3$ denote the local directions of which the first two are tangent and the third one is normal to the mid-surface $S_m$ of the symmetric-thickness piezoelectric structure. It can be noticed that the above condition is expressed by the top and bottom local displacements, $u'^t$ and $u'^b$ that can be described with $u'^t = u'$ for $x \in S_t$ and $u'^b = u'$ for $x \in S_b$, where $S_t$ and $S_b$ are the top and bottom surfaces within the symmetric-thickness structure. Note also that the first and second terms of the right hand-side of the above condition represent the local displacements of the mid-surface and the displacements due to rotation of the normal. The latter displacements change linearly along the normal as $t$ is the structure thickness, while $x'_3 \in (-\frac{t}{2}, \frac{t}{2})$ measures the distance from the mid-surface $S_m$.

The second assumption says that the third local displacements of the top and bottom surfaces are the same. Because of this no elongation is possible in this direction, namely:

$$u'^t_3 - u'^b_3 = 0, \qquad x \in V, \tag{26}$$

where $V$ is defined as in (25). One can express the above condition by the global displacements with use of $u'_3 = \theta_{i3}u_i$ and $u'^t_3 = u'_3$ for $x \in S_t$, $u'^b_3 = u'_3$ for $x \in S_b$.

### 3.2. Transition Piezoelectric Models

Three piezoelectric transition models (classical, modified, and enhanced) will be presented. In each of these three models, continuity of the displacement and electric potential fields, on the boundaries between the transition and basic models, is maintained due to three-dimensional approach which lies in application of the three-dimensional displacements and three-dimensional potential for any model of the piezoelectric structure. The hierarchical or first-order models are consistently treated as constrained three-dimensional models.

The classical transition model is characterized with the mechanical field of the three-dimensional model. Only on the boundary with the first-order model, the plane stress assumption is valid. In the case of the modified transition model, the stress field changes gradually from of three-dimensional one (on the boundary with the three-dimensional domain) to the plane stress one (on the boundary with the first-order model). The same stress field description is valid also for the enhanced transition model.

The enhanced transition model is characterized with the gradually changing assumption of no elongation of the lines perpendicular to the mid-surface of the domain. This assumption changes from its total validity on the boundary with the first-order model to its lack on the boundary with the three-dimensional model. Note that, in the cases of the classical and modified models, this assumption is valid only on the boundary with the first-order model. Apart from this boundary, this assumption does not apply.

### 3.2.1. The Classical Approach

Stress State

The mechanical and electric fields within the classical transition piezoelectric model are defined as in the case of the three-dimensional piezoelctricity characterized with the constitutive Equation (16), completed with the definitions (13), (18), and (19).

Kinematic Assumptions

In this classical transition model, the kinematic assumption of the first-order mechanical theory consisting in deformation of the normals to the mid-surface into normals during deformation and the lack of elongation of these normals are valid on the boundary $R$ between the transition and first-order theories, i.e.,

$$u'_j = \frac{1}{2}(u'^b_j + u'^t_j) + \frac{x'_3}{t}(u'^t_j - u'^b_j), \quad x \in \overline{V} \cap R \equiv R$$

$$0 = u'^t_3 - u'^b_3, \quad x \in \overline{V} \cap R \equiv R,$$

$$(27)$$

where $\overline{V} = V \cup \partial V$, $\partial V \equiv S$. As in the case of the first-order model, the above conditions can be expressed by the global displacements with use of the same relations as before. Note that the mid-surface within the transition domain may only be defined on the boundary $R$ between the first-order and transition models.

### 3.2.2. the Modified Transition Model

The Assumed Stresses

The second approach is based on the transition character of the plane stress assumption. This assumption is valid on the boundary with the first-order piezoelectric model. It does not hold, however, on the boundary with the hierarchical higher-order or three-dimensional piezoelectric model, i.e., three-dimensional stress state is present on this boundary. Between these two boundaries, the stress state is intermediate, namely:

$$\sigma_{33} = D\left\{ v\varepsilon_{11} + v\varepsilon_{22} + (1-v)\left\langle \alpha\varepsilon_{33} + (1-\alpha)\left[\frac{-v}{1-v}(\varepsilon_{11}+\varepsilon_{22}) + \frac{C_{33}}{D(1-v)}E_3\right]\right\rangle\right\} - C_{33}E_3.$$

$$(28)$$

Above, the function $\alpha = \alpha(S_m) \in \langle 0, 1\rangle$, with $S_m$ being the mid-surface of the thick- or thin-walled part of the transition domain, is a blending function equal to 1 at the boundary with the three-dimensional model, and 0 at the boundary with the first-order model. Consequently, the definition (28) becomes identical with (20) for $\alpha = 1$, and with (22) for $\alpha = 0$.

Taking the above equation into account in the third row of the first relation (16) leads to: $\sigma = \sigma_1 - \sigma_2$, where the first component equals:

$$\sigma_1 = \begin{bmatrix} D_{11} & D_{12} & D_{13} & 0 & 0 & 0 \\ D_{21} & D_{22} & D_{23} & 0 & 0 & 0 \\ D_{31} & D_{23} & D_{33} & 0 & 0 & 0 \\ 0 & 0 & 0 & D_{44} & 0 & 0 \\ 0 & 0 & 0 & 0 & D_{55} & 0 \\ 0 & 0 & 0 & 0 & 0 & D_{66} \end{bmatrix} \begin{bmatrix} \varepsilon_{11} \\ \varepsilon_{22} \\ \alpha\varepsilon_{33} + (1-\alpha)\frac{-v}{1-v}(\varepsilon_{11}+\varepsilon_{22}) \\ \varepsilon_{12} \\ \varepsilon_{23} \\ \varepsilon_{33} \end{bmatrix}, \quad (29)$$

while the second one is equal to:

$$\sigma_2 = \begin{bmatrix} 0 & 0 & C_{13}-(1-\alpha)\frac{\nu}{1-\nu}C_{33} \\ 0 & 0 & C_{23}-(1-\alpha)\frac{\nu}{1-\nu}C_{33} \\ 0 & 0 & \alpha C_{33} \\ 0 & 0 & 0 \\ 0 & C_{52} & 0 \\ C_{61} & 0 & 0 \end{bmatrix} \begin{bmatrix} E_1 \\ E_2 \\ E_3 \end{bmatrix}. \tag{30}$$

In addition, the second equation (16) can be divided into two components, i.e., $d = d_1 + d_2$, where the first component is:

$$d_1 = \begin{bmatrix} 0 & 0 & 0 & 0 & 0 & C_{61} \\ 0 & 0 & 0 & 0 & C_{52} & 0 \\ C_{13}-(1-\alpha)\frac{\nu}{1-\nu}C_{33} & C_{23}-(1-\alpha)\frac{\nu}{1-\nu}C_{33} & \alpha C_{33} & 0 & 0 & 0 \end{bmatrix} \begin{bmatrix} \varepsilon_{11} \\ \varepsilon_{22} \\ \varepsilon_{33} \\ \varepsilon_{12} \\ \varepsilon_{23} \\ \varepsilon_{33} \end{bmatrix}, \tag{31}$$

while the second component reads

$$d_2 = \begin{bmatrix} \gamma_{11} & 0 & 0 \\ 0 & \gamma_{22} & 0 \\ 0 & 0 & \gamma_{33}+(1-\alpha)\dfrac{C_{33}^2}{D(1-\nu)} \end{bmatrix} \begin{bmatrix} E_1 \\ E_2 \\ E_3 \end{bmatrix}. \tag{32}$$

Note that, when $\alpha = 1$, the relations (29)–(32) transform into Equations (16)–(19), while, for $\alpha = 0$, into Equations (23) and (24).

The Assumed Kinematics

As it comes to the kinematic assumptions within the modified transition element, they are the same as in the case of the classical elements, i.e., the conditions (27) are valid on the boundary $R$ between the transition and first-order models.

### 3.2.3. The Enhanced Transition Model

Stress State

In the case of the enhanced transition model of piezoelectricity, constitutive relations are the same as in the modified model. This means that (29) and (30), as well as (31) and (32), hold.

Displacement assumptions

The starting point for defining the kinematic assumption within this type of the transition model is the definition of the displacement field of the 3D-based hierarchical shell models [13,14]. This field can be expressed with use of the local or global displacements,

$u'$ or $u$, which are related by the transformation matrix $\theta^T$, i.e., $u' = \theta^T u$. The local displacement components are equal to:

$$
\begin{aligned}
u'_j &= \theta_{ij}u_i = \sum_{l=0}^{I} f_l(x'_3)\theta_{ij}u^l_i = \sum_{l=0}^{I} f_l(x'_3)u'^l_j \\
&= f_0(x'_3)u'^0_j + \sum_{l=1}^{I-1} f_l(x'_3)u'^l_j + f_I(x'_3)u'^I_j \\
&= \left(\frac{1}{2} - \frac{x'_3}{t}\right)u'^0_j + \left(\frac{1}{2} + \frac{x'_3}{t}\right)u'^I_j + \sum_{l=1}^{I-1} f_l(x'_3)u'^l_j \\
&= \frac{1}{2}(u'^0_j + u'^I_j) + \frac{x'_3}{t}(u'^I_j - u'^0_j) + \sum_{l=1}^{I-1} f_l(x'_3)u'^l_j, \quad x \in V,
\end{aligned}
\tag{33}
$$

where $V$ is the symmetric-thickness domain. In the first line above, $u'^l_j, l = 1, 2, ..., I$, are the functions describing changes of displacements in the longitudinal directions. In the case of the symmetric-thickness geometry, these function are the same for each $l$. The functions $f_l(x'_3), l = 1, 2, ..., I$ stand for the polynomial functions in the direction normal to the mid-surface. Along this direction the coordinate $x'_3 \in (-\frac{t}{2}, \frac{t}{2})$ is employed. The polynomial functions in this direction span the respective polynomial space of order $I$. Hence, the quantity $I$ represents the hierarchical model order. It is worth mentioning that with $I \to \infty$ one reaches three-dimensional elasticity description of the above displacement field.

Following our propositions from Reference [14,16], in the second and third lines of (33), we define the first and the last functions, $f_0(x'_3)$ and $f_I(x'_3)$, as linear ones and the rest functions $f_l(x'_3), l = 1, 2, ..., I - 1$ as polynomials of order $I$. After some arithmetic manipulations, one can reach the final form (line) of (33).

In the case of the enhanced transition piezoelectric model, its transverse displacement field is assumed to change from the state of no elongation of the lines perpendicular to the mid-surface to the state of lack of such an assumption. In the same manner, the order of the transverse displacement field is assumed to change from $I = 1$ to an arbitrary value $I$. So as to implement such changes, the blending function $\alpha$, now playing the role of a gradually switching function, is applied, namely:

$$
u'_j = \frac{1}{2}(u'^0_j + u'^I_j) + \frac{x'_3}{t}(u'^0_k - u'^I_k) + \alpha\frac{x'_3}{t}(u'^0_3 - u'^I_3) + \alpha\sum_{l=1}^{I-1} f_l(x'_3)u'^l_j, \quad x \in V, \tag{34}
$$

where $V$ is the symmetric-thickness domain or the domain with the mid-surface defined on $R$ at least, while $j = 1, 2, 3$ and $k = 1, 2$. The later index $k$ corresponds to the first two, longitudinal directions, while the index 3 is used for the transverse direction. It can be noticed that with $\alpha \equiv 1$, (34) becomes (33) which characterizes the hierarchical or three-dimensional models, while, for $\alpha \equiv 0$, (34) becomes (25) and (26) which correspond to the first-order model. In this case, $l = 0$ and $l = 1$ refer to the bottom ($l \equiv b$) and top ($l \equiv t$) surfaces.

## 4. Algorithms of the Transition Piezoelectric Elements

The starting point for our finite element considerations is the variational Formulations (8)–(11). Derivation of the corresponding finite element formulation needs discretization of the domain $V$ and into finite element subdomains $V_e$ and introduction of polynomial interpolation of the displacements $u$ and electric potential $\phi$ (see Reference [11,12]) with use of the global vectors of the nodal, displacement $q^{hpq}$ and potential $\varphi^{h\pi\rho}$, degrees of freedom (dofs). The applied interpolations are suited for the $hpq$- and $h\pi\rho$-adaptive approximations of the displacements and potential fields, with $h$ being the common element size for both fields and independent longitudinal and transverse approximation orders $p$, $q$ and $\pi$, $\rho$ within the mechanical and electric fields, respectively. All these lead to

$$K_M q^{hpq} - K_C \varphi^{h\pi\rho} = F_V + F_S,$$
$$K_C^T q^{hpq} + K_E \varphi^{h\pi\rho} = F_Q, \tag{35}$$

where the components of the first equation correspond to: the first equation (9), the second Equation (9) and the first Equation (11), while the components of the second line result from: the second definition (11), the first definition (10) and the second definition (10). The global (of the entire domain) terms $K_M$, $K_C$, $K_E$ represent the global stiffness matrix, the global characteristic matrix of piezoelectricity, and the global characteristic matrix of dielectricity. The global vectors of nodal mass and surface forces are denoted as $F_V$ and $F_S$, while the global vector of the surface charges as $F_Q$.

### 4.1. Normalized Geometry of the Transition Elements

For three (classical, modified, enhanced) transition piezoelectric models two principal geometries of the prismatic elements has to be introduced in order to be able to join the basic elements corresponding to the three-dimensional or hierarchical models with the first-order model of piezoelectricity.

The example of the first transition geometry I, presented in Figure 1, includes one vertical edge (vertex nodes: $a_1, a_4$) corresponding to the 3D-based first-order shell theory [13,54] within the mechanical field. The other part of this field corresponds to the 3D-based hierarchical shell [13,14] or three-dimensional elasticity models [13,53]. Above, the following linear vertex ($a_2, a_3, a_5, a_6$) nodes and generalized: horizontal mid-edge ($a_7, a_8$, ..., $a_{12}$), vertical mid-edge ($a_{13}, a_{14}$), mid-base ($a_{15}, a_{16}$), mid-side ($a_{17}, a_{18}, a_{19}$) and middle ($a_{20}$) nodes are applied within the geometry and displacement fields. For the element I, the electric field corresponds to the hierarchical symmetric-thickness or three-dimensional dielectricity models [11,16]. The linear vertex ($b_1, b_2, ..., b_6$) nodes, the generalized: horizontal mid-edge ($b_7, b_8, ..., b_{12}$), vertical mid-edge ($b_{13}, b_{14}, b_{15}$), mid-base ($b_{16}, b_{17}$), mid-side ($b_{18}, b_{19}, b_{20}$) and middle ($b_{21}$) nodes characterize the approximated geometry and electric potential fields of the element.

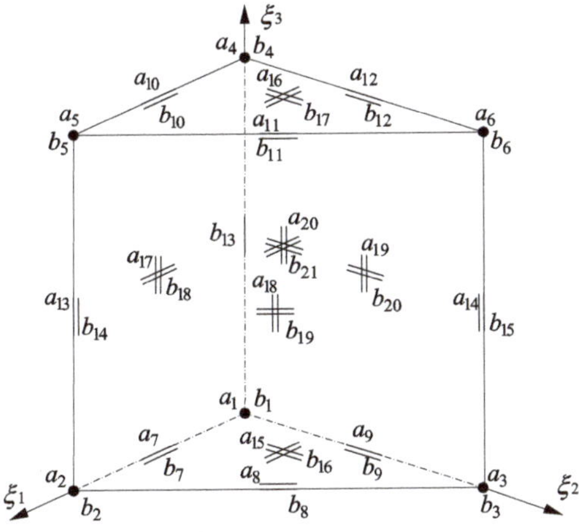

**Figure 1.** An exemplary principal geometry I of the transition elements.

It is worth mentioning that the generalized (varying-order) nodes of both types may contain multiple degrees of freedom along the edges, within the bases and sides, and in the interior of the element. The related finite element approximations within the proposed element are all based on the general rules proposed by Demkowicz and others [1,2].

In the example of the second transition geometry II, shown in Figure 2, one deals with two vertical edges (nodes $a_1, a_4$ and $a_2, a_5$) and the side (nodes $a_7, a_{10}$) between these edges, all corresponding to the first-order mechanical model. The rest of the element (nodes $a_3, a_6, a_8, a_9, a_{11}, a_{12}, a_{13}, a_{14}, a_{15}, a_{16}, a_{17}, a_{18}$) conforms to the hierarchical shell model or the three-dimensional model of elasticity. These mechanical models are coupled with the symmetric-thickness hierarchical or three-dimensional models of dielectricity. The same nodes $b_1, b_2, ..., b_{21}$, as for the geometry I, are employed in this case.

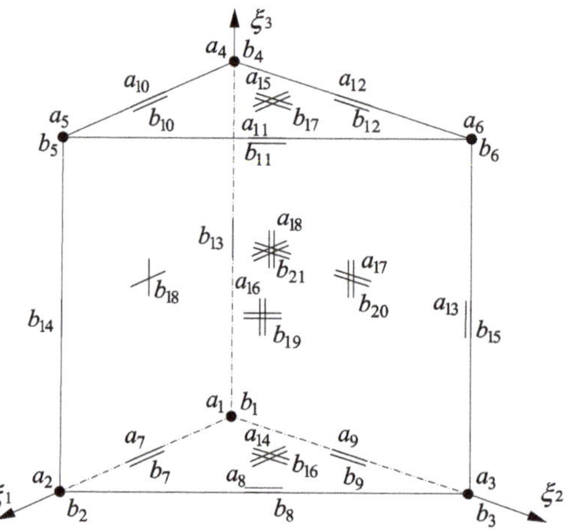

**Figure 2.** An exemplary principal geometry II of the transition elements.

*4.2. Element Characteristic Matrices and Vectors*

Replacement of the global quantities introduced in (24) by the corresponding sums of the element (or local) quantities leads to

$$\sum_e \left( \overset{e}{k}_M \overset{e}{q}^{hpq} - \overset{e}{k}_C \overset{e}{\varphi}^{h\pi\rho} \right) = \sum_e \left( \overset{e}{f}_V + \overset{e}{f}_S \right),$$
$$-\sum_e \left( \overset{e}{k}_C^T \overset{e}{q}^{hpq} + \overset{e}{k}_E \overset{e}{\varphi}^{h\pi\rho} \right) = -\sum_e \overset{e}{f}_Q, \tag{36}$$

where $e$ is the element number, while $\overset{e}{q}^{hpq}$ and $\overset{e}{\varphi}^{h\pi\rho}$ stand for the element, displacements and potential, dofs vectors.

The element stiffness matrix is defined in the standard way, namely:

$$\overset{e}{k}_M = \int_0^1 \int_0^1 \int_0^{-\xi_2+1} \overset{e}{B}^T D \overset{e}{B} \det(J)\, d\xi_1 d\xi_2 d\xi_3, \tag{37}$$

with $\overset{e}{B}$ relating strains $\varepsilon$ with the element displacement dofs $\overset{e}{q}^{hpq}$ and, hence, named the element strain-displacement matrix. The matrix $D$ represents the matrix of elasticities for the isotropy case under consideration, while the term $\det(J)$ is the determinant of the Jacobian matrix transforming the global coordinates $x$ into normalized ones $\xi = [\xi_1, \xi_2, \xi_3]^T$. Note that the applied normalized prismatic geometry of the element $e$ can be characterized with $V_e = \{(\xi_1, \xi_2, \xi_3) \in [0, -\xi_2 + 1] \times [0, 1] \times [0, 1]\}$. In the case of the thick- or thin-walled geometry, the first two normalized directions coincide with the longitudinal shell directions and the third normalized direction is the same as the transverse shell direction.

The definition of the element characteristic matrix of dielectricity reads

$$\overset{e}{k}_E = \int_0^1 \int_0^1 \int_0^{-\xi_2+1} \overset{e}{b}^T \gamma \overset{e}{b} \det(J)\, d\xi_1 d\xi_2 d\xi_3, \tag{38}$$

where $\overset{e}{b}$ allows expressing the electric field $E$ with the element electric potential dofs $\overset{e}{\varphi}^{h\pi\rho}$. It is called the element field-potential matrix. The matrix $\gamma$ is the constitutive constant matrix of orthotropic dielectricity under constant strain.

The element characteristic matrix of piezoelectricity coupling the mechanical and electric fields is defined as follows:

$$\overset{e}{k}_C = \int_0^1 \int_0^1 \int_0^{-\xi_2+1} \overset{e}{B}^T C \overset{e}{b} \det(J)\, d\xi_1 d\xi_2 d\xi_3, \tag{39}$$

with $C$ being the constitutive matrix of orthotropic piezoelectricity under constant strain.

The element nodal forces vector due to the mass loading $f$ is defined in the standard way, i.e.,

$$\overset{e}{f}_V = \int_0^1 \int_0^1 \int_0^{-\xi_2+1} \overset{e}{N}^T f \det(J)\, d\xi_1 d\xi_2 d\xi_3, \tag{40}$$

with $\overset{e}{N}$ standing for the standard shape functions matrix corresponding to the vectorial displacement field.

In addition the element nodal forces due to the surface loading $p$ is defined in the standard way, for the bases and lateral sides, respectively:

$$\overset{e}{f}_S = \int_0^1 \int_0^{-\xi_2+1} \overset{e}{N}^T p \operatorname{cof}(J)\, d\xi_1 d\xi_2$$
$$\overset{e}{f}_S = \int_0^1 \int_0^1 \overset{e}{N}^T p \operatorname{cof}(J)\, d\eta_i d\xi_3. \tag{41}$$

Above, $\eta_i$, $i = 1, 3$ is identical to the directions of the triangle edges, i.e., either $\xi_1$ or $\xi_2$ or $1 - \xi_1 - \xi_2$, while the coefficient $\operatorname{cof}(J)$ transforms the surface element of the part $dS_e$ of the domain surface $S$ from the global coordinates to the normalized ones.

Finally, the element nodal charges due to the charge density $c$ are equal to

$$\overset{e}{f}_Q = \int_0^1 \int_0^{-\xi_2+1} \overset{e}{n}^T c \operatorname{cof}(J)\, d\xi_1 d\xi_2$$
$$\overset{e}{f}_Q = \int_0^1 \int_0^1 \overset{e}{n}^T c \operatorname{cof}(J)\, d\eta_i d\xi_3 \tag{42}$$

for the element bases and sides, respectively, with $\overset{e}{n}$ denoting the standard shape function matrix suitable for the finite element approximation of the scalar electric potential field.

In the next subsection, we will present the changes that have to be implemented in the above finite element formulation depending on the applied transition model.

### 4.3. Strain-Displacement Matrix

Three cases of the classical, modified, and enhanced transition elements corresponding to the analogous transition models are of interest in this section.

### 4.3.1. The Transition Element Based on the Classical Model

The strain-displacement matrix for the classical transition model is assumed in the form corresponding to the three-dimensional elasticity and the hierarchical shell models, for the three-dimensional and thick- or thin-walled geometries, respectively. This means that the three-dimensional strain and stress state is applied for this classical model. This state is characterized by

$$\varepsilon = \overset{e}{\boldsymbol{B}}\,\overset{e}{\boldsymbol{q}}^{hpq}$$ (43)

and reflects the relation between the local strains $\varepsilon$ and global nodal displacement dofs $\overset{e}{\boldsymbol{q}}^{hpq}$. The local strain directions are coincident with two tangent (the first two or two longitudinal) and one normal (the third or transverse) directions within the structure.

The strain-displacement matrix $\overset{e}{\boldsymbol{B}}$ will also be denoted as $\overset{e}{\boldsymbol{B}}{}^{1} \equiv \overset{e}{\boldsymbol{B}}$ in our further considerations. The form of this matrix is derived from the standard approach applied to five-parameter thick-shell elements of the second order [63]. Here, we adopt it to the needs of the 3D-based hierarchical shell model. This form reads:

$$\overset{e}{\boldsymbol{B}} = [..., \boldsymbol{B}_k, ...], \quad \overset{e}{\boldsymbol{B}}{}^{T} = [..., \boldsymbol{B}_k^{T}, ...]^{T}.$$ (44)

The above blocks $\boldsymbol{B}_k \equiv \boldsymbol{B}_k^{1}$ of the strain-displacement matrix correspond to the element displacement dofs at the element nodes and are composed of the following components:

$$\boldsymbol{B}_k = \begin{bmatrix} B_{k,11} & B_{k,12} & B_{k,13} \\ B_{k,21} & B_{k,22} & B_{k,23} \\ B_{k,31} & B_{k,32} & B_{k,33} \\ B_{k,41} & B_{k,42} & B_{k,43} \\ B_{k,51} & B_{k,52} & B_{k,53} \\ B_{k,61} & B_{k,62} & B_{k,63} \end{bmatrix}.$$ (45)

Any term $B_{k,mn}$, $m = 1, 2, .., 6$, $n = 1, 2, 3$ of each block $\boldsymbol{B}_k$, where $m$ corresponds to each of six components of the strain vector and $n$ to each global direction of the displacement dof vector, can be determined as equal to:

$$B_{k,1n} = \beta_{k,1}\theta_{1n}; \quad B_{k,4n} = \beta_{k,1}\theta_{2n} + \beta_{k,2}\theta_{1n}$$

$$B_{k,2n} = \beta_{k,2}\theta_{2n}; \quad B_{k,5n} = \beta_{k,2}\theta_{3n} + \beta_{k,3}\theta_{2n}$$ (46)

$$B_{k,3n} = \beta_{k,3}\theta_{3n}; \quad B_{k,6n} = \beta_{k,3}\theta_{1n} + \beta_{k,1}\theta_{3n},$$

with the terms $\theta_{pn}$ representing components of the transformation matrix $\theta$ which converts the local Cartesian directions $p = 1, 2, 3$ of the displacement dofs into the global ones. The terms $\beta_{k,p}$ can be calculated from

$$\beta_{k,p} = P_{k,1}A^{1p} + P_{k,2}A^{2p} + P_{k,3}A^{3p},$$ (47)

where the quantities $P_{k,r} = \partial N_k / \partial \zeta_r$ are the partial derivatives, with respect to the normalized directions $r = 1, 2, 3$, of the shape function $N_k$ from the corresponding shape function matrix. The auxiliary quantities $A_{rp}$ are equal to:

$$A^{rp} = j^{r1}\theta_{1p} + j^{r2}\theta_{2p} + j^{r3}\theta_{3p}, \tag{48}$$

with the terms $j^{rn}$ representing terms of the inverse Jacobian matrix $J^{-1}$ and $\theta_{np}$ being the components of the transposed transformation matrix $\theta^T$. The latter matrix transforms derivatives with respect to the global Cartesian directions into derivatives with respect to the local Cartesian directions. The terms of the transformation matrices can be obtained from:

$$\theta = \begin{bmatrix} \theta_{11} & \theta_{12} & \theta_{13} \\ \theta_{21} & \theta_{22} & \theta_{23} \\ \theta_{31} & \theta_{32} & \theta_{33} \end{bmatrix} = \begin{bmatrix} u_1 & v_1 & w_1 \\ u_2 & v_2 & w_2 \\ u_3 & v_3 & w_3 \end{bmatrix}, \tag{49}$$

where the unit vectors components $u_p = U_p/U,\ v_p = V_p/V,\ w_p = W_p/W,\ p = 1,2,3$ can be obtained from the vectors $U$, $V$ and $W$ equal to:

$$W = \begin{bmatrix} \partial x_1/\partial \xi_1 \\ \partial x_2/\partial \xi_1 \\ \partial x_3/\partial \xi_1 \end{bmatrix} \times \begin{bmatrix} \partial x_1/\partial \xi_2 \\ \partial x_2/\partial \xi_2 \\ \partial x_3/\partial \xi_2 \end{bmatrix}, \quad U = \begin{bmatrix} \partial x_1/\partial \xi_1 \\ \partial x_2/\partial \xi_1 \\ \partial x_3/\partial \xi_1 \end{bmatrix}, \quad V = W \times U, \tag{50}$$

where the partial derivatives $\partial x_n/\partial \xi_r = j_{nr},\ n,r = 1,2,3$ are the terms of the Jacobian matrix $J$.

### 4.3.2. The Modified and Enhanced Transition Elements

In the case of these two types of the transition element, the constitutive relations (29) and (30) hold within the coupled mechanical field. The strain and stress state changes from the three-dimensional one to the plane stress state. This can be expressed with the following relation

$$\overset{e}{\varepsilon} = \overset{e}{B} q^{hpq} = \left[ \alpha \overset{e}{B^1} + (1-\alpha) \overset{e}{B^2} \right] q^{hpq}, \tag{51}$$

where the components $\overset{e}{B^1}$ and $\overset{e}{B^2}$ correspond to the three-dimensional and plane stress states. The first component is defined in accordance with (44)–(50). The second component is divided into blocks corresponding to the displacement dofs at the element nodes, namely

$$\overset{e}{B^2} = [..., B_k^2, ...]. \tag{52}$$

Due to the plane stress assumption (21), any block $k$ of the second component needs the following modification:

$$B_k^2 = \begin{bmatrix} B_{k,11} & B_{k,12} & B_{k,13} \\ B_{k,21} & B_{k,22} & B_{k,23} \\ \frac{-\nu}{1-\nu}(B_{k,11}+B_{k,21}) & \frac{-\nu}{1-\nu}(B_{k,12}+B_{k,22}) & \frac{-\nu}{1-\nu}(B_{k,13}+B_{k,23}) \\ B_{k,41} & B_{k,42} & B_{k,43} \\ B_{k,51} & B_{k,52} & B_{k,53} \\ B_{k,61} & B_{k,62} & B_{k,63} \end{bmatrix}, \tag{53}$$

where the terms $B_{k,mn},\ m = 1,2,4,5,6,\ n = 1,2,3$ are defined in accordance with (46)–(50).

Because of the modified definition (51) of the strain vector $\varepsilon$, also any block $k$ of the strain-displacement matrix has to be modified in accordance with

$$B_k = \alpha B_k^1 + (1-\alpha) B_k^2 =$$

$$
\begin{bmatrix}
B_{k,11} & B_{k,12} & B_{k,13} \\
B_{k,21} & B_{k,22} & B_{k,23} \\
\alpha B_{k,31}+(1-\alpha)\frac{-\nu}{1-\nu}(B_{k,11}+B_{k,21}) & \alpha B_{k,32}+(1-\alpha)\frac{-\nu}{1-\nu}(B_{k,12}+B_{k,22}) & \alpha B_{k,33}+(1-\alpha)\frac{-\nu}{1-\nu}(B_{k,13}+B_{k,23}) \\
B_{k,41} & B_{k,42} & B_{k,43} \\
B_{k,51} & B_{k,52} & B_{k,53} \\
B_{k,61} & B_{k,62} & B_{k,63}
\end{bmatrix}
\tag{54}
$$

The blending function is defined in accordance with $\alpha = \alpha(\xi_1, \xi_2, \frac{1}{2})$, i.e., as a function of the two normalized directions, $\xi_1$ and $\xi_2$, tangent to the mid-surface defined with $\xi_3 = \frac{1}{2}$. Note that, for $\alpha \equiv 1$ and $\alpha \equiv 0$, the definition (54) transforms into (45) and (53), suitable for the basic (three-dimensional and first-order) piezoelectric elements, respectively.

### 4.4. Field-Potential Matrix

For any type of the transition models: (16), (23), (24), and (30) plus (32), the definition of the electric field vector $E$ remains unchanged. Because of this also the field-potential matrix has the same form for the classical, modified, and enhanced transition elements. This form corresponds to the three-dimensional or hierarchical symmetric-thickness dielectric models. For these models, the following relation holds:

$$E = \overset{e}{b}\, \overset{e}{\varphi}{}^{h\pi\rho} . \tag{55}$$

Above, the locally defined electric field $E$ is expressed with the scalar potential dofs $\overset{e}{\varphi}{}^{h\pi\rho}$. The form of the electric field-potential matrix under consideration is

$$\overset{e}{b} = [..., b_l, ...], \qquad \overset{e}{b}{}^{T} = [..., b_l^{T}, ...]^{T}, \tag{56}$$

where the blocks $b_l$ of the field-potential matrix correspond to the electric potential dofs of the element. The number of these blocks may be different to the number of the displacement dofs as the electric potential field is $h\pi\rho$-interpolated, while the displacement field is $hpq$-interpolated. The form of these blocks is:

$$b_l = \begin{bmatrix} b_{l,1} \\ b_{l,2} \\ b_{l,3} \end{bmatrix}. \tag{57}$$

The terms $b_{l,p}$, where $p = 1,2,3$ are the local Cartesian directions, are equal to

$$b_{l,p} = p_{l,1} A^{1p} + p_{l,2} A^{2p} + p_{l,3} A^{3p}. \tag{58}$$

Note that the quantities $p_{l,r} = \partial n_l / \partial \xi_r$, $r = 1,2,3$ are the derivatives of the terms $n_l$ of the shape function matrix with respect to the normalized directions $r = 1,2,3$. The shape function $n_l$ corresponds to the electric potential dof $l$ at the element node. The auxiliary terms $A_{rp}$, $r, p = 1,2,3$ can be calculated from

$$A^{rp} = j^{r1}\theta_{1p} + j^{r2}\theta_{2p} + j^{r3}\theta_{3p}, \tag{59}$$

i.e., analogously as in the case of the strain-displacement matrix. As before, the terms $\theta_{np}$ belong to the transformation matrix $\theta$ which transforms the local Cartesian directions $p = 1,2,3$ into global Cartesian ones $n = 1,2,3$. The quantities $j^{rn}$ are the terms of the inverse Jacobian matrix $J^{-1}$ again.

### 4.5. Constitutive Constant Matrices

In the case of the classical, modified, and enhanced transition elements, the constitutive, elasticity constant matrix from (37) takes the same form (13).

The classical transition element requires the dielectricity and piezoelectricity constant matrices, in (38) and (39), in the form defined by (18) and (19), respectively. As far as the modified and enhanced transition elements are concerned, one has to apply, in relations (38) and (39), the definitions included in (32) and (30), respectively.

### 4.6. Blending Functions

As said above, blending functions describe the linear change of the stress state within transition elements from the three-dimensional to plane stress state. This change manifests itself by the presence of function $\alpha = \alpha(\xi_1, \xi_2)$ in the strain definition of the modified and enhanced transition elements and in the piezoelectric and dielectric constant matrices for these two transition elements. Note that $\xi_i$, $i = 1, 2, 3$ are the normalized directions within an element and $\xi_3$ defines location of the reference surface within the element.

The blending function corresponding to the transition element of the geometry I is presented in Figure 3. The function is represented by the triangle spanned at the vertices $a_1, a_5, a_6$ and takes the value of 0 for the vertical edge 1 corresponding to the mechanical field of the first-order in the transverse direction $\xi_3$ and to the electric field of an arbitrary order in this direction. The blending function value of 1 corresponds to the mechanical and electric fields of arbitrary orders for the vertical edges 2 and 3. This function can be expressed with the affine coordinates $\lambda_1$, $i = 1, 2, 3$ of the triangular element base or the normalized coordinates $\xi_1$ and $\xi_2$ as

$$\alpha = 1 - \lambda_i, \quad i = 1, 2, 3, \tag{60}$$

where $\lambda_1 = 1 - \xi_1 - \xi_2$, $\lambda_2 = \xi_1$ and $\lambda_3 = \xi_2$.

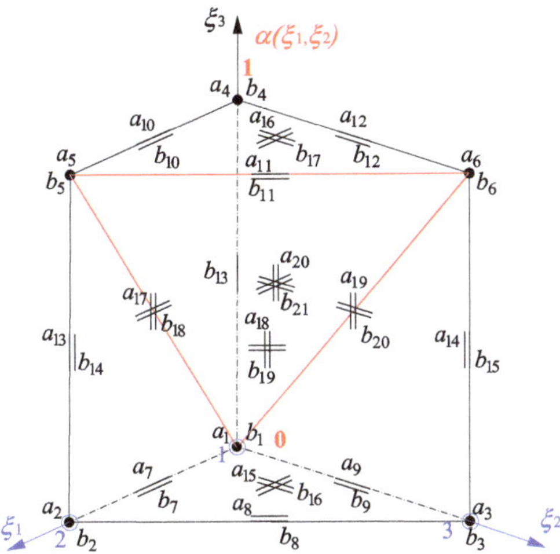

**Figure 3.** The blending function for an example of the principal transition geometry I.

The geometry II of the transition element and the respective blending function are presented in Figure 4. The function is represented by the triangle spanned at the vertices $a_1, a_2, a_6$ and equals 0 for two edges 1 and 2 corresponding to the mechanical field of the first-order in the transverse direction. This value is equal to 1 for the vertical edge 3 where the mechanical field is of arbitrary order in the transverse direction $\xi_3$. The order of the electric field is arbitrary for all three vertical edges. The function can be expressed in the following way:

$$\alpha = \lambda_i, \quad i = 1, 2, 3. \tag{61}$$

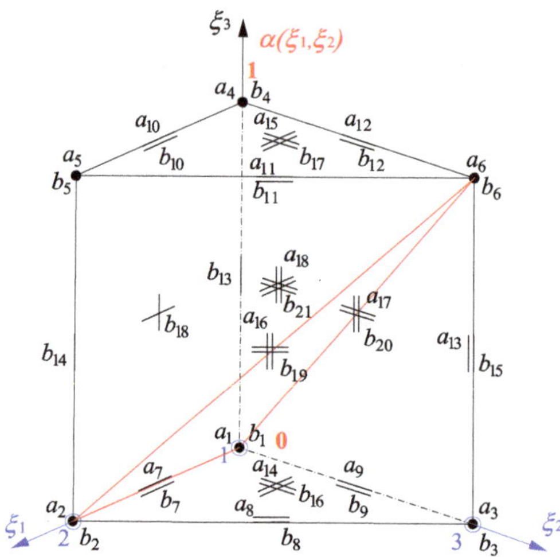


**Figure 4.** The blending function for an example of the principal transition geometry II.

It should be noted that due to introduction of the linear blending functions into the vectors and matrices from (29)–(32), the polynomial order the integrands of (37)–(39) increases. This needs the respective increase of the number of the Gauss points in the numerical integration of these integrands.

*4.7. Shape Function Matrices*

The shape function matrices $\overset{e}{N}$ corresponding to the displacement field of the transition piezoelectric elements are of the standard form for any type of the element. This form reads

$$\overset{e}{N} = [..., N_k, ...], \quad \overset{e}{N}{}^T = [..., N_k^T, ...]^T, \tag{62}$$

where the blocks $N_k$ are diagonal and correspond to the vectorial displacement dofs $k$ at the element node:

$$N_k = \begin{bmatrix} N_k & 0 & 0 \\ 0 & N_k & 0 \\ 0 & 0 & N_k \end{bmatrix}. \tag{63}$$

In the case of the electric potential, the following standard form of the shape function matrix $\overset{e}{n}$ is valid for the piezoelectric transition element of any type:

$$\overset{e}{n} = [..., n_l, ...], \quad \overset{e}{n}{}^T = [..., n_l^T, ...]^T, \tag{64}$$

where the blocks $n_l$, corresponding to the scalar electric potential dof $l$ at the element node, reduce to:

$$n_l = \begin{bmatrix} n_l \end{bmatrix}. \tag{65}$$

It is worth noticing that the forms (44)–(45) and (64)–(65) corresponding to three proposed transition models are exactly the same as in the case of the three-dimensional and hierarchical models of piezoelectricity.

*4.8. Kinematic Constraints within the Elements*

In this section, the algorithms for imposition of the constraints of no elongation of the normals to the mid-surface, and for application of the constraints imposed on the generalized (varying-order) displacement dofs, as well, will be presented.

4.8.1. The Classical and Modified Elements

Stiffness Matrix Modification

In this case the constraints of no elongation of the normals are assumed on the boundary $R_e$ between the first-order and transition elements. These constraints are imposed with the penalty method where the parameter $P$ is employed. It represents the inverse of the penalty parameter and is a number sufficiently large to overwrite the stiffness matrix terms with the constraint terms and to replace the finite element equation with the constraint equation. Our approach takes advantage of the presentation of the method given in Reference [1,2]. Imposition of the constraints under consideration needs the following modification of the stiffness matrix:

$$\overset{e}{k}_M = (\overset{e}{R}{}^{-1})^T \left( \overset{e}{R}{}^T \overset{e}{k}_M \overset{e}{R} + Z^T \overset{e}{k}_P Z \right) \overset{e}{R}{}^{-1}, \tag{66}$$

where $\overset{e}{k}_P$ is called the penalty matrix. Its terms include the mentioned parameter $P$.

Note that no modification of the force vectors $\overset{e}{f}_V + \overset{e}{f}_S$ due to penalty method is required as one deals with zero constraints in the paper, e.g., the constraints of no elongation of the normals to the mid-surface. The same refers to other constraints presented in this paper.

The arithmetic operator matrices $\overset{e}{R}$ and $\overset{e}{R}{}^{-1}$ of the element $e$ from (66) are capable of converting the top and bottom displacement dofs into their mean sums and differences, and conversely, respectively. The form of the first operator is related to the division of the element dofs into two groups: the pairs of top and bottom displacement dofs $q^1$ and all other displacement dofs $q^2$, namely:

$$\overset{e}{q}{}^{hpq} = \begin{bmatrix} q^1 \\ q^2 \end{bmatrix}, \tag{67}$$

where the two blocks consist of the following component sub-blocks:

$$q^1 = [..., q_k, ..., q_l, ...]^T, \quad q^2 = [..., q_m, ...]^T. \tag{68}$$

The top and bottom dofs, $k$ and $l$, are associated with the nodes belonging to $R_e = R|_{V_e}$ only. In the case of the transition element geometry I (Figure 3), the top dofs $k$ are associated with the node $a_4$, the bottom dofs $l$ with the node $a_1$, the other dofs $m$ with the nodes $a_2, a_3, a_5, a_6, a_7, ..., a_{18}$. In the case of the geometry II, one has the following assignments: the top dofs $k$ are associated with the nodes $a_4, a_5, a_{10}$, the bottom dofs $l$ with the nodes $a_1, a_2, a_7$, the other dofs $m$ with the nodes $a_3, a_6, a_8, a_9, a_{11}, a_{12}, a_{13}, ..., a_{20}$.

Modifying Operators

Following the above assignments, one defines

$$\overset{e}{R} = \begin{bmatrix} R^1 & 0 \\ 0 & R^2 \end{bmatrix}. \tag{69}$$

Within the block $\boldsymbol{R}^1$, the following overlapping sub-blocks can be distinguished

$$\boldsymbol{R}^1 = \begin{bmatrix} & \vdots & & \vdots & \\ \cdots & \boldsymbol{R}_{kk} & \cdots & \boldsymbol{R}_{kl} & \cdots \\ & \vdots & & \vdots & \\ \cdots & \boldsymbol{R}_{lk} & \cdots & \boldsymbol{R}_{ll} & \cdots \\ & \vdots & & \vdots & \end{bmatrix}. \tag{70}$$

Sub-blocks $\boldsymbol{R}_{kk}$, $\boldsymbol{R}_{kl}$ are responsible for summation of the top and bottom displacement dofs, while $\boldsymbol{R}_{lk}$ and $\boldsymbol{R}_{ll}$ for subtraction of these dofs. In turn, the block $\boldsymbol{R}^2$ possesses the following structure based on the identity sub-blocks $\boldsymbol{I}$

$$\boldsymbol{R}^2 = \operatorname{diag}\left[..., \boldsymbol{R}_m, ...\right], \quad \boldsymbol{R}_m \equiv \boldsymbol{I} = \operatorname{diag}\left[1, 1, 1\right]. \tag{71}$$

This means that neither the summation nor the subtraction is performed for the dofs $m$.

One more remark concerns the sums and differences of displacements present in the constraints (27). In the case of the symmetric-thickness geometry, these sums define mid-surface displacements. The first two differences, after division by the thickness $t$, give rotations of the normal to the mid-surface, while the third difference determines a transverse elongation of the corresponding normal. Note that only the latter difference needs to be constrained (equal to zero) in accordance with the last Equation (27). Because of this, multiplication of the penalty stiffness by the diagonal matrix $\boldsymbol{Z}$ is necessary. This matrix possesses the following structure:

$$\overset{e}{\boldsymbol{Z}} = \begin{bmatrix} \boldsymbol{Z}^1 & \boldsymbol{0} \\ \boldsymbol{0} & \boldsymbol{Z}^2 \end{bmatrix}. \tag{72}$$

The component block $\boldsymbol{Z}^1$ is composed of the diagonal blocks $\boldsymbol{Z}_k$ and $\boldsymbol{Z}_l$ corresponding to the degrees of freedom $k$ and $l$:

$$\boldsymbol{Z}^1 = \operatorname{diag}\left[..., \boldsymbol{Z}_k, ..., \boldsymbol{Z}_l, ...\right], \quad \boldsymbol{Z}_k = \operatorname{diag}\left[0, 0, 0\right], \quad \boldsymbol{Z}_l = \operatorname{diag}\left[0, 0, 1\right], \tag{73}$$

where the degrees of freedom $k$ and $l$ correspond to the sums and differences of the top and bottom displacement dofs. The second component block has the simpler form

$$\boldsymbol{Z}^2 = \operatorname{diag}\left[..., \boldsymbol{Z}_m, ...\right], \quad \boldsymbol{Z}_m = \operatorname{diag}\left[0, 0, 0\right], \tag{74}$$

which results from the fact that, for the degrees of freedom $m$, other than the sums and differences of the displacement dofs, the penalty terms are not necessary.

Penalty Stiffness

The respective penalty stiffness $\overset{e}{\boldsymbol{k}}_P$, present in (66), is defined in the directions coincident with the constraint directions of which the first two are tangent and the third one is perpendicular to the mid-surface. The stiffness form corresponds to the displacement dofs that represent the mentioned sums and differences of the top and bottom dofs. The definition of the penalty stiffness reads:

$$\overset{e}{\boldsymbol{k}}_P = \int_{R_e} \overset{e}{\boldsymbol{N}}{}^T \boldsymbol{n} \boldsymbol{P} \boldsymbol{n}^T \overset{e}{\boldsymbol{N}} \operatorname{cof}(\boldsymbol{J}) \, d\xi_3 d\eta_i. \tag{75}$$

The penalty parameter has to be larger than the terms of $\overset{e}{\boldsymbol{k}}_M$ and the terms of $\overset{e}{\boldsymbol{k}}_C$, i.e., $P \gg \max\{(k_M)_{ij}; \, i, j = 1, 2, ..., n_M\}$, where $(k_M)_{ij}$ represent terms of the stiffness matrix and $n_M$ stands for the corresponding number of displacement dofs. Additionally, one has $P \gg$

$\max\{(k_C)_{ij}; i = 1, 2, ..., n_M; j = 1, 2, ..., n_E\}$, with $(k_C)_{ij}$ being terms of the piezoelectricity matrix and $n_E$ denoting the corresponding number of electric potential dofs.

The above form (75) of the penalty matrix is suitable for the geometry I of the classical and modified transition elements. In the above equation, $R_e$ stands for the element side being the subset of the boundary $R$ between the first-order and transition model: $R_e \subset R$. The normal vector $n$ is defined as perpendicular to the mid-surface $S_m$, i.e., $n \perp S_m$. It also approximately satisfies the condition $n \in R_e$ – approximately due to the finite element approximation of the element geometry. Note that two tangential directions and the third normal direction are coincident with the local directions present in the constraint Equation (27). On the element level, they are defined by the relations (49) and (50). Note also that the presence of the vector $n$ in the above definition results in imposition of the penalty terms in the third (normal) direction only, i.e., the constraints are applied in this direction only. The shape function matrix terms are non-zero on the boundary $R_e$. Hence, the non-zero penalty terms appear for the element nodes placed on this boundary.

In the case of the classical and modified transition elements of the geometry II, the above definition has to take into consideration the fact that the constraints (27) are imposed on the element vertical edge $L_e$ only. The edge can be characterized with: $L_e \subset R$. The normal vector is defined analogously as in the case of the geometry I, i.e., $n \perp S_m$ and approximately $n \in L_e$. With these remarks in mind, one can write

$$\overset{e}{k}_P = \int_{L_e} \overset{e}{N^T} n P n^T \overset{e}{N} \, \text{der}(J) \, d\xi_3, \tag{76}$$

where, due to one dimensional integration, the value of $\text{der}(J)$ is applied for transformation of the global directions into the normalized direction $\xi_3$. As previously, the constraints concern the differences of the top and bottom displacement dofs. The non-zero penalty terms appear for the element nodes and the corresponding dofs placed on the boundary $L_e$, in the transverse direction only.

### 4.8.2. The Enhanced Transition Element

Two types of constraints are present in the case of the corresponding transition model. The first type are the gradually changing constraints of no elongation of the normals to the mid-surface, represented by the last but one term of (34). The considered type of the element requires also the additional constraints corresponding to the gradual change of the transverse order of approximation from $q \equiv I = 1$ to an arbitrary value $q \equiv I$, where $I$ is the hierarchical model order, and represented by the last term of (34). As the constraints of both types are fully active on the boundary with the first-order model ($\alpha = 1$) and gradually switched off ($\alpha \to 0$) towards the boundary with the hierarchical or three-dimensional model, the nodal (collocation) values of the complement of the switching functions to unity $1 - \alpha$ will appear in the penalty stiffness definitions so as to activate the penalty terms.

### The Modified Stiffness Matrix

The relations (66)–(68) are still valid for the enhanced transition elements and constraints of no elongation of the normals to the mid-surface $S_m$. However, the degrees of freedom $k$ and $l$ corresponding to the top and bottom displacements, and the other dofs $m$, as well, are associated with different nodes now because the constraints of no elongation of the lines normal to the mid-surface are applied in the entire element volume $V_e$. In the case of the geometry I, the top dofs $k$ are associated with the nodes $a_4, a_5, a_6, a_{10}, a_{11}, a_{12}, a_{16}$ the bottom dofs $l$ with the nodes $a_1, a_2, a_3, a_7, a_8, a_9, a_{15}$, and the other dofs $m$ with the nodes $a_{13}, a_{14}, a_{17}, a_{18}, a_{19}, a_{20}$. The geometry II needs the following assignments: the top dofs $k$ correspond to the nodes $a_4, a_5, a_6, a_{10}, a_{11}, a_{12}, a_{15}$, the bottom dofs $l$ to the nodes $a_1, a_2, a_3, a_7, a_8, a_9, a_{14}$, and the other dofs $m$ to the nodes $a_{13}, a_{16}, a_{17}, a_{18}$. These assignments hold also for the constraints related to the change of the transverse approximation order. In this case, the stiffness matrix modification is described with the relation

$$\overset{e}{k}_M = \overset{e}{k}_M + \mathbf{Z}^T \overset{e}{k}_P \mathbf{Z}. \tag{77}$$

Comparing (66) and (77), one can notice that there are no matrices $\overset{e}{R}$ and $\overset{e}{R}^{-1}$ present in the above relation. This means that the replacement of the top and bottom dofs with their sums and differences and inversely is not necessary for the constraints of the second type.

The Modifying Operators

In the case of the constraints of no elongation of the normals, the definitions of the operators $\overset{e}{R}$ and $\mathbf{Z}$, (69)–(71) and (72)–(79), respectively, are still valid. However, the new assignments of the top $k$, bottom $l$, and other $m$ dofs to the element nodes for geometries I and II have to be taken into account while calculating the corresponding operator matrices. In the case of the constraints of the second type, which concern the other dofs $m$ only, the following definition is appropriate

$$\overset{e}{\mathbf{Z}} = \begin{bmatrix} \mathbf{0} & \mathbf{0} \\ \mathbf{0} & \mathbf{Z}^2 \end{bmatrix}. \tag{78}$$

As it can be seen, the first component block of (78) is $\mathbf{Z}^1 \equiv \mathbf{0}$. The second component block has the form

$$\mathbf{Z}^2 = \operatorname{diag}[..., \mathbf{Z}_m, ...], \quad \mathbf{Z}_m = \operatorname{diag}[1, 1, 1], \tag{79}$$

which results in appearance of the penalty terms for all dofs $m$ in all three global directions.

The Penalty Matrices

The constraints of no elongation of the normals require the following definition of the penalty stiffness matrix:

$$\overset{e}{k}_P = \int_0^1 \int_0^1 \int_0^{-\xi_2+1} \overset{e}{\mathbf{N}}^T (\mathbf{I} - \boldsymbol{\alpha}^T) \boldsymbol{n} P \boldsymbol{n}^T (\mathbf{I} - \boldsymbol{\alpha}) \overset{e}{\mathbf{N}} \det(\mathbf{J}) \, d\xi_1 d\xi_2 d\xi_3, \tag{80}$$

where normal vector is $\boldsymbol{n} \perp S_m$, with $S_m$ being the mid-surface within the symmetric-thickness geometry. The presence of this vector results in application of the constraints in the normal direction only. In turn, the presence of the switching function $\mathbf{I} - \boldsymbol{\alpha}$ guarantees the gradual change of the penalty terms. It employs the identity matrix $\mathbf{I}$ and the matrix of nodal values of the blending function:

$$\boldsymbol{\alpha} = [..., \boldsymbol{\alpha}_m, ...], \quad \boldsymbol{\alpha}^T = [..., \boldsymbol{\alpha}_m^T, ...]^T, \tag{81}$$

and the nodal blocks $\boldsymbol{\alpha}_m$ are diagonal and correspond to the dofs $m$, other than the top and bottom ones:

$$\boldsymbol{\alpha}_m = \begin{bmatrix} \alpha_m & 0 & 0 \\ 0 & \alpha_m & 0 \\ 0 & 0 & \alpha_m \end{bmatrix}, \tag{82}$$

and where $\alpha_m \in \langle 0, 1 \rangle$. Its value depends on the longitudinal location of the node, the dof $m$ is associated with. In other words, $\alpha_m = \alpha(\xi_{1,m}, \xi_{2,m})$, where $\xi_{1,m}$ and $\xi_{2,m}$ represent first two normalized coordinates for the dof $m$.

In the case of the constraints of the second type one has

$$\overset{e}{k}_P = \int_0^1 \int_0^1 \int_0^{-\xi_2+1} \overset{e}{\mathbf{N}}^T (\mathbf{I} - \boldsymbol{\alpha}^T) P (\mathbf{I} - \boldsymbol{\alpha}) \overset{e}{\mathbf{N}} \det(\mathbf{J}) \, d\xi_1 d\xi_2 d\xi_3. \tag{83}$$

These constraints also change gradually and are applied in all three global directions for degrees of freedom $m$ of the element. The definitions (81) and (82) of the switching function remain unchanged.

It should stressed that the above definitions (80) and (83) are valid for both the transition elements acting between the piezoelectric elements based on the first-order shell theory and the elements based on either the hierarchical shell theory or the three-dimensional theory. The difference is that, in the first case, the entire surface $S_m$ (characterized with $\xi_3 = \frac{1}{2}$) represents the shell mid-surface, while, in the second case, it represents the reference surface only (defined with $\xi_3 = \frac{1}{2}$). The reference surface becomes the mid-surface only on the boundary with the first-order model $R \cap S_m$.

## 5. Numerical Tests

Two aspects of performance of the proposed transition piezoelectric elements will be addressed in this section. The first one is the ability to remove high stress gradients between the 3D-based transition and 3D-based basic elements in the chosen model problems. The mechanical models of the basic elements conform to the 3D-based first-order shell theory and the 3D-based hierarchical shell or three-dimensional theory. The electric field is characterized with the same dielectric model conforming to either three-dimensional or 3D-based hierarchical symmetric-thickness theory.

The second aspect will be comparison of the solution convergence for the same model problems as in the first aspect with three versions of the transition elements employed. The model problems will concern the bending-dominated plate and shell and membrane-dominated shell. These problems manifest different and numerically demanding stress and strain states. Explanation of the differences between these three cases of the mechanical fields can be found in Reference [53]. The electric field and electric displacement states will be typical for piezoelectricity.

It should be stressed that our model problems presented below serve the numerical assessment of the presented algorithms. Because of that the loads and charges are assumed so as the mechanical and electric parts of the potential electro-mechanical energy (or potential co-energy) are of the same order – the case very demanding from the numerical point of view as the sign of the electro-mechanical potential energy may change in the convergence studies where the error of the energy is displayed as a function of the number of dofs in the problem. Note that, in the two technologically important actuating and sensing modes of piezoelectric operation, one part of the co-energy dominates the other one. Because of this, such modes are less demanding numerically.

### 5.1. Model Problems and Methodology

#### 5.1.1. Model Structures and Their Geometry

Three piezoelectric domains corresponding to the bending-dominated plate, bending-dominated half-cylindrical shell and a half of the membrane-dominated cylindrical shell are presented in Figures 5–7. The plate dimensions are $l \times l$, the length of both shells is equal to $l$, while the circumferential length of a half of the cylinder is $l \approx 2\pi R$ with $R$ being the radius of the mid-surface of the shells. All the structures are of the same thickness $t$. The following values are set in our tests: $l = 3.1415 \times 10^{-2}$ m, $R = 1.0 \times 10^{-2}$ m, $t = 0.03 \times 10^{-2}$ m.

#### 5.1.2. Material, Load, Charge and Boundary Conditions

All three structures are made of the same piezoelectric material of material constants typical for piezoceramics [57]. The Young modulus is $E = 0.5 \times 10^{11}$ N/m$^2$, the Poisson ratio equals $\nu = 0.294$, the dielectricity constant under constant stress is $\delta = 0.1593 \times 10^{-7}$ F/m, while the piezoelectric constants under constant stress are equal to $d_{13} = d_{23} = -0.15 \times 10^{-9}$ C/N, $d_{33} = 0.3 \times 10^{-9}$ C/N, and $d_{52} = d_{61} = 0.5 \times 10^{-9}$ C/N.

The plate and bending-dominated shell are loaded with the vertical traction of the value $p = -0.4 \times 10^6$ N/m$^2$, while the membrane-dominated shell with the internal pressure of the same value $p$ acting in the outward direction. All the structures are electrically charged on their upper (or outward) surface where the surface charge density is $c = 0.2 \times 10^{-1}$ C/m$^2$. On the lower (or inward) surfaces the charge density is set to 0.

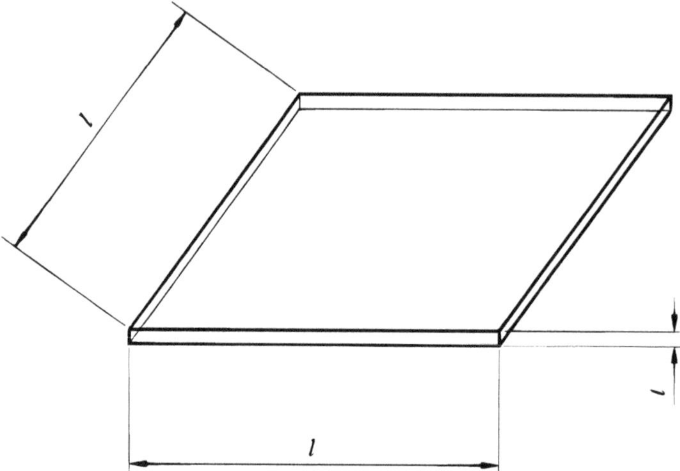

**Figure 5.** The domain of the bending-dominated piezoelectric plate.

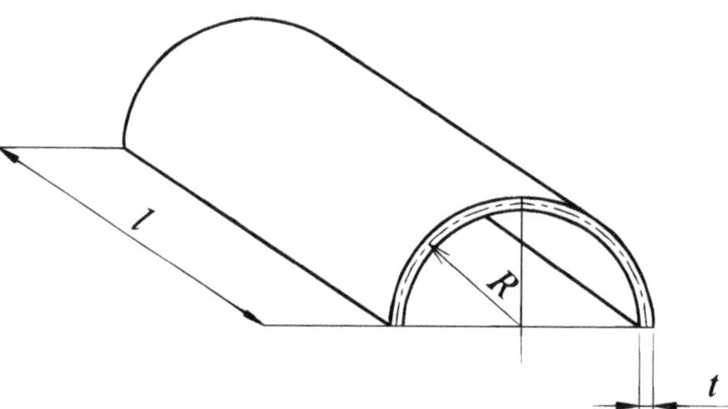

**Figure 6.** The domain of the bending-dominated piezoelectric half-cylindrical shell.

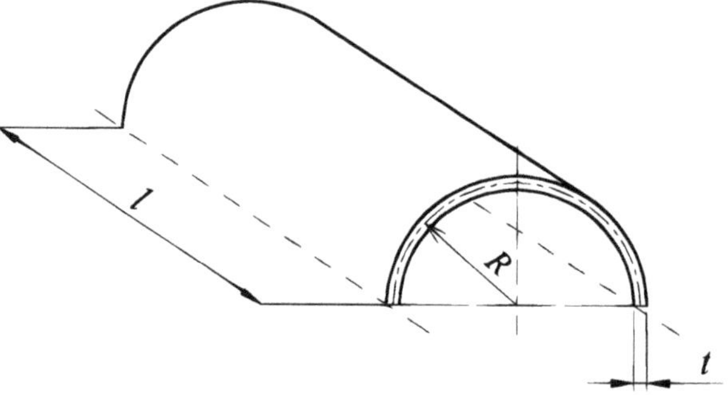

**Figure 7.** A half of the membrane-dominated piezoelectric shell domain.

The mechanical boundary conditions are as follows. The plate is clamped along its four edges. The bending-dominated shell has two straight edges clamped and two curved edges free. In the case of the membrane-dominated shell, there is no rotation along the curved edges. The electrical boundary conditions assume grounding along: four edges of the plate, along two straight and two curved edges of the half-cylindrical shell, and two curved edges of the cylindrical shell.

### 5.1.3. Discretization and Methodology

Due to symmetry of the geometry, loading, electric charge, and boundary conditions, only quarters of the bending-dominated domains and an octant of the membrane-dominated domain are taken into account in or numerical tests and presented in the figures below. The numerical settings are as follows. We apply the meshes $4 \times 4 \times 2$ of the prismatic elements in the displayed quarter or octant parts of the structures in our tests. The longitudinal order of approximation within the mechanical and electric field is the same and equal $p = \pi = 1, 2, \ldots, 8$. The transverse orders of approximation of both fields are equal to $q = \rho = 2$ in the hierarchical shell zone of the domains, $q = \rho = 1$ in the first-order shell zone, and the changing (from 2 to 1) order in the transition zone constituting a layer composed of couples of prismatic transition elements. The number of the first-order and hierarchical (three-dimensional) elements changes from the entirely hierarchical shell structure to the shell structure composed of the first-order shell elements only. The intermediate (or mixed) case includes a single layer of piezoelectric transition elements. The exemplary cases for the plate and shells are presented in Figures 8–10.

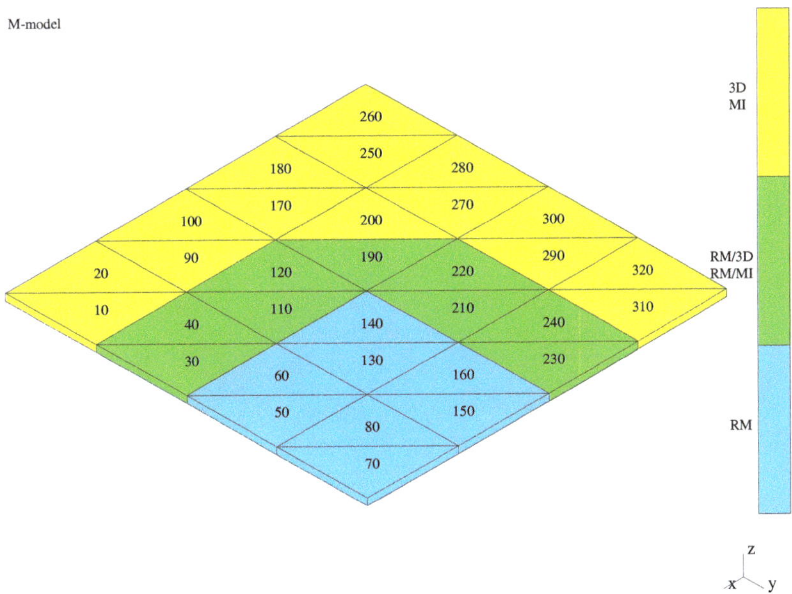


**Figure 8.** Mixed models of the bending-dominated plate.

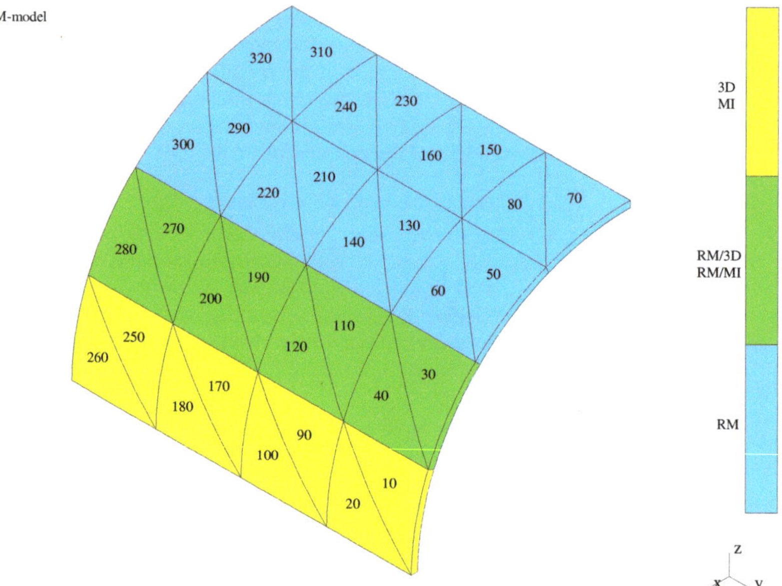


**Figure 9.** Mixed models of the bending-dominated shell.

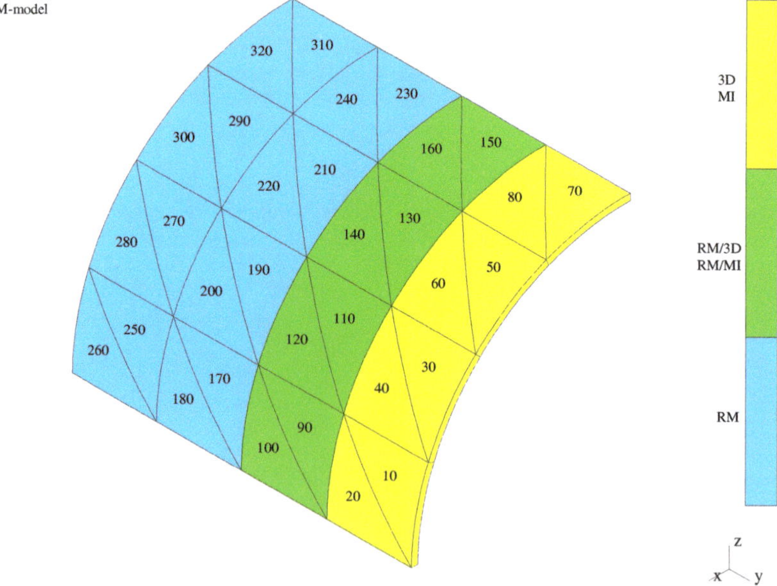


**Figure 10.** Mixed models of the membrane-dominated shell.

The presented meshes are composed of one layer of hierarchical elements (yellow and marked as MI in the figure), one layer of the transition elements (green and marked as RM/MI), and two layers of the first-order elements (blue and marked as RM in the figure).

Our parametric studies are limited to changes of the longitudinal approximation orders $p \equiv \pi$, the ratios of the numbers of elements of various types, and the types of the transition elements for three model problems. The mentioned ratios are equal to $r = 1.0, 0.5, 0.0$ and represent the number of the first-order elements divided by the sum of the transition and hierarchical elements along the characteristic longitudinal dimensions of the structures.

### 5.2. Ability to Remove High Stress Gradients

In order to present the ability of three (classical, modified, and enhanced) types of the transition elements to remove the high stress gradients between the transition elements and the neighboring basic (hierarchical and first-order) elements, exemplary ($p \equiv \pi = 6$) distributions of the global normal stress $\sigma_{33}$ (marked as *szz*) for three model problems and three types of the transition elements are presented. These distributions correspond to the mixed models from Figures 8–10. The stress distributions for the plate problem, with the classical, modified and transition elements employed, are presented in Figures 11–13, respectively. Figures 14–16 display the analogous three stress distributions for the bending-dominated shell. The last figures, Figures 17–19, correspond to the membrane-dominated problem with three types of the transition elements applied. Note that, in the figures, the stress values are written in numerical form not in analytical form due to capabilities of the numerical code used.

In the case of the plate and membrane problems, high solution gradients are visible when the classical transition elements are employed. These gradients are reduced for the mixed models with the modified and enhanced elements applied. In the case of the bent shell, high gradients are not visible for the classical elements as the displayed global normal stress component is a combination of the transverse and longitudinal stresses.

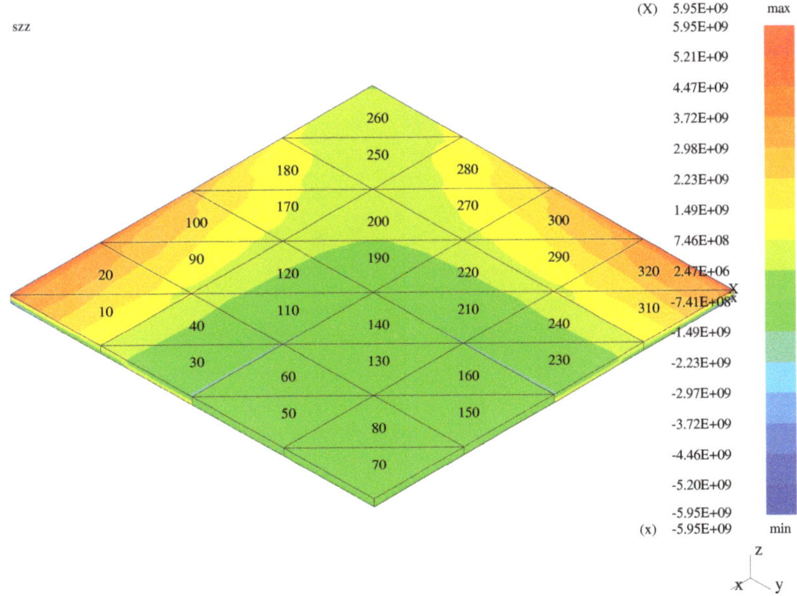


**Figure 11.** Transverse stress—the bent plate with the classical transition elements.

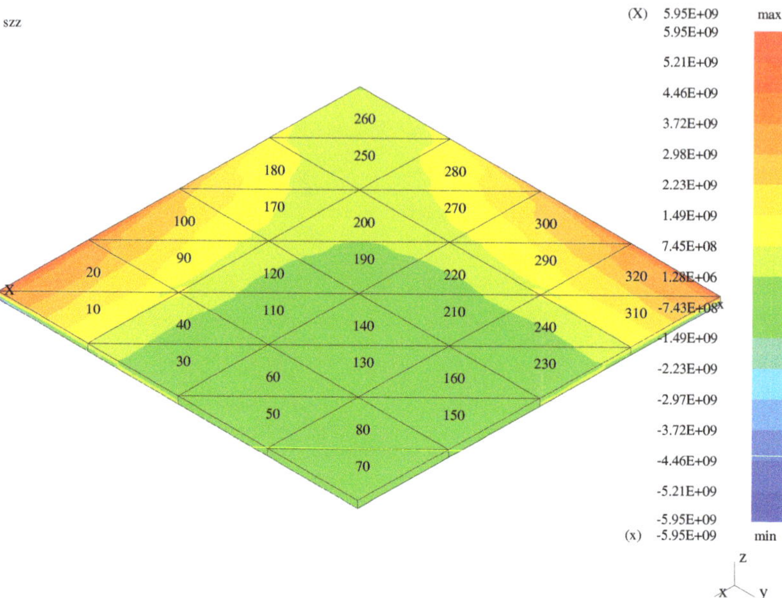

**Figure 12.** Transverse stress—the bent plate with the modified transition elements.

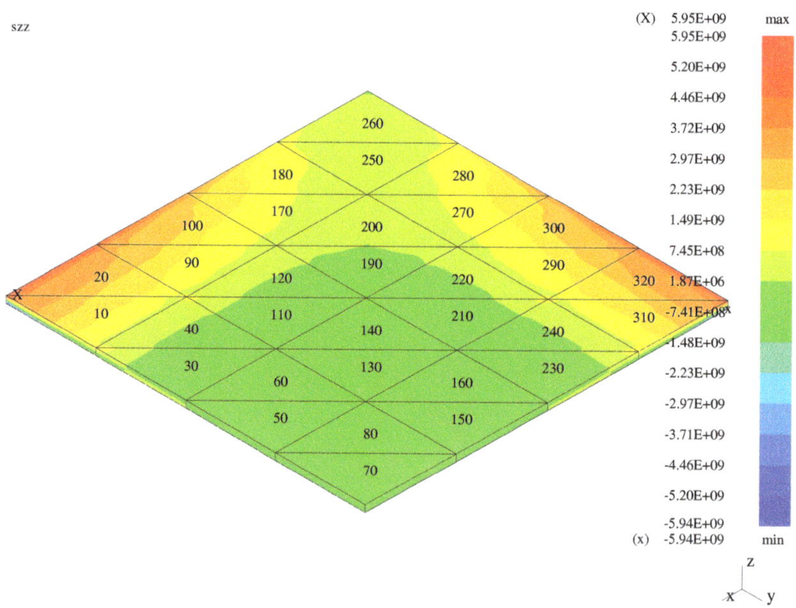

**Figure 13.** Transverse stress—the bent plate with the enhanced transition elements.

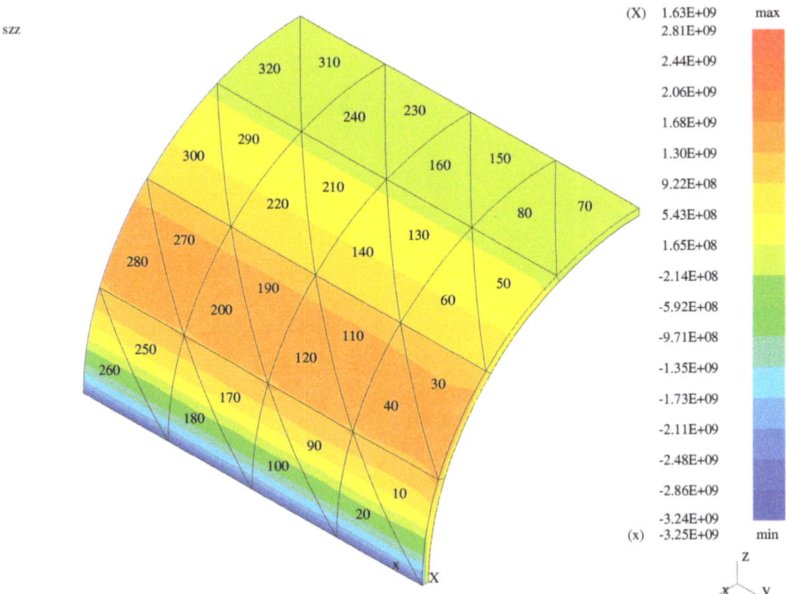

**Figure 14.** Third normal stress—the bent shell with the classical transition elements.

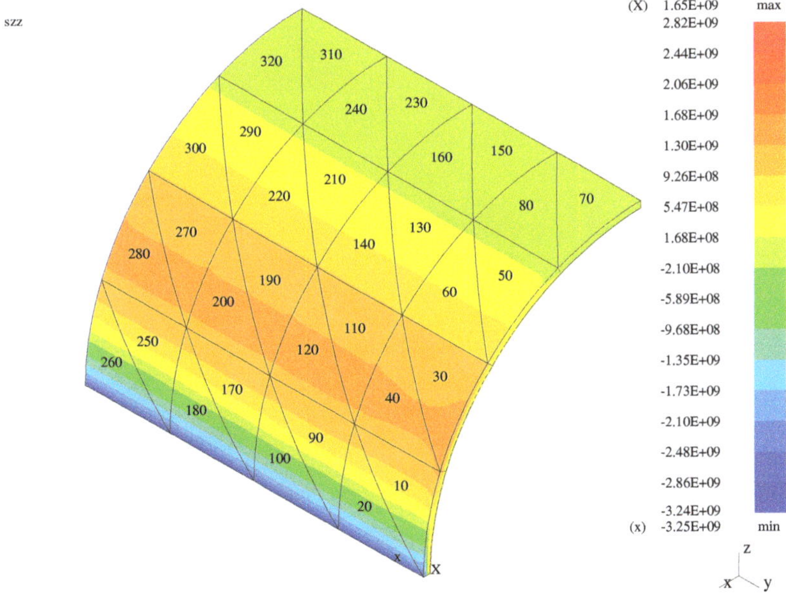

**Figure 15.** Third normal stress—the bent shell with the modified transition elements.

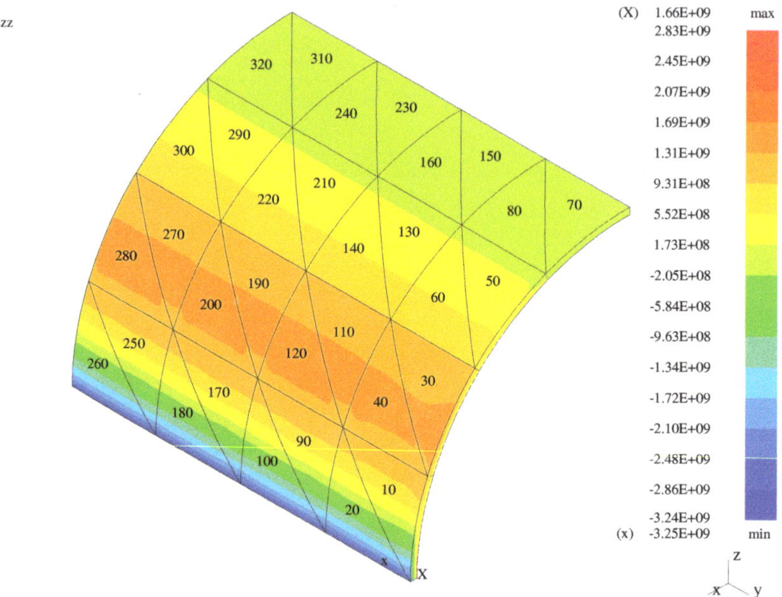

**Figure 16.** Third normal stress—the bent shell with the enhanced transition elements.

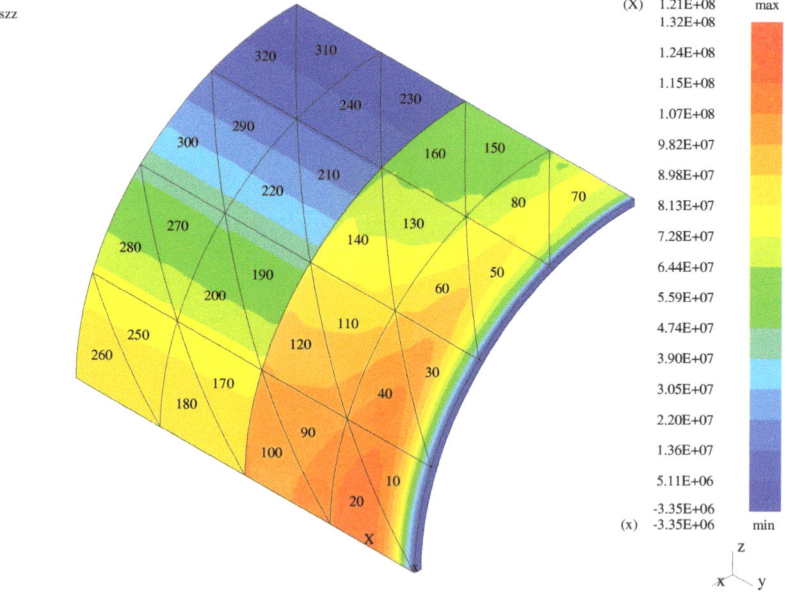

**Figure 17.** Third normal stress—the membrane with the classical transition elements.

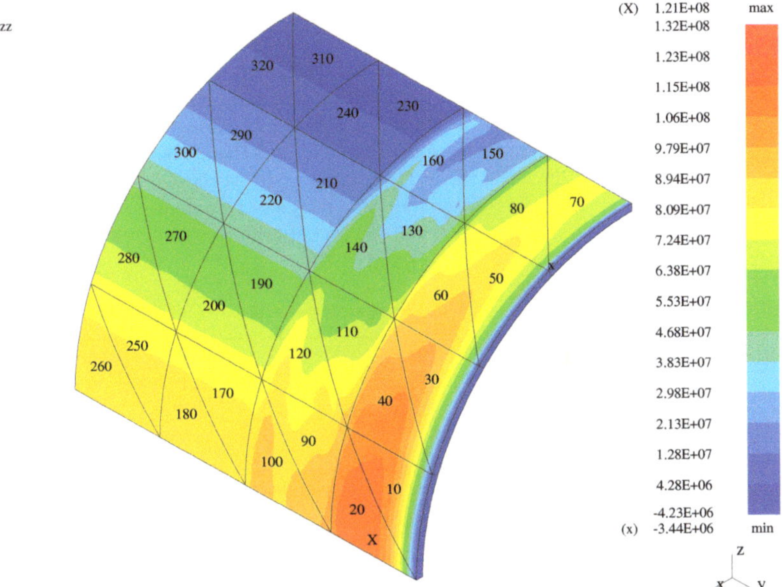

**Figure 18.** Third normal stress—the membrane with the classical transition elements.

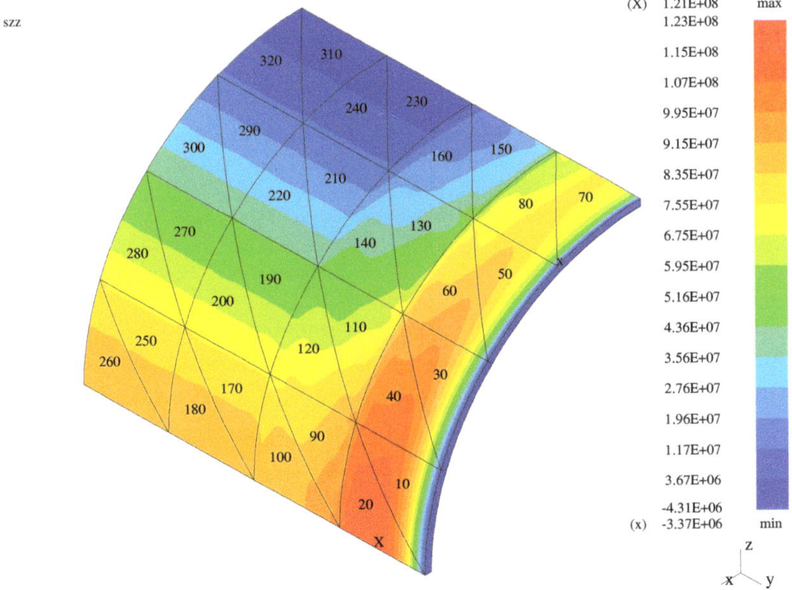

**Figure 19.** Third normal stress—the membrane with the enhanced transition elements.

In order to quantify the presented graphical results concerning stress gradients, the Table 1 is presented. Therein, gradients (jumps/differences) of the global stress $\sigma_{33}$ between the face centers of the neighboring elements are displayed for three model structures. Five models for each of three structures are included: the entirely hierarchical (MI) shell model ($r = 0.0$), three mixed (RM/MI) models ($r = 0.5$) with the classical, modified, and enhanced (respectively, TR1, TR2, TR3) transition models, and entirely first-order (RM) model ($r = 1.0$). In the case of the bending-dominated (bent) plate, the chosen elements 31 and 24 correspond to 3D model and TR1 or TR2 or TR3 model. Both elements 24 and 23 correspond to either TR1 or TR2 or TR3 model, while the elements 23 and 16 to either TR1 or TR2 or TR3 model and RM model. In the case of the bending-dominated (bent) shell, the corresponding couples of elements are: 25 and 28, 28 and 27, and 27 and 30. In the case of the membrane (membrane-dominated shell), the corresponding couples are: 6 and 13, 13 and 14, 14 and 21.

**Table 1.** Summary of stress gradients [N/m$^2$] for three model problems.

| Structure Type | Stress Gradient * | Structure Model | | | | |
|---|---|---|---|---|---|---|
| | | MI | TR1 | TR2 | TR3 | RM |
| bent plate | figure no. | – | Figure 11 | Figure 12 | Figure 13 | – |
| | el. 31, 24 | $-0.0085 \times 10^7$ | $0.0064 \times 10^7$ | $1.6533 \times 10^7$ | $2.0853 \times 10^7$ | $0.0000 \times 10^{-12}$ |
| | el. 24, 23 | $0.0001 \times 10^7$ | $0.0025 \times 10^7$ | $0.0312 \times 10^7$ | $0.2602 \times 10^7$ | $-0.2096 \times 10^{-12}$ |
| | el. 23, 16 | $0.0001 \times 10^7$ | $1.0996 \times 10^7$ | $1.1658 \times 10^{-7}$ | $4.0159 \times 10^{-7}$ | $3.7411 \times 10^{-12}$ |
| bent shell | figure no. | – | Figure 14 | Figure 15 | Figure 16 | – |
| | el. 25, 28 | $-0.1342 \times 10^8$ | $-0.1348 \times 10^8$ | $-0.2991 \times 10^8$ | $-0.2661 \times 10^8$ | $0.0286 \times 10^8$ |
| | el. 28, 27 | $-0.0313 \times 10^8$ | $-0.0327 \times 10^8$ | $-0.0519 \times 10^8$ | $-0.0109 \times 10^8$ | $0.0051 \times 10^8$ |
| | el. 27, 30 | $-0.0256 \times 10^8$ | $0.8547 \times 10^8$ | $1.2494 \times 10^8$ | $1.3887 \times 10^8$ | $0.0173 \times 10^8$ |
| membrane shell | figure no. | – | Figure 17 | Figure 18 | Figure 19 | – |
| | el. 6, 13 | $-0.0251 \times 10^7$ | $-0.0505 \times 10^7$ | $2.1664 \times 10^7$ | $2.1448 \times 10^7$ | $-0.0053 \times 10^7$ |
| | el. 13, 14 | $-0.0002 \times 10^7$ | $0.0035 \times 10^7$ | $-0.0434 \times 10^7$ | $-0.0048 \times 10^7$ | $0.0043 \times 10^7$ |
| | el. 14, 21 | $-0.0013 \times 10^7$ | $4.4918 \times 10^7$ | $0.0335 \times 10^7$ | $0.0068 \times 10^7$ | $-0.0008 \times 10^7$ |

* The presented result values correspond to $p \equiv \pi = 6$.

In the case of the plate, the third normal stress $\sigma_{33}$ corresponds to the transverse normal stress. In the entirely hierarchical (MI) shell model, the observed values of gradients are small and due to finite element discretization. In the entirely first-order (RM) model, the jumps are extremely small (correspond to the numerical zero), as the plane stress assumption holds in all elements. In the case of the mixed TR1 model, the jump between the transition and first-order elements 23 and 16 results from the three-dimensional stress state and plane stress state in the neighboring elements. Such jumps are diminished to numerical zero for the cases of the TR2 and TR3 models, as the plane stress assumption holds on both sides of the neighboring faces of the transition and first-order elements. In both shell cases, the presented jumps of $\sigma_{33}$ result from both discretization and plane stress assumption. The exact analysis for the transverse direction requires taking jumps for the stress components $\sigma_{11}$ and $\sigma_{13} = \sigma_{31}$ into account. Such an analysis (not revealed here) leads to exactly the same observations as in the case of the plate. For the membrane case, where the contribution of $\sigma_{33}$ to the transverse normal stress is higher than the influence of two other stress components, the stress jumps in $\sigma_{33}$ between elements 14 and 21 are smaller in the case of the TR3 model than in the cases of the TR2 and TR1 models. In the bent shell case, the jumps in $\sigma_{33}$ between elements 27 and 30 grow from TR1 through TR2 to TR3, but this leads to zero transverse normal stress values in both neighboring elements for the TR2 and TR3 models. Demonstration of this needs inclusion of the stress components $\sigma_{11}$ and $\sigma_{13} = \sigma_{31}$ again.

### 5.3. Solution Convergence of the Problems

In the convergence studies, the logarithm of the approximation error in potential energy is presented as a function of the logarithm of the number of dofs $N$ of the problem. The corresponding curves are $p$-convergence ones as the number of dofs in the problems increase due to the change of the longitudinal order of approximation $p$ within elements. In order to present the solution convergence for the three model problems with three types of the transition elements applied, three threesomes of pictures are displayed. The first three pictures (Figures 20–22) correspond to the bending-dominated plate problem, the second threesome (Figures 23–25) to the bending-dominated shell problem, while the last three drawings (Figures 26–28) to the membrane-dominated shell. In the figures, the first, second and third drawings correspond to the cases of the classical, modified, and enhanced transition elements applied.

The mentioned approximation error is defined as the difference of two energy measures of which $E$ corresponds to the numerical solution under consideration and the reference solution $E_r$ plays a role of the exact solution, namely:

$$energy\ error = |E_r - E|. \tag{84}$$

The energy measures represent the sum of absolute values of the mechanical and electrical parts of the electro-mechanical potential energy, i.e.,

$$E = |B(\boldsymbol{u}, \boldsymbol{u}) - C(\boldsymbol{u}, \phi)| + |-b(\phi, \phi) - C(\phi, \boldsymbol{u})|. \tag{85}$$

Due to stationarity of the solutions $\boldsymbol{u}$ and $\phi$, resulting from (8), the potential energy can be expressed with the components of the electro-mechanical co-energy (strain, electric field, and coupling energies) corresponding to the bilinear forms $B$, $b$ and $C$ from (9)–(11). The solutions for displacements $\boldsymbol{u}$ and electric potential $\phi$ are approximated with use of the global nodal vectors of the displacement dofs $q^{hpq}$ and potential dofs $\varphi^{h\pi p}$ obtained from solution of the problem (35).

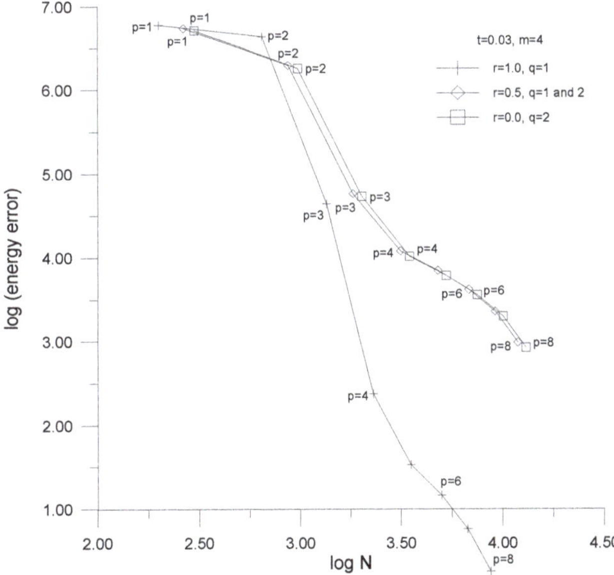


**Figure 20.** Convergence curves—the plate problem, the classical transition elements employed.

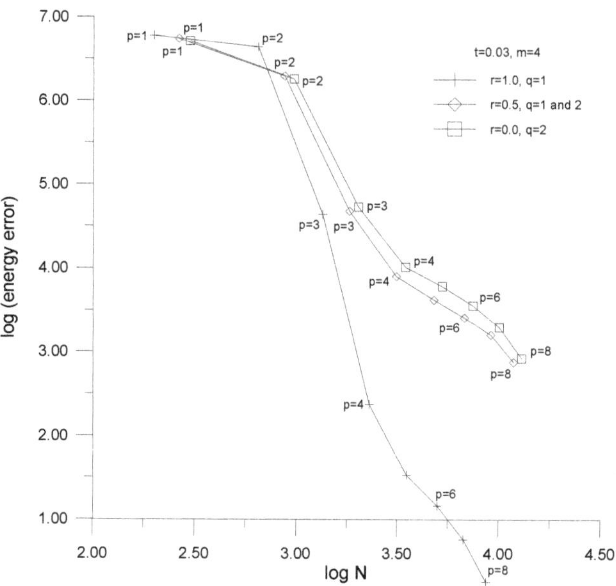

**Figure 21.** Convergence curves—the plate problem, the modified transition elements applied.

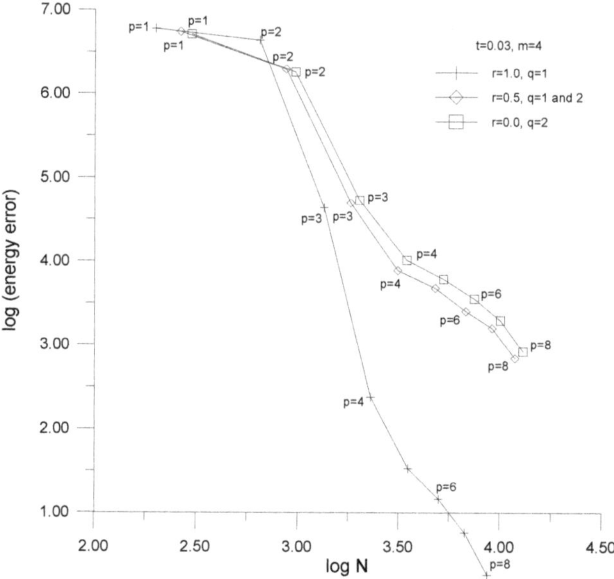

**Figure 22.** Convergence curves—the plate problem, the enhanced transition elements applied.

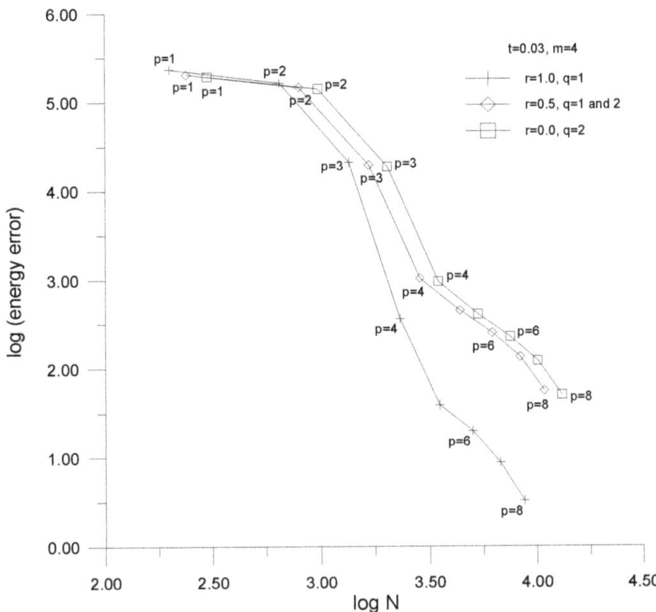


**Figure 23.** Convergence curves—the bending-dominated shell with the classical transition elements.

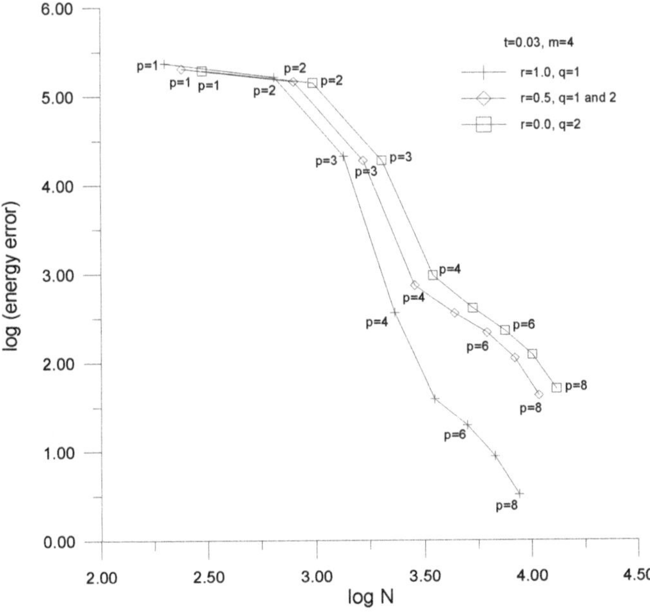


**Figure 24.** Convergence curves—the bending-dominated shell with the modified transition elements.

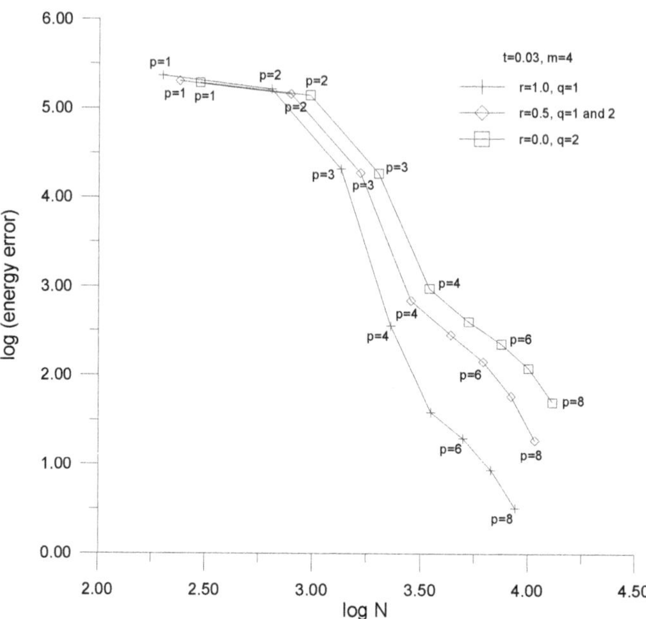

**Figure 25.** Convergence curves—the bending-dominated shell with the enhanced transition elements.

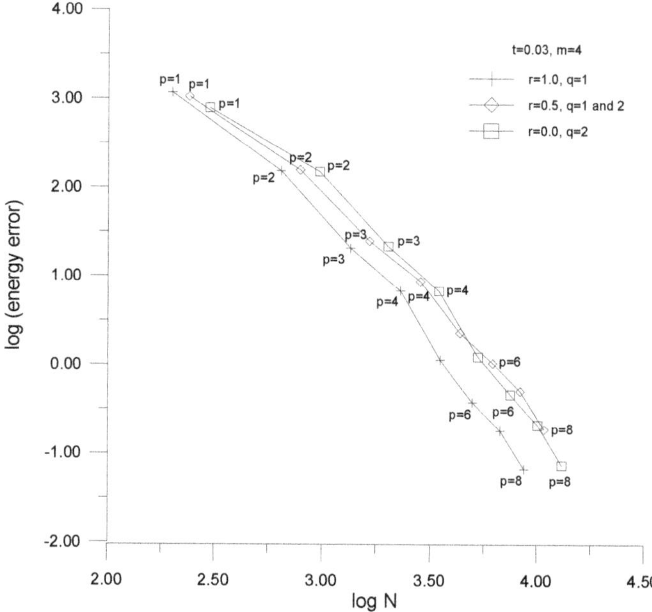

**Figure 26.** Convergence curves—the membrane shell with the classical transition elements.

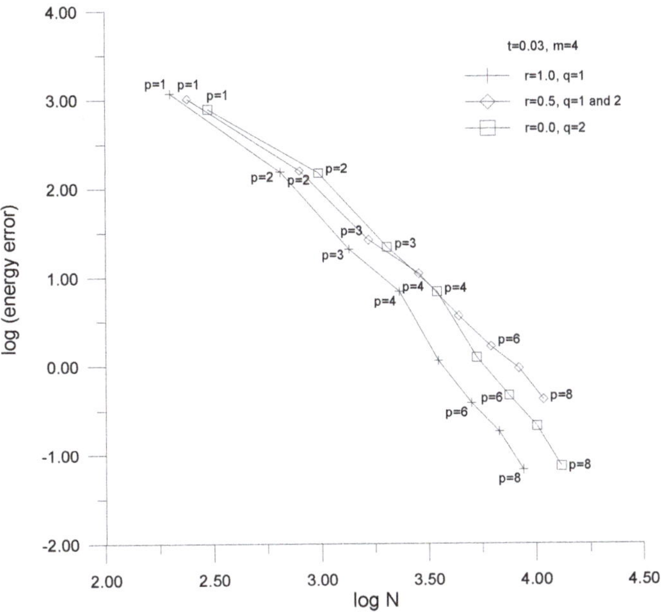

**Figure 27.** Convergence curves—the membrane shell with the modified transition elements.

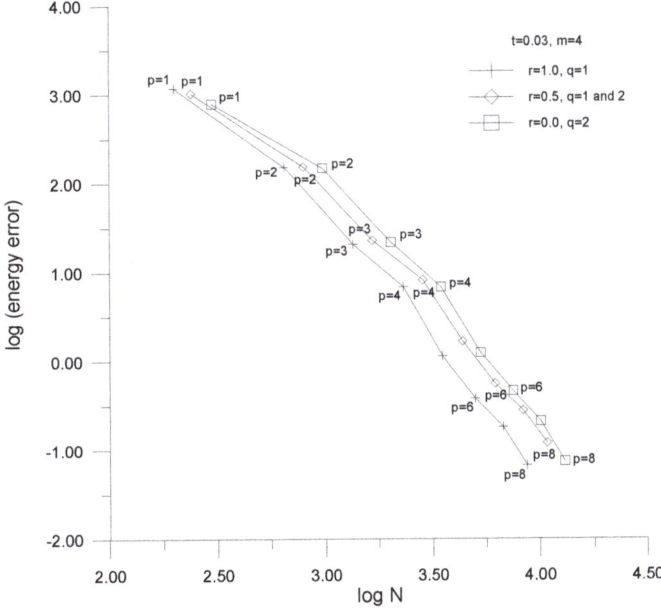

**Figure 28.** Convergence curves—the membrane shell with the enhanced transition elements.

The proposed measure of the approximation error was applied as it guarantees monotonicity of the energy approximation error with changing value of $p \equiv \pi$, while the simple sum of both parts of the energy (equal to the electro-mechanical co-energy) may not. Note that such monotonicity is a must in the case of error-controlled adaptivity suggested in Reference [11,12].

In order to calculate the errors, the exact values $E_r$ of the energy measures are replaced with their best numerical approximations obtained from the so-called over-killed meshes, i.e., the meshes for which the longitudinal order of approximation is equal to $p = \pi = 9$.

In the case of two bending-dominated problems the convergence for the enhanced models is slightly higher than for the modified transition elements employed in the mixed models. Additionally, the latter elements produce slightly better convergence than the classical ones. In the case of the membrane-dominated problem, only the enhanced transition elements deliver convergence curves between the curves obtained for the basic models. The confirmation is presented in the Table 2, where the absolute values $\log(E_r - E)$ of the error and relative error values $(E_r - E)/E_r$ (both for $p = 8$, i.e., $E \equiv E_8$) are included. In addition, the averaged convergence rates are given and equal to $-\log(\delta_8/\delta_7)/\log(N_8/N_7)$, $\delta_8 = E_r - E_8$, $\delta_7 = E_r - E_7$, where $E_8$, $N_8$ and $E_7$, $N_7$ represent the energies and numbers of dofs for $p = 8$ and $p = 7$.

**Table 2.** Error and convergence summary for three model problems.

| Problem Type | Result Quantity * | Model Type | | | | |
|---|---|---|---|---|---|---|
| | | MI | TR1 | TR2 | TR3 | RM |
| bending-dominated plate | figure no. | Figures 20–22 | Figure 20 | Figure 21 | Figure 22 | Figures 20–22 |
| | dofs number $N$ [1] | 13,068 | 11,912 | 11,912 | 11,912 | 8712 |
| | $\log(E_r - E)$ [N·m] | 2.922 | 2.983 | 2.879 | 2.803 | 0.253 |
| | $(E_r - E)/E_r$ [%] | 0.0149 | 0.0162 | 0.0127 | 0.0116 | 0.00003 |
| | $-\log(\delta_8/\delta_7)/\log(N_8/N_7)$ | 3.330 | 3.301 | 2.867 | 3.789 | 4.504 |
| bending-dominated shell | figure no. | Figures 23–25 | Figure 23 | Figure 24 | Figure 25 | Figures 23–25 |
| | dofs number $N$ [1] | 13,068 | 10,824 | 10,824 | 10,824 | 8712 |
| | $\log(E_r - E)$ [N·m] | 1.701 | 1.749 | 1.626 | 1.478 | 0.511 |
| | $(E_r - E)/E_r$ [%] | 0.0046 | 0.0051 | 0.0039 | 0.0018 | 0.0003 |
| | $-\log(\delta_8/\delta_7)/\log(N_8/N_7)$ | 3.411 | 3.393 | 3.741 | 4.429 | 3.788 |
| membrane-dominated shell | figure no. | Figures 26–28 | Figure 26 | Figure 27 | Figure 28 | Figures 26–28 |
| | dofs number $N$ [1] | 13,068 | 10,824 | 10,824 | 10,824 | 8712 |
| | $\log(E_r - E)$ [N·m] | −1.131 | −0.727 | −0.379 | −0.923 | −1.174 |
| | $(E_r - E)/E_r$ [%] | 0.0000083 | 0.0000211 | 0.0000471 | 0.0000134 | 0.0000076 |
| | $-\log(\delta_8/\delta_7)/\log(N_8/N_7)$ | 3.832 | 3.857 | 3.125 | 3.268 | 4.036 |

* The presented result values correspond to $p \equiv \pi = 8$.

## 6. Conclusions

Three types of the transition piezoelectric elements joining the basic piezoelectric elements are possible. The joined basic elements are the three-dimensional piezoelectric elements (or hierarchical symmetric-thickness piezoelectric elements) and the first-order symmetric-thickness piezoelectric elements.

The first (classical) transition element guarantees continuity of the displacement and electric potential fields between the basic elements. It also guarantees that the assumption of no elongation of the normals to the mid-surface holds on the boundary between the transition elements and the first-order elements.

The second (modified) transition element, additionally, guarantees the gradual change from the plane stress to three-dimensional stress state between the basic elements of the first-order and three-dimensional (or hierarchical) character, respectively.

The third (enhanced) transition element, additionally, assures the gradual change of the assumption of no elongation of the normals to the mid-surface between the basic elements. The latter assumption holds on the boundary with the first-order element and is gradually switched off towards the boundary with the three-dimensional (or hierarchical) element. On this boundary, this assumption is completely off.

The gradual change of the stress state from the plane to three-dimensional is based on introduction of the blending functions that determine contributions of the basic models to the transition model. Implementation of this idea on the element level needs modifications of the strain vector, and the piezoelectric and dielectric constant matrices as well.

The gradual change of the assumption of no elongation of the normals needs utilization of the blending functions, which play the role of gradually switching functions. On the element level, the nodal values of the complement (to unity) of the gradually switching functions appear in the penalty stiffness matrix responsible for the gradual change of the constraints.

Comparing the stress distributions within the transition zones for three model problems, one can notice that high stress gradients appear on the boundary between the classical transition elements and the first-order elements. In the case of the enhanced transition elements, the gradual change of the stress state can be observed. In the case of the modified transition element, the stress distribution is intermediate with respect to the previous two distributions.

It can be observed that the discrete models of the three analyzed model structures, with three types of the transition elements employed, produce the convergence curves of slightly higher convergence for the case of the modified elements in comparison to the classical elements. The highest convergence, however, is obtained in the case of the enhanced transition elements.

It can be concluded that the enhanced transition element should be recommended in the analysis of mixed models of piezoelectrics, as this element gives the same or better convergence than the other transition models, and it removes the high stress gradients between the classical transition elements and the first-order piezoelectric elements.

Even though all three transition models can be applied to error-controlled adaptivity due to monotonicity of the applied error measure, the enhanced transition elements deliver convergence curves with the error level and convergence rates just between the curves for the basic models.

The presented research needs the following next steps. Firstly, the error estimation methods and adaptive procedures for the new transition elements have to be prepared and verified. Next, application of the proposed transition models to parametric and/or adaptive analysis of electro-mechanical systems should be performed and critical factors influencing quality of such an analysis should be determined.

**Author Contributions:** Conceptualization, G.Z. and M.Z.; methodology, G.Z. and M.Z.; writing and verifying software, G.Z. and M.Z.; validation, G.Z. and M.Z.; formal analysis, G.Z. and M.Z.; investigation, G.Z. and M.Z.; resources, G.Z. and M.Z.; writing—original draft preparation, G.Z. and M.Z.; writing—review and editing, G.Z. and M.Z.; projects administration, G.Z. and M.Z.; funding acquisition, G.Z. and M.Z. All authors have read and agreed to the published version of the manuscript.

**Funding:** This research was partly funded by Scientific Research Committee Poland under research grants 7 T07A 004 14 and 5 T07A 040 24 and by National Science Center Poland under research grant N N504 515340.

**Institutional Review Board Statement:** Not applicable.

**Informed Consent Statement:** Not applicable.

**Conflicts of Interest:** The authors declare no conflict of interest.

# References

1. Demkowicz, L. *Computing with hp-Adaptive Finite Elements. Vol. 1. One- and Two-Dimensional Elliptic and Maxwell Problems*; Chapman & Hall/CRC: Boca Raton, FL, USA, 2007.
2. Demkowicz, L.; Kurtz, J.; Pardo, D.; Paszyński, M.; Rachowicz, W.; Zdunek, A. *Computing with hp-Adaptive Finite Elements. Vol. 2. Three-Dimensional Elliptic and Maxwell Problems with Applications*; Chapman & Hall/CRC: Boca Raton, FL, USA, 2008.

3. Auricchio, F.; Bisegna, P.; Lovadina, C. Finite element approximation of piezoelectric plates. *Int. J. Numer. Meth. Eng.* **2001**, *50*, 1469–1499.

4. Bisegna, P.; Maceri, F. A consistent theory of thin piezoelectric plates. *J. Intell. Mater. Syst. Struct.* **1996**, *7*, 372–389.

5. Lammering, R.; Yang, F. A Four-node Finite Element for Piezoelectric Shell Structures in Convective Co-ordinates. *Proc. Appl. Math. Mech.* **2006**, *6*, 237–238.

6. Lammering, R.; Mesecke-Rischmann, S. Multi-field variational formulations and related finite elements for piezoelectric shells. *Smart Mater. Struct.* **2003**, *12*, 904–913.

7. Braess, D.; Kaltenbacher, M. Efficient 3D-Finite-Element-Formulation for Thin Mechanical and Piezoelectric Structures. *Int. J. Numer. Meth. Eng.* **2008**, *73*, 147–161.

8. Botta, F.; Cerri, G. Wave propagation in Reissner-Mindlin piezoelectric coupled cylinder with non-constant electric field through the thickness. *Solids Struct.* **2007**, *44*, 6201–6219.

9. Madureira, A.L. Hierarchical modeling of piezoelectric plates. *Appl. Math. Model.* **2012**, *36*, 3555–3569.

10. Carrera E., Boscolo M., Robaldo A. Hierarchic multilayered plate elements for coupled multifield problems of piezoelectric adaptive structures: Formulation and numerical assessment. *Arch. Comput. Methods Eng.* **2007**, *14*, 384–430.

11. Zboiński, G. Problems of Hierarchical Modelling and *hp*-Adaptive Finite Element Analysis in Elasticity, Dielectricity and Piezoelectricity. In *Perusal of the Finite Element Method*; Petrova, R., Ed. InTech: Rijeka, Croatia, 2016; Chapter 1, pp. 1–29.

12. Zboiński, G. Adaptive modeling and simulation of elastic, dielectric and piezoelectric problems. In *Finite Element Method—Simulation, Numerical Analysis and Solution Techniques*; Pacurar, R., Ed.; InTech: Rijeka, Croatia, 2018; Chapter 7, pp. 157–192.

13. Zboiński, G. Hierarchical modeling and Finite Element Approximation for Adaptive Analysis of Complex Structures. D.Sc. Thesis, Institute of Fluid Flow Machinery, Gdańsk, Poland, 2001. (In Polish)

14. Zboiński, G. Adaptive *hpq* finite element methods for the analysis of 3D-based models of complex structures. Part 1. Hierarchical modeling and approximation. *Comput. Methods Appl. Mech. Eng.* **2010**, *199*, 2913-2940.

15. Zboiński, G. Hierarchical models for adaptive modelling and analysis of coupled electro-mechanical systems. In *Recent Advances in Computational Mechanics*; Łodygowski, T., Rakowski, J., Litewka, P., Eds.; CRC Press: London, UK, 2014; pp. 339–334.

16. Zboiński, G. 3D-based hierarchical models and *hpq*-approximations for adaptive finite element method of Laplace problems as exemplified by linear dielectricity. *Comput. Math. Appl.* **2019**, *78*, 2468–2511.

17. Krommer, M.; Irschik, H. A Reissner-Mindlin-type plate theory including the direct piezoelectric and the pyroelectric effect. *Acta Mech.* **2000**, *141*, 51–69.

18. Kant, T.; Shiyekar, S.M. Cylindrical bending of piezoelectric laminates with a higher order shear and normal deformation theory. *Comput. Struct.* **2008**, *86*, 1594–1603.

19. Bisegna, P.; Caruso, G.; Maceri, F. A layer-wise Reissner-Mindlin-type model for the vibration analysis and suppression of piezoactuated plates. *Comput. Struct.* **2001**, *79*, 2309–2319.

20. Kanber, B.; Bozkurt, O.Y. Finite element analysis of elasto-plastic plate bending problems using transition rectangular plate elements. *Acta Mech. Sin.* **2006**, *22*, 355–365.

21. Liao, L.; Yu, W. An electromechanical Reissner-Mindlin model for laminated piezoelectric plates. *Compos. Struct.* **2009**, *88*, 394–402.

22. Quoc, T.H.; Tham, V.V.; Tu, T.M. Active vibration control of a piezoelectric functionally graded carbon nanotube-reinforced spherical shell panel. *Acta Mech.* **2021**, *232*, 1005–1023.

23. Kapuria, S.; Ahmad, A. A coupled efficient layerwise finite element model for free vibration analysis of smart piezo-bonded laminated shells featuring delaminations and transducer debonding. *Int. J. Mech. Sci.* **2021**, *194*, 106195.

24. Zielińska, M. Elastic Structures Analysis by the Adaptive Finite Element Method with Use of the Solid-to-Shell Transition Elements. Ph.D. Thesis, Faculty of Technical Sciences, University of Warmia and Mazury, Olsztyn, Poland, 2016. (In Polish)

25. Surana, K.S. Transition finite elements for three-dimensional stress analysis. *Int. J. Numer. Meth. Eng.* **1980**, *15*, 991–1020.

26. Ahmad, S. Curved Finite Elements in the Analysis of Solid, Shell and Plate Structures. Ph.D. Thesis. University College of Swansea, Swansea, UK, 1969.

27. Surana, K.S. Lumped Mass Matrices with Non-Zero Interia for General Shell and Axi-Symmetric Shell Elements. *Int. J. Numer. Meth. Eng.* **1978**, *12*, 1635–1650.

28. Surana, K.S. Geometrically nonlinear formulation for the axi-symmetric transition finite element. *Comput. Struct.* **1983**, *17*, 243–255.

29. Surana, K.S. Geometrically non-linear formulation for the three dimensional solid-shell transition finite elements. *Comput. Struct.* **1982**, *15*, 549–566.

30. Surana, K.S. Three dimensional solid-shell transition finite elements for heat conduction. *Comput. Struct.* **1987**, *26*, 941–950.

31. Bathe, K.J.; Bolourchi, S. A geometric and material nonlinear plate and shell element. *Comput. Struct.* **1980**, *11*, 23–48.

32. Gmür, T.C.; Schorderet, A.M. A set of three-dimensional solid to shell transition elements for structural dynamics. *Comput. Struct.* **1993**, *46*, 583–591.

33. Liao, C.L.; Reddy, J.N.; Engelstad, S.P. A solid-shell transition element for geometrically nonlinear analysis of laminated composite structures. *Int. J. Number. Meth. Eng.* **1988**, *26*, 1843–1854.

34. Cofer, W.F.; Will, K.M. A three-dimensional, shell-solid transition element for general nonlinear analysis. *Comput. Struct.* **1991**, *38*, 449–462.

35. Dávila, C.G. Solid-to-shell transition elements for the computation of interlaminar stresses. *Comput. Syst. Eng.* **1994**, *5*, 193–202.
36. Gong, Y. Local/global structural analysis by transition elements. *Comput. Struct.* **1988**, *30*, 831–836.
37. Stein, E.; Rüter, M. Error-cotrolled adaptive modelling and finite element analysis in the context of European structural codes. In *International Conference on Adaptive Modeling and Simulation ADMOS 2003*; Wiberg, N.-E., Diez, P., Eds.; CIMNE: Barcelona, Spain, 2003; CD-Rom 1-28.
38. Stein, E.; Ohnimus, S. Equilibrium method for postprocessing and error estimation in the finite element method. *Comput. Assist. Mech. Eng. Sci.* **1997**, *4*, 645–666.
39. Niekamp, R.; Stein, E. An object-oriented approach for parallel two- and three-dimensional adaptive finite element computations. *Comput. Struct.* **2002**, *80*, 317-328.
40. Gupta, A.K. A finite element for transition from a fine to a coarse grid. *Int. J. Numer. Meth. Eng.* **1978**, *12*, 35–45.
41. Huang, F.; Xie, X. A modified nonconforming 5-node quadrilateral transition finite element. *Adv. Appl. Math. Mech.* **2011**, *2*, 784–797.
42. Wan, K. Transition Finite Elements for Mesh Refinement in Plane and Plate Bending Analyses. Master's Thesis. Mechanical Engineering Department, The University of Hong Kong, Hong Kong, China, 2004.
43. Jeyachandrabose C.; Kirkhope J. Construction of transition finite elements for the plane triangular family. *Comput. Struct.* 1984, *18*, 1127–1134.
44. Lo S.H.; Wu, D.; Sze, K.Y. Adaptive meshing and analysis using transitional quadriateral and hexahedral elements. *Finite Elem. Anal. Des.* **2010**, *46*, 2–16.
45. Guzelbey, I.H.; Kanber, B. A practical rule for the derivation of transition finite elements. *Int. J. Numer. Meth. Eng.* **2000**, *47*, 1029–1056.
46. Kim, H.-S.; Hong, S.-M. Formulation of transition elements for the analysis of coupled wall structures. *Comput. Struct.* 1995, *57*, 333–344.
47. Lim, J.-K.; Song, T.-H.; Seog, B.-H. Formulation Method for Solid-to-Beam Transition Finite Elements. *KSME Int. J.* **2001**, *15*, 1499–1506.
48. Garusi, E.; Tralli, A. A hybrid stress-assumed transition element for solid-to-beam and plate-to-beam connections. *Comput. Struct.* **2002**, *80*, 105–115.
49. Stein, E.; Rüter, M.; Ohnimus, S. Adaptive finite element analysis and modelling of solids and structures. Findings, problems and trends. *Int. J. Numer. Meth. Eng.* **2004**, *60*, 103–138.
50. Hauptmann, R.; Schweizerhof, K. A systematic development of solid-shell element formulations for linear and non-linear analyses employing only displacement degrees of freedom. *Int. J. Numer. Meth. Eng.* **1998**, *42*, 49–69.
51. Zboiński, G.; Ostachowicz, W. An algorithm of a family of 3D-based solid-to-shell transition $hpq/hp$-adaptive finite elements. *J. Theor. Appl. Mech.* **2000**, *38*, 791–806.
52. Zielińska, M.; Zboiński, G. $hp$-Adaptive finite element analysis of thin-walled structures with use of the shell-to-shell transition elements. In *Recent Advances in Computational Mechanics*; Łodygowski, T., Rakowski, J., Litewka, P., Eds.; CRC Press: London, UK, 2014; pp. 63–72.
53. Zboiński, G. Application of the three-dimensional triangular-prism $hpq$ adaptive finite element to plate and shell analysis. *Comput. Struct.* **1997**, *65*, 497–514.
54. Zboiński, G.; Jasiński, M. 3D-based $hp$-adaptive first order shell finite element for modeling and analysis of complex structures—Part 1. The model and the approximation. *Int. J. Numer. Meth. Eng.* **2007**, *70*, 1513–1545.
55. Szabó, B.A.; Sahrmann, G.J. Hierarchic plate and shell models based on $p$-extension. *Int. J. Numer. Meth. Eng.* **1988**, *26*, 1855-1881.
56. Oden, J.T.; Cho, J.R. Adaptive $hpq$ finite element methods of hierarchical models for plate- and shell-like structures. Comput. Methods Appl. Mech. Eng. **1996**, *136*, 317–345.
57. Premount, A. *Mechatronics. Dynamics of Electromechanical and Piezoelectric Systems*; Springer: Dordrecht, The Netherlands, 2006.
58. Zboiński, G. Analytical aspects of the 3D-based hierarchical models of three-dimensional, transition and symmetric-thickness piezoelectrics as applied in adaptive FEM. In *Shell Structures. Theory and Applications*; Pietraszkiewicz, W., Witkowski, W., Eds.; CRC Press: London, UK, 2017; Volume 4, pp. 304–306.
59. Ieşan, D. Reciprocity, uniqueness and minimum principles in the linear theory of piezoelectricity. *Int. J. Eng. Sci.* **1990**, *28*, 1139–1149.
60. Cimatti, G. The piezoelectric continuum. *Ann. Metematica Pura Appl.* **2004**, *183*, 495–514.
61. Babuška, I. Error-bounds for finite element method. *Numer. Math.* **1971**, *16*, 322–333.
62. Lax, P.D.; Milgram, A.N. Parabolic equations. In *Contributions to the Theory of Partial Differential Equations, Annalsof Mathematics Studies 33*; Princeton University Press: Princeton, NJ, USA,1954; pp. 167–190.
63. Hinton, E.; Owen, D.R.J. *Finite Element Software for Plates and Shells*; Pineridge Press: Swansea, UK, 1984.

 *applied sciences*

*Article*

# A Posteriori Detection of Numerical Locking in *hpq*-Adaptive Finite Element Analysis

**Łukasz Miazio [1] and Grzegorz Zboiński [1,2,*]**

[1] Faculty of Technical Sciences, The University of Warmia and Mazury, Oczapowskiego 11, 10-719 Olsztyn, Poland; lukasz.miazio@uwm.edu.pl
[2] Institute of Fluid Flow Machinery, Polish Academy of Sciences, Fiszera 14, 80-231 Gdańsk, Poland
[*] Correspondence: zboi@imp.gda.pl; Tel.: +48-58-522-5219

Received: 29 September 2020; Accepted: 17 November 2020; Published: 20 November 2020

**Abstract:** The proposed detection algorithms are assigned for the *hpq*-adaptive finite element analysis of the solid mechanics problems affected by the locking phenomena. The algorithms are combined with the *M*- and *hpq*-adaptive finite element method, where *M* is the element model, *h* denotes the element size parameter, and *p* and *q* stand for the longitudinal and transverse approximation orders within an element. The applied adaptive scheme is extended with the additional step where the locking phenomena are a posteriori detected, assessed and resolved. The detection can be applied to shear, membrane, or shear–membrane locking phenomena. The removal of the undesired influence of the numerical locking on the problem solution is based on *p*-enrichment of the mesh. The detection algorithm is also enriched with the locking assessment algorithm which is capable of determination of the optimized value of *p* which is sufficient for the phenomena removal. The detection and assessment algorithms are based on a simple sensitivity analysis performed locally for the finite elements of the thin-walled domain. The sensitivity analysis lies in comparison of the element solutions corresponding to two values of the order *p*, namely current and potentially eliminating the locking. The local solutions are obtained from the element residual method. The elaborated algorithms are original, relatively simple, extremely reliable, and highly effective.

**Keywords:** solid mechanics; finite elements; *hp*-adaptivity; numerical locking; detection; assessment; resolution; equilibrated residual method; sensitivity analysis; *p*-enrichment

---

## 1. Introduction

This paper concerns application of the algorithms for detection, assessment and resolution of numerical locking in the *hpq*-adaptive finite element elastic analysis of thin-walled structures or complex structures which include thin-walled, solid, and transition parts. We consider all cases when the influence of the locking phenomenon on the problem solution is significant. We focus on the theoretical and methodological aspects such as the idea and justification of the elaborated algorithms for a posteriori detection and assessment of the phenomenon. In addition, the necessary modification of the applied *hpq*-adaptive algorithms is of our interest. The employed model- and *hpq*-adaptive method allows for different element size *h*, different element longitudinal and transverse orders of approximation, *p* and *q*, and different element model *M* in each finite element. The paper also presents application of the introduced detection, assessment, and resolution algorithms in the *hpq*-adaptive analysis of structural elements. These algorithms are investigated in the contexts of their generality, reliability, and effectiveness.

### 1.1. State-of-the-Art Issues

We address two specific issues dealt with the locking phenomena. The first one concerns some basic research on the nature of this phenomenon, while the second issue is the existing methods of removal of the numerical consequences of the phenomenon.

#### 1.1.1. Theoretical and Numerical Research of Locking Phenomena

The numerical locking phenomena concern thin-walled structures in which the true solution is characterized by bending strains dominance over shear and/or membrane strains. The phenomenon does not appear in the case of the membrane strains dominance. If, due to poor discretization of the problem, the shear and/or membrane strains are not equal to zero, as it results from the thin-walled theories for the thickness $t$ tending to 0, then the shear, membrane, or shear–membrane strain energy numerically dominates over the bending energy, leading to numerical over-stiffening of the structure, which in turn results in too low (zero or almost zero) values of displacements (compare the work [1]). This phenomenon is called the shear, membrane, or shear–membrane locking and is typical for the displacement finite element method.

Theoretical and numerical studies of the locking phenomena concern one- and two-dimensional problems, including beams, arcs, plates, and shells. Different kinds of locking are investigated: volume (Poisson's) locking present in nearly incompressible materials ($\nu \rightarrow 0.5$), deformational locking present in bending-dominated thin-walled structures within the displacement formulation of the finite element method, and finally the trapezoidal locking present in hybrid-stress finite elements. The deformational locking, which is the subject of this work, may be shear (plates), membrane, or shear–membrane (shells). The significant exemplary theoretical and numerical research results concerning deformational locking are presented in [1–6], respectively. These works refer to the first-order, higher-order, and hierarchical models of plates and shells.

#### 1.1.2. Overcoming the Locking Phenomena

We limit this survey to the methods related to the shear and/or membrane locking. This type of locking results from the thick- or thin-walled character of the plate and shell structures.

The first method of overcoming the locking is based on application of the mixed or hybrid formulations of the finite element method instead of the displacement formulation. This leads to elements of the class $C^1$ instead of the class $C^0$. The elements of this group are usually of low-order and may need stabilization. The examples of the elements of this group are presented in [7–9]. More recent examples of the mixed and hybrid finite elements resistant to locking are published in [10,11].

The second approach takes advantage of the so-called reduced or reduced selective numerical integration, sometimes enriched with the stabilization matrix which removes deformation modes of zero energy. The reduced integration consists in integration of the stiffness matrix with the numerical integration parameters as for the elements described with the polynomial interpolation of one order lower. In the selective version of the reduced integration the lower order is applied to a part of the stiffness matrix, responsible for the locking, i.e., the part corresponding to shear and/or membrane strain energy. The examples of application of this approach to plate and shell elements can be found in [12–14]. The recent works dealing with the reduced and reduced selective integration concern either the isogeometric analysis or the standard finite element methods, for example, in [15,16].

The third way to overcome the phenomenon lies in introduction of the discrete Kirchhoff constraints into the elements of the class $C^0$. The method is directed towards removal of the shear locking and requires that the Kirchhoff constraints are imposed on the selected points or lines within $C^0$ element. The prominent examples of plate or thick shell elements of this type can be found in [17,18]. Recent examples of the discrete Kirchhoff and Kirchhoff-Love constraints are presented in the works [19,20].

The fourth method consists in application of the consistent (interdependent) fields of the transverse displacement and rotations, one order higher in the case of the mentioned displacement. The method leads to different numbers of unknowns of both types within an element. The surplus displacement degrees of freedom (dofs) are removed from the model based on the condition of zero transverse shear strains and/or zero membrane strain condition. The prominent examples of this method of the locking removal can be found in [21,22].

The fifth approach is based on the assumed shear or membrane strains consistent with the interpolated transverse displacement at some points. In this method, bending strains result from the interpolated displacement field, while the transverse shear and/or membrane (in-plane) strains possess the assumed form resulting from the interpolation based on some chosen points. The significant works, leading to the current state of this method in relation to quadrilateral plate and shell elements, are found in [23–25]. Two versions of the presented approach, based on either the enhanced assumed strains or assumed natural strains, are still being developed, for example, in [26,27].

The sixth method lies in application of higher-order elements conforming to the displacement finite element formulation. The examples of application of such elements in the case of the classical (non-adaptive) finite element methods, can be found in [28,29]. In the non-adaptive methods, plate or shell elements conforming to the first-order or higher-order theories are applied. The fixed longitudinal order of $h$-approximation up to the fifth order is usually applied within elements of this type. The adaptive quadrilateral elements, conforming to hierarchical approximations and higher-order shell models, are used in [30]. The hexahedral elements corresponding to three-dimensional elasticity, equipped with independent transverse and longitudinal approximations of the higher order, and assigned for plate and shell analysis, are proposed in [31] and applied in $hp$-adaptive version in [32], for example. The recent applications of the higher-order models and approximations to locking removal are presented in the works [33,34]. These proposals are not consistent with the hierarchical approach.

Let us conclude the above survey of the methods of overcoming the locking phenomena in the context of needs of the $hpq$-adaptive method for complex structures analysis. Firstly, it should be noticed that only low-order longitudinal approximations and first-order plate and shell models are possible in the cases of the mixed and hybrid methods, the uniform or selective reduced integration, and the consistent field method. In the case of the discrete Kirchhoff constraints, only shear locking can be removed effectively. Additionally, this method of removal leads to large variability of elements. In the case of the assumed strains approach, the claimed generalization of the method for high-order in-plane approximations has not been proved in practice. For the reduced integration and assumed stress method, there are problems with their application to triangular or prismatic elements. In the case of the discrete Kirchhoff constraint and consistent field methods of removal, one deals with different number and location of translational and rotational degrees of freedom. Note that the higher-order elements are free from all these defects. Due to our earlier choices concerning the applied hierarchical models and $hpq$-approximations, the displacement formulation of such elements becomes our obvious choice.

### 1.1.3. Detection and Assessment of the Locking

The basic method of detection and assessment of the locking phenomena is the a priori theoretical analysis of the numerical solutions of the model problems potentially suspected to be a subject of locking (see Section 1.1.1 of this literature survey). Another interesting method of detection and assessment of the locking in the one- and two-dimensional elements is proposed in [35,36]. It is based on application of the finite difference operators corresponding to the problem local formulation. In [37–39], the numerical methods of detection and assessment of the phenomenon are proposed. These methods are based on the sensitivity analysis, i.e., two or a sequence of local problems are solved for each element of the potentially affected domain.

*1.2. The Applied Methodology*

Two issues are addressed here. The first one deals with the best choice of the method of effective detection and assessment of the locking phenomena. The second issue is related to the numerical methods of removing the phenomena.

It results from the above literature survey that the available knowledge on the locking phenomena allows understanding of the nature and sources of appearance of the phenomena. The accumulated knowledge on the phenomenon allows also for a priori determination of the solution convergence of the problems where the phenomena appear. The main difficulty in the direct application of these results in the numerical analysis of any arbitrary thin-walled structure is that the available results concern the specific model problems which may differ to the arbitrary problem under consideration.

The second conclusion from the literature survey is that the most effective way of removing the locking phenomena lies in application of the higher-order longitudinal $p$-approximation of the displacement field in the analyzed thin-walled structure or a thin-walled part of the complex structure.

It also results from the literature that some detection methods of the phenomena exist. Among these methods, the approach proposed in [37–39] seems to be best suited for adaptive analysis. This approach is based on the same numerical techniques that are applied in the error-controlled adaptivity.

The main feature of the proposed a posteriori detection, assessment and removal of the locking phenomena is that the adaptation process requires four steps, instead of the standard three steps of the error-controlled $hp$-adaptivity proposed in [40]. The additional step of adaptation, which lies in initial mesh modification, incorporates not only the automatic removal of the locking phenomena but also the automatic resolution of the boundary layers [38]. The main idea standing behind the additional adaptation step is to move the numerical solution, obtained with use of $hp$-approximations [41,42], to the asymptotic convergence range. Within this range, the standard $h$- and $p$-adaptation steps can be made based on the $hp$-convergence theorem and upper-bounding values of the global error estimates from the equilibrated residual method [43,44].

Finally, it should be noted that our detection and assessment tools are based on sensitivity analysis, not on the estimated error values themselves. Thanks to this, requirements concerning the error estimation can be relaxed.

*1.3. Novelty of the Paper*

The novelty of this work consists in the new algorithms for a posteriori detection and assessment of the numerical locking phenomena. This refers to shear, membrane, or shear–membrane locking. With these new algorithms, one is able to detect the presence of the phenomenon and assess its strength so that the adequate numerical means can be used to remove the phenomenon. The proposed numerical means of the removal consist in introduction of the new adaptation step, called the modification one, into the existing three-step, model- and $hpq$-adaptive finite element procedure for analysis of complex structures. The new step employs the mentioned detection and assessment algorithms and performs modification of the initial mesh through $p$-enrichment. The numerical cost of this new step is low as the detection and assessment is performed on the initial, usually coarse mesh.

The applied adaptive method [38,45,46], the new algorithms are incorporated in, takes advantage of the hierarchical models proposed in [31,47], hierarchical $hp$-approximations elaborated in [41,42], a posteriori error estimation from [30,43,44], and error-controlled adaptive procedure given in [40].

## 2. Preliminaries

Two issues are addressed in this section. The first one deals with the model problems considered in this research. The second one is presentation of the nature of the locking phenomena.

### 2.1. Model Problems

Let us consider a wide range of problems of linear elasticity covered by the standard local (strong) formulation

$$\left.\begin{array}{c} \sigma^{ij}_{,j} + f^i = 0 \\ \sigma^{ij} = D^{ijkl}\varepsilon_{kl} \\ \varepsilon_{kl} = \frac{1}{2}(u_{k,l} + u_{l,k}) \end{array}\right\}, \quad x \in V \tag{1}$$

composed of the equilibrium, constitutive, and geometrical equations, respectively. Above, $i, j, k, l = 1, 2, 3$, while the smooth components $f^i \in L^2(V)$ stand for the known body loading vector, with $V$ representing volume of the body. The tensor components $\sigma^{ij}$ and $\varepsilon_{kl}$ stand for stresses and strains, while the unknown vector components $u_k$ denote displacements. The terms $D^{ijkl}$ are components of the fourth-order tensor of elasticities. The elasticities are symmetric $D^{ijkl} = D^{jikl} = D^{ijlk} = D^{klij}$ and satisfy the strong and uniform ellipticity condition: $D^{ijkl}\zeta_{ij}\zeta_{kl} \geq \alpha\zeta_{ij}\zeta_{kl}$, $\zeta_{ij} = \zeta_{ji}$, where $\zeta_{ij}$ is any real-valued tensor of the second rank and $\alpha$ is a positive constant. Note that such elasticities may correspond to three-dimensional stress and strain states or to the constrained, plane stress or plane strain, states. Note also that, for the specific case of isotropy, the elasticities can be expressed by the Young's modulus $E$ and Poisson's ratio $v$.

The set (1) has to be completed with the displacement and stress boundary conditions of the form

$$\begin{aligned} \sigma^{ij}n_j &= p^i, \quad x \in S_P \\ u_i &= w_i, \quad x \in S_W \end{aligned} \tag{2}$$

where $i, j = 1, 2, 3$, while the smooth components $p^i \in L^2(S)$ and the terms $w_i$ stand for the known values of stress vector components and displacement vector components on the parts $S_P$ and $S_W$ of the surface $S \equiv \partial V$ of the body $V$, with $S = S_P \cup S_W$ and $S_P \cap S_W = \varnothing$. The vector components $n_j$ represent unit outward normal to the surface part $S_P = S_T \cup S_B$ composed of the top $S_T$ and bottom $S_B$ surfaces of the body. The way they are determined is explained below.

Equation (1) is valid within the body volume $V$. Such a volume may represent thick- or thin-walled structures or complex structures containing such parts. Due to the model character of the considerations of this section, we limit ourselves to the first case. The corresponding thin-walled domain of the volume $V$ is sufficiently smooth (Lipschitzian or smoother), open, and bounded region. As we apply the 3D Cartesian description of the problem, i.e., we employ Cartesian coordinates $x$, the following explicit (curvilinear) and implicit (Cartesian) definitions of the volume $V$ are appropriate:

$$V = \left\{ \eta \in R^3 : (\eta^1, \eta^2) \in S_M, \eta^3 \in (-t/2, t/2) \right\} \tag{3}$$

$$\equiv \left\{ x(\eta) \in R^3 : x = F(\eta), (\eta^1, \eta^2) \in S_M, \eta^3 \in (-t/2, t/2) \right\}$$

where $F$ is the reversible map converting the curvilinear coordinates $\eta$ into the Cartesian ones $x$. The function $t = t(\eta^1, \eta^2)$ measures the symmetric thickness in the direction $\eta^3$ normal to the mid-surface $S_M$. The mid-surface $S_M$, where $\eta^3 = 0$, of the thin-walled body can be any sufficiently smooth (Lipschitzian or smoother), two-dimensional, open and bounded region. Formal definition of the thin-walled body geometry also needs introduction of the lateral part $S_L$ of the body boundary $S$ as well as the upper (top) $S_T$ and lower (bottom) $S_B$ parts of this boundary. More details on the applied definition of the thin-walled geometry can be found in [48].

The considered model problem, described by the local (strong) formulation (1), completed with the boundary conditions (2), can also be presented in the weak variational form:

$$B(\boldsymbol{u}, \boldsymbol{v}) = L(\boldsymbol{v}) \tag{4}$$

where the admissible displacement vector is $\boldsymbol{v} \in \mathcal{V}$, while the corresponding space reads $\mathcal{V} = \{\boldsymbol{v} \in (H^1(V))^3 : \boldsymbol{v} = \boldsymbol{0} \text{ on } S_W\}$. The solution function for displacements is $\boldsymbol{u} \in \boldsymbol{w} + \mathcal{V}$, with $\boldsymbol{w}$ standing for the lift of the Dirichlet data (see [41]). This lift is consistent with the second Equation (2). The bilinear form $B(\boldsymbol{v}, \boldsymbol{u})$ and linear form $L(\boldsymbol{v})$ from the above functional represent the virtual strain energy and the virtual work of the external body and surface loadings, $\boldsymbol{f}$ and $\boldsymbol{p}$, respectively, i.e.,

$$
\begin{aligned}
B(\boldsymbol{u}, \boldsymbol{v}) &= \int_V \varepsilon^T(\boldsymbol{v}) \boldsymbol{D} \, \varepsilon(\boldsymbol{u}) \, dV \\
L(\boldsymbol{v}) &= \int_V \boldsymbol{v}^T \boldsymbol{f} \, dV + \int_{S_P} \boldsymbol{v}^T \boldsymbol{p} \, dS
\end{aligned}
\tag{5}
$$

Above, $\boldsymbol{D}$ stands for the matrix representation of the elasticity constant tensor present in the second Equation (1), while $\varepsilon$ is the six-component vectorial representation of the strain tensor defined with the third Equation (1).

Let us introduce now the finite element approximation of the variational functional (4). The approximation is based on the general rules of $hp$ approximations [41], applied to 3D-based hierarchical shell models of order $q$ presented in [38,45]. The approximated variational functional reads

$$B(\boldsymbol{u}^{hpq}, \boldsymbol{v}^{hpq}) = L(\boldsymbol{v}^{hpq}) \tag{6}$$

where the approximated admissible displacements $\boldsymbol{v}^{hpq} \in \mathcal{V}^{hpq}$ belong to the space $\mathcal{V}^{hpq} = \{\boldsymbol{v}^{hpq} \in (H^1(V))^3 : \boldsymbol{v}^{hpq} = \boldsymbol{0} \text{ on } S_W\}$. Additionally, $\boldsymbol{u}^{hpq} \in \boldsymbol{w}^{hpq} + \mathcal{V}^{hpq}$, with the approximated values $\boldsymbol{w}^{hpq}$ of the lift $\boldsymbol{w}$.

Note that the approximated variational formulation (6) can also be expressed in the language of finite elements

$$\boldsymbol{K} \boldsymbol{q}^{hpq} = \boldsymbol{F} \tag{7}$$

where $\boldsymbol{K}$ and $\boldsymbol{F}$ are the global stiffness matrix and the global forces vector, while $\boldsymbol{q}^{hpq}$ stands for the displacement dofs vector corresponding to the solution field $\boldsymbol{u}^{hpq}$ of displacements.

### 2.2. Locking Phenomena

It was demonstrated in a numerous works (see [49,50], for example) that the three-dimensional elasticity description of the strain energy $U \equiv \frac{1}{2}B(\mathbf{u}, \mathbf{u})$ of the thin-walled body $V$ tends to the following limit value when the thickness of the body tends to zero, $t \to 0$:

$$
\begin{aligned}
U \;=\;& \frac{1}{2}\int_V \sigma^{ij}\varepsilon_{ij}\,dV \equiv \frac{1}{2}\int_V [\sigma^{\alpha\beta}\varepsilon_{\alpha\beta} + \sigma^{33}\varepsilon_{33} + 2\sigma^{3\beta}\varepsilon_{3\beta}]\,dV \\[4pt]
\to\;& \frac{1}{2}t\int_{S_M}\frac{E}{1-\nu^2}[(1-\nu)\gamma_{\alpha\beta}\gamma_{\alpha\beta}+\nu\gamma_{\alpha\alpha}\gamma_{\beta\beta}]\,dS_M + \frac{1}{2}t\int_{S_M}\frac{Ek}{1+\nu}(\gamma_{\alpha 3}\gamma_{\alpha 3}+\gamma_{3\beta}\gamma_{3\beta})\,dS_M \\[4pt]
+\;& \frac{1}{2}t^3\int_{S_M}\frac{E}{12(1-\nu^2)}[(1-\nu)\kappa_{\alpha\beta}\kappa_{\alpha\beta}+\nu\kappa_{\alpha\alpha}\kappa_{\beta\beta}]\,dS_M \\[4pt]
=\;& U_m + U_s + U_b
\end{aligned}
\tag{8}
$$

Above, the three-dimensional components of the stress and strain tensors, $\sigma^{ij}$ and $\varepsilon_{ij}$, $i, j = 1, 2, 3$, of the three-dimensional theory of elasticity can be expressed in the thin limit with the two-dimensional longitudinal strain components of the mid-surface $S_M$, $\gamma_{\alpha\beta}$, $\alpha, \beta = 1, 2$; transverse strain components, $\gamma_{\alpha 3}$ and $\gamma_{3\beta}$, $\alpha, \beta = 1, 2$; and the components of the tensor of curvature variation, $\kappa_{\alpha\beta}$, $\alpha, \beta = 1, 2$. Additionally, in the thin limit, the three-dimensional isotropic elasticity constants can be replaced with the isotropic plane stress constants of the first-order shell theory, expressed by the Young's modulus $E$ and the Poisson's ratio $\nu$ and resulting from the plane stress assumption $\sigma_{33} = 0$. In addition, the kinematic condition of no elongation of the normals to the mid-surface, $\gamma_{33} = 0$, comes into play. The thin limit strain energy can then be approximated with the limit values of the membrane, shear, and bending parts, $U_m$, $U_s$, $U_b$, of the strain energy $U$. The quantity $k$ is the shear strain correction factor, equal to $5/6$ and resulting from the applied first-order model which leads to false (constant) transverse-shear stresses.

Depending on the geometry, loading and kinematic boundary conditions, one may distinguish between the bending-dominated problems, when

$$
U_b \gg U_m + U_s
\tag{9}
$$

and membrane-dominated or shear–membrane-dominated problems, where

$$
\begin{aligned}
U_m &\gg U_b, \; U_s \cong 0 \\
U_m + U_s &\gg U_b
\end{aligned}
\tag{10}
$$

respectively. Then, it results from (9) that, in the analytical solution of the bending-dominated problems, the shear and membrane strains should be equal or very close to zero, i.e.,

$$
\gamma_{13} = \gamma_{31} \cong 0, \; \gamma_{23} = \gamma_{32} \cong 0
\tag{11}
$$

for the plate and shell structures, and

$$
\gamma_{11} \cong 0, \; \gamma_{12} \cong 0, \; \gamma_{22} \cong 0
\tag{12}
$$

for the shell structures where the coupling of the membrane and bending strains exists. Note that the decoupling of the membrane and bending strains shown by the limit expression of the relation (8) exists only in the thin limit ($t \rightarrow 0$) for the first-order shells, while for the first-order plates it holds for any thickness $t$.

The shear and shear–membrane lockings are purely numerical phenomena which happen for poor discretizations and result from insufficiently accurate approximation of the analytical constraints (11) and/or (12). If such inaccurate approximation occurs then the shear and/or membrane strains are too large and the shear and membrane strain energies grow. These energies dominate over the bending strain energy, leading to numerical over-stiffening of the structure, which in turn gives too small (locked) values of displacements of the numerical solution. Remembering that $U^{hpq} \equiv \frac{1}{2}B(u^{hpq}, u^{hpq}) \equiv \frac{1}{2}(q^{hpq})^T K q^{hpq}$, one has

$$Kq^{hpq} = F, \quad F_I = \text{const}, \; K_{IJ} \uparrow \; \Rightarrow q_J^{hpq} \downarrow \tag{13}$$

where $K_{IJ}$, $F_I$ are components of the global stiffness matrix and the global forces vector, with $I, J = 1, 2, \ldots, N$ and $N$ being the global number of degrees of freedom (dofs). The components $q_J^{hpq}$ represent the global vector of unknown displacement dofs $q^{hpq}$ corresponding to the approximated field of displacements $u^{hpq}$.

### 3. Locking Detection, Assessment, and Resolution

The proposed ideas of locking detection, assessment, and resolution are based on application of three existing numerical techniques. The first one is the equilibrated residual method of error estimation [43,44], here applied to solution of the local problems of two types. The second idea is sensitivity analysis based on comparison of the solutions to two local problems which differ with the longitudinal orders of approximation. The results of such a comparison can be used for the locking detection or assessment. If the locking is detected, then the mesh can be modified through the increase of the longitudinal order of approximation, by means of the standard $p$-enrichment technique [41,42].

The mentioned three techniques can be combined with the three-step $hp$-adaptive procedure controlled by the approximation error [51,52] or the automatic, iterative $hp$-adaptation driven by the interpolation error [41,42,53].

#### 3.1. The Idea and Algorithm of a Posteriori Phenomenon Detection

The idea of a posteriori detection of the numerical locking was originally proposed by Zboiński [38]. Here, we develop and verify this idea. It consists in solution of two local problems for each element of the thick- or thin-walled part of the structure. The solutions to these problems are obtained from the equilibrated residual method. These two problems differ with the longitudinal order of approximation $p$. In the first problem, it is equal to its current value from the global problem under consideration. In the second local problem, its value corresponds to the problem potentially free of locking. For two mentioned solutions, the strain energy norms are calculated and compared. This corresponds to sensitivity analysis where the sensitivity of the solutions of the local problems to the change of the longitudinal order of approximation is assessed. Once the locking is detected, one may modify the mesh by increasing the current order of the longitudinal approximation to its value corresponding to the problems free of locking. The similar approach is used also for determination of the optimized value of the longitudinal approximation order which corresponds to the minimum value sufficient for removal of the phenomenon.

#### 3.1.1. Solutions from the Equilibrated Residual Method

Let us recall now the equilibrated residual method of a posteriori error estimation, invented by Ainsworth and Oden [43]. All relations of this section are taken from the cited work. Here, we apply this method to the estimation of the numerical solutions of the 3D-based hierarchical shell problems [46]. The principal relation of the method is [44]

$$
\begin{aligned}
B(e, v) &= B(u^{HPQ} - u^{hpq}, v) = B(u^{HPQ}, v) - B(u^{hpq}, v) \\
&= L(v) - B(u^{hpq}, v), \quad \forall v \in \mathcal{V}^{HPQ}(V)
\end{aligned}
\tag{14}
$$

Above, $u^{HPQ}$ stands for either the exact solution of the problem or sufficiently accurate approximation of such a solution. The quantity $e$ is the error vector corresponding to the assessed numerical solution $u^{hpq}$. The kinematically admissible displacements $v \equiv v^{HPQ}$ belong to the space $\mathcal{V}^{HPQ}$ defined in analogy to the spaces introduced for (4) and (6). The discretization parameters $H$, $P$, and $Q$ are the counterparts of $h$, $p$, and $q$ from the relation (6). In the case of the exact solution ($u^{HPQ} \equiv u$), one deals with $1/H, P, Q \to \infty$, while in the case of the approximation, one deals with the finite values of the discretization parameters $H$, $P$ and $Q$.

After noticing that $e = u^{HPQ} - u^{hpq}$ and introduction of this definition into the error functional (14), the latter simplifies to:

$$
B(u^{HPQ}, v) = L(v), \quad \forall v \in \mathcal{V}^{HPQ}(V)
\tag{15}
$$

This way one searches for the approximation of the exact solution (displacements) of the problem instead of the problem error itself. However, once the displacements are determined, the error can be calculated from the above error definition. The presented approach is applied in the works [30,46], for example.

Let us now follow the work [43] and divide the domain $V$ into $E$ subdomains $V_e$, corresponding to finite elements $e = 1, 2, \ldots, E$. The global functional (15) and its left- and right-hand side components can now be presented as a sum of the element contributions:

$$
\begin{aligned}
B(u^{HPQ}, v) &= \sum_{e=1}^{E} \overset{e}{B}(\overset{e}{u}^{HPQ}, \overset{e}{v}) \\
L(v) &= \sum_{e=1}^{E} \overset{e}{L}(\overset{e}{v}) = \sum_{e=1}^{E} \overset{e}{L}(\overset{e}{v}) - \sum_{e=1}^{E} \overset{e}{\beta}(\overset{e}{v}) \\
&= \sum_{e=1}^{E} \left[ \overset{e}{L}(\overset{e}{v}) + \int_{S_e \backslash (S_P \cup S_W)} \overset{e}{v}^T \langle \overset{e}{r}(u^{hpq}) \rangle \, dS_e \right]
\end{aligned}
\tag{16}
$$

where the element admissible displacements are defined as the global admissible displacements projected onto elements $V_e$: $\overset{e}{v} \equiv v|_{V_e}$. As demonstrated by Ainsworth and Oden [43], the sum of the auxiliary element functionals $\overset{e}{\beta}(\overset{e}{v})$ is equal to 0, as the internal load consistency condition must hold.

Above, the vectors $\langle \overset{e}{r}(u^{hpq}) \rangle$ represent the interelement loading due to equilibrated stresses. These stresses are defined in [43,44], for example. Their definition is as follows

$$
\langle \overset{e}{r}(u^{hpq}) \rangle = \overset{fe}{\alpha} \overset{e}{r}(u^{hpq}) + \overset{ef}{\alpha} \overset{f}{r}(u^{hpq})
\tag{17}
$$

where

$$
\begin{aligned}
\overset{e}{r}(u^{hpq}) &= H(\overset{e}{v}) \, \overset{e}{\sigma}(u^{hpq}), \\
\overset{f}{r}(u^{hpq}) &= H(\overset{e}{v}) \, \overset{f}{\sigma}(u^{hpq}),
\end{aligned}
\tag{18}
$$

and

$$H(\overset{e}{\nu}) = \begin{bmatrix} \nu_1 & 0 & 0 & \nu_2 & 0 & \nu_3 \\ 0 & \nu_2 & 0 & \nu_1 & \nu_3 & 0 \\ 0 & 0 & \nu_3 & 0 & \nu_2 & \nu_1 \end{bmatrix} \tag{19}$$

The components of the unit normal vector $\overset{e}{\nu} = [\nu_1, \nu_2, \nu_3]^T$ are defined on the element surface $S_e$. The vectorial six-component representations $\overset{e}{\sigma}$ and $\overset{f}{\sigma}$ of the stress tensors have to be determined for the element $e$ and any of its neighbours $f$. The terms $\overset{fe}{\alpha}$ and $\overset{e}{r}$ stand for the splitting function diagonal matrices and stress vectors. These matrices are: $\overset{fe}{\alpha} = \mathrm{diag}[\alpha_1, \alpha_2, \alpha_3]$, with $\overset{fe}{\alpha} = 1 - \overset{ef}{\alpha}$, $1 = \mathrm{diag}[1, 1, 1]$ and directional components $\alpha_m$, $m = 1, 2, 3$. The algorithms for calculation of the splitting functions in the case of the internally unconstrained and constrained (e.g., by the Reissner-Mindlin kinematic constraints) are described in [43,44,46], respectively. The alternative is replacement of the equilibrated stresses with their averaged counterparts: $\overset{fe}{\alpha} = \overset{ef}{\alpha} = \mathrm{diag}[\frac{1}{2}, \frac{1}{2}, \frac{1}{2}]$.

Comparing relations (15) and (16) one can notice that the minimization of the global energy functional can be replaced with the minimization of local (element) functionals (see [43,44,46]). The element functionals read

$$\overset{e}{u}{}^{HPQ} \in \overset{e}{\mathcal{V}}{}^{HPQ}(V_e): \quad \overset{e}{B}(\overset{e}{u}{}^{HPQ}, \overset{e}{v}) = \overset{e}{L}(\overset{e}{v}) + \int_{S_e \backslash (P \cup Q)} \overset{e}{v}{}^T \langle \overset{e}{r}(u^{hpq}) \rangle \, dS_e, \quad \forall \, \overset{e}{v} \in \overset{e}{\mathcal{V}}{}^{HPQ}(V_e) \tag{20}$$

where $\overset{e}{u}{}^{HPQ}$ is the element solution function and $\overset{e}{\mathcal{V}}{}^{HPQ}$ denotes the space of kinematically admissible element displacements $\overset{e}{v} \equiv v^{HPQ}$. As shown in [43,44], the solution to the local problems (20) exists and is either unique or unique up to rigid body motions, for Dirichlet and Neumann boundary value problems, respectively. The mixed boundary value problems are also possible. Note that one deals with the Dirichlet problems for clamped elements adjacent to the external boundary of the structure, and with the Neumann problems for elements not adjacent to this boundary, namely for elements from the interior of the structure.

### 3.1.2. Check on Bending-Dominance of the Solution

Check on bending-dominance of the solution to the global problems of the type (6) was proposed by Zboiński [32]. Such a check is necessary as the locking phenomena are present only in the bending-dominated problems. In the membrane-dominated problems the phenomena do not appear. In addition, in the case of the mixed dominance, the locking may appear.

### Strain Energy Components

In this work, we apply the proposed approach to the local solutions from the equilibrated residual method and to their sum. To apply this approach, one has to take into consideration that the strain energy approximation from (8) corresponds to the structures of the infinitely small thickness, for which the first-order theory can approximate the three-dimensional description. As we intend to check the locking phenomena in the structures of the finite thickness, described with the hierarchical models corresponding to higher-order shell theories, the decomposition of the three-dimensional strain energy into component energies typical for the first-order models can only be done approximately. Let us start with the following decomposition of the strain vector into its longitudinal, transverse and shear parts

$$\varepsilon = [\varepsilon_l, \varepsilon_n, \varepsilon_s]^T \tag{21}$$

where $\varepsilon_l = [\dots, \varepsilon_{\alpha\beta}, \dots]^T$, $\varepsilon_n = [\varepsilon_{33}]^T$, $\varepsilon_s = [\dots, \varepsilon_{3\alpha}, \varepsilon_{\alpha3}, \dots]^T$ and $\alpha, \beta = 1, 2$ are the local directions tangent to longitudinal directions of the thin-walled structure, and the index 3 corresponds to the local transverse direction. The analogous decomposition of the stress vector $\sigma$ is also possible. Such decompositions lead to the division of the density $v$ of the total strain energy into the following components:

$$
\begin{aligned}
v &= \sigma^{ij}\varepsilon_{ij} \\
&= \sigma^{\alpha\beta}\varepsilon_{\alpha\beta} + \sigma^{33}\varepsilon_{33} + 2\sigma^{3\alpha}\varepsilon_{3\alpha} \\
&= v_l + v_n + v_s
\end{aligned}
\tag{22}
$$

where $i, j = 1, 2, 3$ stand for the global Cartesian directions. In the above relation, the symmetries: $\sigma^{3\alpha} = \sigma^{\alpha3}$ and $\varepsilon_{3\alpha} = \varepsilon_{\alpha3}$ were taken into account. The terms $v_l$, $v_n$, $v_s$ represent longitudinal, transverse, and shear components, respectively, of the density function. Consequently, the strain energy can also be divided into the corresponding energy components $U_l$, $U_n$, $U_s$:

$$
\begin{aligned}
U &= \int_V v \, dV \\
&= U_l + U_n + U_s
\end{aligned}
\tag{23}
$$

Our calculations of the strain components of (21) on the element level correspond to the 3D-based hierarchical shell formulation proposed and developed in [38,45]. In this formulation local strains (defined in two longitudinal and the third transverse directions) and the global displacement dofs are applied. Because of that, the relation defining local strains $\varepsilon$ (or their components) by the product of the element strain-displacement matrix $\overset{e}{B}$ (or its components) and the element displacement dofs vector $\overset{e}{q}^{hpq}$ can be written as follows

$$
\begin{aligned}
\varepsilon &= \overset{e}{B}\,\overset{e}{q}^{hpq} \\
&= [\overset{e}{B_l}, \overset{e}{B_n}, \overset{e}{B_s}]^T \overset{e}{q}^{hpq} \\
&= [\overset{e}{B_l}\,\overset{e}{q}^{hpq}, \overset{e}{B_n}\,\overset{e}{q}^{hpq}, \overset{e}{B_s}\,\overset{e}{q}^{hpq}]^T \\
&= [\varepsilon_l, \varepsilon_n, \varepsilon_s]^T
\end{aligned}
\tag{24}
$$

where the longitudinal $\overset{e}{B_l}$, transverse $\overset{e}{B_n}$, and shear $\overset{e}{B_s}$ blocks of the strain-displacement matrix include the latter matrix rows corresponding to the in-plane, normal and shear strain components: $\varepsilon_l$, $\varepsilon_n$, $\varepsilon_s$. Note that the vector of displacement dofs is related to the displacement field $\overset{e}{u}^{hpq}$ with the standard interpolation formula of the element $e$: $\overset{e}{u}^{hpq} = \overset{e}{N}\overset{e}{q}^{hpq}$, where $\overset{e}{N}$ stands for the shape function matrix of the element.

Because of the 3D-based shell models applied in our research, further decomposition of the strain energy will be performed approximately. The exact decomposition is easy for the conventional shell models with the degrees of freedom defined on the mid-surface. Then, the dofs corresponding to the even and odd powers of the thickness contribute to the membrane and bending strains, respectively. In the 3D-based formulation, the dofs are defined along (or through) the thickness, and extracting their membrane and bending contributions needs much more complex procedure explained in [32]. The approximate procedure requires replacement of the top and bottom displacements of the symmetric thin-walled structure with their sums and differences which contribute to the mid-surface displacements and rotations. This allows for distinguishing the following components of the in-plane strains

$$\varepsilon_{\alpha\beta} \approx \gamma_{\alpha\beta} + \kappa_{\alpha\beta} \tag{25}$$

where $\gamma_{\alpha\beta}$ stand for the mid-surface strains (membrane strains) and $\kappa_{\alpha\beta}$ are the three-dimensional counterparts of the change of curvature tensor (bending strains). With the above division, the in-plane density $v_l$ of strain energy can be decomposed into three parts (membrane, bending, and coupling ones):

$$v_l \approx v_m + v_b + v_c \tag{26}$$

that contain products of the membrane strains, bending strains, and their combination, respectively. Consequently, one also has

$$U_l \approx U_m + U_b + U_c \tag{27}$$

Calculation of the two components of the right-hand side of (25) on the element level needs adequate transformation of the numerical representation of element longitudinal strains $\varepsilon_l$, defined as a matrix product of the corresponding part of the strain-displacement matrix and the vector of displacement dofs. For the 3D-based hierarchical shell formulation, presented in [38,45], this transformation reads

$$
\begin{aligned}
\varepsilon_l \ &= \ \overset{e}{B}_l \overset{e}{q}^{hpq} \\[4pt]
&= \ [B_t, B_b, B_o] \times [q_t, q_b, q_o]^T \\[4pt]
&= \ [B_t + B_b, B_t - B_b, B_o] \times [\tfrac{1}{2}(q_t + q_b), \tfrac{1}{2}(q_t - q_b), q_o]^T \\[4pt]
&= \ [B_s, B_\phi, B_o] \times [q_s, q_\phi, q_o]^T = B_s\, q_s + B_\phi\, q_\phi + B_o\, q_o \\[4pt]
&= \ \gamma + \kappa + r \\[4pt]
&\approx \ \gamma + \kappa
\end{aligned}
\tag{28}
$$

where the sub-blocks $B_t$, $B_b$, and $B_o$ correspond to the top $q_t$, bottom $q_b$, and all other $q_o$ displacement dofs of the element dofs vector $\overset{e}{q}^{hpq}$. Location of the dofs of three types is on the top and bottom surfaces of the shell element, and apart from these two surfaces, respectively. In turn, the sub-blocks $B_s$, $B_\phi$ correspond to the mid-surface displacement dofs $q_s = \tfrac{1}{2}(q_t + q_b)$ and rotational dofs $q_\phi = \tfrac{1}{2}(q_t - q_b)$. The approximate character of the performed transformation results from neglecting the mentioned other dofs in the resultant decomposition into the membrane and bending strain contributions.

The Criterion

In the criterion for detection of the dominance of bending strains, the solutions of two local problems from the equilibrated residual method are applied. The first solution corresponds to the following local discretization parameters: $H = h$, $P_1 = p$, and $Q = q$. In the second problem, set to be free of the locking phenomena, one has: $H = h$, $P_2 = p_{max} = 8$, and $Q = q$. The value of $p_{max} = 8$ is taken from the research of Pitkaranta [4] who demonstrated that for such a value the membrane locking disappears in shell structures regardless of the thickness. For this value also the shear locking in plate structures disappears. It is well known that this type of locking is weaker than the membrane one in shells (compare remarks in [5,32], for example). The proposed criterion reads

$$
\begin{cases}
\sum_{e=1}^{E} \overset{e}{U}_m(\boldsymbol{u}^{HP_1Q}) + \sum_{e=1}^{E} \overset{e}{U}_s(\boldsymbol{u}^{HP_1Q}) \;\geq\; \sum_{e=1}^{E} \overset{e}{U}_b(\boldsymbol{u}^{HP_1Q}) \\[2mm]
\sum_{e=1}^{E} \overset{e}{U}_m(\boldsymbol{u}^{HP_2Q}) + \sum_{e=1}^{E} \overset{e}{U}_s(\boldsymbol{u}^{HP_2Q}) \;<\; \sum_{e=1}^{E} \overset{e}{U}_b(\boldsymbol{u}^{HP_2Q})
\end{cases}
\tag{29}
$$

The first condition detects dominance of the membrane and shear strains over the bending strains in the local problems corresponding to the global problem under consideration, while the second condition reflects the change of the dominance in the local problems. The form of the criterion corresponds to the shell structures, where coupling between shear, membrane and bending strains exists. In the case of plate structures, the coupling between the membrane strains and the shear and bending strains disappears and because of this the membrane strains have to be removed from the above criterion.

Fulfillment of the above criterion means that the true nature of the problem is bending-dominated and that the locking phenomenon appears in the assessed global problem. As a consequence, the locking has to be removed through increase of the approximation order $p$. The above criteria are necessary to confirm the results from the detection and assessment tools presented in the next sections, as those tools are not capable of distinguishing between the membrane- (and/or) shear-dominated problems and the bending-dominated ones.

### 3.1.3. Sensitivity Analysis of the Local Solutions

The main purpose of the sensitivity analysis performed in this section is detection of the situation that the assessed global problem solution differs to the solution of the problem potentially free of locking to such an extent that the difference may suggest the presence of the locking phenomena. Note that the theoretical ratio of the true bending-dominated solution and locked membrane-dominated (and/or shear-dominated) solution is $C \cdot t^2$, with $C$ being a constant. This ratio results from the proportionality of the membrane (and/or shear) part of the strain energy and the bending part of this energy to $t$ and $t^3$, respectively, as shown in (8).

To compare the above two solutions, the following two local problems for each element of the plate or shell structure, or such parts of complex structures, has to be solved:

$$
\overset{e}{B}(\boldsymbol{u}^{HP_1Q}, \overset{e}{\boldsymbol{v}}^{HP_1Q}) - \overset{e}{L}(\overset{e}{\boldsymbol{v}}^{HP_1Q}) - \int_{S_e\setminus(P\cup W)} (\boldsymbol{v}^{HP_1Q})^T \, \langle \overset{e}{\boldsymbol{r}}(\boldsymbol{u}^{hpq}) \rangle \, dS_e \;=\; 0
\tag{30}
$$

where $P_1 = p$, $H = h$ and $Q = q$ correspond to the global problem under consideration, and:

$$
\overset{e}{B}(\boldsymbol{u}^{HP_2Q}, \overset{e}{\boldsymbol{v}}^{HP_2Q}) - \overset{e}{L}(\overset{e}{\boldsymbol{v}}^{HP_2Q}) - \int_{S_e\setminus(P\cup W)} (\boldsymbol{v}^{HP_2Q})^T \, \langle \overset{e}{\boldsymbol{r}}(\boldsymbol{u}^{hpq}) \rangle \, dS_e \;=\; 0
\tag{31}
$$

with $P_2 = p_{max} = 8$, $H = h$, and $Q = q$ corresponding to the problem potentially free of locking.

The following criterion is set to assess sensitivity of the solution to the change of the longitudinal approximation order from $P_1$ to $P_2$:

$$\sum_{e=1}^{E} \frac{1}{2} \overset{e}{B}(\overset{e}{u}^{HP_1Q}, \overset{e}{u}^{HP_1Q}) < a \sum_{e=1}^{E} \frac{1}{2} \overset{e}{B}(\overset{e}{u}^{HP_2Q}, \overset{e}{u}^{HP_2Q}) \tag{32}$$

where $a$ is the coefficient which determines the ability of the adaptive method to overcome the locking phenomena with the standard $hp$-adaptive procedure. In the case of the applied procedure based on Texas three-step strategy [40] and error estimation based on the equilibrated residual method [43,44], the reasonable (verified numerically) value is suggested to be equal to $a = 0.1$. Fulfillment of the above criterion suggests the possibility of locking appearance. This possibility has to be confirmed by the criterion (29) of bending-dominance.

### 3.1.4. The Detection Algorithm

The algorithm of detection of the locking phenomena introduced into the standard three-step adaptive strategy proposed by Oden [40] is presented in Figure 1. The original part of the standard adaptation includes three steps: initial ($i = 1$), intermediate ($i = 3$) called also $h$-step, and final ($i = 4$) called also the $p$-step. Either local $h$-refinement or local $p$-enrichment is performed within each element of the latter two steps, based on the estimated approximation error values obtained from the equilibrated residual method of Ainsworth and Oden [43,44]. The modification of the initial mesh ($i = 2$) is performed as the second step when necessary. This additional step is composed of two stages: the detection of the phenomenon, and the changes in the mesh which consist in the global $p$-enrichment performed within all elements of the thin- or thick-walled structure or such a part of the complex structure.

The details of the additional step modifying the initial mesh are presented in Figure 2. At first, we search for the thick- or thin-walled parts of the generally complex structure. Such a structure can be composed of thick- and thin-walled parts, solid parts, and transition parts joining the previous two types of geometries. Of course, simple geometries composed of a single thin- or thick-walled part are also possible. Then, for each element of such a thin- or thick-walled part, the interelement stresses on the element boundary have to be calculated. Subsequently, for two local problems described with Equations (30) and (31), the terms of these equations are generated and the equations are solved. Next, one has to sum the element energies from (32) and compare two sums for two types of the local problems. Finally, if the criterion for locking detection is met, one has to confirm this result with the criterion for bending dominance detection (29). If it is fulfilled, then the modification of the initial mesh is performed, which is based on $p$-enrichment of all elements of the thin- or thick-walled part of the structure. Such elements of the modified structure are all equipped with the longitudinal order of approximation equal to $p = p_{max} = 8$. Next, the standard steps, $i = 3$ and $i = 4$, of $hp$-adaptation can be performed.

It should be underlined that calculation of the interelement stresses, performed in the above algorithm, is skipped in this paper as it is performed in the standard way presented in [43,44,46,54,55].

We also apply the standard $p$-enrichment algorithm performed at the modification stage of the additional adaptation step ($i = 2$). Such modification does not require any additional actions in comparison to the standard $p$-adaptation [41,42].

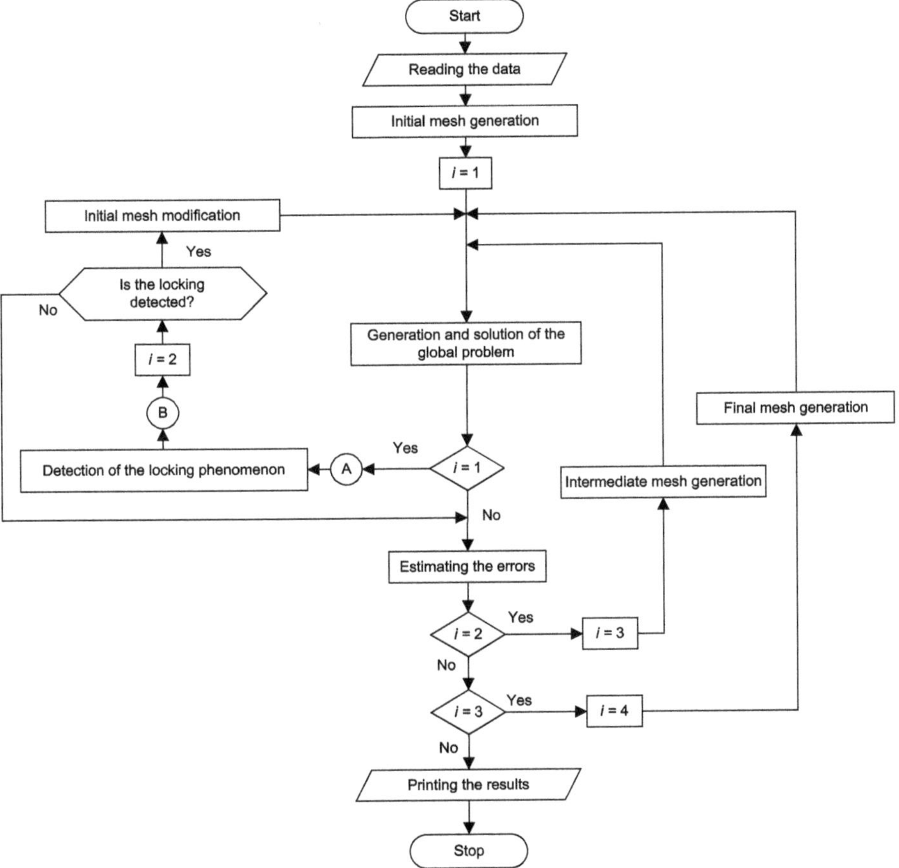


**Figure 1.** Three-step adaptive procedure extended with the fourth modification step.

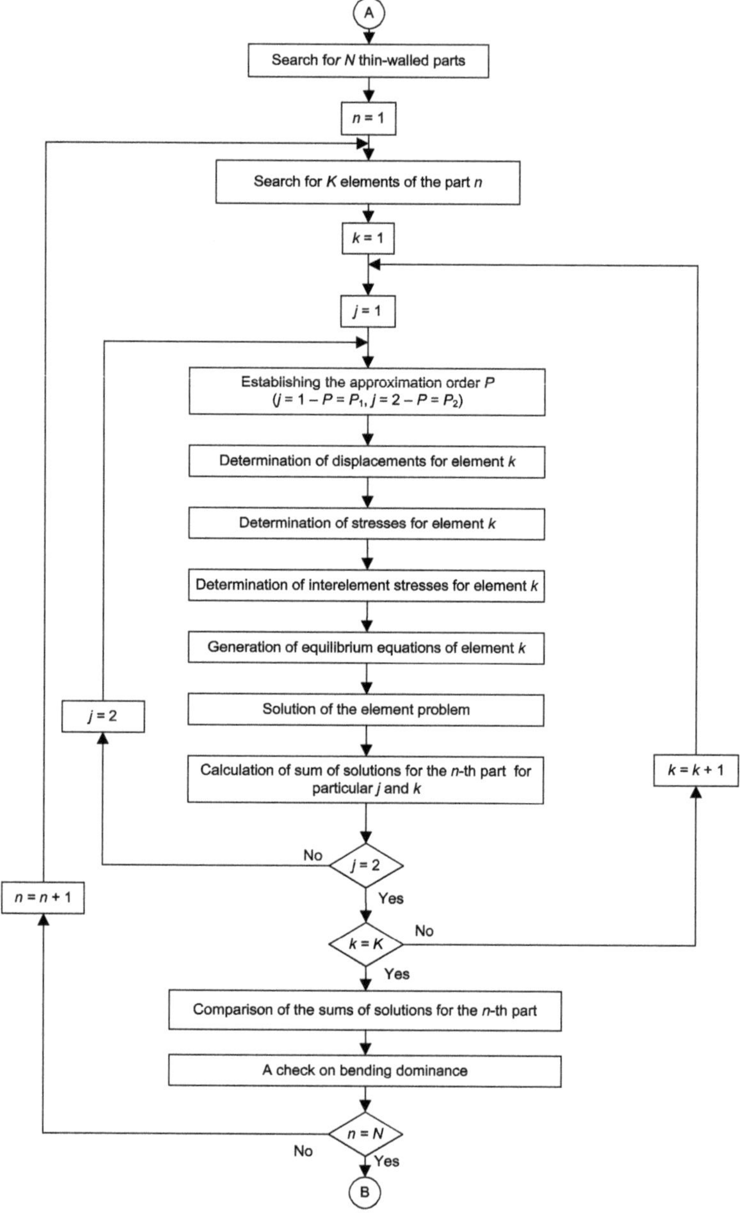



**Figure 2.** Flow diagram of the overall algorithm for the locking detection.

### 3.2. Calculation of the Optimized Value of p

The idea presented in this section lies in making an attempt to assess the locking phenomena strength. This strength is measured with the $p$ value necessary to remove the locking phenomenon. The criterion (32) assumes that the value of $p = 8$ is necessary for the removal.

This value corresponds to the coarsest mesh possible, composed of one rectangle (hexahedra) or two triangles (prisms). However, depending on the structure thickness $t$ and the initial global discretization characterized by the value of $h$, the removal may require much lower (more optimal) value of the longitudinal order $p$. Moreover, in such circumstances, $p = p_{max} = 8$ leads to too rich discretizations. Note also that setting $p = 8$ in the modified mesh reduces further $hp$-adaptation to $h$-adaptation only as the maximum value of the longitudinal order of approximation $p = p_{max} = 8$ is applied in the modified mesh. We address both situations in numerical tests presented in the next sections of the paper.

#### 3.2.1. The Idea

The idea lies in solution of the sequence of the local problems for each element $e$ of a thin- or thick-walled structure or such a part of a complex structure. The sequence can be characterized with

$$\overset{e}{B}(\overset{e}{u}^{HP_jQ}, \overset{e}{v}^{HP_jQ}) - \overset{e}{L}(\overset{e}{v}^{HP_jQ}) - \int_{S_e\backslash(P\cup W)} (\overset{e}{v}^{HP_jQ})^T \langle \overset{e}{r}(u^{hpq}) \rangle \, dS_e = 0 \tag{33}$$

where $P_j = P_1, \ldots, P_{max} - 1$, with $P_1 = p$ and $P_{max} = 8$. The solutions of the above sequence are compared with the solution to the problem potentially free of locking phenomena, i.e.,

$$\overset{e}{B}(\overset{e}{u}^{HP_2Q}, \overset{e}{v}^{HP_2Q}) - \overset{e}{L}(\overset{e}{v}^{HP_2Q}) - \int_{S_e\backslash(P\cup W)} (\overset{e}{v}^{HP_2Q})^T \langle \overset{e}{r}(u^{hpq}) \rangle \, dS_e = 0 \tag{34}$$

where $P_2 = P_{max} = 8$. The comparisons are based on the summary criterion

$$\sum_{e=1}^{E} \frac{1}{2} \overset{e}{B}(\overset{e}{u}^{HP_kQ}, \overset{e}{u}^{HP_kQ}) \geq a \sum_{e=1}^{E} \frac{1}{2} \overset{e}{B}(\overset{e}{u}^{HP_2Q}, \overset{e}{u}^{HP_2Q}) \tag{35}$$

where, as above, one may set $a = 0.1$. For all particular orders of approximation $P_k$, fulfilling the above criterion, the adaptive algorithm can obtain the global adapted solution with the standard $hp$-adaptive algorithm. We choose the solution $P_l$ fulfilling (35), such that:

$$\sum_{e=1}^{E} \frac{1}{2} \overset{e}{B}(\overset{e}{u}^{HP_lQ}, \overset{e}{u}^{HP_lQ}) = \min_{P_k} \left\{ \ldots, \sum_{e=1}^{E} \frac{1}{2} \overset{e}{B}(\overset{e}{u}^{HP_kQ}, \overset{e}{u}^{HP_kQ}), \ldots \right\} \tag{36}$$

The solution corresponding to $P_l$ is enough and the most optimal for removal of the phenomenon. Note that for $P_l = P_1$ removal of the locking is not necessary, as the standard $hp$-adaptation can be performed without changing the longitudinal order of approximation. This order is equal to its value from the assessed global problem, i.e., $P_1 = p$. For $P_l = P_{max}$, one obtains the same result as for the detection criterion (32). Note that the search for the optimized value of $p$ should be performed if and only if the bending dominance criterion (29) is fulfilled.

#### 3.2.2. The Algorithm

The structure of the algorithm remains the same as for the locking detection algorithm described in Section 3.1.4. In the adaptive algorithm presented in Figure 1, the only difference is that the modified mesh is generated with the optimized value of the longitudinal order of approximation within each element of the thin- or thick-walled part of the structure. This optimized value replaces the maximum value $p_{max} = 8$ applied in the case of the detection without optimization. In the algorithm presented in Figure 2, apart from the problem (31) or (34), the sequence of the local problems (33) is solved instead of the problem (30). After obtaining the solutions, the summary criteria (35) and (36) are checked, and the optimized value of the longitudinal order of approximation is established. If the optimized value is greater than the current value from the initial mesh, the confirmation from the criterion for the detection of bending dominance (29) has to be checked. If it is met, then the mesh is modified by increasing the longitudinal order in any element of the thick- or thin-walled part of the structure. Finally, the standard $hp$-adaptation ($i = 3$ and $i = 4$) can be performed.

## 4. Verification and Utilization of the Proposed Tools

Our numerical verification of the proposed numerical tools for detection and optimization of the longitudinal order of approximation is based on comparison of the results obtained from the detection and optimization tools and the results from the corresponding global solution of the model problems under consideration. The model problems concern all possible situations of locking existence or not, i.e., shear locking, membrane-shear locking, and the lack of locking. The performed comparisons should confirm that the detection and optimization based on the local, element solutions can replace the corresponding global analysis.

### 4.1. Model Problems

It is well known from the literature presented in the state-of-the-art section that in the case of the bending-dominated plates, the shear locking may appear, depending on the plate thickness and the applied discretization. In the case of the bending-dominated shells, the shear–membrane locking is possible. On the other hand, in the case of the membrane-dominated shells, the locking phenomena do not appear. This is the reason for our choice of three model problems introduced below.

The first problem concerns a bending-dominated square plate. The plate longitudinal dimensions are equal to $l = 3.1415 \times 10^{-2}$ m, while its basic thickness is $t = 0.03 \times 10^{-2}$ m. A symmetric quarter of the plate can be seen in the Section 5.2.1. The plate is clamped along its edges. The vertical surface traction of the value equal to $p = 4.0 \times 10^6$ N/m$^2$ is applied to the upper surface of the plate. The traction acts downwards. The plate, as well as the next two shell structures, is made of steel. For this material the Young's modulus is $E = 2.1 \times 10^{11}$ N/m$^2$, while the Poisson's ratio equals $\nu = 0.3$.

The second model problem is a bending-dominated half-cylindrical shell. Its length is equal to $l = 3.1415 \times 10^{-2}$ m, while its semi-circle circumference is $\pi R \approx 3.1415 \times 10^{-2}$ m, where $R = 1.0 \times 10^{-2}$ m is the radius of the shell middle surface. The shell thickness is $t = 0.03 \times 10^{-2}$ m. A symmetric quarter of the shell is displayed in Section 5.2.3. The shell is clamped along its straight edges, and the curved edges are free. The shell is loaded with the vertical traction of the value $p = 4.0 \times 10^6$ N/m$^2$, directed downwards.

The third example is a symmetric half of a membrane-dominated cylindrical shell. The shell dimensions are analogous as for the bending-dominated shell, i.e., $l = 3.1415 \times 10^{-2}$ m, $\pi R \approx 3.1415 \times 10^{-2}$ m, with $R = 1.0 \times 10^{-2}$ m, and $t = 0.03 \times 10^{-2}$ m. A symmetric octant of the shell is presented in Section 5.2.4. The symmetry boundary conditions are applied along the straight edges and one (left) curved edge of the octant, while there are no rotations along the other (right) curved edge. The shell is loaded with the internal pressure $p = 4.0 \times 10^6$ N/m$^2$ acting outwards.

### 4.2. Local Problems Solutions Versus Global Solutions

In our tests, we applied the results from the algorithms for the detection of the bending dominance. We calculated sums of elemental strain energy components present in the criterion (29). In the case of the plate problem, the shear and bending energy components were computed, while, in the shell examples, the sum of shear and membrane components and the bending component of the energy were determined. The assessed global plate problem was characterized with $p = 1$, while in the shell examples the assessed global problems were characterized with $p = 2$. Then, the energy components were calculated for the sequence of problems (30) characterized with $P_1 = p, p+1, \ldots, 8$. The averaged values of the interelement stresses were used in the local problems definition. The transverse order of approximation corresponded to the second-order hierarchical shell model ($Q = q = 2$) for all examples. The used mesh was characterized with $H = h = l/2$.

The above results from the sequence of the local problems were compared with the results from the global problems, where $q = 2$ and $h = l/2$. In the case of the plate, the following longitudinal orders of approximation were applied $p = 1, 2, \ldots, 8$ in the global problems, while in the case of two shell examples, the values of $p = 2, 3, \ldots, 8$ were taken into account. The results of the comparisons are presented in Figures 3–8 for the plate and bending- and membrane-dominated shells, respectively. The figures present the ratios of the shear (the plate) or the sum of shear and membrane (the shells) energies to the entire strain energy and the ratios of the bending energy to the entire strain energy obtained from the local problems (top) and the global problems (bottom). Values of these ratios are presented versus the longitudinal order of approximation.

Comparison of the top and bottom figures for the model problems leads to the conclusion that the detection tools based on solution of the element local problems give practically the same results as the solutions to the global problems, both qualitatively and quantitatively. In both bending-dominated problems, the change of the membrane dominance resulting from the locking phenomena to bending dominance due to removal of the locking is perfectly indicated. In addition, the membrane-dominated problem has been perfectly identified, as shown by the third couple of figures—the membrane energy component dominates over the bending one for all values of the longitudinal order of approximation.

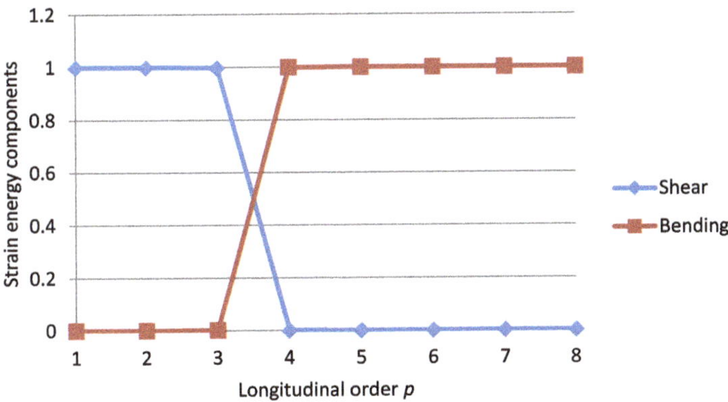


**Figure 3.** The sum of local energies—the bending-dominated plate.

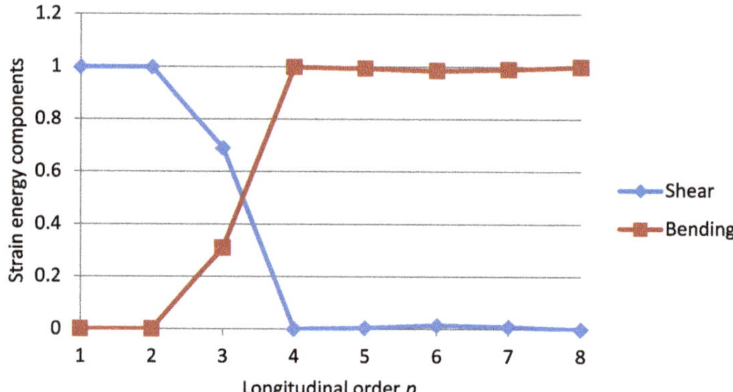

**Figure 4.** The global energy—the bending-dominated plate.

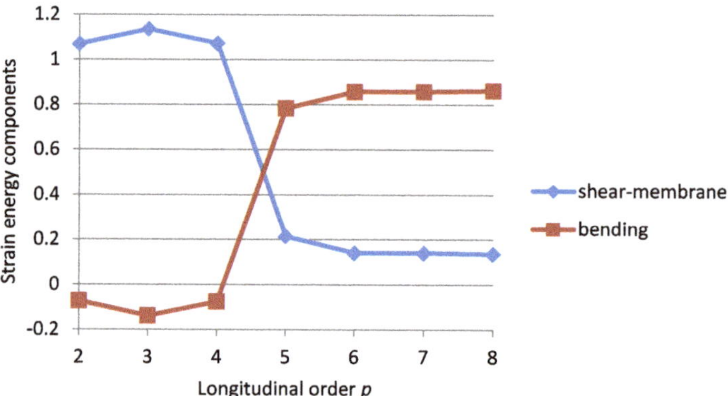

**Figure 5.** The sum of local energies—the bending-dominated shell.

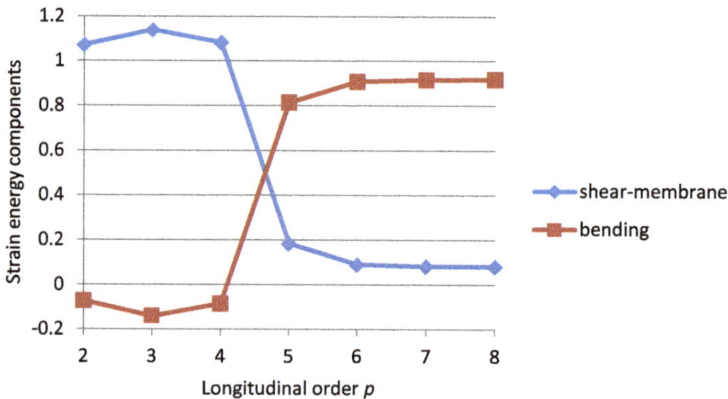

**Figure 6.** The global energy—the bending-dominated shell.

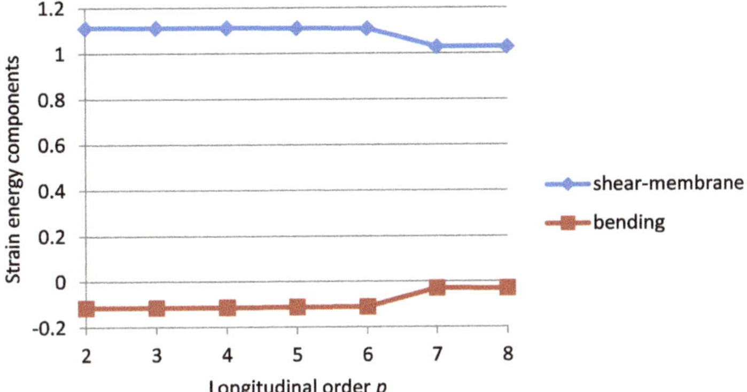

**Figure 7.** The sum of local energies—the membrane-dominated shell.

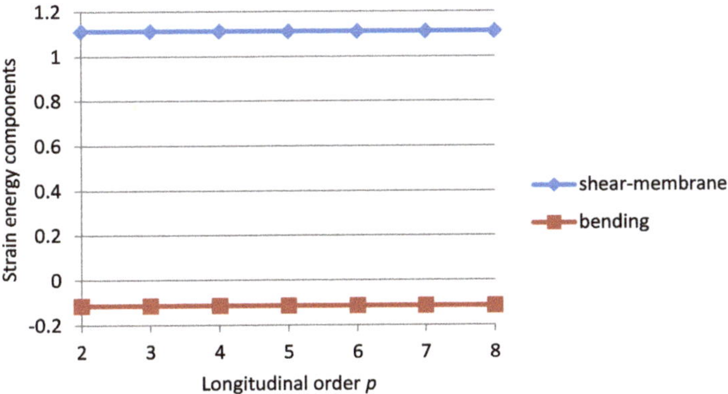

**Figure 8.** The global energy—the membrane-dominated shell.

## 5. Effectivity of the Method in Model Problems

In this section, we present examples of application of the algorithms for detection and/or assessment of the locking phenomena, and the standard mesh modification algorithms as well, in the adaptive analysis of model problems. The adaptive procedure is derived from the three-step strategy [40] composed of the global solution and error estimation for the initial, intermediate (or *h*-) and final (or *p*-) steps. The global problems are based on 3D-based hierarchical modelling and approximations for complex structures [45]. Our models and approximations are related, based or derived from [30,31,41,42], respectively. The error estimation is performed using the equilibrated residual method [43,54], adjusted to the 3D-based formulation of the hierarchical models and approximations [46]. The three-step strategy is completed with the modification step performed after the initial one. The algorithms from the previous sections are used within this additional step. The averaged stresses are used instead of the equilibrated ones in the local problems.

### 5.1. Problems and Methodology

The three model problems in Section 4.1 are applied again. Two model bending-dominated plate examples are considered. In the first one, the data are exactly the same as in the mentioned section. In the second plate problem, we solve the plate of the same dimensions $l/2$ and $t$ as for a quarter of the plate from the first plate. In the second problem, all four lateral sides of the plate are clamped. In the first problem, two of them correspond to symmetry boundary conditions, and only two sides are clamped. In the second problem, one deals with the most coarse mesh possible and the lowest longitudinal order of approximation possible. As a result, the strongest possible locking appears for the assumed plate thickness. The third example is the bending-dominated shell from Section 4.1. The only difference is the shell thickness which is now equal to $t = 0.003 \times 10^{-2}$ m. The last example concerns the membrane-dominated shell. In this example, all data are exactly the same as in Section 4.1.

### 5.2. Numerical Examples

In the analysis of the locking phenomena, the following data were treated as independent: the problem type (the type of the strain dominance), and the structure length $l$ and thickness $t$ resulting in the thinness ratio $l/t$. In the numerical analysis, also the initial mesh data, the relative value of the target admissible error $\gamma_T$, and the ratio $\gamma_I/\gamma_T$ of the intermediate (after $h$-step) error to the target (after $p$-step) error were treated as independent quantities.

As results of the analysis, we present the meshes corresponding to three performed courses of adaptation: standard $hp$-adaptation and such adaptations preceded by the modification of the initial mesh with the increased longitudinal order of approximation $p$, equal either to the maximum or optimized value. The results are completed with the adaptive convergence curves corresponding to these three courses of adaptation. The convergence curves present the approximation error as a function of the number $N$ of degrees of freedom (dofs). The absolute error is defined as a negative difference of the total strain energy $U$ corresponding to the global solution under consideration and the reference energy $U_r$ replacing the unknown exact value. The energies are calculated in accordance with the strain energy definition from Section 2. Due to the exponential character of $hp$-convergence, the curves are plotted as $\log(U_r - U)$ versus $\log N$. The reference energy $U_r$ is obtained numerically from calculations performed on over-killed meshes with the global discretization parameters equal to: $p = 9$, $q = 2$, $m = 9$, where $m = l/2h$. Apart from the absolute error values, the relative error values $(U_r - U)/U_r$ are also presented and discussed.

#### 5.2.1. A Quarter of a Bending Dominated Plate

**Data**

The dimensions of a quarter of the plate are such that the thinness ratio of the plate is $l/t = 3.1415 \times 10^{-2}/0.03 \times 10^{-2}$. The initial mesh is coarse. Its data are as follows: the longitudinal approximation order $p = 1$, the transverse approximation order $q = 2$, and the element size $h = l/2$. This mesh is shown in Figure 9. In the error analysis, the target error $\gamma_T = 0.01$ and the ratio $\gamma_I/\gamma_T = 3$.

**Results**

Three different courses of the adaptation are presented for this example. The first one corresponds to the standard $hp$-adaptivity composed of the $h$- and $p$-adaptation steps only. The final mesh for this first course is presented in Figure 10. The second course is the $hp$-adaptation preceded by the modification step based on the detection algorithm from Section 3.1.3, where, after the detection, the modification of the initial mesh is performed with the maximum setting $p = 8$ (Figure 11). No further automatic $h$-adaptation is performed by the adaptivity control algorithm. In addition, no further $p$-adaptation can be performed due to the maximum value of $p$ applied in the modification step after the detection. The last course is the $hp$-adaptation performed after the initial mesh modification based on setting the longitudinal order of approximation to its optimized value $p = 4$ (Figure 12) obtained from the algorithm of Section 3.2.1. The final mesh completing this course of adaptation is presented in Figure 13. The not presented intermediate mesh possesses the same division pattern as the final mesh but the order of approximation is uniform and taken from the modified mesh ($p = 4$).

The adaptive convergence curves for three described cases are presented in Figure 14. In the case of the standard $hp$-adaptivity, the curve consists of two sections and three points corresponding to the initial, intermediate ($h$-adapted) and final ($hp$-adapted) meshes. The influence of the shear locking is visible in the first section of the curve—this section is almost horizontal. In the case of the $hp$-adaptivity performed after detection of the locking and based on the modified value of $p = 8$, two points of the convergence curve correspond to the initial and modified meshes. The locking has been removed—the only section of the curve is not horizontal. In the case of the $hp$-adaptation performed after the detection and optimization of the value of $p$, the convergence curve consists of four points (the initial, modified, intermediate, and final meshes) and three sections. The first not horizontal section of the curve reflects locking removal, while the next two sections correspond to $h$-refinement and $p$-enrichment.

Finally, it is worth mentioning that the automatic choice of the program may be the two courses with the initial mesh modification. These two automatic courses correspond to either the assumed maximum or determined optimized value of $p$ applied in the modification step. The enforced course corresponds to the standard $hp$-adaptation.

The relations between the number $N$ of degrees of freedom (dofs) and the absolute $U_r - U$ and relative $(U_r - U)/U_r$ errors are summarized in Table 1 for the consecutive points of the three mentioned convergence curves. The mesh figure numbers corresponding to these points are also indicated.

**Table 1.** Result summary—a quarter of the bending-dominated plate.

| Adaptive Method | Result Quantity | Initial | Modified | Intermediate | Final |
|---|---|---|---|---|---|
| | | | Mesh Type | | |
| standard $hp$-adaptivity | mesh figure no. | Figure 9 | – | – | Figure 10 |
| | dofs number $N$ [1] | 36 | – | 225 | 9648 |
| | $\log(U_r - U)$ [N/m] | 0.459481 | – | 0.452821 | $-2.352724$ |
| | $(U_r - U)/U_r$ [%] * | 99.9 | – | 98.4 | 0.15 |
| standard $hp$ after detection | mesh figure no. | Figure 9 | Figure 11 | – | – |
| | dofs number $N$ [1] | 36 | 729 | – | – |
| | $\log(U_r - U)$ [N/m] | 0.459481 | $-1.43770$ | – | – |
| | $(U_r - U)/U_r$ [%] * | 99.9 | 1.27 | – | – |
| standard $hp$ after optimization | mesh figure no. | Figure 9 | Figure 12 | – | Figure 13 |
| | dofs number $N$ [1] | 36 | 531 | 3438 | 7704 |
| | $\log(U_r - U)$ [N/m] | 0.459481 | $-0.297250$ | $-1.531051$ | $-2.354113$ |
| | $(U_r - U)/U_r$ [%] * | 99.9 | 17.5 | 1.02 | 0.15 |

* admissible relative error value $\gamma_T = 1.0$ %.

Discussion

The shear locking has been detected in both cases which include modification of the initial mesh based on the maximum and optimized values of the longitudinal order $p$. One can see that the standard $hp$-adaptation leads to the final error value below the desired admissible value of $\gamma_T$. The same refers to the $hp$-adaptation preceded by the modification based on the optimized value of the longitudinal order of approximation $p = 4$. In the case of the $hp$-adaptation performed after the modification of the mesh based on the fixed maximum value of $p = 8$, the admissible error value has not been reached, even though the error has been diminished. The reason is the discrete character of the possible $h$-adaptation. For the estimated error level, this adaption has not been performed.

**Figure 9.** A quarter of a bending-dominated plate—initial mesh.

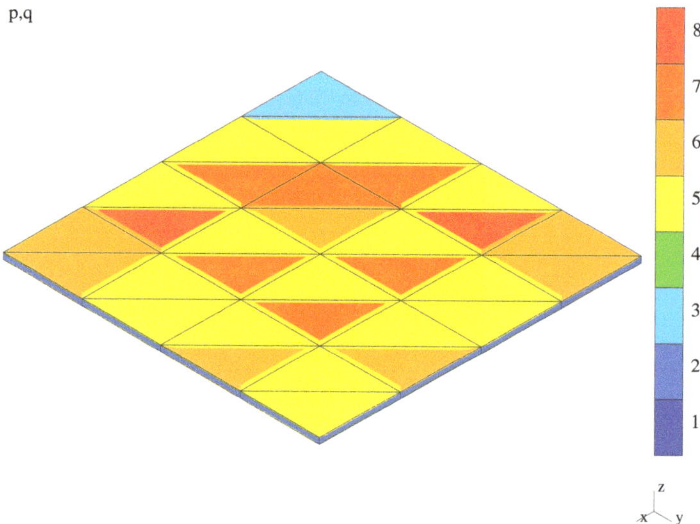

**Figure 10.** A quarter of a bending-dominated plate—$hp$-adapted (final) mesh.

**Figure 11.** A quarter of a bending-dominated plate—after simple detection, no *hp*-adaptation.

**Figure 12.** A quarter of a bending-dominated plate—mesh after optimized modification.

**Figure 13.** A quarter of a bending-dominated plate—after optimization and *hp*-adaptation.

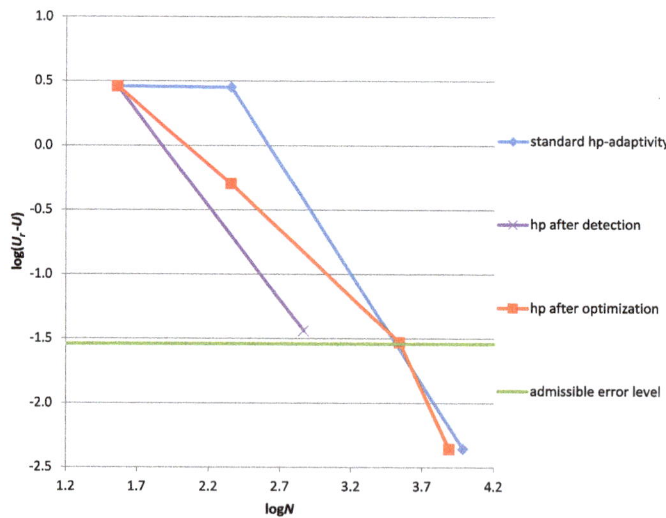


**Figure 14.** A quarter of a bending-dominated plate—convergence for three adaptation cases.

### 5.2.2. A Bending-Dominated Plate

#### Data

In this example, the entire plate is considered. Thus, the thinness ratio is equal to $l/2t = 1.57075 \times 10^{-2}/0.03 \times 10^{-2}$. The discretization parameters of the coarse initial mesh, i.e., the element longitudinal and transverse approximation orders and the element size, are $p = 1$, $q = 2$, and $h = l/2$, respectively. These parameters can be seen in Figure 15. The data for the error analysis are assumed as: the admissible target error $\gamma_T = 0.02$ and the ratio of the admissible intermediate to admissible target errors $\gamma_I/\gamma_T = 3$.

**Figure 15.** A bending-dominated plate—initial mesh.

#### Results

As in the previous test, three courses of adaptation are of our interest. The first course corresponds to the standard $hp$-adaptivity where the initial mesh is $h$- and then $p$-adapted. The $hp$-adapted final mesh is shown in Figure 16. The next course of the adaptation is based on the $hp$-adaptation which is preceded by the modification of the initial mesh. This modification (Figure 17) lies in setting the longitudinal order of approximation as equal to the maximum possible value removing the locking, i.e., $p = 8$, after the phenomenon has been detected by means of the algorithm of Section 3.1.3. As in the previous example, no further adaptive actions have been performed. The $h$-division has not been made due to the estimated approximation error level and discretized character of performance of the adaptivity control algorithm. The $p$-enrichment has not been possible as the maximum value of $p = 8$ has already been applied in the modified mesh. The third course of adaptation consists of the $hp$-adaptation following the modification of the initial mesh by adopting the optimized value ($p = 5$) of the longitudinal order of approximation (Figure 18). This value was established by the algorithm from Section 3.2.1. The final mesh for this adaptation is presented in Figure 19. The not displayed intermediate mesh has the same division pattern as the final mesh and the longitudinal approximation order as in the modified mesh, i.e., $p = 5$.

**Figure 16.** A bending-dominated plate—*hp*-adapted (final) mesh.

**Figure 17.** A bending-dominated plate—mesh after simple detection, no *hp*-adaptation.

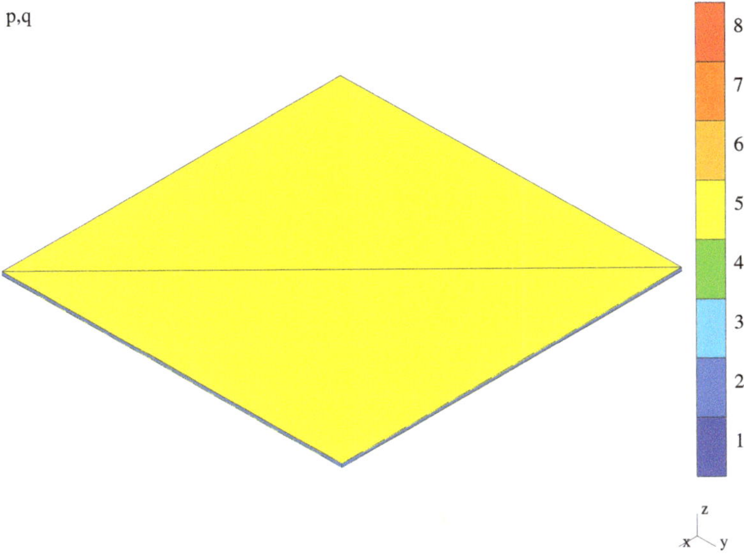



**Figure 18.** A bending-dominated plate—mesh after optimized modification.

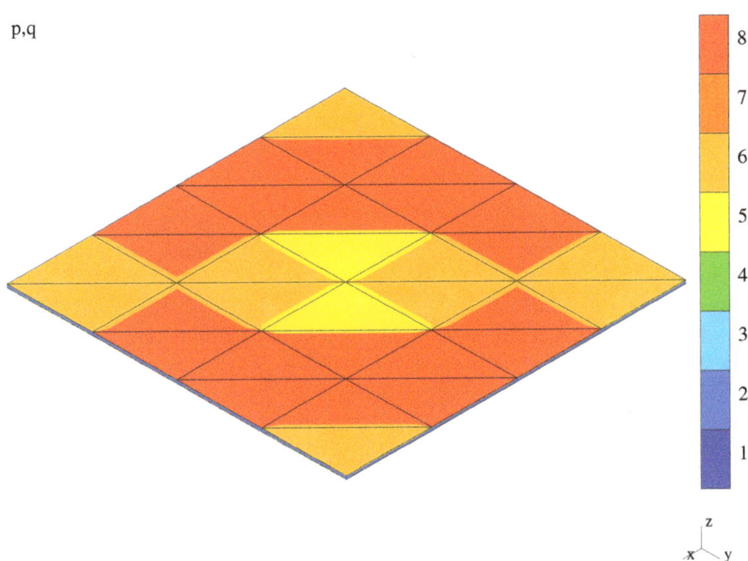



**Figure 19.** A bending-dominated plate—mesh after optimization and *hp*-adaptation.

The convergence curves corresponding to the performed adaptations are displayed in Figure 20. In the case of the standard *hp*-adaptivity, the curve consists of three points and two section. These three points correspond to three meshes generated within this course: the initial, intermediate (or $h$-adapted), and final (or $hp$-adapted) ones. The presence of the shear locking is reflected by the first horizontal section of the curve. In the case of the $hp$-adaptivity performed after detection of the locking and applying the modified maximum value of the longitudinal order of approximation, i.e., $p = 8$, the convergence curve consists of two points and one section. These two points correspond to the initial and modified meshes. In the only section of the curve that is not horizontal, he locking has been removed. In the third course of adaptation, based on the modification of the initial mesh by the optimized value of the approximation order $p = 5$, the convergence curve consists of three sections and four points—the initial, modified, intermediate, and final meshes have been generated. It can be seen that the locking has been removed as the first section is not horizontal.

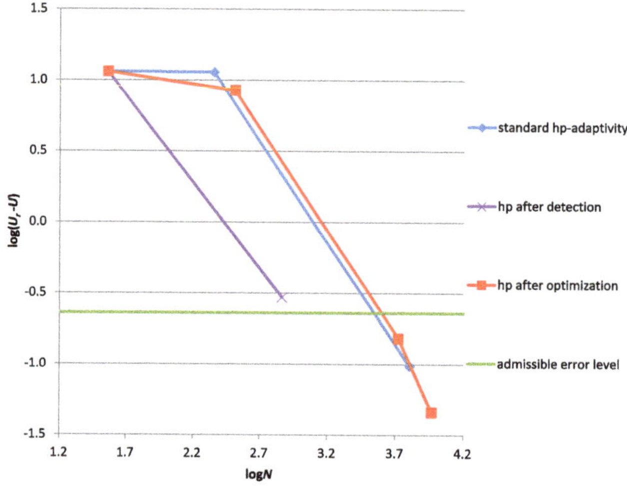


**Figure 20.** A bending-dominated plate—convergence for three adaptation cases.

It should be stressed that the modes including modification step are the automatic choice of the adaptive algorithm, while the standard $hp$-adaptivity is the enforced mode.

In Table 2, the relations between the number $N$ of degrees of freedom and the absolute and relative errors, respectively, $U_r - U$ and $(U_r - U)/U_r$, are presented for the consecutive points of the convergence curves of three adaptation modes. The mesh figure numbers corresponding to these points are also indicated in the table.

Discussion

Firstly, as expected, the shear locking has been detected. Secondly, the adaptation based on modification of the initial mesh by the optimized value of the longitudinal order of approximation leads to final error value below the admissible error value, in contrast to the modification based on the maximum value of the longitudinal order of approximation $p = 8$. Again, the discrete values of parameters controlling the adaptation process have resulted in no further adaptation after the initial mesh modification. One can see, however, that the standard $hp$-adaptation leads to the final error level below the admissible value again. We show in the next example that this is not the rule.

**Table 2.** Result summary—the bending-dominated plate.

| Adaptive Method | Result Quantity | Mesh Type | | | |
|---|---|---|---|---|---|
| | | Initial | Modified | Intermediate | Final |
| standard $hp$-adaptivity | mesh figure no. | Figure 15 | – | – | Figure 16 |
| | dofs number $N$ [1] | 36 | – | 225 | 6381 |
| | $\log(U_r - U)$ [N/m] | 1.060969 | – | 1.059175 | $-1.010223$ |
| | $(U_r - U)/U_r$ [%] * | 100. | – | 99.6 | 0.85 |
| standard $hp$ after detection | mesh figure no. | Figure 15 | Figure 17 | – | – |
| | dofs number $N$ [1] | 36 | 729 | – | – |
| | $\log(U_r - U)$ [N/m] | 1.060969 | $-0.527871$ | – | – |
| | $(U_r - U)/U_r$ [%] * | 100. | 2.58 | – | – |
| standard $hp$ after optimization | mesh figure no. | Figure 15 | Figure 18 | – | Figure 19 |
| | dofs number $N$ [1] | 36 | 324 | 5265 | 9279 |
| | $\log(U_r - U)$ [N/m] | 1.060969 | 0.927962 | $-0.817269$ | $-1.338752$ |
| | $(U_r - U)/U_r$ [%] * | 100. | 73.6 | 1.32 | 0.40 |

* admissible relative error value $\gamma_T = 2.0$ %.

### 5.2.3. A Quarter of a Bending-Dominated Shell

Data

This example concerns a symmetric quarter of the half-cylindrical shell. The thinness ratio for this structure is equal to $l/t = 3.1415 \times 10^{-2}/0.003 \times 10^{-2}$. We assume the following discretization parameters: the longitudinal order of approximation $p = 2$, the transverse one $q = 2$, and the element size $h = l/2$. The shell quarter and its initial discretization is illustrated in Figure 21. The error analysis is based on the following assumptions: $\gamma_T = 0.007$ and $\gamma_I/\gamma_T = 3$.

**Figure 21.** A quarter of a bending-dominated shell—initial mesh.

Results

As in the previous two bending-dominated examples, also here three courses of the adaptation are performed. The first adaptation is the standard procedure composed of $h$- and $p$-steps. The mesh after $hp$-adaptation can be seen in Figure 22. The second and third types of the adaptation are composed of the $hp$-adaptation which follows the modification of the initial mesh. In the second type, the modification is based on the maximum possible value of the longitudinal order of approximation $p = 8$ (the corresponding figure is not displayed). In the case of the second type of adaptation, the modified mesh has been $h$-adapted further (see Figure 23). In the third type, the modification takes advantage of the optimized value of $p = 5$ (Figure 24). In the second type, the $p$-adaptation has not been possible as the maximum value of $p = 8$ has already been applied. In the third type of adaptation, the modified mesh has been $hp$-adapted. The final mesh is presented in Figure 25. The not revealed intermediate mesh possesses the same division pattern as the final mesh and the uniform order of approximation as in the modified mesh, namely $p = 5$.

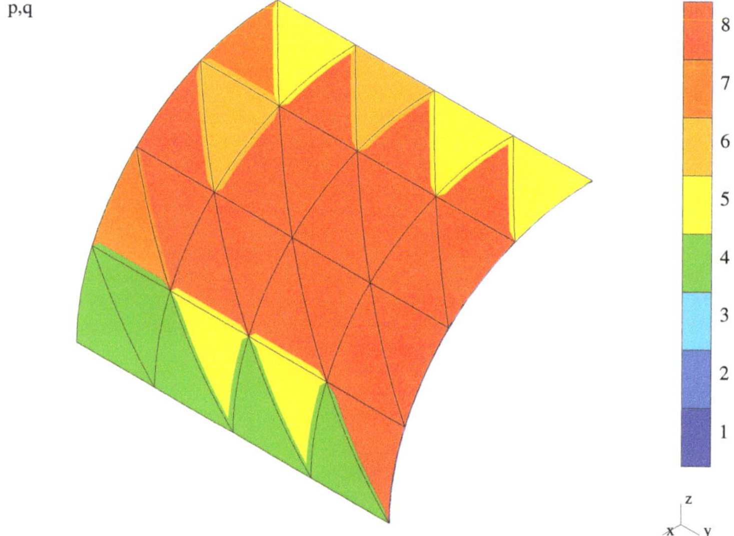

**Figure 22.** A quarter of a bending-dominated shell—$hp$-adapted (final) mesh.

Three convergence curves resulting from the described adaptations are presented in Figure 26. The standard $hp$-adaptivity convergence curve consists of two sections and three points indicating the estimated error level for the initial, intermediate, and final meshes. In the case of the $hp$-adaptation following the modification based on the maximum value of $p = 8$, the curve consists of two sections and three points corresponding to the initial, modified, and $h$-adapted (intermediate) meshes. The shear–membrane locking has been removed—the first section of the curve is not horizontal. In the last case of the $hp$-adaptivity following the modification based on the optimized value of $p = 5$, the curve is composed of three sections and four points—the initial, modified, intermediate, and final meshes have been generated. The locking has not been removed from the modified mesh, however the value of $p = 5$ has appeared sufficient for removal of the locking from the $h$-adapted (intermediate) mesh—the second section of the convergence curve is not horizontal.

**Figure 23.** A quarter of a bending-dominated shell—after simple detection and *h*-adaptation

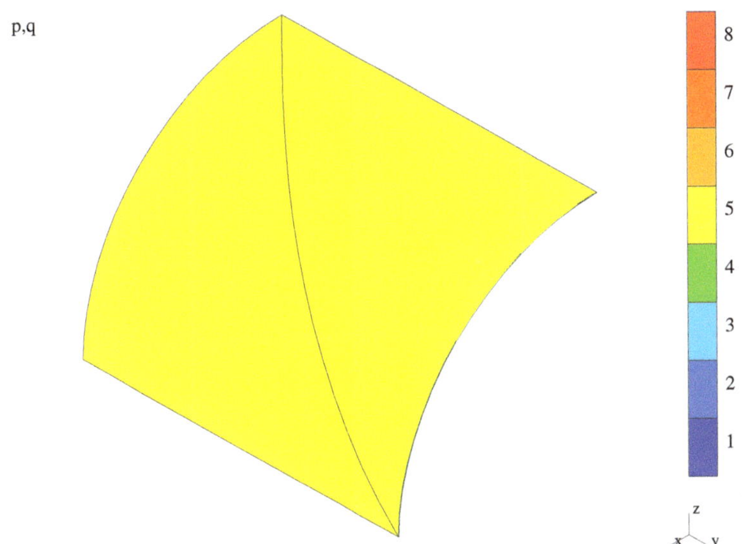


**Figure 24.** A quarter of a bending-dominated shell—mesh after optimized modification.

Note that the adaptations including the described mesh modifications, based on either the maximum ($p = 8$) or optimized ($p = 5$) values of the longitudinal order of approximation, may be performed automatically by the program, while the standard $hp$ course has to be enforced by a user.

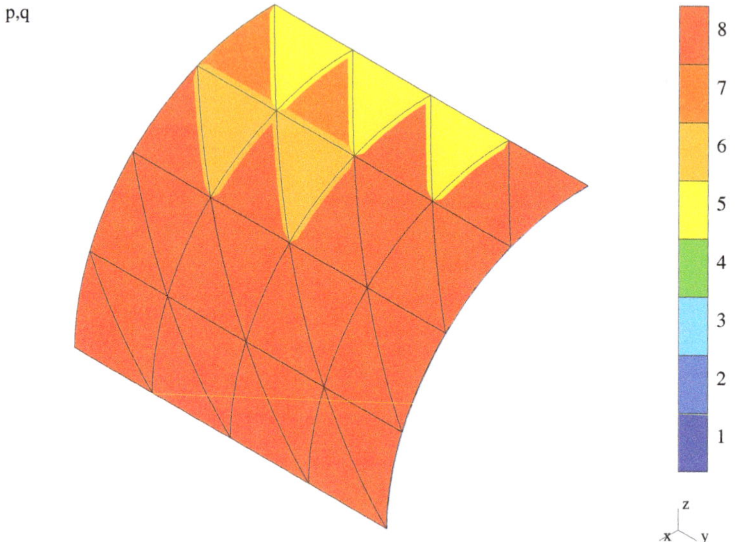



**Figure 25.** A quarter of a bending-dominated shell—after optimization and *hp*-adaptation.

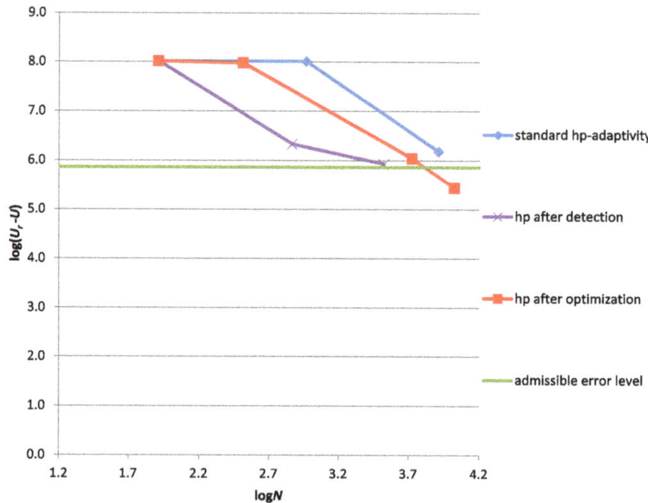



**Figure 26.** A quarter of a bending-dominated shell—convergence for three adaptation cases.

As in the previous numerical examples, the absolute and relative error values, respectively, $\log(U_r - U)$ and $(U_r - U)/U_r$, are presented in Table 3 versus the number $N$ of degrees of freedom (dofs). For the points of three adaptive convergence curves, the corresponding mesh figure numbers are included in the table.

**Table 3.** Result summary—a quarter of the bending-dominated shell.

| Adaptive Method | Result Quantity | Mesh Type | | | |
| --- | --- | --- | --- | --- | --- |
| | | Initial | Modified | Intermediate | Final |
| standard $hp$-adaptivity | mesh figure no. | Figure 21 | – | – | Figure 22 |
| | dofs number $N$ [1] | 81 | – | 918 | 8145 |
| | $\log(U_r - U)$ [N/m] | 8.015996 | – | 8.014184 | 6.184208 |
| | $(U_r - U)/U_r$ [%] * | 99.9 | – | 99.5 | 1.47 |
| standard $hp$ after detection | mesh figure no. | Figure 21 | – | Figure 23 | – |
| | dofs number $N$ [1] | 81 | 729 | 3294 | – |
| | $\log(U_r - U)$ [N/m] | 8.015996 | 6.330842 | 5.930839 | – |
| | $(U_r - U)/U_r$ [%] * | 99.9 | 2.06 | 0.82 | – |
| standard $hp$ after optimization | mesh figure no. | Figure 21 | Figure 24 | – | Figure 25 |
| | dofs number $N$ [1] | 81 | 324 | 5265 | 10,512 |
| | $\log(U_r - U)$ [N/m] | 8.015996 | 7.978775 | 6.039104 | 5.449802 |
| | $(U_r - U)/U_r$ [%] * | 99.9 | 91.73 | 1.05 | 0.27 |

* admissible relative error value $\gamma_T = 0.7$ %.

Discussion

The anticipated shear–membrane locking has been detected in two adaptation modes including modification of the initial mesh. In the case of the standard $hp$-adaptation, the admissible error value has not been achieved. This is because the estimated error level and the discretization parameters in the adapted meshes have not been determined accurately due to the locking. In the adaptation with modification based on the maximum $p$ possible, the admissible error value has not been achieved. This is because only further $h$-adaptation has been performed and further $p$-adaptation has not been possible as the maximum value of $p = 8$ has already been adopted. However, the admissible error value has been confidently achieved in the adaptation with the modification of the initial mesh based on the optimized value of $p$.

5.2.4. An Octant of a Membrane-Dominated Shell

Data

A symmetric octant of the cylindrical shell and its initial discretization, based on $p = 2$, $q = 2$, and $h = l/2$, are presented in Figure 27. A very coarse initial mesh can be seen in this figure. The thinness ratio equals $l/t = 3.1415 \times 10^{-2}/0.03 \times 10^{-2}$. The error data controlling the $hp$-adaptation are the admissible relative error value on the final mesh $\gamma_T = 0.01$ and the ratio of the admissible error values on the intermediate and final meshes $\gamma_I/\gamma_T = 2$.

Results

As expected, the automatic tools for detection and/or assessment of the locking have not indicated the appearance of the phenomenon. Because of this, the automatic mode of adaptation corresponds to standard $hp$-approach. The final, $hp$-adapted mesh is presented in Figure 28. Note that the intermediate ($h$-adapted) mesh is not presented but it can easily be obtained from the results in Figures 27 and 28 by taking approximation orders $p$ and $h$-division pattern from these two figures, respectively. It can be seen in the presented figure that, in the vicinity of the curved edge with the rotations constrained, the higher error level and the resultant higher approximation orders are present due to the local bending along the constrained boundary. The membrane and bending strain energies are of the same order along this boundary. In the rest of the shell octant, the membrane strains dominate over bending ones and the error level is lower and evenly distributed. Because of this, the resultant approximation orders are also lower and evenly distributed.

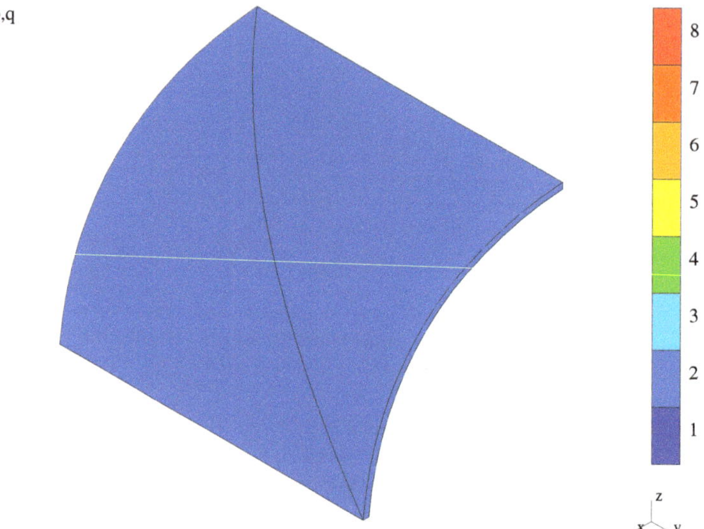

**Figure 27.** An octant of a membrane-dominated shell—initial mesh.

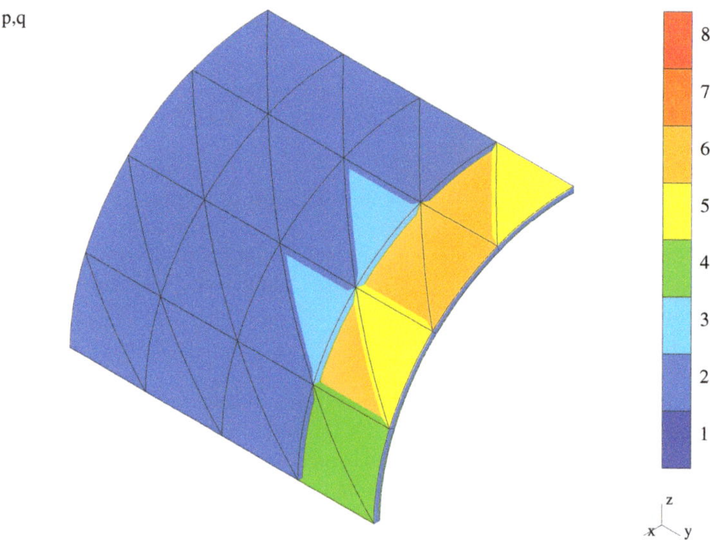

**Figure 28.** An octant of a membrane-dominated shell—$hp$-adapted (final) mesh.

The convergence curve corresponding to the standard three-step $hp$-adaptivity is shown in Figure 29. The first point on the curve corresponds to the initial mesh. The second point of the curve is for the intermediate mesh, while the third point for the final mesh. The admissible error value was confidently achieved in the corresponding three steps.

Finally, in Table 4, the summarizing results of the absolute error $U_r - U$ and the relative error $(U_r - U)/U_r$ are presented versus the number $N$ of degrees of freedom (dofs) for three points of the convergence curve. The mesh figure numbers corresponding to these points are also included in the presented table.

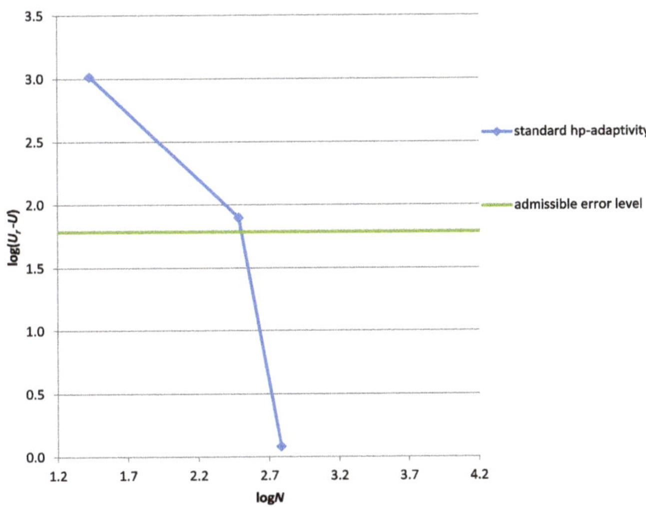



**Figure 29.** An octant of a membrane-dominated shell—convergence for $hp$-adaptation.

**Table 4.** Result summary—an octant of the membrane-dominated shell.

| Adaptive Method | Result Quantity | Mesh Type | | | |
|---|---|---|---|---|---|
| | | Initial | Modified | Intermediate | Final |
| standard $hp$-adaptivity | mesh figure no. | Figure 27 | – | – | Figure 28 |
| | dofs number $N$ [1] | 27 | – | 306 | 606 |
| | $\log(U_r - U)$ [N/m] | 3.017413 | – | 1.898823 | 0.086034 |
| | $(U_r - U)/U_r$ [%] * | 16.9 | – | 1.29 | 0.02 |

* admissible relative error value $\gamma_T = 1.0\,\%$.

Discussion

It can seen that the locking detection and/or assessment tools are capable of not only detecting the phenomena but also recognizing the problems where the phenomenon is not present. Such a result of the detection is consistent with theory presented in the literature (compare [32]). In situations such as this, performance of the standard $hp$-approach is enough to achieve the admissible error value as shown in this example.

### 5.2.5. Generalizations

Based on the above representative examples, and other analogous examples performed by the authors and not presented here due to their qualitative and quantitative similarity, one can state that:

- The adaptation mode based on standard $hp$-adaptivity may fail when applied to the problems where the locking phenomena are present.
- The adaptation mode which allows modification of the initial mesh with the maximum possible longitudinal order of approximation may lead to underestimation of the final error value.
- The course of adaptation with the modification of the initial mesh by means of the optimized value of the longitudinal order of approximation is more effective in achieving the admissible error value than the previous two courses of adaptation.

## 6. Conclusions

The proposed tools for detection and/or assessment of the locking phenomena are capable of the detection of the shear locking and shear–membrane locking as well. These tools are also capable of recognizing the problems without locking phenomena present.

The removal of the locking phenomena by modification of the initial mesh, based on the optimized value of the longitudinal order of approximation $p$, can effectively lead to the admissible error value in the final mesh. This may be possible neither in the case of mesh modification based on the maximum value of $p$ nor in the case of the standard $hp$-adaptivity.

The elaborated tools are perfectly suited to the adaptation based on approximation error control employing, e.g., the equilibrated residual method of error estimation and the three-step $hp$-adaptive strategy. One may also consider application of these tools as the first step of the adaptation based on iterative diminishing the interpolation error.

**Author Contributions:** Conceptualization, Ł.M. and G.Z.; methodology, Ł.M. and G.Z.; writing and verifying software, Ł.M. and G.Z.; validation, Ł.M. and G.Z.; formal analysis, Ł.M. and G.Z.; investigation, Ł.M. and G.Z.; resources, Ł.M. and G.Z.; writing–original draft preparation, Ł.M. and G.Z.; writing–review and editing, Ł.M. and G.Z.; projects administration, Ł.M. and G.Z.; funding acquisition, Ł.M. and G.Z. Both authors have read and agreed to the published version of the manuscript.

**Funding:** This research was partly funded by Scientific Research Committee Poland under research grants 7 T07A 004 14 and 5 T07A 040 24.

**Conflicts of Interest:** The authors declare no conflict of interest.

## References

1. Babuška, I.; Suri, M. Locking effects in finite element approximation of elasticity problems. *Numer. Math.* **1992**, *62*, 439–463. [CrossRef]
2. Suri, M. Analytical and computational assessment of locking in the $hp$ finite element method. *Comput. Methods Appl. Mech. Eng.* **1996**, *133*, 347–371. [CrossRef]
3. Schwab, C.; Suri, M. Locking and boundary layer effects in the finite element approximation of the Reissner-Mindlin plate model. *Proc. Symp. Appl. Math.* **1994**, *48*, 367–371.
4. Pitkäranta, J. The problem of membrane locking in finite element analysis of cylindrical shells. *Numer. Math.* **1992**, *61*, 523–542. [CrossRef]
5. Cho, J.R.; Oden, J.T. Locking and boundary layer in hierarchical models for thin elastic structures. *Comput. Methods Appl. Mech. Eng.* **1997**, *149*, 33–48. [CrossRef]
6. Stolarski, H.; Belytschko, T. Shear and membrane locking in curved $C^0$ elements. *Comput. Methods Appl. Mech. Eng.* **1983**, *41*, 279–296. [CrossRef]

7.  Pinsky, P.M.; Jasti R.V. A mixed finite element formulation for Reissner-Mindlin plates based on the use of bubble function. *Int. J. Numer. Methods Eng.* **1989**, *28*, 1677–1702. [CrossRef]

8.  Kang, D.S.; Pian, T.H.H. A 20-dof hybrid stress general shell element. *Comput. Struct.* **1988**, *30*, 789–794. [CrossRef]

9.  Ayad, R.; Dhatt, G.; Batoz, J.L. A new hybrid-mixed variational approach for Reissner-Mindlin plates. The MiSP model. *Int. J. Numer. Methods Eng.* **1998**, *42*, 1149–1179. [CrossRef]

10. Cohen, G.; Grob, P. Mixed Higher Order Spectral Finite Elements for Reissner-Mindlin Equations. *SIAM J. Sci. Comput.* **2007**, *29*, 986–1005. [CrossRef]

11. Guo, Y.H.; Yu, G.Z.; Xie, X.P. Uniform analysis of a stabilized hybrid finite element method for Reissner-Mindlin plates. *Sci. China Math.* **2103**, *56*, 1727–1742. [CrossRef]

12. Zienkiewicz, O.C.; Too, J.; Taylor, R.L. Reduced integration technique in general analysis of plates and shells. *Int. J. Numer. Methods Eng.* **1971**, *3*, 275–290. [CrossRef]

13. Jacquotte, O.P.; Oden, J.T. Analysis of hourglass instabilities and control in underintegrated finite element methods. *Comput. Methods Appl. Mech. Eng.* **1984**, *44*, 339–363. [CrossRef]

14. Reese, S.; Wriggers, P. A stabilization technique to avoid hourglassing in finite elasticity. *Int. J. Numer. Methods Eng.* **2000**, *48*, 79–109. [CrossRef]

15. Adam, C.; Bouabdallah, S.; Zarroug, M.; Maitournam, H. Improved numerical integration for locking treatment in isogeometric structural elements. Part II: Plates and shells. *Comput. Methods Appl. Mech. Eng.* **2105**, *284*, 106–137. [CrossRef]

16. Abed-Meraim, F.; Combescure, A. A physically stabilized and locking-free formulation of the (SHB8PS) solid-shell element. *Eur. J. Comput. Mech.* **2007**, *16*, 1037–1072. [CrossRef]

17. Wempner, G.; Oden, J.T.; Kross, D.A. Finite element analysis of thin shells. *J. Eng. Mech. Div. (ASCE)* **1968**, *94*, 1273–1294.

18. Crisfield, M.A. A four-noded thin-plate bending element using shear constraints–A modified version of Lyon's element. *Comput. Methods Appl. Mech. Eng.* **1983**, *38*, 93–120. [CrossRef]

19. Areias, P.M.A.; Song, J.H.; Belytschko, T. A finite strain quadrilateral shell element based on discrete Kirchhoff-Love constraints. *Int. J. Numer. Methods Eng.* **2005**, *64*, 1166–1206. [CrossRef]

20. Kapuria, S.; Kulkarni, S.D. An improved discrete Kirchhoff quadrilateral element based on third order zigzag theory for static analysis of composite and sandwich plates. *Int. J. Numer. Methods Eng.* **2007**, *69*, 1948–1981. [CrossRef]

21. Carpenter, N.; Belytschko, T.; Stolarski, H. Improvements in 3-node triangular shell element. *Int. J. Numer. Methods Eng.* **1986**, *23*, 1643–1667. [CrossRef]

22. Naganarayana, B.P.; Prathap, G.; Dattaguru, B.; Ramamurty, T.S. A field-consistent and variationally correct representation of transverse shear strains in the nine-noded plate element. *Comput. Methods Appl. Mech. Eng.* **1992**, *97*, 355–374. [CrossRef]

23. MacNeal, R.H. Derivation of element stiffness matrices by assumed strain distribution. *Nuclear Eng. Des.* **1982**, *70*, 3–12. [CrossRef]

24. Stolarski, H.; Chiang, M.Y.M. Assumed strain formulation for triangular $C^0$ plate elements based on a weak form of the Kirchhoff constraints. *Int. J. Numer. Methods Eng.* **1989**, *28*, 2323–2338. [CrossRef]

25. Militello, C.; Felippa, C.A. A variational justification of the assumed natural strain formulation of finite elements–II. The $C^0$ four node plate element. *Comput. Struct.* **1990**, *34*, 438–444. [CrossRef]

26. Bui, Q.V.; Papeleux, L.; Ponthot, J.P. Numerical simulation of springback using enhanced assumed strain elements. *J. Mater. Process. Technol.* **2004**, *153–154*, 314–318. [CrossRef]

27. Reinoso, J.; Paggi, M.; Linder, C. Phase field modeling of brittle fracture for enhanced assumed strain shells at large deformations: formulation and finite element implementation. *Comput. Mech.* **2017**, *59*, 981–1001. [CrossRef]

28. Dasgupta, S.; Sengupta, D. A high-order triangular plate bending element revisited. *Int. J. Numer. Methods Eng.* **1990**, *30*, 419–430. [CrossRef]

29. Scapolla, T.; Della Croche, L. On robustness of hierarchic finite elements for Reissner-Mindlin plates. *Comp. Methods Appl. Mech. Eng.* **1992**, *101*, 43–60. [CrossRef]

30. Oden, J.T.; Cho, J.R. Adaptive *hpq* finite element methods of hierarchical models for plate- and shell-like structures, *Comput. Methods Appl. Mech. Eng.* **1996**, 136, 317–345. [CrossRef]

31. Szabó, B.A.; Sahrmann, G.J. Hierarchic plate and shell models based on *p*-extension. *Int. J. Numer. Methods Eng.* **1988**, 26, 1855–1881. [CrossRef]

32. Zboiński, G. Application of the three-dimensional triangular-prism *hpq* adaptive finite element to plate and shell analysis. *Comput. Struct.* **1997**, *65*, 497–514. [CrossRef]

33. Echter, R.; Bischoff, M. Numerical efficiency, locking and unlocking of NURBS finite elements. *Comput. Methods Appl. Mech. Eng.* **2010**, *199*, 374–382. [CrossRef]

34. Özdemir, Y.I. Development of a higher order finite element on a Winkler foundation. *Finite Elem. Anal. Des.* **2012**, *48*, 1400–1408. [CrossRef]

35. Rakowski, J. A critical analysis of quadratic beam finite elements. *Int. J. Numer. Methods Eng.* **1991**, *31*, 949–966. [CrossRef]

36. Rakowski, J. A new methodology of evaluation of $C^0$ bending finite elements. *Comput. Methods Appl. Mech. Eng.* **1991**, *91*, 1327–1338. [CrossRef]

37. Zboiński, G.; Ostachowicz, W.; Krawczuk, M. *Modifications of Adaptive Procedures for Analysis of Complex Structures in the Case of the Improper Solution Limit, Locking and Boundary Layer*; Report No. 204/2000; Institute of Fluid Flow Machinery: Gdańsk, Poland, 2000. (In Polish)

38. Zboiński, G. Hierarchical Modeling and Finite Element Approximation for Adaptive Analysis of Complex Structures. D.Sc. Thesis, Institute of Fluid Flow Machinery, Gdańsk, Poland, 2001. (In Polish)

39. Zboiński, G. Numerical tools for a posteriori detection and assessment of the improper solution limit, locking and boundary layers in analysis of thin walled structures. In *Adaptive Modeling and Simulation 2005. Proceeding of the Second International Conference on Adaptive Modeling and Simulation*; Wiberg, N.-E., Diez, P., Eds.; CIMNE: Barcelona, Spain, 2005; pp. 321–330.

40. Oden, J.T. Error estimation and control in computational fluid dynamics. The O. C. Zienkiewicz Lecture. In *Proc. Math. of Finite Elements–MAFELAP VIII*; Brunnel Univ.: Uxbridge, UK, 1993; pp. 1–36.

41. Demkowicz, L. *Computing with hp-Adaptive Finite Elements. Vol. 1. One- and Two-Dimensional Elliptic and Maxwell problems*; Chapman & Hall/CRC: Boca Raton, FL, USA, 2007.

42. Demkowicz, L.; Kurtz, J.; Pardo, D.; Paszyński, M.; Rachowicz, W.; Zdunek, A. *Computing with hp-Adaptive Finite Elements. Vol. 2. Three-Dimensional Elliptic and Maxwell Problems with Applications*; Chapman & Hall/CRC: Boca Raton, FL, USA, 2008.

43. Ainsworth, M.; Oden, J.T. *A Posteriori Error Estimation in Finite Element Analysis*; Wiley: New York, NY, USA, 2000.

44. Ainsworth, M.; Oden, J.T.; Wu, W. A posteriori error estimation for *hp* approximation in elastostatics. *Appl. Numer. Math.* **1994**, *14*, 23–55. [CrossRef]

45. Zboiński, G. Adaptive *hpq* finite element methods for the analysis of 3D-based models of complex structures. Part 1. Hierarchical modeling and approximation. *Comput. Methods Appl. Mech. Eng.* **2010**, *199*, 2913–2940. [CrossRef]

46. Zboiński, G. Adaptive *hpq* finite element methods for the analysis of 3D-based models of complex structures. Part 2. A posteriori error estimation. *Comput. Methods Appl. Mech. Eng.* **2013**, *267*, 531–565. [CrossRef]

47. Cho, J.R.; Oden, J.T. A priori error estimations of *hp*-finite element approximations for hierarchical models of plate- and shell-like structures. *Comput. Methods Appl. Mech. Eng.* **1996**, *132*, 135–177. [CrossRef]

48. Zboiński, G. 3D-based hierarchical models and *hpq*-approximations for adaptive finite element method of Laplace problems as exemplified by linear dielectricity. *Comput. Math. Appl.* **2019**, *78*, 2468-2511. [CrossRef]

49. Pietraszkiewicz, W. *Finite Rotations and Lagrangean Description in the Non-Linear Theory of Shells*; Polish Scientific Publishers: Warsaw, Poland, 1979.

50. Ciarlet, P.G. *Plates and Junctions in Elastic Multi-Structures*; Springer: Berlin, Germany, 1990.

51. Rachowicz, W.; Oden, J.T.; Demkowicz, L. Towards a universal *hp* adaptive finite element strategy. Part 3. Design of *h-p* meshes. *Comp. Methods Appl. Mech. Eng.* **1989**, *77*, 181–212. [CrossRef]

*Appl. Sci.* **2020**, *10*, 8247

52.    Oden, J.T. The best FEM. *Finite Elem. Anal. Des.* **1990**, *7*, 103–114. [CrossRef]
53.    Rachowicz, W.; Pardo, D.; Demkowicz, L. Fully automatic *hp*-adaptivity in three dimensions. *Comput. Methods Appl. Mech. Eng.* **2006**, *195*, 4816–4842. [CrossRef]
54.    Ainsworth, M.; Oden, J.T. A posteriori error estimators for second order elliptic systems: Part 1. Theoretical foundations and a posteriori error analysis. *Comput. Math. Appl.* **1993**, *25*, 101–113. [CrossRef]
55.    Ainsworth, M.; Oden, J.T. A posteriori error estimators for second order elliptic systems: Part 2. An optimal order process for calculating self-equilibrating fluxes. *Comput. Math. Appl.* **1993**, *26*, 75–87. [CrossRef]

**Publisher's Note:** MDPI stays neutral with regard to jurisdictional claims in published maps and institutional affiliations.

MDPI

St. Alban-Anlage 66

4052 Basel

Switzerland

Tel. +41 61 683 77 34

Fax +41 61 302 89 18

www.mdpi.com

*Applied Sciences* Editorial Office

E-mail: applsci@mdpi.com

www.mdpi.com/journal/applsci